Communications in Computer and Information Science 2969

Series Editors

Rationale
The CCIS series is devoted to the publication of proceedings of computer science conferences. Its aim is to efficiently disseminate original research results in informatics in printed and electronic form. While the focus is on publication of peer-reviewed full papers presenting mature work, inclusion of reviewed short papers reporting on work in progress is welcome, too. Besides globally relevant meetings with internationally representative program committees guaranteeing a strict peer-reviewing and paper selection process, conferences run by societies or of high regional or national relevance are also considered for publication.

Topics
The topical scope of CCIS spans the entire spectrum of informatics ranging from foundational topics in the theory of computing to information and communications science and technology and a broad variety of interdisciplinary application fields.

Information for Volume Editors and Authors
Publication in CCIS is free of charge. No royalties are paid, however, we offer registered conference participants temporary free access to the online version of the conference proceedings on SpringerLink (http://link.springer.com) by means of an http referrer from the conference website and/or a number of complimentary printed copies, as specified in the official acceptance email of the event.

CCIS proceedings can be published in time for distribution at conferences or as post-proceedings, and delivered in the form of printed books and/or electronically as USBs and/or e-content licenses for accessing proceedings at SpringerLink. Furthermore, CCIS proceedings are included in the CCIS electronic book series hosted in the SpringerLink digital library at http://link.springer.com/bookseries/7899. Conferences publishing in CCIS are allowed to use our online conference service (Meteor) for managing the whole proceedings lifecycle (from submission and reviewing to preparing for publication) free of charge.

Publication process
The language of publication is exclusively English. Authors publishing in CCIS have to sign the Springer CCIS copyright transfer form, however, they are free to use their material published in CCIS for substantially changed, more elaborate subsequent publications elsewhere. For the preparation of the camera-ready papers/files, authors have to strictly adhere to the Springer CCIS Authors' Instructions and are strongly encouraged to use the CCIS LaTeX style files or templates.

Abstracting/Indexing
CCIS is abstracted/indexed in DBLP, Google Scholar, EI-Compendex, Mathematical Reviews, SCImago, Scopus. CCIS volumes are also submitted for the inclusion in ISI Proceedings.

How to start
To start the evaluation of your proposal for inclusion in the CCIS series, please send an e-mail to ccis@springer.com

Hamid R. Arabnia · Leonidas Deligiannidis ·
Farzan Shenavarmasouleh ·
Farid Ghareh Mohammadi · Soheyla Amirian
Editors

Emerging Trends in Data Science and Cyber Warfare

12th International Conference on Computational Science and Computational Intelligence, CSCI 2025
Las Vegas, NV, USA, December 3–5, 2025
Proceedings, Part III

Springer

Editors
Hamid R. Arabnia
University of Georgia
Athens, GA, USA

Leonidas Deligiannidis
Wentworth Institute of Technology
Boston, MA, USA

Farzan Shenavarmasouleh
Medialab Inc.
Lawrenceville, GA, USA

Farid Ghareh Mohammadi
Verify Radiology Images Consultants
Houston, TX, USA

Soheyla Amirian
Pace University
New York, NY, USA

ISSN 1865-0929 ISSN 1865-0937 (electronic)
Communications in Computer and Information Science
ISBN 978-3-032-28459-4 ISBN 978-3-032-28460-0 (eBook)
https://doi.org/10.1007/978-3-032-28460-0

This Springer imprint is published by the registered company Springer Nature Switzerland AG
The registered company address is: Gewerbestrasse 11, 6330 Cham, Switzerland

Preface

It is our great pleasure to present this collection of research papers from the 12th International Conference on Computational Science and Computational Intelligence (CSCI 2025). This volume, titled "Emerging Trends in Data Science and Cyber Warfare", comprises selected papers accepted in the Research Tracks on Big Data and Data Science (CSCI-RTBD); and Cyber Warfare, Cyber Defense, & Cyber Security (CSCI-RTCW). The conference was held in Las Vegas, Nevada, USA, from December 3 to December 5, 2025.

The CSCI 2025 International Conference brought together papers from a diverse array of communities, including researchers from universities, corporations, and government agencies. Accepted papers are published by Springer Nature, and the proceedings showcase solutions to key challenges in various critical areas of Computational Science and Computational Intelligence.

Computational Science (CS) is the study of addressing problems that are impossible to solve (or difficult to solve) without computers. CS can be considered to be the bridge between computer science and other sciences. The field is interdisciplinary by nature and includes the use of advanced computing capabilities to understand and solve complex problems. In short, CS is the science of using computers to do science. Computational Intelligence (CI) is the study of computational methods that exhibit intelligence. These methods adapt to changing environments and changing goals. There is a significant overlap between the fields of CI and Artificial Intelligence (AI). However, there is also a difference: AI techniques often involve top-to-bottom methods (i.e., methods to reach solutions are imposed from the top) whereas CI techniques often involve bottom-up methods (i.e., solutions emerge from unstructured beginnings). An important part of CI includes a set of Nature-inspired computational approaches to address complex problems for which traditional methods are infeasible. Computational Science and Computational Intelligence both share the same objective: finding solutions to difficult problems. However, the methods to reach the solutions are different. The main objective of the CSCI Conference is to facilitate increased opportunities for cross-fertilization across CS and CI.

Considering the above broad outline, the CSCI 2025 International Conference was composed of the following focused Research Tracks:

Artificial Intelligence (CSCI-RTAI); Big Data and Data Science (CSCI-RTBD); Computational Science (CSCI-RTCS); Computational Intelligence (CSCI-RTCI); Computational Biology (CSCI-RTCB); Cyber Warfare, Cyber Defense, & Cyber Security (CSCI-RTCW); Signal & Image Processing, Computer Vision & Pattern Recognition (CSCI-RTPC); Smart Cities and Smart Mobility (CSCI-RTSC); Education - CS & CE (CSCI-RTED); Health Informatics and Medical Systems (CSCI-RTHI); Mobile Computing, Wireless Networks, & Security (CSCI-RTMC); Software Engineering (CSCI-RTSE); Internet of Things & Internet of Everything (CSCI-RTOT); Social Network Analysis, Social Media, & Mining (CSCI-RTNA); Cloud Computing and Data Centers

(CSCI-RTCC); Parallel & Distributed Computing (CSCI-RTPD); and Large Language Models (LLM). The scope of each track can be found at: https://www.american-cse.org/csci2025/topics and https://www.american-cse.org/LLM2025/topics

An important objective of the CSCI 2025 International Conference and its associated Research Tracks was to foster opportunities for cross-fertilization between the fields of Computational Science and Computational Intelligence. The CSCI Conference is deeply committed to promoting diversity and eliminating discrimination, both in its role as a conference organizer and as a service provider. Our goal is to create an inclusive culture that respects and values differences, promotes dignity, equality, and diversity, and encourages individuals to reach their full potential. We are also dedicated, wherever possible, to organizing a conference that represents the global community. We sincerely hope that we have succeeded in achieving these important objectives.

The Steering Committee and the Program Committee would like to extend their gratitude to all the authors who submitted papers for consideration. CSCI 2025 received submissions from 48 countries, with approximately 48% of them coming from outside the United States. Each submitted paper underwent a rigorous peer-review process, with at least two experts (an average of 2.5 referees per paper) evaluating the submissions based on originality, significance, clarity, impact, and soundness. In cases where reviewers' recommendations were contradictory, a program committee member was tasked with making the final decision, often consulting additional referees for further guidance. The CSCI Conference followed the guidelines of COPE (Committee on Publication Ethics):

- Typical submissions underwent a single-blind peer review process, in which the authors remain unaware of the identities of the reviewers, while the reviewers are informed of the authors' identities.
- Papers authored by one or more members of the program committee, including co-chairs, were subjected to a double-blind peer review process, ensuring that neither the authors nor the reviewers were aware of each other's identities or affiliations.

The Research Track on Big Data and Data Science (CSCI-RTBD) received 98 submissions, of which 19 papers were accepted, resulting in a paper acceptance rate of 19%; we selected only 13 papers for this volume. The Research Track on Cyber Warfare, Cyber Defense, & Cyber Security (CSCI-RTCW) received 184 submissions, of which 39 papers were accepted, resulting in a paper acceptance rate of 21%; we selected only 27 papers for this volume. A total of 40 papers are included in this volume.

We are deeply grateful to the many colleagues who contributed their time and effort to organizing the CSCI 2025 Conference. In particular, we extend our thanks to the members of the Program Committee, the Steering Committee, and the referees. We would also like to express our appreciation to the primary sponsor of the conference, the American Council on Science & Education. The list of members of the Program Committee for each track can be found at: https://www.american-cse.org/csci2025/committees

We extend our heartfelt gratitude to all the speakers and authors for their valuable contributions. We would also like to thank the following individuals and organizations for their support: the staff at the Luxor Hotel (an MGM property, conference/meeting department) and the staff of Springer for their assistance in various aspects of the event.

We are pleased to present the proceedings of CSCI 2025 (selected papers of CSCI-RTBD and CSCI-RTCW Research Tracks). These proceedings represent a collection of outstanding research contributions that reflect the diversity and depth of work in Data Science and Cyber Warfare.

December 2025

Hamid R. Arabnia
Leonidas Deligiannidis
Farzan Shenavarmasouleh
Farid Ghareh Mohammadi
Soheyla Amirian

Organization

Steering Committee – Co-chairs (CSCI 2025)

Hamid R. Arabnia	University of Georgia, USA
Leonidas Deligiannidis	Wentworth Institute of Technology, USA
Fernando G. Tinetti	Universidad Nacional de La Plata, Argentina
Quoc-Nam Tran	Southeastern Louisiana University, USA

Publication Co-chairs and Co-editors

"Emerging Trends in Artificial Intelligence and Applications"
Research Track: Artificial Intelligence (CSCI-RTAI)
Part of Proceedings of The 2025 International Conference on Computational Science and Computational Intelligence (CSCI)

Hamid R. Arabnia	University of Georgia, USA
Leonidas Deligiannidis	Wentworth Institute of Technology, USA
Farzan Shenavarmasouleh	Medialab Inc., USA
Farid Ghareh Mohammadi	Verify Radiology Images Consultants, LLC, USA
Soheyla Amirian	Pace University, USA

Members of Steering Committee (CSCI 2025)

Babak Akhgar	Sheffield Hallam University, UK
Abbas M. Al-Bakry	University of Information Technology & Communications, Iraq
Nizar Al-Holou	University of Detroit Mercy, USA
Soheyla Amirian	Pace University, USA
Hamid R. Arabnia	University of Georgia, USA
Rajab Challoo	Texas A&M University-Kingsville, USA
Chien-Fu Cheng	Tamkang University, Taiwan
Hyunseung Choo	Sungkyunkwan University, South Korea
Kevin Daimi	University of Detroit Mercy, USA
Leonidas Deligiannidis	Wentworth Institute of Technology, USA
Eman M. El-Sheikh	University of West Florida, USA
Mary Mehrnoosh Eshaghian-Wilner	University of California, Los Angeles, USA

David L. Foster	Kettering University, USA
Ching-Hsien (Robert) Hsu Chung	Asia University, Taiwan
Farid Ghareh Mohammadi	Verify Radiology Images Consultants, LLC, USA
James J. (Jong Hyuk) Park	SeoulTech, South Korea
Mohammad S. Obaidat	University of Jordan, Jordan
Gerald Schaefer	Loughborough University, UK
Farzan Shenavarmasouleh	Medialab Inc., USA
Fernando G. Tinetti	Universidad Nacional de La Plata, Argentina
Quoc-Nam Tran	Southeastern Louisiana University, USA
Shiuh-Jeng Wang	Central Police University, Taiwan
Layne T. Watson	Virginia Polytechnic Institute & State University, USA
Chao-Tung Yang	Tunghai University, Taiwan
Mary Yang	University of Arkansas for Medical Sciences, USA

Program Committee – Computational Science and Computational Intelligence (CSCI 2025)

Afrand Agah	West Chester University of Pennsylvania, USA
Bharat Bhushan Agarwal	IFTM University, India
Omaima Nazar Ahmad AL	Al-Zaytoonah University of Jordan, Jordan
Wasim A. Al – Hamdani	Kentucky State University, USA
Ismail Khalil Al Ani	Ittihad University, United Arab Emirates
Mehran Asadi	Lincoln University, USA
Travis Atkison	University of Alabama, USA
Azita Bahrami	IT Consult, USA
Mehdi Bahrami	Fujitsu Laboratories of America, Inc., USA
P. Balasubramanian	Nanyang Technological University, Singapore
Petra Saskia Bayerl	Erasmus University Rotterdam, The Netherlands
Jane M. Binner	University of Birmingham, UK
Juan-Vicente Capella-Hernandez	Universitat Politècnica de València, Spain
Juan Jose Martinez Castillo	Universidad Nacional Abierta, Venezuela
Rui Chang	Mount Sinai School of Medicine, USA
Dongsheng Che	East Stroudsburg University of Pennsylvania, USA
Jianhung Chen	Chung Hua University, Taiwan
Mu-Song Chen	Da-Yeh University, Taiwan
Xin (Thomas) Chen	University of Hawaii, USA
Steve C. Chiu	Idaho State University, USA
Mark Yul Chu	University of Texas Rio Grande Valley, USA

Jose Alfredo F. Costa	Universidade Federal do Rio Grande do Norte, Brazil
Arianna D'Ulizia	National Research Council of Italy, Italy
Zhangisina G. Davletzhanovna	Central Asian University, Kazakhstan
Wesley Deneke	Western Washington University, USA
Noel De Palma	Grenoble Alpes University, France
Mohsen Doroodchi	University of North Carolina at Charlotte, USA
Levent Ertaull	California State University, East Bay, USA
Mahmood Fazlali	Shahaid Beheshti University, Iran
Boyuan Feng	University of California, Santa Barbara, USA
George A. Gravvanis	Democritus University of Thrace, Greece
Gheorghe Grigoras	"Gheorghe Asachi" Technical University of Iaşi, Romania
Ray Hashemi	Georgia Southern University, USA
Houcine Hassan	Universitat Politècnica de València, Spain
Abdolreza Hatamlou	Islamic Azad University, Iran
Bing He	Cisco System Inc., USA
Henry Hexmoor	Southern Illinois University at Carbondale, USA
Gahangir Hossain	Texas A&M University - Kingsville, USA
Guofeng Hou	AQR Capital Management, USA
Ren-Junn Hwang	Tamkang University, Taiwan
Naseem Ibrahim	Penn State Erie, The Behrend College, USA
Rabia Jafri	King Saud University, Saudi Arabia
Shahram Javadi	Azad University, Iran & University of Lisbon, Portugal
Young-Sik Jeong	Dongguk University, South Korea
Aleksandr Katkow	University of Computer Sciences & Skills, Poland
Byung-Gyu Kim	Sun Moon University, South Korea
Taihoon Kim	University of Tasmania, Australia
Sang-Wook Kim	Hanyang University, South Korea
Dattatraya Vishnu Kodavade	D.K.T.E Society's Textile & Engineering Institute, India
Elena B. Kozerenko	Russian Academy of Sciences, Moscow, Russia
Guoming Lai	Sun Yat-sen University, China
Hyo Jong Lee	Chonbuk National University, South Korea
Bo Liu	NEC Labs China, China
Eleanor Lockley	Sheffield Hallam University, UK
Jianbing Ma	Bournemouth University, UK
Julius Marpaung	Northeastern University, USA
Andrew Marsh	HoIP Telecom Ltd, UK
Juan Martinez	Universidad Gran Mariscal de Ayacucho, Venezuela

Praveen Meduri	California State University, Sacramento, USA
Ali Mostafaeipour	California State University, Fullerton, USA
Houssem Eddine Nouri	University of Tunis, Tunisia
Michael B. O'Hara	KB Computing, LLC, USA
Robert Ehimen Okonigene	Ambrose Alli University, Nigeria
Funminiyi Olajide	Nottingham Trent University, UK
Satish Penmatsa	University of Maryland - Eastern Shore, USA
Saman Parvaneh	Philips Research North America, USA
Ponalagusamy	National Institute of Technology, Tiruchirappalli, India
Laura L. Pullum	Oak Ridge National Laboratory, USA
Xuewei Qi	University of California, Riverside, USA
Junfeng Qu	Clayton State University, USA
Shahram Rahimi	University of Alabama, USA
Jaime Raigoza	California State University, Chico, USA
Arvind Ramanathan	Oak Ridge National Laboratory, USA
Iman M. Rezazadeh	University of California, Davis, USA
Om Prakash Rishi	University of Kota, India
Cristian Rodriguez Rivero	Universidad Nacional de Córdoba, Argentina
Seyed Roosta	Albany State University, USA
K. Martin Sagayam	Karunya Institute of Technology and Sciences, India
P. Sanjeevikumar	University of South-Eastern Norway, Norway
Benaoumeur Senouci	University of Southern Denmark, Denmark
Zhefu Shi	Microsoft Corporation, USA
Jawed Siddiqi	Sheffield Hallam University, UK
Akash Singh	IBM Corporation, USA
Anthony Skjellum	Auburn University, USA
Omer Muhammet Soysal	Southeastern Louisiana University, USA
Helman Stern	Ben-Gurion University of the Negev, Israel
Jonathan Z. Sun	University of Southern Mississippi, USA
Rahman Tashakkori	Appalachian State University, USA
Predrag Tosic	Microsoft, USA
Quoc-Nam Tran	Southeastern Louisiana University, USA
Jesus Vigo-Aguiar	University of Salamanca, Spain
Patrick Wang	East China Normal University, China & Northeastern University, USA
Weiqiang Wang	Opera Solutions LLC, USA
Yin Wang	Lawrence Technological University, USA
Alicia Nicki Washington	Duke University, USA
Wei Wei	Xi'an University of Technology, China
Yong Wei	University of North Georgia, USA

Santoso Wibowo	Central Queensland University, Australia
Alexander Woehrer	Vienna Science and Technology Fund, Austria
Bernd E. Wolfinger	University of Hamburg, Germany
Jay Xiong	NSEsoftware, LLC, USA
Hyun Yoe	Sunchon National University, South Korea
Jane You	Hong Kong Polytechnic University, China
Ismail Yusuf	Lamintang Education and Training Center, Indonesia
Wei Zhong	University of South Carolina Upstate, USA

Contents

Poster Papers

Big Data, Data Science, and Applications

Accelerating HNSW in RAG: A Dynamic Cache Method Based on Topic-Biased Access Patterns

Rei Masuda[1](✉), Kazuma Iwamoto[1], Hinata Yokoyama[1], Kazuaki Ando[2], and Hitoshi Kamei[3]

[1] Graduate School of Science for Creative Emergence, Kagawa University, Takamatsu, Japan
s24g357@kagawa-u.ac.jp

[2] Faculty of Engineering and Design, Kagawa University, Takamatsu, Japan

[3] Integrated Center for Informatics, Kagawa University, Takamatsu, Japan

Abstract. HNSW (Hierarchical Navigable Small World) is a method widely used for vector search in RAG (Retrieval-Augmented Generation). It consumes vast amounts of memory at scale. To reduce memory usage, there are two methods. One is memory-mapped file I/O (`mmap`) to avoid loading data into memory. The other method is quantization, e.g., Product Quantization (PQ). However, these methods cause response time degradation or lower accuracy. To mitigate these problems, we focused on the "topic bias" of the search process in RAG. Topic bias refers to category locality in search queries from users. We previously proposed a method called Tiered Cache-HNSW (TC-HNSW), which utilizes the topic bias to narrow down vector data to cache it efficiently. The proposed method identifies frequently searched topic categories from recent history and dynamically loads the corresponding vector data groups from disk into memory as a cache. We implemented TC-HNSW in open-source software and evaluated it using a categorized Wikipedia dataset. The results showed that TC-HNSW shortens response time by approximately 30% to 60% compared to the conventional method, especially when a large dataset is applied and high-accuracy settings are configured (high *ef* values). The contribution of this study is the demonstration of a practical cache method for TC-HNSW that exploits the topic bias to shorten response time under memory constraints without search accuracy degradation.

Keywords: RAG · HNSW · Vector Search · Topic Bias · Tiered Cache

1 Introduction

In recent years, the response quality of Large Language Models (LLMs) such as GPT-4 has improved significantly [1]. A method called Retrieval-Augmented Generation (RAG) [2] has also become popular [3]. RAG is an architecture designed to achieve more accurate responses by using external knowledge. RAG often uses vector storage, such as Qdrant, which employs a graph-based Approximate Nearest Neighbor Search (ANNS) algorithm as a search backend.

Hierarchical Navigable Small World (HNSW) [4] is one of the most commonly used methods to implement ANNS in vector search. HNSW consumes a large amount of

H. R. Arabnia et al. (Eds.): CSCI 2025, CCIS 2969, pp. 3–18, 2027.
https://doi.org/10.1007/978-3-032-28460-0_1

memory as the scale increases, especially when the stored data becomes large in RAG. This makes it impossible to operate within a realistic memory capacity. Thus, conventional methods use memory-mapped file I/O (`mmap`) to avoid loading data from disk. Another method to reduce memory size is quantization, e.g., product quantization (PQ) [5]. However, these methods involve a trade-off between response time and accuracy. Therefore, it becomes a challenge to achieve high accuracy and shorten response time for search in RAG.

To address this problem, we propose a method called Tiered Cache-HNSW (TC-HNSW), which focuses on the "topic bias" of search processes in RAG, which means the category locality for search queries from users. Our proposed method leverages this bias for caching; crucially, it does not alter the HNSW graph structure, thereby maintaining search accuracy while shortening response time. To clarify this topic bias, we analyzed the trend of search queries in our previous study [6]. We implemented the proposed method, which utilizes the topic bias, in Qdrant [7], one of the most widely used vector databases in RAG. TC-HNSW performs dynamic hierarchical caching as required. This method shortens the response time compared to conventional methods while maintaining accuracy, without loading the entire graph into memory. In this paper, we present the design of TC-HNSW and evaluate its effectiveness using a dataset of categorized English Wikipedia articles.

2 Background

2.1 RAG Overview

Retrieval-Augmented Generation (RAG) is an architecture that enhances the knowledge of Large Language Models (LLMs) by retrieving information from external document databases. Figure 1 shows the architecture of RAG. When a user inputs a search query, the retrieval module in RAG first obtains related information from an external knowledge source stored in a vector storage. Then, it provides the search results to the LLM as part of the prompt, and the LLM then generates an answer. This process enables the LLM to generate answers based on the retrieved context. As a result, it can reduce factually incorrect answers (a phenomenon known as hallucination). Thus, RAG allows LLMs to produce more accurate responses based on external knowledge [8].

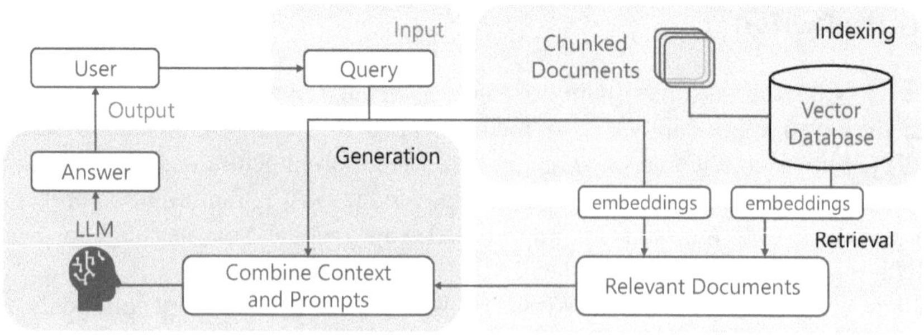

Fig. 1. Architecture of RAG

Knowledge sources used in RAG, such as internal documents, are first chunked (divided into chunks of a certain length). Then, an embedding model converts them into high-dimensional vector representations. Finally, they are stored in a vector storage. In the search phase, the user query is also vectorized. Then, an Approximate Nearest Neighbor Search (ANNS) is performed in this vector storage.

Therefore, the overall response time and answer quality of RAG fundamentally depend on the search performances which are the response time and accuracy of the vector storage used in the search phase. This study focuses on HNSW, which is widely and efficiently used as the search backend in RAG.

2.2 Approximate Nearest Neighbor Search

Nearest Neighbor Search (NNS) is a method used for searching for similar information, and it plays a basic role in various application domains, including information retrieval, pattern recognition, data mining, machine learning, and recommendation systems. However, the dataset sizes handled by NNS are rapidly increasing. In this situation, exact NNS cannot meet efficiency or cost requirements. For this reason, research on Approximate NNS (ANNS) has been actively conducted [9], because ANNS improves efficiency by relaxing accuracy constraints. RAG also uses vector storage that implements ANNS as a component for searching embedded documents [10].

ANNS finds the approximate nearest neighbors for a query from a high-dimensional dataset. ANNS algorithms are classified into four main types based on their index structure: hash-based, tree-based, quantization-based, and graph-based [9]. Recently, graph-based structures have been getting attention as an effective choice for ANNS.

A typical graph-based ANNS is Hierarchical Navigable Small World (HNSW) [4]. HNSW is an extension of NSW. Figure 2 shows the architecture of HNSW. HNSW employs a hierarchical graph with multiple layers to achieve efficient nearest neighbor search. Each layer is a subset of data points. Upper layers contain fewer data points while lower layers contain more. Upper layers have long-distance links (links to far data points), while lower layers have more short-distance links (links to near data points). In the upper layer, the algorithm performs a rough and quick search, while in the lower layer, it searches more accurately. This allows for fast and accurate searches even for large-scale datasets. Furthermore, in HNSW, the parameter *ef* can be specified to control the search. It represents the number of neighbor nodes evaluated during the graph traversal. A larger *ef* value improves search accuracy but also increases the CPU and I/O load. This makes response time longer. Conversely, reducing this value degrades search accuracy but reduces the CPU and I/O load. In applications such as RAG, which use many high-dimensional vectors, search accuracy is particularly important. Therefore, a large *ef* value is often used.

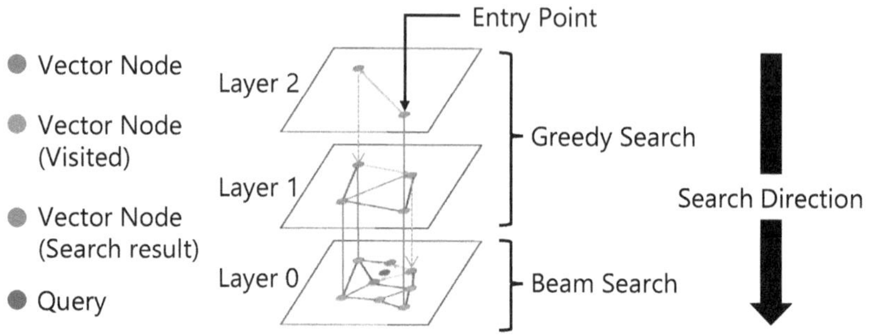

Fig. 2. Architecture of HNSW

2.3 Challenges

As RAG becomes more popular, the demand for local graph search algorithms will increase to handle organization-specific data. These systems will be required to perform larger-scale, higher-accuracy searches with a shorter response time. Conventional search methods need to load the entire graph into memory to ensure consistent response times. Vector search in RAG needs to hold a large amount of high-dimensional vector data. Therefore, it is not feasible to store all of this in memory.

To mitigate this problem, large-scale vector storage often uses memory-mapped file I/O (`mmap`). This method loads only the necessary data into memory by using the OS page cache and accessing files on disk to avoid loading all of the data into memory. However, when using this method in a large-scale environment, response times tend to increase. It also increases the overall response time of the system. Our previous research also showed that response times significantly increase as the amount of data stored in vector storage when using `mmap` in memory-limited environments [6].

Another method to reduce the size of high-dimensional vector data is also being studied, and Product Quantization (PQ) [5] is typically applied. This method reduces the amount of vector data for the graph loaded into memory through quantization. This allows large-scale RAG to run even with a realistic amount of memory. Meanwhile, reducing the dataset size with PQ always degrades accuracy.

When running RAG with large-scale data using the existing methods described above, the response time becomes longer and the search accuracy degrades, or both may occur. Therefore, it becomes a challenge to reduce the response time of HNSW search and to maintain its search accuracy in RAG. To address this challenge, we previously proposed a new search method called TC-HNSW [6]. This method focuses on the topic bias of searches in RAG. It adds an efficient caching method, using category classification and history, to the conventional HNSW search algorithm. In this paper, we clarify the specific design of TC-HNSW and verify its effectiveness through practical operation by implementation.

3 Proposed Method

3.1 Design Overview

TC-HNSW leverages the characteristics of RAG search within organizations. We assume that these characteristics are as follows:

1. Search topics are biased towards certain categories within an organization.
2. Frequently searched topics change over time.

As for characteristic (1), which relates to topic bias, consider the use-case of an HR department in a company. It is expected that topics such as employee rosters and evaluation criteria would be searched frequently. We expect this to be very different for other departments. On the other hand, in the development department of the same company, many topics would be related to development, such as product design for customers, internal documentation, and QA progress.

The characteristic (2) is based on the hypothesis that topic bias changes over time. For example, in the case of the HR department, the topics are still related to HR. At the end of the year, topics related to employee evaluations and promotions might be searched more frequently. Around the new fiscal year, topics related to new employee training and hiring information might be searched more frequently.

Therefore, we propose a method to cache vector data in groups likely to be searched next, based on recent search frequencies. To achieve efficient caching through prefetching, we categorize vectors by their searched topics and control the cache at the category level. Furthermore, these vector data are managed in a tiered cache, where they are moved among tiers based on their reference frequency. Owing to this tiered cache, the issued I/O for comparing vector data in HNSW are effectively distributed. The more frequently accessed vector data are placed in main memory as cache instead of accessing storage. This prevents an increase in response time even for large datasets. Figure 3 shows the concept of TC-HNSW.

In TC-HNSW, vector data are cached on the vector storage, such as Qdrant. When the graph structure is re-constructed, the search accuracy varies. This leads to confusion for users of RAG. Therefore, TC-HNSW does not change the HNSW graph structure. Instead, it shortens the response time while managing search accuracy by changing the arrangement of vector data on the disk side.

3.2 Design Specifications

TC-HNSW is realized by adding a novel cache method to the existing HNSW search algorithm based on memory-mapped files. We assume that categories are already defined at the dataset stage and are labeled. Multiple categories can be assigned to a single data item if necessary. Figure 4 shows the system configuration of TC-HNSW.

A new cache management module is implemented within the HNSW graph management module for cache control. Temporary management data is kept in memory to accelerate processes of HNSW. Figure 4 A (Lookup Table) is the Category-Node ID Conversion Table. One category has multiple node IDs, and some node IDs may overlap. This table maintains these relationships. When a category ID is provided, it returns a list of node IDs belonging to it to Fig. 4 C (Cache Management Control Code). When a node ID is provided, it returns the corresponding category IDs. Figure 4 B is the Cache Management Data Structure, which consists of two main parts. One is a method to hold the history of node IDs from past searches. The other is a method to represent the current cache state. The method to represent the current cache state holds a list of category IDs to be cached. During periodic caching (based on the number of searches), it gets a list of node IDs for a category ID using the Category-Node ID Conversion Table. It passes this list of node IDs to the data management module to replace the cache contents. Figure 4 C is the Cache Control Implementation. The cache is replaced based on the reference frequency at the category level. It counts the number of references for each category based on the node ID reference history and prioritizes the category with the most references as the next cache target. It also checks the current cache state from the cache management data structure. If data is already in the cache, it is marked to avoid unnecessary cache updates.

The caching and vector data retrieval processes are implemented in the data management module. The control code in data management module receives the list of node IDs to be cached, which is determined by the cache management module. It reads the vector data specified by those node IDs into the memory cache. During a search, this module is used to read the actual data for data comparison. At this time, if the target data is in the cache, the module returns it directly. If it is not in the cache, access to the conventional memory-mapped file occurs.

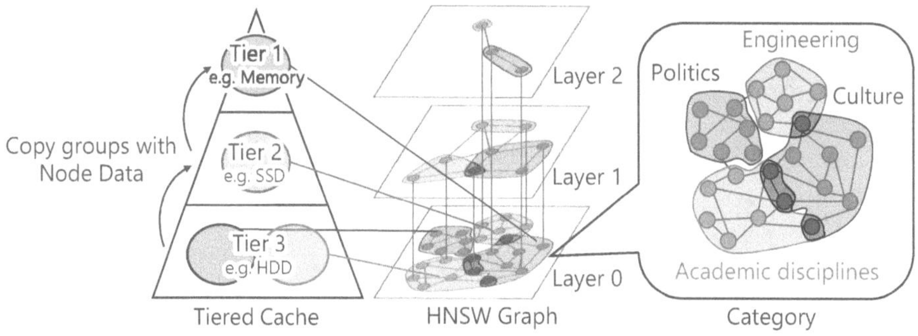

Fig. 3. Concept of the proposed method

The blue region in Fig. 4 represents data stored in the lowest tier of the tiered cache, such as SSD or HDD. The orange region corresponds to data placed in the cache tier of the tiered cache, such as memory. The bottom layer of the tiered cache holds all the vector data and the complete graph structure of the HNSW graph. In this method, a part of the HNSW vector data is copied from the bottom layer of the tiered cache to the cache tier. In this way, it achieves search with maintained accuracy and response time using a realistic amount of memory, even for large datasets.

The cache replacement frequency can be specified. In this experiment, we set it to once every 100 queries. In this paper, we evaluate the change in response time with simple implementation. Thus, the tiered cache has a two-layer structure consisting of memory and SSD. Generally, the size of vector datasets in RAG use cases ranges from tens to hundreds of gigabytes, TC-HNSW assigns a 2-byte category ID per node. Thus, the metadata overhead is limited to tens or hundreds of megabytes, which is negligible compared to the total size.

4 Evaluation

4.1 Experimental Setup

In this paper, we evaluate the effectiveness of TC-HNSW. We implement TC-HNSW in Qdrant [7] and construct categorized English Wikipedia articles as a dataset, utilizing the mdx platform [11]. First, we measure CPU and I/O utilization for both the conventional method and TC-HNSW to confirm that the cache function works as intended. While caching is in progress, it performs a large amount of memory reading. I/O utilization may increase, and CPU utilization may decrease. Meanwhile, after caching vector data with TC-HNSW, I/O utilization decreases, and CPU utilization becomes higher than that of the conventional method. This behavior indicates that searches are performed using the data cached by TC-HNSW. Next, to further evaluate the effectiveness of TC-HNSW, we evaluate the response time per query for both the conventional method and TC-HNSW. If the TC-HNSW cache is effective, the response time is shorter than the conventional method. For the cache operation check and response time comparison experiments, we conduct experiments in the following three patterns.

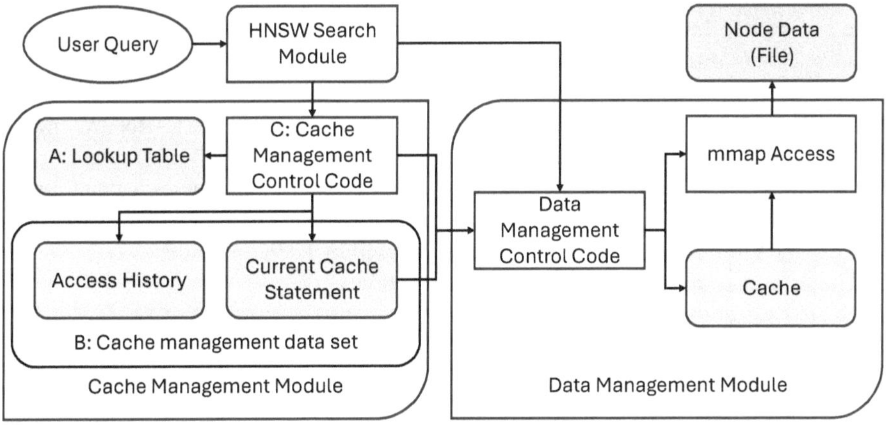

Fig. 4. System configuration

1. **Conventional Qdrant + Random Queries:** HNSW search in `mmap` mode. Queries are random.
2. **Conventional Qdrant + Specific Category Queries:** HNSW search in `mmap` mode. Queries are randomly selected from a specific category.
3. **Qdrant with TC-HNSW + Specific Category Queries:** HNSW search with the TC-HNSW cache method. Queries are randomly selected from a specific category.

For pattern 1, we randomly select 5,000 queries from the entire dataset. For patterns 2 and 3, we randomly select 5,000 queries from a specific category. The queries in pattern 1 simulate a scenario in which users from multiple departments access the system at the same time, and their requests are processed together. The queries in patterns 2 and 3 simulate a scenario for RAG search where the topic distribution is biased, which happens when the search is limited to a specific department. This is the scenario in which TC-HNSW is designed to be used.

In the TC-HNSW cache operation check experiment, we use dataset sizes of 1M and 6M. We use an *ef* value of 512. For patterns 2 and 3, we use the Engineering category. The `sar` command is used to measure resource utilization. For CPU utilization, we use the sum of `%usr` and `%sys`, and for I/O utilization, we use `%util`.

In the response time comparison, we change the dataset size from 1M to 6M. We also experiment with four *ef* values for each condition: 64, 128, 256, and 512. These values are commonly used, including in the official Qdrant benchmark [12]. In patterns 2 and 3, we use the following three specific categories: Engineering, Academic disciplines and Politics. These three categories are randomly selected. They are chosen from the 35 initial root categories in the Wikipedia dataset, which are described later.

4.2 Wikipedia Dataset

For this experiment, we use a dataset based on the English Wikipedia [13] which we categorized using our own criteria. In Wikipedia, one article has multiple categories to improve management and search efficiency. The categories are organized hierarchically and subdivided by topic, management criteria, and the number of articles. However, as noted by Stefano Faralli et al. [14], Wikipedia categories are a directed cyclic graph. Thus, we cannot directly apply them to TC-HNSW's category division.

Therefore, we created a dataset using a new division method under specific conditions. This allows us to apply such categories of information to TC-HNSW. In this method, we use categories belonging to "Main_topic_classifications" [15] in the English Wikipedia as roots. From these points, we recursively get child categories and repeat merging and splitting. In this way, we group related information into the same category and keep the number of articles in each category within a certain range.

The dataset creation algorithm operates as follows:

1. We start processing with one arbitrary category from "Main_topic_classifications" as the root. The result of this process does not affect the processing of other root categories.
2. Count the number of articles in the root category. If the number of articles exceeds 10,000 or there are no valid child categories, finalize the state of this root category.
3. If the number of articles in the root category is less than 10,000, recursively get its child categories and add them to the root category.
4. To prevent loops within the root category, we create a visited list and exclude categories that have already been visited from further processing.
5. Repeat steps 2–4 until the state of the root category is finalized.
6. If the root category state is finalized, valid child categories exist, and the total number of articles is below the threshold, make the current child category a new, independent root category.

7. Repeat the process from step 2 for the new independent root category.

We perform the above process for all 35 categories. In this experiment, we use data from May 1, 2025, and collect articles until the total number of articles after deduplication exceeded 2M. We chunked the collected articles into 512 tokens and embedded them, resulting in a total of approximately10M. We create subsets with sizes from 1M to 6M (in steps of 1M) by randomly sampling from this dataset.

In the experiment, we use the root categories from this process as root categories in TC-HNSW. We use article titles as search queries, and article bodies as the search targets stored in the vector storage. We embed article titles and bodies into 1024-dimensional vectors using bge-m3 [16].

4.3 Experimental Environment

Table 1. Specifications of experimental server

	Qdrant Server	Query Server
CPU	Ryzen7 5825U	Ryzen7 3700X
Memory	DDR4–3200 64GB	DDR4–3200 128GB
Platform	LXC, Docker Container	LXC, Docker Container
Storage	NVMe SSD 1TB	NVMe SSD 1TB
Network	1Gbps LAN	
Qdrant	Single node v1.9.5	Single node v1.9.5
OS	Ubuntu 24.04 LTS	Ubuntu 24.04 LTS
Kernel	Linux 6.8	Linux 6.8

Table 1 shows the experimental environment. The experiments are conducted using two physical servers. One server runs a benchmark tool that sends queries, and the other server runs Qdrant. Each service runs in a Docker environment within an LXC container. The two physical machines are connected via a 1Gbps network, and the benchmark searches are performed over this network. We use a modified version of the official Qdrant benchmark tool [12] adapted to work with TC-HNSW. On the Qdrant side, we limit the memory size for the container to 8 GB. For TC-HNSW, we set the cache size to 3 GB, and the cache is updated once every 100 queries.

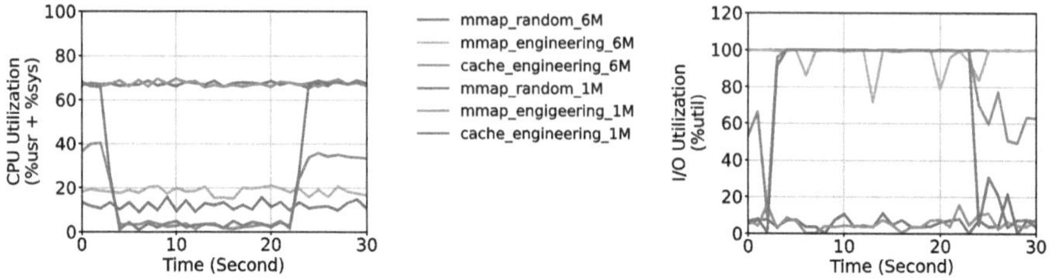

A. CPU utilization (`%usr + %sys`) **B.** I/O utilization (`%util`)

Fig. 5. Resource utilization in the conventional method and TC-HNSW

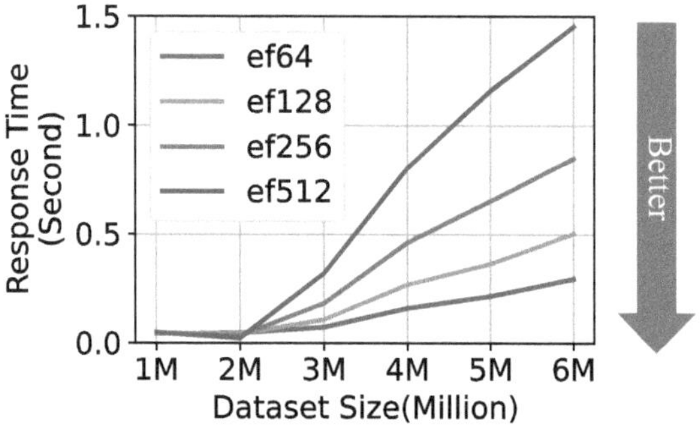

Fig. 6. Response time per query (`mmap` + Random)

4.4 Experimental Results

Cache Operation Check

First, the results of the cache operation check experiment are presented below. Figure ure5 shows the CPU and I/O utilization over 30-s period. These results correspond to *ef*512 with dataset sizes 1M and 6M, for both the conventional method and TC-HNSW. Data was collected for three experiments: conventional (`mmap` + random), conventional with category (`mmap` + Engineering), and TC-HNSW (cache + Engineering). Figure ure5A shows CPU utilization around the 200th query, when caching occurs. Figure ure5B shows I/O utilization at the same time as Fig. 5A. For the 1M dataset, the conventional method remained stable at 65%-70% CPU utilization, and the I/O utilization was approximately 10%. In TC-HNSW, the CPU utilization ranged from 65% to 75% before suddenly dropping to approximately 3%. It then rose back to 65%-75%. The I/O utilization for TC-HNSW temporarily reached 100%. After around 22 s, it settled down to approximately 10%.

For the 6M dataset, the conventional method (mmap + random) was at 10%-15% CPU utilization. The conventional method restricted to the Engineering category was slightly higher, at approximately 20%. The I/O utilization for both methods was nearly 100%. For TC-HNSW, the CPU utilization started at 40%, then dropped to 3% (similar to the 1M dataset). After approximately 20 s, it increased to around 35%. The I/O utilization for TC-HNSW started at 60%, then reached 100% for a while. After that, it decreased again to 50%-80%. From these results, we can see a correlation between CPU and I/O utilization. Furthermore, compared to the conventional method, TC-HNSW is characterized by large changes in CPU and I/O utilization over time.

Response Time Results

Next, we show the experimental results for response time per query for the conventional method and TC-HNSW. Figures 6, 7, 8, 9, 10, 11 and 12 illustrate changes in response time for each parameter. The horizontal axis is the dataset size, and the vertical axis is the response time per query in seconds. In Fig. 6, for the conventional Qdrant with random queries, the response time remained within 0.05 s up to 2M, regardless of the *ef* value. However, after 3M, the response time increased sharply. With *ef*512 and the 6M dataset, the result was over 1.4 s. For other *ef* values, the response time also increased significantly after 3M. In Fig. 7, for the conventional Qdrant with queries limited to the Engineering category, the response time decreased after 3M compared to Fig. 6. Regarding other categories, Fig. 9 (Politics) showed consistent improvement in response time, similar to Engineering. However, Fig. 8 (Academic disciplines) exhibited response times similar to those of random queries.

In Figs. 10, 11 and 12, for searches using TC-HNSW, the response time decreased and increased occasionally. In Fig. 10 (Engineering), the response time was slower than in Fig. 7 for dataset sizes up to 3M, across all *ef* values. However, at 4M, *ef*512 became faster. At 5M and above, *ef*256 and *ef*512 were faster than the corresponding results in Fig. 7. Additionally, in Fig. 12 (Politics), for *ef*512, the increase in response time was smaller for dataset sizes 4M and above. On the other hand, in Fig. 11 (Academic disciplines), the response time was slower than that of the conventional method in Fig. 8 across all dataset sizes and *ef* values. From the experimental results, we can see that TC-HNSW significantly reduces the response time compared to the conventional method for specific categories, high *ef* values, and larger dataset sizes.

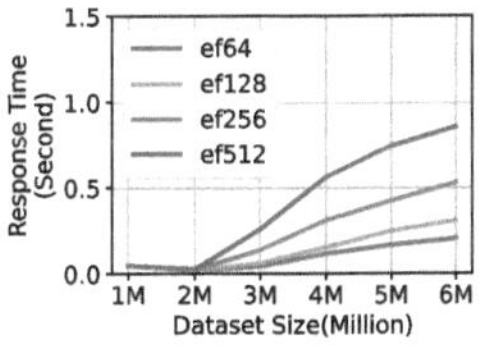

Fig. 7. Mmap + Engineering.

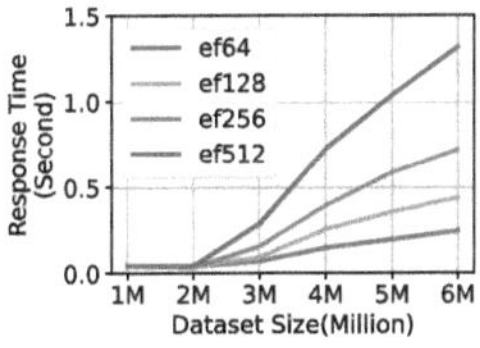

Fig. 8. Mmap + Academic disciplines.

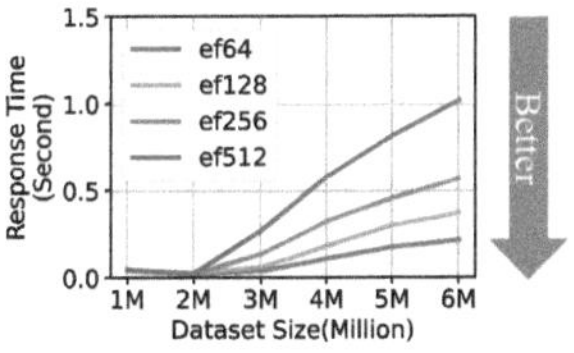

Fig. 9. Mmap + Politics.

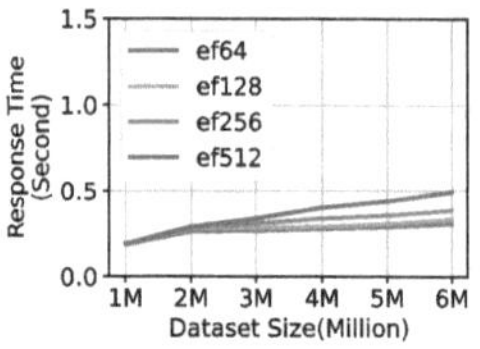

Fig. 10. Cache + Engineering

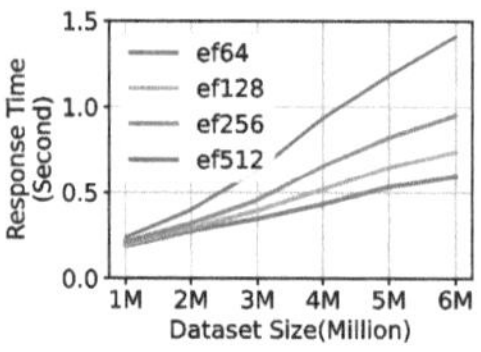

Fig. 11. Cache + Academic disciplines

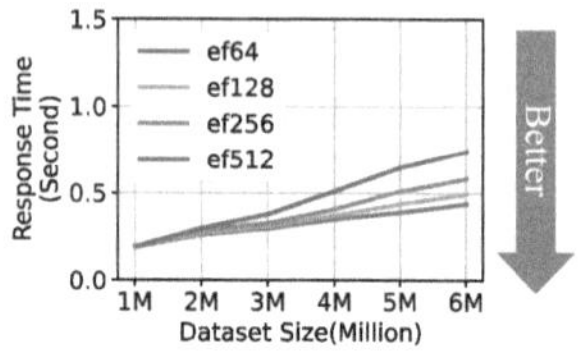

Fig. 12. Cache + Politics

5 Discussion

From the CPU and I/O utilization trends shown in Fig. 5, we observed that TC-HNSW has times when I/O utilization increases significantly and when CPU utilization increases significantly. When vector data are read from disk or a cache miss occurs, I/O utilization increases while CPU utilization decreases due to I/O waiting. Conversely, when a cache hit occurs or when reading vector data is unnecessary, I/O utilization decreases and CPU utilization increases. Therefore, the results in Fig. 5 indicate that TC-HNSW reads data from disk into memory and preferentially uses the cache in memory during search. In other words, TC-HNSW operates as designed, and the cache effect is clearly observed.

Next, focusing on the response time, the results of the conventional method (`mmap` + random queries) in Fig. 6 show that the response time increases significantly to 1.4 s as the *ef* value increased. In applications, such as RAG, it is necessary to search a sufficient amount of relevant information for the LLM from high-dimensional vectors within a short time. In other words, it is desirable to shorten the response time to 60% at high *ef* values, such as 256 and 512. Conversely, parameters such as *ef*64 sacrifice accuracy to ensure short response times (approximately 0.05 s). Therefore, they are a compromise value for the applications. In this context, TC-HNSW can reduce the response time to approximately one-third of the conventional method at high *ef* values (from 1.4 s to 0.5 s), as shown in Fig. 10 and 12, demonstrating its effectiveness in shortening response time without sacrificing the search accuracy provided by a given *ef* value.

On the other hand, as shown in Fig. 11 (Academic disciplines), the effect of TC-HNSW was not observed even at high *ef* values. This is because this category lacked topic bias. As shown in Fig. 8, the effect of shortening response time by restricting the category was minimal even with the conventional method. The response time for other categories using the conventional method, shown in Fig. 7 and 9, was shorter than that for random queries in Fig. 8. Therefore, Academic disciplines category contains various topics in the dataset, and the search process is similarly to that of random queries.

Focusing on dataset size, the results of the conventional method (mmap + random queries) in Fig. 6 show that the response time increased significantly to 0.4 s after exceeding 3M. This is because a dataset size of up to 2M fits entirely within the page cache of our experimental environment, resulting in minimal disk usage. Since this experiment aims to evaluate the cache effect of TC-HNSW, it is difficult to compare with such a small dataset size. Additionally, as shown in Fig. 10, for dataset sizes of 3M and 4M, it is possible that the 3 GB cache size we set was excessive. When experimenting with dataset sizes of 5M or 6M and categories with topic bias, a significant reduction in response time was observed at high *ef* values, such as *ef*512 and *ef*256. Therefore, by selecting appropriate dataset sizes and parameters suitable for RAG, it is possible to leverage the cache effect of TC-HNSW and shorten the response time to one-third that of the conventional method.

On the other hand, our study has several limitations. First, our evaluation relies solely on a dataset derived from a single source, specifically, that is the Wikipedia dataset. Second, the effectiveness of TC-HNSW is highly dependent on the quality of the categorization. Finally, we used fixed values for key parameters such as the cache size (3 GB) and update frequency (every 100 queries). We will aim to improve the performance by dynamically adjusting these parameters could potentially yield further performance improvements.

6 Related Work

Graph-based Approximate Nearest Neighbor Search (ANNS) is widely used as a search backend for RAG systems. However, it has problems such as large memory consumption at scale and I/O bottlenecks when using disks. This research aims to shorten the response time of HNSW search under memory constraints. Existing studies can be classified into the following three main approaches.

Graph Structure Optimization
Research like NSG [17] and Vamana (used in DiskANN [18]) improves the graph construction algorithm itself and proposes graph structures with higher search efficiency. This aims to shorten search paths and reduce calculation costs. Unlike these approaches, our research does not change the existing HNSW graph structure. Instead, we improve performance through data management and cache methods on top of it.

I/O Optimization for ANNS
HM-ANN [19] places the HNSW's hierarchical structure according to the speed hierarchy of heterogeneous memory. DiskANN [18] and GoVector [20] assume the use of secondary storage, such as SSDs. They aim to reduce delays from disk access by improving data placement and optimizing I/O access patterns. As a method using quantized vectors, Yamane et al. (Chimera-VDB) [21] proposed "Mixed-Precision" approach based on the HNSW graph structure. Their method keeps important vectors in high precision to achieve high recall while greatly reducing storage usage. Our study also aims to improve I/O efficiency, which is a shared objective among these studies. However, our approach is different because we focus on the "topic bias," which is a unique access pattern in RAG. We use category information to dynamically select and place cache targets.

Cache Strategies in the Search Process
Several methods have also been proposed to improve caching during the search process. They aim to increase the cache hit rate and reduce access to slower storage. Graph reordering [22] optimizes the placement of graph nodes in memory to improve CPU cache efficiency. PROXIMITY [23] uses the semantic similarity between queries in RAG. It caches and reuses the vector search results (document ID lists) for similar

queries. This reduces the number of queries sent to the vector database itself. In contrast, our research focused on caching the vector data itself. Our key feature is moving cache contents dynamically between storage tiers based on category information and access frequency. This is more specialized for RAG use case of topic bias.

Our research does not modify the existing HNSW graph structure. Instead, we use the topic bias of search queries within RAG. We aim to improve vector search performance under limited memory conditions by adopting a dynamic hierarchical caching mechanism based on category information. This represents a fundamentally different approach from the related works mentioned above.

7 Conclusion

In this paper, we addressed the response time problem of HNSW vector search in RAG use cases. The conventional method, relying on the page cache with memory-mapped files, suffers from a major issue: as the data size increases, disk I/O becomes a bottleneck, leading to significant increases in response time, especially at high *ef* values. To address this issue, we focused on the "topic bias" observed in RAG use cases. We proposed TC-HNSW, which incorporates a dynamic hierarchical cache method. The proposed method analyzes the reference frequency of categories from recent search history and proactively places vector groups that are likely to be searched next into the memory cache. This effectively reduces disk access. Furthermore, by managing the cache hierarchically and using higher-performance memory for the upper tiers, frequently accessed categories can be processed more efficiently. Evaluation experiments confirmed that TC-HNSW significantly reduces response time compared with the conventional `mmap`-based method for specific datasets.

This study demonstrates the effectiveness of a practical caching method that uses RAG-specific search patterns. It shortens the response time of HNSW search, which must run with `mmap` under memory constraints, without sacrificing search accuracy. Future challenges include evaluating the proposed method using additional datasets and optimizing parameter settings.

Acknowledgments. In this research, we used "mdx: a platform for building data-empowered society". This research was also supported by "Joint Usage/Research Center for Interdisciplinary Largescale Information Infrastructures (JHPCN)" in Japan (Project ID: jh251003).

References

1. Tomlinson, K., Jaffe, S., Wang, W., Counts, S., Suri, S.: Working with AI: measuring the applicability of generative AI to occupations. https://arxiv.org/abs/2507.07935. Last accessed 17 Oct 2025
2. Lewis, P., et al.: Retrieval-augmented generation for knowledge-intensive NLP tasks. In: Advances in neural information processing systems 33, pp. 9459–9474. Curran Associates, Inc. (2020)
3. Sharma, C.: Retrieval-augmented generation: a comprehensive survey of architectures, enhancements, and robustness frontiers. https://arxiv.org/abs/2506.00054. Last accessed 17 Oct 2025

4. Malkov, Y.A., Yashunin, D.A.: Efficient and robust approximate nearest neighbor search using hierarchical navigable small world graphs. IEEE Trans. Pattern Anal. Mach. Intell. **42**(4), 824–836 (2020)
5. Jégou, H., Douze, M., Schmid, C.: Product quantization for nearest neighbor search. IEEE Trans. Pattern Anal. Mach. Intell. **33**(1), 117–128 (2011)
6. Masuda, R., Iwamoto, K., Ando, K., Kamei, H.: Tiered Cache-HNSW: using hierarchical caching system in HNSW. In: 2025 1st International Conference on Consumer Technology (ICCT-Pacific), pp. 1–4 (2025)
7. Qdrant: What is Qdrant?. https://qdrant.tech/documentation/overview/. Last accessed 01 Oct 2025
8. Gao, Y., et al.: Retrieval-Augmented generation for large language models: a survey. https://arxiv.org/abs/2312.10997. Last accessed 17 Oct 2025
9. Wang, M., Xu, X., Yue, Q., Wang, Y.: A comprehensive survey and experimental comparison of graph-based approximate nearest neighbor search. Proc. VLDB Endow. **14**(11), 1964–1978 (2021)
10. Patil, R., Boit, S., Gudivada, V., Nandigam, J.: A survey of text representation and embedding techniques in NLP. IEEE Access **11**, 36120–36146 (2023)
11. Suzumura, T., et al.: mdx: a cloud platform for supporting data science and cross-disciplinary research collaborations. In: 2022 IEEE International Conference on Dependable, Autonomic and Secure Computing (DASC), pp. 1–7. IEEE (2022)
12. Qdrant: vector-db-benchmark. https://github.com/qdrant/vector-db-benchmark. Last accessed 17 Oct 2025
13. Wikipedia: Main Page. https://en.wikipedia.org/wiki/Main_Page. Last accessed 26 Oct 2025
14. Faralli, S., Finocchi, I., Ponzetto, S.P., Velardi, P.: Efficient pruning of large knowledge graphs. In: Proceedings of the Twenty-Seventh International Joint Conference on Artificial Intelligence, IJCAI-18, pp. 4055–4063 (2018)
15. Wikipedia: Category:Main topic classifications. https://en.wikipedia.org/wiki/Category:Main_topic_classifications. Last accessed 19 Oct 2025
16. Chen, J., Xiao, S., Zhang, P., Luo, K., Lian, D., Liu, Z.: M3-Embedding: multi-linguality, multi-functionality, multi-granularity text embeddings through self-knowledge distillation. In: Findings of the Association for Computational Linguistics: ACL 2024, pp. 2318–2335 (2024)
17. Fu, C., Xiang, C., Wang, C., Cai, D.: Fast approximate nearest neighbor search with the navigating spreading-out graph. Proc. VLDB Endow. **12**(5), 461–474 (2019)
18. Jayaram Subramanya, S., Devvrit, F., Simhadri, H.V., Krishnawamy, R., Kadekodi, R.: DiskANN: fast accurate billion-point nearest neighbor search on a single node. In: Advances in Neural Information Processing Systems 32, pp. 14695–14705. Curran Associates, Inc. (2019)
19. Ren, J., Zhang, M., Li, D.: HM-ANN: Efficient Billion-Point nearest neighbor search on heterogeneous memory. In: Larochelle, H., Ranzato, M., Hadsell, R., Balcan, M.F., Lin, H. In: Advances in Neural Information Processing Systems 33, pp. 10672–10684. Curran Associates, Inc. (2020)
20. Zhou, Y., et al.: GoVector: An I/O-Efficient caching strategy for high-dimensional vector nearest neighbor search. https://arxiv.org/abs/2508.15694. Last accessed 17 Oct 2025
21. Yamane, N., Zielewski, M.R., Nakamura, T., Suganuma, T.: Chimera-VDB: mixed-precision vector database with HNSW index for RAG-LLM. In: Proceedings of the 16th ACM SIGOPS Asia-Pacific Workshop on Systems, pp. 61–67. ACM, New York (2025)

22. Coleman, B., Segarra, S., Smola, A., Shrivastava, A.: Graph reordering for cache-efficient near neighbor search. In: Advances in neural information processing systems 35, pp. 27797–27810. Curran Associates, Inc. (2022)
23. Bergman, S.A., et al.: Leveraging approximate caching for faster retrieval-augmented generation. In: Proceedings of the 5th Workshop on Machine Learning and Systems, pp. 66–73. ACM, New York (2025)

Using Embeddings to Train and Build Privacy-Preserving Machine Learning Models and Applications

Pierce Buckner-Wolfson[1] and Dalmo Cirne[2(✉)]

[1] Wesleyan University, Middletown, CT 06459, USA
pbucknerwolf@wesleyan.edu
[2] Streams Group, LLC., Longmont, CO 80503, USA
contact@dalmocirne.com

Abstract. Protecting data privacy is a critical responsibility for application developers. However, without data, it becomes impossible to build entire categories of products. This paper proposes an innovative method for training Machine Learning (ML) models using only embeddings (derived data). Embeddings represent the original data as multidimensional vectors and, as such, can be plotted and clustered in hyperspace, enabling effective solutions for problems such as anomaly detection, search, sentiment analysis, recommendation, and graph prediction, without requiring access to the raw data. Using only derived representations of the data and their clustering patterns, this approach preserves privacy while allowing responsible application development. Many machine learning algorithms such as Neural Networks, Gradient Boosting, and K-Nearest Neighbors are well-suited tools, providing cost-effective and computationally efficient alternatives to Large Language Models (LLMs). The approach proposed here balances data-driven innovation that is compliant with strict privacy requirements while unlocking the space for the development of powerful applications.

Keywords: Embeddings · Machine Learning · Privacy

1 Introduction

A central challenge in modern machine learning is balancing the field's dependence on vast datasets with the critical need for safeguarding user privacy. As ML models become integral to sensitive domains like finance and healthcare, the need for tools to protect this data has led to the emergence of Privacy-Preserving Machine Learning (PPML) [1].

A commonly used privacy-preserving technique is to analyze data at an aggregate level, such as looking at distributions or summary statistics. Although this approach is inherently safe, it sacrifices the fine-grained individual-level detail required to build robust predictive systems.

H. R. Arabnia et al. (Eds.): CSCI 2025, CCIS 2969, pp. 19–30, 2027.
https://doi.org/10.1007/978-3-032-28460-0_2

With these concerns in mind, this paper proposes the use of embeddings as a practical and effective solution. Embedding sensitive data acts as a proxy layer, transforming it into high-dimensional vectors that are not suitable for human interpretation. Moreover, these vectors preserve the original relational patterns in the raw data, making them highly useful for computation. By operating on these vector representations, ML models can learn from the underlying structure of the data to perform tasks like *anomaly detection*, by identifying embeddings outside typical clusters, *recommendation*, by suggesting the typical values in the cluster, or *classification*, by observing patterns. All without accessing the raw sensitive inputs.

- **Anomaly detection**
  - General purpose anomaly detection engine (i.e., risk score calculator).
  - Text, tags, and non-continuous data are clustered by semantic similarity.
  - Simple fraud detection.
- **Recommendations**
  - Similar and frequently used items are suggested during data entry.
  - Resource management (assets and resource forecasting).
- **Classification**
  - Extract keywords from text and use them to categorize content (e.g., procurement, expenses, accounts payable, categorize memos).

These embedding patterns can be effectively leveraged by a number of efficient machine learning models, including Gradient Boosting [2], Neural Networks [3], and K-Nearest Neighbors [4]. Although an LLM may solve similar problems when augmented with techniques such as Retrieval Augmented Generation (RAG) [5] and Low-Rank Adaptation (LoRA) [6], the models proposed here offer a more scalable and computationally efficient pathway for many real-world applications.

This paper empirically evaluates the embedding-based approach on a real-world financial task: loan default prediction. The results demonstrate that this method is highly competitive with a baseline model trained on raw data, validating its use as a foundational step for building effective and more responsible PPML systems.

2 Methodology

2.1 System Architecture

The architecture diagram for the proposed privacy-preserving machine learning system is shown in Fig. 1. This blueprint serves as the foundation for the methodology detailed in this paper. The workflow, designed to prevent data exposure, proceeds in several distinct stages:

1. First, embeddings are generated within the same secure environment as the source database. This is a critical design choice, as it ensures that sensitive data is processed without requiring new access permissions.

2. Next, only the abstract embeddings are streamed (e.g., using a message queue like Kafka [7]) to the separate machine learning environment. The raw data never leaves its secure enclave.
3. Finally, models are trained on these embeddings, and the resulting endpoints are published as a service, allowing them to be used by applications while preserving privacy.

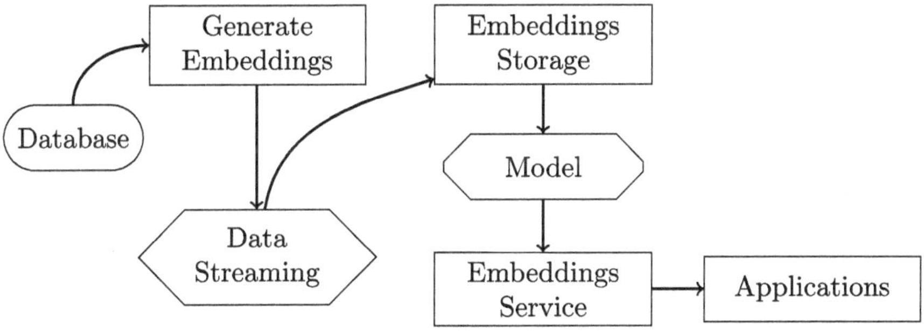

Fig. 1. Architecture diagram

The architecture diagram illustrates the high-level, complete end-to-end system for a production environment. This paper covers it only partially, focusing on validating the core components required to make such a system viable (i.e., generate embeddings, model training, embeddings service, and sample application). The implementation of the core Embedding Service is available on GitHub [8].

The following three sections (Data Pipeline, Embeddings Pipeline, and Model Framework and Architecture) describe each stage in details.

2.2 Data Pipeline

This study uses the publicly available Lending Club (LC) dataset [9], chosen specifically because it has undergone minor preprocessing. Other datasets, particularly in the credit fraud space where privacy concerns are of high importance, often obscure raw features through techniques like Principal Component Analysis (PCA) [10], the LC dataset provides the rich, mixed-type features (e.g., raw text, numerical values) that are essential for developing and testing a privacy-preserving technique that operates on original data. The only significant anonymization is the usage of k-anonymity [11] to mask the exact zip code, for example: 809xx.

The dataset contains records for approximately 2 million accepted loans. For this classification task, all loans with a "Current" status were filtered out, resulting in a final working dataset of approximately 1.4 million completed loans. Exploratory analysis revealed a significant class imbalance, with defaulted loans representing 21.2% of outcomes. This imbalance is addressed in the Model Training section.

A multistage feature selection process was employed to ensure methodological rigor. First, an initial pruning was performed, where features with over 60% missing values and those with high inter-feature correlation were removed, exceptions were made for pairs where both features demonstrated high predictive importance with the target variable, such as loan sub-grade and interest rate. Second, to ensure a realistic simulation of a real-world loan application scenario, any features containing "leaky" information (i.e., data that would not be available at the time of origination) were discarded.

Table 1. Selected Features and Descriptions

Feature Name	Description
`loan_amnt`	The total amount of the loan applied for by the borrower.
`int_rate`	The interest rate assigned to the loan.
`sub_grade`	The loan sub-grade assigned by Lending Club.
`title`	The loan title provided by the borrower (e.g., "Business").
`emp_title`	The job title provided by the borrower.
`emp_length`	The borrower's employment length in years.
`addr_state`	The state where the borrower resides.
`zip_code`	The first three digits of the borrower's zip code.
`dti`	Borrower's ratio of total monthly debt to monthly income.
`fico_range_high`	The upper bound of the borrower's FICO credit score.
`earliest_cr_line`	The month and year of the borrower's first credit line.
`revol_util`	Percent of revolving credit the borrower is using.
`mort_acc`	The number of mortgage accounts on record.
`avg_cur_bal`	The average current balance on all of the borrower's accounts.
`num_op_rev_tl`	The number of open revolving credit lines.

From the remaining 68 candidates, a final set of 15 features (shown on Table 1) was selected based on a feature importance analysis using a preliminary XGBoost model [12] and a review of prior literature [13–15].

2.3 Embeddings Pipeline

This paper introduces a novel embeddings pipeline designed to unlock inaccessible private data for machine learning. The pipeline leverages the privacy-preserving nature of embeddings by transforming raw data into abstract, high-dimensional vectors. This transformation yields representations that are not directly human-interpretable, yet retain the essential relational patterns required for building powerful predictive models. Crucially, the proposed pipeline is specifically designed to handle the mixed-type datasets common in real-world records by employing distinct components to process textual and numerical features.

The specific implementation of this pipeline for the experiments is detailed below.

For the experiments, a pre-trained sentence-transformer model was used to generate the embeddings. Given the large scale of the dataset, computational efficiency and cost were primary design constraints. Therefore, the all-MiniLM-L6-v2 (MiniLM) [16,17] model was selected, which offers a strong balance between high embedding quality and efficient processing speed. (While more powerful models exist on the MTEB [18] and FinMTEB leaderboards [19], MiniLM represents a practical and robust choice.)

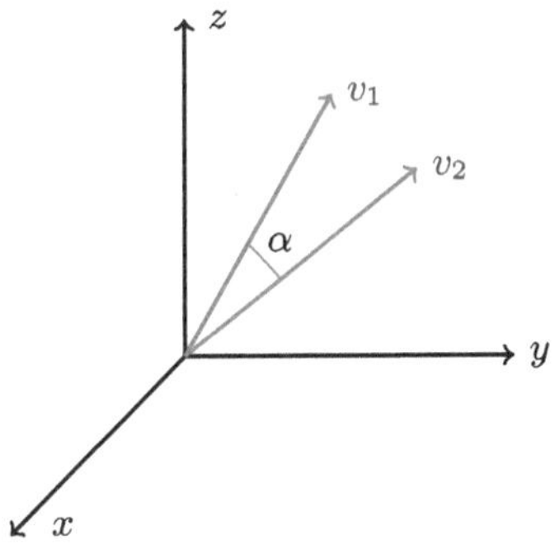

Fig. 2. Vector space.

While transformer-based embeddings models excel at capturing semantic meaning, they are not designed to preserve the mathematical properties of numerical data. To address this limitation, the pipeline explores incorporating a specialized technique for numeracy-preserving embeddings, Deterministic Independent of Corpus Embeddings (DICE) [20]. DICE transforms numbers such that the cosine similarity (1) between their vectors reflects their distance; the closer they are, the more likely for them to be similar (Fig. 2).

$$\cos \alpha = \frac{v_1 \cdot v_2}{|v_1|\,|v_2|} \tag{1}$$

The length of the vectors v_1 and v_2 must be greater than zero ($|v_n| > 0$). Otherwise, a division by zero will lead to an undetermined value.

2.4 Model Framework and Architecture

To assess the performance of the proposed pipeline, a number of predictive models were considered based on their demonstrated success in prior analysis of the LC dataset. Consistent with the literature, XGBoost [21] and Neural Network

[22] models were selected as subjects for this classification task. Given the sensitivity of these models to their configuration, an important goal was to ensure fairness across experiments. Thus, a consistent training and evaluation protocol was established. It was developed and validated on a subset of the data before being applied to full experiments.

To model the complex relationships within the data, the neural network used is a standard feed-forward network, a Multi-Layer Perceptron (MLP). The architecture consists of two hidden layers and a single output neuron suitable for a binary classification task. The design and parameter decisions are based upon that of the previous literature [22].

Consistent with common deep learning practices for high-dimensional data, the design employed a funnel structure [23]. The network's input layer is determined by the dimensionality of each data pipeline's feature vector (see Table 2 for a breakdown of feature vector sizes). The first hidden layer consists of 512 neurons, designed to learn a dense, compressed representation of the wide input. The second hidden layer further distills these patterns with 256 neurons. The final output layer consists of a single neuron to produce the classification logit. As an activation function, we selected the Rectified Linear Unit (ReLU) over older alternatives like tanh (see [22]), as it is a modern standard known for its computational efficiency and ability to mitigate the vanishing gradient problem [24].

Experimental Design. The experimental process was designed to isolate the impact of the embeddings pipeline by comparing three distinct data representation methods.

1. **Non-Private Baseline**: Models trained on the raw features. Textual features were processed using a TF-IDF vectorizer [25], and categorical features were one-hot encoded [26].
2. **Text-Only Embeddings**: Models trained on a dataset where all features, including numerical ones, were first cast to strings and then embedded using the all-MiniLM-L6-v2 transformer (384 dimensions).
3. **Hybrid Embeddings**: Models trained where text features were processed by the transformer and numerical features were processed using DICE embeddings (32 dimensions).

Each representation method results in a distinct input feature dimensionality, summarized in Table 2.

Model Training. For model training and optimization, the data was partitioned into 3 sets for training (64%), validating (16%), and testing (20%). The intent is to ensure fair tuning and final evaluation. The validation set was utilized to implement early stopping for both models to prevent overfitting [27] and to determine the optimal number of training iterations (epochs). In addition, class imbalance was addressed by using a class weighting strategy to prevent

Table 2. Input Vector Dimensions by Representation Method

Representation	Feature Composition	Input Size
Text-Only	15 features × 384-dim transformer embeddings	5760
Hybrid	(8 numerical × 32-dim DICE) + (7 text × 384-dim)	2944
Non-Private Baseline	Raw features (TF-IDF + one-hot encoded)	2514

either model from developing a bias toward the majority (negative) class. Thus, a parameter *pos_weight* was incorporated for both models. It is calculated as the ratio of the number of negative samples to the number of positive samples in the training data (2).

$$pos\_weight = \frac{\text{Number of negative samples}}{\text{Number of positive samples}} \tag{2}$$

This method increases the penalty for miss-classifying the minority (positive) class, pushing the models to learn the minority's distinguishing features more effectively, resulting in a more robust model.

The MLP was trained using the Adam optimizer, selected for its adaptive learning rate and robust performance, which aligns with findings in prior literature [22]. To handle the class imbalance present in the dataset, we used a BCEWithLogitLoss function, which combines a sigmoid activation with binary cross-entropy loss for numerical stability and allows for the addition of the *pos_weight* parameter.

To prevent overfitting, a major issue, and improve generalization, the training procedure incorporated two regularization techniques. First, Batch Normalization was applied after each linear transformation to stabilize and accelerate the training process. Second, a Dropout rate of 0.3 was applied to the hidden layers, forcing the network to learn more robust feature representations.

Evaluation Protocol. Model evaluation utilized a comprehensive set of metrics including Precision, Recall, F1-score [28], and the Area Under the Receiver Operating Characteristic Curve (ROC AUC) [29]. The primary metric for comparing the overall discriminative power of the different experimental pipelines (e.g., Non-Private Baseline vs. Hybrid Embeddings) was ROC AUC. Given the significant class imbalance, this threshold-independent metric provides the most robust measure of a model's intrinsic ability to distinguish between positive and negative classes.

Additionally, to provide insight into each model's practical utility at an optimal decision point, threshold-dependent metrics (Precision, Recall, and F1-score) are also reported. To ensure a fair comparison, these metrics were calculated using a classification threshold that was individually optimized for each model. This optimal threshold was determined by finding the point on the Precision-Recall curve that maximized the F1-score, thereby representing the model's best achievable balance between precision and recall.

3 Experimental Results

The following are the performance results of the XGBoost and Neural Network models across the three data representation pipelines, evaluated on the held-out test set. Detailed classification reports are in Tables 4 and 6 and summaries are provided in Tables 3 and 5.

3.1 XGBoost Performance

For the XGBoost model, the embedding-based pipelines demonstrated a modest but consistent improvement in discriminative ability over the non-private baseline. As summarized in Table 3, the hybrid embedding pipeline achieved the highest performance with a ROC AUC score of 0.722. This surpassed both the Text-Only pipeline (0.721) and the Non-Private Baseline (0.717).

Table 3. XGBoost Models Overall Performance Summary

Data Pipeline	ROC AUC Score
Non-Private Baseline	0.717
Text-Only Embeddings	0.721
Hybrid Embeddings	**0.722**

Although the ROC AUC scores indicate a stronger underlying signal from the embedded representations, this did not translate to significant gains in the threshold-dependent metrics. As shown in the detailed report in Table 1, the F1-score for the positive (minority) class remained stable at 0.45 across all three pipelines. This suggests that while the model's ability to rank candidates improved with embeddings, the task of cleanly separating the classes at a single decision point remains challenging.

Table 4. Detailed Classification Report for XGBoost Models on the Test Set

Data Pipeline	Class	Precision	Recall	F1-score	Support
Non-Private Baseline	Negative (0)	0.87	0.69	0.77	215748
	Positive (1)	0.35	0.63	0.45	58165
Text-Only Embeddings	Negative (0)	0.87	0.68	0.77	215748
	Positive (1)	0.35	0.64	0.45	58165
Hybrid Embeddings	Negative (0)	0.88	0.66	0.75	215748
	Positive (1)	0.34	0.67	0.45	58165

3.2 Neural Network Performance

In contrast to the XGBoost model, the Neural Network achieved its peak performance when trained on the Text-Only embeddings, yielding the highest ROC AUC of 0.718, outperforming both the hybrid pipeline (0.717) and the Non-Private Baseline (0.715), as shown in Table 5. This suggests that the neural network, which processes all input features simultaneously, struggled to effectively integrate the two heterogeneous embedding types. In terms of threshold-dependent metrics like F1-score the Neural Network had stable performance across pipelines, similar to XGBoost.

Table 5. Neural Network Models Overall Performance Summary

Data Pipeline	ROC AUC Score
Non-Private Baseline	0.715
Text-Only Embeddings	**0.718**
Hybrid Embeddings	0.717

The dense, 384-dimensional text embeddings contain a stronger and more complex signal than the 32-dimensional numerical DICE embeddings. For the neural network, concatenating these may have caused the less informative numerical features to act as noise, degrading the learning process and making it difficult for the network to capitalize on the patterns in the dominant text features.

Table 6. Detailed Classification Report for Neural Network Models on the Test Set

Data Pipeline	Class	Precision	Recall	F1-score	Support
Non-Private Baseline	Negative (0)	0.88	0.66	0.75	215748
	Positive (1)	0.34	0.66	0.45	58165
Text-Only Embeddings	Negative (0)	0.88	0.64	0.74	215748
	Positive (1)	0.34	0.67	0.45	58165
Hybrid Embeddings	Negative (0)	0.87	0.70	0.78	215748
	Positive (1)	0.36	0.61	0.45	58165

Conversely, XGBoost's tree-based architecture is more robust to this feature heterogeneity. Its inherent feature selection mechanism at each split allowed it to selectively leverage the most predictive features from either the text or numerical embeddings at any given decision point, effectively ignoring the noisier inputs when necessary.

4 Conclusion

This research addresses the critical challenge of privacy in machine learning by introducing a pipeline that utilizes embeddings to enable model development on sensitive data. While the field of PPML has established powerful tools like Differential Privacy (DP) [30] and Homomorphic Encryption [31], these methods often involve well-documented trade-offs, such as significant utility loss or high computational overhead. This study explores a complementary approach, focusing on embeddings as a privacy-preserving foundation.

In conclusion, this paper demonstrates that embedding-based pipelines are a viable and effective strategy for building predictive models on sensitive data, enabling the development of privacy-preserving applications. This approach was validated by the highly competitive performance between the embedding-based models and the non-private baseline. The top-performing model, an XGBoost classifier using hybrid embeddings, achieved a ROC AUC score of 0.72, which is comparable to both our baseline and results from previous literature [21,22].

This key finding suggests that abstracting raw data into high-dimensional embeddings successfully preserves the underlying predictive patterns necessary for effective modeling. A notable secondary result was the varied response of different model architectures to embedding strategies, highlighting the importance of matching the data representation to the chosen model.

The successful validation of this pipeline opens two key avenues for future work: enhancing model performance and formalizing privacy guarantees.

First, to improve predictive utility, more advanced embedding models need to be explored, such as state-of-the-art transformers from the MTEB leaderboard or domain-specific models like those on the FinMTEB leaderboard.

Second, a critical direction is to provide formal mathematical privacy guarantees. While it is known that applying Differential Privacy (DP) directly to raw, high-dimensional data can degrade model utility; the embedding pipeline presents a new opportunity. By analyzing the practical reversibility of embeddings to better understand the inherent privacy they provide [32]. A reduced amount of calibrated noise can be applied to the abstract embeddings rather than the raw features, making it possible to achieve a strong privacy guarantee with a much lower impact on model performance. Future work will investigate this trade-off, aiming to combine the privacy-preserving embedding approach with DP to create a framework that is both highly accurate and formally private.

References

1. Joshi, J., Xu, R., Baracaldo, N.: Privacy-Preserving Machine Learning: Methods, Challenges and Directions. arXiv preprint arXiv:2108.04417 (2021). https://doi.org/10.48550/arXiv.2108.04417
2. Biau, G., Cadre, B.: Optimization by gradient boosting. arXiv eprint arXiv:1707.05023 (2017). https://doi.org/10.48550/arXiv.1707.05023

3. Gabrié, M., Ganguli, S., Lucibello, C., Zecchina, R.: Neural networks: from the perceptron to deep nets. arXiv eprint arXiv:2304.06636 (2023). https://doi.org/10.48550/arXiv.2304.06636
4. Cunningham, P., Delany, S.: K-Nearest neighbour classifiers - a tutorial. ACM Comput. Surv. **54**, New York, NY, USA (2021). https://doi.org/10.1145/3459665
5. Lewis, P., et al.: Retrieval-augmented generation for knowledge-intensive NLP tasks. In: NIPS'20: Proceedings of the 34th International Conference on Neural Information Processing Systems, pp. 9459–9474, Vancouver, BC, Canada (2020). https://dl.acm.org/doi/10.5555/3495724.3496517
6. Yang, M., et al.: Low-Rank Adaptation for Foundation Models: A Comprehensive Review arXiv, (2024). https://doi.org/10.48550/arXiv.2501.00365
7. Apache Kafka. https://kafka.apache.org. Accessed 05 Sep 2025
8. Buckner-Wolfson, P.: Embeddings-Service. Version 1.0.0, GitHub (2025). https://github.com/PierceBW/Embeddings-Service. Accessed 05 Sep 2025
9. Lending Club Dataset. https://www.kaggle.com/datasets/wordsforthewise/lending-club. Accessed 02 Sep 2025
10. Credit Card Fraud Detection. https://www.kaggle.com/datasets/mlg-ulb/creditcardfraud. Accessed 02 Sep 2025
11. Sweeney, L.: k-anonymity: a model for protecting privacy. In: International Journal of Uncertainty, Fuzziness and Knowledge-based Systems, vol. 10, pp. 557–570. World Scientific Publishing Co., Inc. USA (2002). https://doi.org/10.1142/S0218488502001648
12. Chen, T., Guestrin, C.: XGBoost: a scalable tree boosting system. In: 22nd ACM SIGKDD International Conference on Knowledge Discovery and Data Mining, pp. 785–794. ACM, San Francisco, California, USA (2016). https://doi.org/10.1145/2939672.2939785
13. Emekter, R., Tu, Y., Jirasakuldech, B., Lu, M.: Evaluating credit risk and loan performance in online Peer-to-Peer (P2P) lending. Appl. Econ. **47**(1), 54–70 (2015). https://doi.org/10.1080/00036846.2014.962222
14. Möllenkamp, N.: Determinants of loan performance in P2P lending, Enschede, The Netherlands (2017). BS thesis
15. Sharar, M.L.: Online peer-to-peer lending: determinants of loan performance. Available at SSRN: 3785323 (2021). https://doi.org/10.2139/ssrn.3785323
16. Wang, W., et al.: MiniLM: Deep self-attention distillation for task-agnostic compression of pre-trained transformers. arXiv preprint arXiv:2002.10957 (2020). https://doi.org/10.48550/arXiv.2002.10957
17. Sentence-Transformers: all-MiniLM-L6-v2 model. https://huggingface.co/sentence-transformers/all-MiniLM-L6-v2. Accessed 02 Sep 2025
18. Enevoldsen, K., et al.: MMTEB: Massive Multilingual Text Embedding Benchmark. arXiv preprint arXiv:2502.13595 (2025). https://doi.org/10.48550/arXiv.2502.13595
19. Tang, Y., Yang, Y.: FinMTEB: Finance Massive Text Embedding Benchmark. arXiv preprint arXiv:2502.10990 (2025). https://doi.org/10.48550/arXiv.2502.10990
20. Carin, L., et al.: Methods for numeracy-preserving word embeddings. In: Proceedings of the 2020 Conference on Empirical Methods in Natural Language Processing (EMNLP), pp. 4742–4753. Association for Computational Linguistics, Online (2020). https://doi.org/10.18653/v1/2020.emnlp-main.384
21. Souadda, L.I., Halitim, A.R., Benilles, B., Oliveira, J.M., Ramos, P.: Optimizing credit risk prediction for peer-to-peer lending using machine learning. Forecasting **7**(3), 35 (2025). https://doi.org/10.3390/forecast7030035

22. Turiel, J.D., Aste, T.: Peer-to-Peer loan acceptance and default prediction with artificial intelligence. R. Soc. Open Sci. **7**(6), 191649 (2020). https://doi.org/10.1098/rsos.191649
23. Hinton, G.E., Salakhutdinov, R.R.: Reducing the dimensionality of data with neural networks. Science **313**(5786), 504–507 (2006). https://doi.org/10.1126/science.1127647
24. Agarap, A.F.: Deep Learning using Rectified Linear Units (ReLU). arXiv preprint arXiv:1803.08375 (2019). https://doi.org/10.48550/arXiv.1803.08375
25. TF-IDF Demystified. https://towardsdatascience.com/tf-idf-demystified-7dd3ef071b24/. Accessed 02 Sep 2025
26. Robust One-Hot Encoding. https://towardsdatascience.com/robust-one-hot-encoding-930b5f8943af/. Accessed 02 Sep 2025
27. Overfitting. https://developers.google.com/machine-learning/crash-course/overfitting/overfitting. Accessed 02 Sep 2025
28. Understanding Precision, Recall, F1-score, and Support in Machine Learning Evaluation. https://medium.com/@nirajan.acharya777/understanding-precision-recall-f1-score-and-support-in-machine-learning-evaluation-7ec935e8512e. Accessed 02 Sep 2025
29. Area Under the Curve (AUC): A Robust Performance Measure of Classification Models. https://medium.com/@bayramorkunor/area-under-the-curve-auc-a-robust-performance-measure-of-classification-models-cbfc3549d8c6. Accessed 02 Sep 2025
30. Danger, R.: Differential Privacy: What is all the noise about? arXiv preprint arXiv:2205.09453 (2022). https://doi.org/10.48550/arXiv.2205.09453
31. Agrawal, J., Nguyen, H.: Fraud detection using machine learning models and encryption techniques. In: Arabnia, H.R., Deligiannidis, L., Amirian, S., et al. (eds.) Artificial Intelligence and Applications, pp. 115–124. Springer Nature Switzerland, Cham (2025). https://doi.org/10.1007/978-3-031-86623-4_9
32. Morris, J.X., Kuleshov, V., Shmatikov, V., Rush, A.M.: Text Embeddings Reveal (Almost) As Much As Text. arXiv preprint arXiv:2310.06816 (2023).https://doi.org/10.48550/arXiv.2310.06816

Multiregional Characterization of Traditional Street Cries in Mexico Using Acoustic Parameters

Agosto de la Gala-Ureña[1](✉), Julio-Alejandro Romero-González[1], M. Florencia Assaneo[2], Diana-Margarita Córdova-Esparza[1], Juvenal Rodríguez-Reséndiz[3], and Ricardo Chaparro-Sánchez[1]

[1] Faculty of Informatics, Autonomous University of Queretaro, 76230 Queretaro, Mexico
crodriguezdelagala27@alumnos.uaq.mx,
{julio.romero,diana.cordova,rchapa}@uaq.mx

[2] Institute of Neurobiology, National Autonomous University of Mexico, Juriquilla, Queretaro, Mexico
fassaneo@inb.unam.mx

[3] Faculty of Engineering, Autonomous University of Queretaro, 76230 Queretaro, Mexico
juvenal@uaq.edu.mx

Abstract. This study aims to quantitatively characterize Mexican street cries (pregones) by exploring a range of acoustic parameters and examining potential prototype distinctions across geographical regions. The samples ($n = 166$) were obtained from the downloadable audio files of the Mexican Sound Map (https://mapasonoro.cultura.gob.mx/). They were classified according to three geographical regions: North–Northeast "NNe" ($n = 13$), Center–West "CW" ($n = 93$) and South–Southeast "SSe" ($n = 60$). A window size of 3 s was selected (NNe=45, CW=102, SSe=86) for the calculation of the selected acoustic parameters: jitter, shimmer, pitch (f_0), the standard deviation of pitch ($f_0 stdv$), the averages of the formants F_i ($i = 1, ..., 4$) and the standard deviation of the differences between these formants $Disp(F_j - F_i)$. A Principal Component Analysis (PCA) was applied, and two groupings were identified: one associated with the vocal tract, represented by three principal components, and another related to the vocal folds, comprising two components. In certain 2D and 3D PC subspaces, a clearer distinction was observed between the cries from the CW and NNe regions, whereas the separation between CW and SSe was less pronounced. When comparing each region against the others, the cries from the NNe region exhibited the greatest degree of differentiation. For classification, the Support Vector Machine (SVM), Random Forest (RF), and K-Nearest Neighbors (KNN) classifiers were employed. The results obtained were expressed in terms of *accuracy* (SVM: 0.76, RF: 0.73, KNN: 0.76), *precision* (SVM: 0.77, RF: 0.74, KNN: 0.75), and *recall* (SVM: 0.76, RF: 0.73, KNN: 0.76).

H. R. Arabnia et al. (Eds.): CSCI 2025, CCIS 2969, pp. 31–47, 2027.
https://doi.org/10.1007/978-3-032-28460-0_3

Keywords: Acoustic adaptation · Street cries · Acoustic parameters · PCA · Kendall correlation

1 Introduction

Street cries (pregones) have been part of the urban soundscape in different regions of Mexico for centuries, becoming an integral element of popular culture and exhibiting distinctive acoustic properties that may vary across geographical areas. However, despite their cultural and social relevance, an objective distinction of potential regional variants has received little attention from a quantitative perspective. To date, street cries have been studied primarily within the social sciences, from historical, socioeconomic, and cultural perspectives [20,21,24]. Sharma [20] even considers them as an acoustic phenomenon, since one consequence of urbanization is the introduction of motorized machines such as vehicles and their increased use, leading to heavier street traffic, higher decibel levels, and forcing street vendors to cry out their products or services at much higher volumes than before in order to be heard above the noise. From this perspective, street cries appear to represent another case within the acoustic adaptation hypothesis, for which there is extensive research across different animal species, including birds, mammals, insects, and anurans [4,5,14]. This possible adaptation is driven by two main objectives: on the one hand, to be heard by as many people as possible, and on the other, to increase sales. However, from a scientific perspective, this phenomenon can be analyzed through acoustic parameters, which have been extensively studied and widely applied in various investigations to characterize phonatory phenomena, such as evaluating voice quality in patients with voice disorders [11], determining whether these parameters are predictive and discriminative in patients with and without depression [22], or identifying significant differences between spoken and sung tasks [13]. The purpose of this work is to provide an approach to the characterization of street cries from different regions of Mexico, aimed at identifying distinctions in their phonation through a set of acoustic parameters (jitter, shimmer, pitch (f_0), the standard deviation of pitch ($f_0 stdv$), the mean values of the formants F_i ($i = 1, ..., 4$) and the standard deviation of the inter-formant differences $Disp(F_j - F_i)$). The analysis of these variables revealed the existence of two blocks of principal components: the first related to vocal tract parameters, and the second associated with vocal fold parameters. This characterization may contribute to a better understanding of the phonatory diversity among regional styles of pregones.

2 Methods

Figure 1 presents the methodological framework implemented in the present study, outlining the main stages of the process from data acquisition to classifier evaluation.

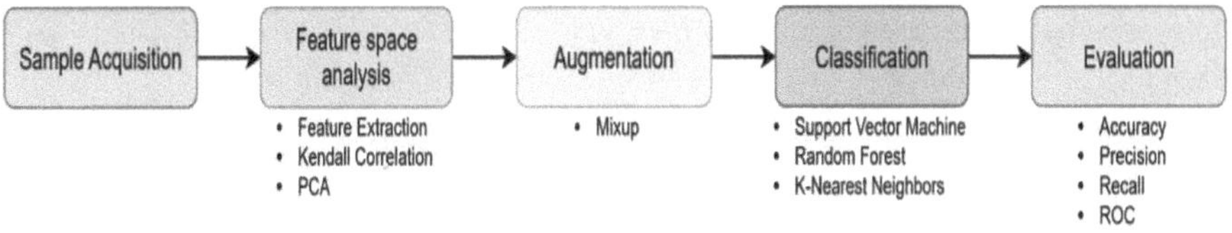

Fig. 1. Methodological framework of the study.

2.1 Sample Acquisition

In this stage, samples ($n = 166$) were obtained from the downloadable audio files of the Mapa Sonoro de México [16] and classified into three geographical regions: North–Northeast (NNe, $n = 13$) which includes the states of Chihuahua, Nuevo León, San Luis Potosí and Sonora; Center–West (CW, $n = 93$), comprising CDMX, Estado de México, Guanajuato, Jalisco and Michoacán; and South–Southeast (SSe, $n = 60$) including Campeche, Chiapas, Guerrero, Oaxaca, Puebla, Tabasco, Veracruz and Yucatán. This classification only considered states for which audio samples were available.

Cleaning and Exclusion Criteria. All selected files were preprocessed using Sound Forge Audio Cleaning LAB 4. The parameters were adjusted as follows: DeHisser at 40% and DeNoiser configured between 60% and 80%. In addition, the noise suppression function of Clipchamp was applied to automatically remove unwanted background noise and improve sound quality. For each selected audio file, sections in which background noise predominated over the street cry were removed, as well as cries in languages other than Spanish.

Cropping. From all the cleaned segments extracted, a window size of 3 s was selected (NNe = 45, CW = 102, SSe = 86), considering that this length could represent the minimum window necessary to distinguish a street cry from other types of speech. It is important to note that more than one window may originate from the same audio file.

2.2 Feature Space Analysis

The analysis of the feature space is essential for the study of street cries (pregones), as it allows transforming the information contained in the acoustic signal into a set of parameters that capture phonatory, temporal, and spectral aspects of the voice.

Feature Extraction. The acoustic parameters used to characterize the audio files were jitter, shimmer, pitch (f_0), the standard deviation of pitch ($f_0 stdv$), the mean values of the formants F_1, F_2, F_3 and F_4, and the standard deviation of the differences between these formants: $Disp(F_2-F_1)$, $Disp(F_3-F_1)$, $Disp(F_4-F_1)$, $Disp(F_3-F_2)$, $Disp(F_4-F_2)$ and $Disp(F_4-F_3)$.

Jitter. It calculates the cycle-to-cycle period perturbations of successive glottal cycles during sustained phonation [12]. There are several ways to calculate jitter; one of them is *local jitter*, defined by the following Eq. 1.

$$jitter\ (local) = \frac{\frac{1}{N-1}\sum_{i=1}^{N-1} |T_i - T_{i-1}|}{\frac{1}{N}\sum_{i=1}^{N} T_i} \cdot 100 \tag{1}$$

where T_i denotes the extracted glottal periods and N the total number of periods [25].

Shimmer. This parameter assesses the perturbation of the amplitude of the sound wave across consecutive periods; that is, it quantifies the variability in intensity or energy between successive cycles of vocal fold vibration during speech [12]. It is calculated as shown in the Eq. 2.

$$shimmer\ (dB) = \frac{1}{N-1}\sum_{i=1}^{N-1} \left| 20 \cdot \log\left(\frac{A_{i+1}}{A_i}\right)\right| \tag{2}$$

where A_i are the extracted amplitudes and N is the number of periods obtained from the fundamental frequency [25].

Fundamental frequency f_0. Also known as *pitch*, it is directly associated with the vibration of the vocal folds and corresponds to the number of times the glottis opens and closes within one second [12]. Vocal folds of greater length tend to vibrate at a slower rate; nevertheless, additional factors including mass, thickness, tension, age, and sex also play a significant role.

Formants. These are the resonant frequencies of the vocal tract that generate regions of higher spectral intensity. They are commonly denoted as F_1, F_2, $F_3, \cdots, F_i$, ordered in ascending frequency. The pattern formed by the first formants (typically the first two) is largely responsible for the perceptual differences among vowels. Since longer vocal tracts exhibit lower resonances, the absolute frequencies of the formants depend on the size of the vocal tract. [1].

The extraction of acoustic features was conducted with Praat software (version 6.4.38), using its default hyperparameter settings (Table 1). The computation of $Disp(F_j - F_i)$ involved extracting the formant frequency values from each frame and calculating the standard deviation of the absolute differences between formant pairs. Instances containing NaN values were excluded. The resulting data were then standardized using z-scores ($\mu = 0, \sigma = 1$) to normalize the acoustic variables and mitigate scale-related differences, thus enabling meaningful comparisons. Class balance was achieved by means of random sampling.

Table 1. Standard hyperparameters in Praat (version 6.4.38) for jitter, shimmer, pitch, and formant parameters.

Acoustic Parameters	Object	Hyperparameters
Jitter	To PointProcess (periodic, cc):	Pitch range (Hz): 75 – 600
	Get jitter (local):	Time range (s): 0, 0 (= all)
		Shortest period (s): 0.0001
		Longest period (s): 0.02
		Maximum period factor: 1.3
Shimmer	To PointProcess (periodic, cc):	Pitch range (Hz): 75 – 600
	Get shimmer (local_dB):	Time range (s): 0, 0 (= all)
		Shortest period (s): 0.0001
		Longest period (s): 0.02
		Max period factor: 1.3
		Max amplitude factor: 1.6
Pitch	To Pitch:	Time range (s): 0
		Pitch range (Hz): 75 – 600
	Get mean:	Time range (s): 0, 0 (= all)
	Get standard deviation:	Time range (s): 0, 0 (= all)
Formants	To Formant (robust):	Time step (s): 0
		Max. number of formants: 5
		Formant ceiling (Hz): 5500 (= adult female)
		Window length (s): 0.025
		Pre-emphasis from (Hz): 50
		Number of std. dev.: 1.5
		Max. iterations: 5
		Tolerance: 1e-06

Kendall's τ Correlation. For an initial exploration of the acoustic parameters, Kendall's correlation coefficients were computed to identify monotonic associations with $|\tau| > 0.6$. Its application is justified when the variables under analysis do not meet the assumption of a bivariate normal distribution, a requirement for the proper use of Pearson's correlation [6]. A notable feature of Kendall's coefficient is that it tends to report lower values compared to Spearman's and Pearson's coefficients when analyzing linear associations of the same strength (in the absence of outliers) [15]. For a set of observations between two variables (x_i, y_i), $i = (1, 2, \cdots, n)$, if a pair of observations (x_i, y_i), (x_j, y_j) with $i < j$ satisfies $x_i > x_j \rightarrow y_i > y_j$ or $x_i < x_j \rightarrow y_i < y_j$, it is considered a concordant pair. Conversely, when $x_i > x_j \rightarrow y_i < y_j$ or $x_i < x_j \rightarrow y_i > y_j$, it is considered a discordant pair. Kendall's correlation coefficient τ can be calculated using equation (3) [3]:

$$\tau = \frac{n_C - n_D}{C_2^n} = \frac{n_C - n_D}{\frac{n(n-1)}{2}} = \frac{2\,(n_C - n_D)}{n(n-1)} \tag{3}$$

where n denotes the number of observations, n_C and n_D represent the numbers of concordant and discordant pairs, respectively, and C_2^n corresponds to the total number of possible pairwise combinations. The coefficient takes values within the range $-1 \leq \tau \leq 1$.

Principal Component Analysis. With the aim of identifying the most relevant acoustic dimensions within the parameters, Principal Component Analysis (PCA) was applied to reduce the complexity of the entire dataset and to explore potential structures that enable improved interpretability and visualization of the results. The objective of the PCA technique is to find a lower-dimensional space, or PCA space, that is used to transform the data $X = x_1, x_2, \cdots, x_N$ from a higher-dimensional space R^M into another of lower dimension R^k, where N denotes the total number of samples or observations and x_i represents the i-th sample or observation [26]. In other words, the dataset is described in terms of new, uncorrelated variables (components). These components are ordered according to the amount of variance they explain. To conduct the principal component analysis, we define the matrix $A_{p \times q} = A$, where p is the number of samples or observations and q is the number of acoustic parameters. Then, we compute $A_c = A - \mu$, where μ is the mean vector of each column or parameter (dimension $1 \times q$). The covariance matrix is obtained as $S_{q \times q} = S = (p-1)^{-1} A_c^T A_c$ [18]. Subsequently, the problem described in equation (4) must be solved:

$$S = V \Lambda V^T \tag{4}$$

where V is an orthogonal $q \times q$ matrix whose columns are the eigenvectors v_i (directions of the principal components), and Λ is a diagonal matrix whose elements λ_i are the eigenvalues, representing the variance explained by each principal component. By convention, the principal components are ordered in decreasing order, that is:

$$S = \sum_{i=1}^{q} \lambda_i v_i v_i^T = \lambda_1 v_1 v_1^T + \cdots + \lambda_q v_q v_q^T \tag{5}$$

where $\lambda_1 \geq \lambda_2 \geq \cdots \geq \lambda_q$ [10,17]. Typically, not all principal components are selected, but only those that account for a specified amount of variance. If the first k PCs ($k \leq q$) explain a given cumulative variance, then the projection of the data onto the k PCs is obtained using equation (6):

$$Z_{p \times k} = A_c V_{q \times k} \tag{6}$$

where Z is the projected data matrix and V is the matrix of the selected eigenvectors. The PCA space is composed of the principal components, which are mutually orthogonal, thereby generating a new coordinate system in which each axis is perpendicular to the others. The percentage of variance explained by principal component j is calculated using equation (7):

$$PC_j = \left(\frac{\lambda_j}{\sum_{i=1}^{q} \lambda_i}\right) \cdot 100 \tag{7}$$

Once the points are projected into the PCA space, the centroid of each class is obtained from the means of its coordinates. Assuming a two-dimensional space, the projections onto PC_x and PC_y allow computing the centroid of class α as indicated in equation (8):

$$C_\alpha = (\bar{x}_\alpha, \bar{y}_\alpha) = \left(\frac{1}{n_\alpha}\sum_{i=1}^{n_\alpha} PC_{x,i}, \frac{1}{n_\alpha}\sum_{i=1}^{n_\alpha} PC_{y,i}\right) \tag{8}$$

where n_α is the number of instances in class α. To calculate the Euclidean distance d_i of each point to the centroid C_α, equation (9) is used:

$$d_i = \sqrt{(PC_{x,i} - \bar{x}_\alpha)^2 + (PC_{y,i} - \bar{y}_\alpha)^2} \tag{9}$$

The separation between classes C_α and C_β is quantified through the Euclidean distance between their centroids, as expressed in equation (10):

$$d\,(C_\alpha, C_\beta) = \sqrt{(\bar{x}_\alpha - \bar{x}_\beta)^2 + (\bar{y}_\alpha - \bar{y}_\beta)^2} \tag{10}$$

2.3 Data Augmentation

According to [27], data augmentation refers to methods for constructing iterative optimization or sampling algorithms through the introduction of unobserved data or latent variables. Nowadays, in order to take full advantage of artificial intelligence methods such as Deep Learning or Machine Learning and their generalization capabilities, a complete and meaningful dataset is required for training. Data augmentation aims to increase the amount of data available for training machine learning models and to make them more accurate and robust [7]. To achieve a sample size that would improve the generalization capacity of the classifiers, the *mixup* augmentation technique was applied.

Mixup. In [9], a comprehensive review of fundamental mixup methods and their applications is presented. In this study, the *Static Linear* method was implemented, which generates a new instance $\hat{x}$ through a linear combination (interpolation) of two vectors x_i and x_j:

$$\hat{x} \;=\; \theta \cdot x_i \;+\; (1-\theta)x_j \tag{11}$$

where θ represents the weight of the combination and is drawn from a beta distribution $B(0.2, 0.2)$. This distribution favors values close to 0 and 1; that is, the probability that θ falls within the intervals $[0, 0.1]$ or $[0.9, 1.0]$ is approximately 0.7 [23]. The rationale is to avoid, as much as possible, the introduction of inconsistent synthetic values and to promote the generation of instances that resemble the original ones.

2.4 Classification

In this stage, the Support Vector Machine (SVM), Random Forest (RF), and K-Nearest Neighbors (KNN) classifiers were compared to evaluate the learning performance of these models and to identify the most efficient one for distinguishing regional street cry classes based on the selected acoustic parameters. For this purpose, the scikit-learn library implemented in Python (version 3.12.7) was employed. To maximize the performance of the classifiers, the GridSearchCV function was applied to perform hyperparameter tuning, using 5-fold cross-validation to identify the hyperparameter combination in each classifier that yielded the highest accuracy. Subsequently, an additional independent cross-validation was employed for post-tuning evaluation. For SVM, the following hyperparameters were explored: $C\{0.1, 1, 10\}$, $kernel\{\text{poly, rbf}\}$ and $\gamma\{\text{scale}\}$. For RF: $n\_estimators\{50, 100\}$, $max\_depth\{4, 8\}$ and $min\_samples\_split\{2, 5\}$. For KNN: $n\_neighbors\{5, 7\}$, $weights\{\text{uniform, distance}\}$ y $p\{1 = \text{Manhattan}, 2 = \text{Euclidean}\}$.

2.5 Evaluation

During the analysis, the performance metrics *accuracy*, *precision*, and *recall* were obtained. For the *precision* and *recall* metrics, the *macro average* was employed. Accuracy is a metric that provides the proportion of correctly predicted classifications over the total number of instances [2,19], that is:

$$accuracy = \frac{TP + TN}{TP + FP + TN + FN} \tag{12}$$

where TP are the positive instances correctly predicted as positive, TN the negative instances correctly predicted as negative, FP the negative instances incorrectly predicted as positive (Type I error), and FN the positive instances incorrectly predicted as negative (Type II error). The precision metric is the fraction of positive predictions that were correct for a class k, that is:

$$precision_k = \frac{TP_k}{TP_k + FP_k} \tag{13}$$

The *macro average precision* for K classes is calculated using equation (14).

$$precision_{macro} = \frac{1}{K}\sum_{k=1}^{K} precision_k \tag{14}$$

Recall (also referred to as *sensitivity*) is the proportion of all truly positive instances that the model correctly predicted for a class k. It is defined as:

$$recall_k = \frac{TP_k}{TP_k + FN_k} \tag{15}$$

The *macro average recall* for K classes is calculated using equation (16).

$$recall_{macro} = \frac{1}{K}\sum_{k=1}^{K} recall_k \tag{16}$$

Another highly useful tool is the ROC curve (Receiver Operating Characteristic) for evaluating performance in binary classification. In this study, it corresponds to one class or region versus the others (*One-vs-Rest*). The ROC curve is a two-dimensional plot of sensitivity versus (1 – *specificity*), where *specificity* denotes the proportion of negative cases correctly classified. The term (1 – *specificity*) is also known as the *false positive rate* (FPR), which identifies the proportion of negative cases classified as positive (17). For this plot, the AUC measure is used, corresponding to the area under the ROC curve, with values ranging from 0 to 1. A value of 0.5 indicates that the test has no discriminative power, representing a completely random case, whereas a value of 1 indicates perfect discriminative ability [2,8,19].

$$FPR = 1 - specificity = 1 - \frac{TN}{FP + TN} = \frac{FP}{FP + TN} \tag{17}$$

3 Results

Feature Space Analysis. After the acquisition and characterization of the samples, according to the procedures described in Sects. 2.1 and 2.2, a single instance from class CW was discarded due to containing NaN values, resulting in a total of 232 instances (NNe = 45, CW = 101, SSe = 86).

After balancing the recordings, 45 instances were obtained for each region. The results of the Kendall correlation matrix for each class are presented in Table 2.

Table 2. Acoustic Parameters with Kendall's Correlation $|\tau| > 0.6$

Region	Parameter	τ	p
Street cries NNe	$F_2 \leftrightarrow F_3$	0.83	8.906e−16
	$Disp(F_3 - F_1) \leftrightarrow Disp(F_3 - F_2)$	0.66	1.220e−10
	$Disp(F_4 - F_1) \leftrightarrow Disp(F_4 - F_2)$	0.68	3.302e−11
	$Disp(F_4 - F_2) \leftrightarrow Disp(F_4 - F_3)$	0.68	5.596e−11
Street cries CW	Jitter↔Shimmer	0.68	5.950e−11
Street cries SSe	$Disp(F_3 - F_1) \leftrightarrow Disp(F_3 - F_2)$	0.61	2.720e−09
	Jitter↔Shimmer	0.67	8.278e−11

In the general matrix (Fig. 2), it was observed that only formants F_2 and F_3 exhibited a positive monotonic association greater than 0.6 ($F_2 \leftrightarrow F_3 : \tau = 0.63$).

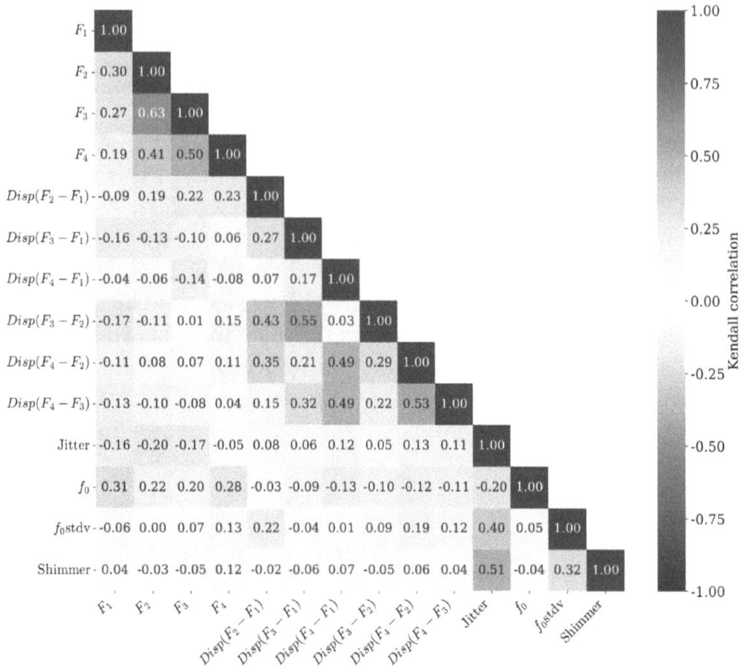

Fig. 2. Kendall correlation matrix of acoustic parameters across all regions. Among them, only formants F_2 and F_3 show a positive monotonic association above 0.6 ($F_2 \leftrightarrow F_3 : \tau = 0.63$).

This may indicate that there is no excessive redundancy among most of the variables when considering the three regions together.

PCA. The number of principal components that explain at least 80% of the cumulative variance was selected, as this is considered adequate to retain most of the structure of the original data. Individually, PC_1 and PC_2 account for the largest proportion of the variance (27.50% and 22.95%, respectively), followed by PC_3 (12.97%), PC_4 (10.24%), and PC_5 (7.37%).

To analyze the relative contribution of each variable, the coefficients (loadings) associated with each principal component were normalized. Parameters with a higher relative weight (≥ 0.7) allow for the distinction of two groupings. PC_1, PC_2, and PC_4 are mainly associated with the vocal tract, as they exhibit high loadings in the parameters $Disp(F_j - F_i)$ and F_i, jointly explaining 60.69% of the cumulative variance (Fig. 3a). In contrast, PC_3 and PC_5 are related to the vocal folds, due to the parameters jitter, shimmer, f_0, and f_{0stdv}, with a cumulative variance of 20.34% (Fig. 3b).

Subsequently, to identify which dimensions in the common PCA space best distinguish the street cry classes, the Euclidean distance between the centroids of each class (Eq. 10) was calculated, considering all possible combinations of PCs in 2D and 3D spaces. To obtain more robust centroids, the instances corresponding

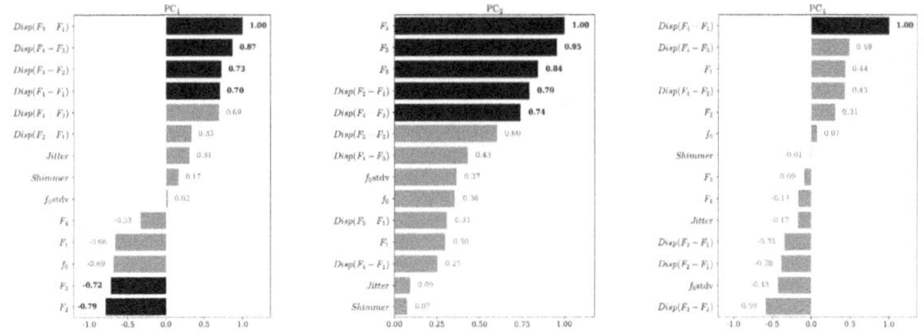

(a) PC_1, PC_2, and PC_4 are predominantly associated with the vocal tract, as indicated by the parameters $Disp(F_j - F_i)$ and F_i (cumulative variance: 60.69%).

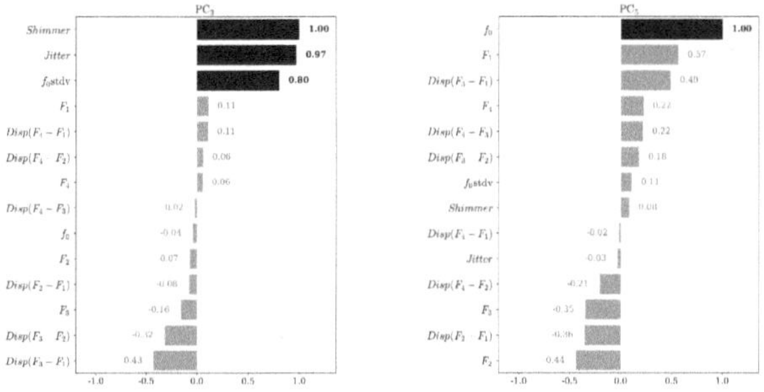

(b) PC_3 and PC_5 are predominantly associated with the vocal folds, as indicated by the parameters jitter, shimmer, f_0, and f_{0stdv} (cumulative variance: 20.34%).

Fig. 3. Principal component groupings associated with the vocal tract and the vocal folds.

to the 10% of points farthest from the initial centroids (five in each class) were removed. Likewise, all combinations were evaluated to determine those in which the largest differences were observed, that is, the greatest distances between the centroids of each region with respect to the others.

The 2D and 3D combinations with the greatest distance between the recalculated centroids were PC_1–PC_2 (*dist*: 6.03) (Fig. 4) and PC_1–PC_2–PC_5 (*dist*: 6.45) (Fig. 4b), respectively. These spaces are those that exhibit the greatest distinction among the street cry regions.

From the contrast of each region against the other two considered as a single set, the best combinations in the PCA space with the greatest centroid distances in 2D were: for NNe: PC_1–PC_2 (*dist*: 2.65), for CW: PC_1–PC_2 (*dist*: 1.89), and for SSe: PC_1–PC_5 (*dist*: 1.03) (Fig. 5); and in 3D: for NNe: PC_1–PC_2–PC_5 (*dist*: 2.74), for CW: PC_1–PC_2–PC_5 (*dist*: 1.92), and for SSe: PC_1–PC_2–PC_5 (*dist*: 1.16) (Fig. 6).

Data Augmentation. Initially, from the total set of instances in each class, all possible unique pairwise combinations were obtained (pairs without repeti-

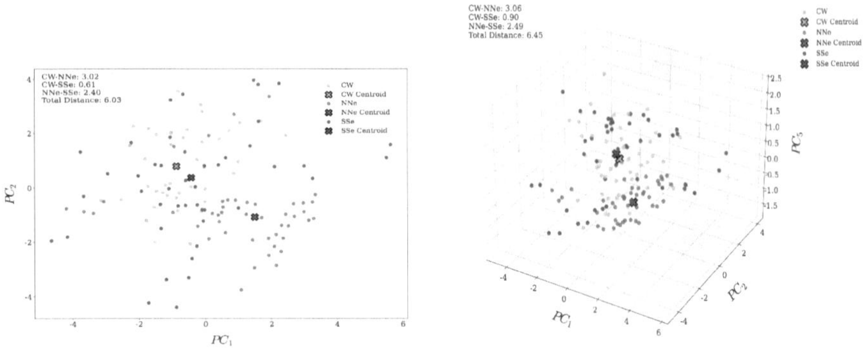

(a) Among the evaluated combinations, PC_1–PC_2 achieved the largest total distance between the recalculated centroids (*dist*: 6.03).

(b) The PC_1–PC_2–PC_5 combination yielded the greatest total distance between the recalculated centroids (*dist*: 6.45).

Fig. 4. Projections onto the 2D and 3D spaces defined by the PCs.

tion: NNe: $C_2^{45} = 990$, CW: $C_2^{101} = 5050$, SSe: $C_2^{86} = 3655$) for the generation of synthetic instances. The distributions of each parameter between the original sample and the augmented sample (original + synthetic instances) were compared in order to verify that the augmentation process did not alter their behavior. For this purpose, the non-parametric Kolmogorov–Smirnov test was applied. Subsequently, their values were normalized together with the original vectors using z-score standardization ($\mu = 0, \sigma = 1$). Finally, the classes were balanced to 300 instances, retaining all original ones and completing the remainder with synthetic instances selected at random. In all comparisons for each parameter between the distributions of the original instances (NNe: 45, CW: 101, SSe: 86) and those including synthetic instances (NNe: 1035, CW: 5151, SSe: 3741), the non-parametric Kolmogorov–Smirnov test yielded p-values greater than 0.05. This implies that the null hypothesis of equal distributions cannot be rejected, suggesting that there is no evidence of statistically significant differences between the original data and the augmented datasets. Consequently, it is valid to increase the sample size by employing synthetic instances in order to enhance the generalization capacity of the classifiers.

Classification and Evaluation. In the hyperparameter tuning of the classification models, the best combinations were determined for the original data (normalized and balanced original instances). For SVM: {*C*: 10, *gamma*: scale, *kernel*: rbf}; RF: {*max_depth*: 8, *min_samples_split*: 2, *n_estimators*: 50}; KNN: {*n_neighbors*: 5, *p*: 2 = Euclidean, *weights*: distance}. For the augmented data (normalized and balanced original plus synthetic instances), the best combinations were: SVM: {*C*: 10, *gamma*: scale, *kernel*: rbf}; RF: {*max_depth*: 8, *min_samples_split*: 5, *n_estimators*: 100}; KNN: {*n_neighbors*: 5, *p*: 2 = Euclidean, *weights*: distance}. The models exhibit satisfactory performance with

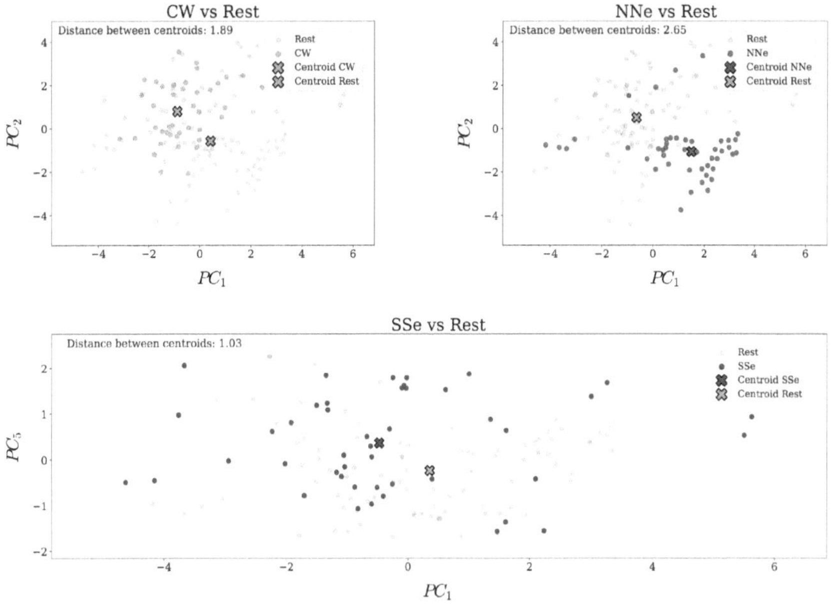

Fig. 5. 2D PCA spaces with the greatest centroid distances (Region vs. Rest).

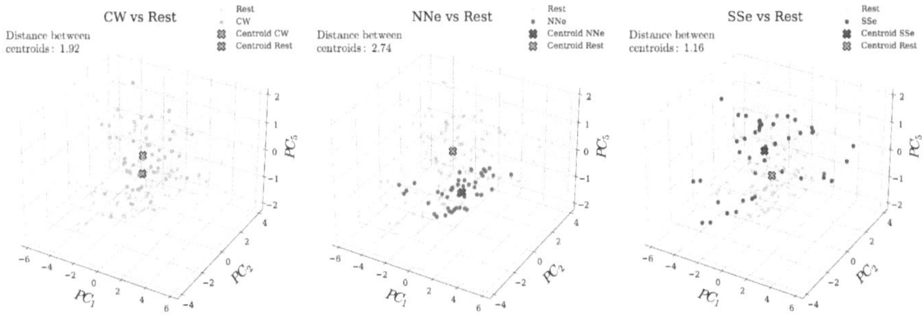

Fig. 6. 3D PCA space with the largest centroid separations (Region vs. Rest).

the original data (≥ 0.7 across the three metrics), while with the augmented data their performance surpasses 0.9, as illustrated in Fig. 7.

In the ROC curves (One-vs-Rest) of the original data, all three classes demonstrated substantial discriminability (AUC > 0.8), with NNe exhibiting the highest degree of separation from the others (Fig. 8).

Since the street cries from the CW and SSe regions are the least distinguishable from each other, a binary ROC curve (CW vs. SSe) was included, with CW set as the positive class, in order to quantify the specific separability between these two classes. All three classifiers achieved AUC > 0.7, i.e., discrimination above chance level (Fig. 9), with KNN yielding the best result (AUC $= 0.8$).

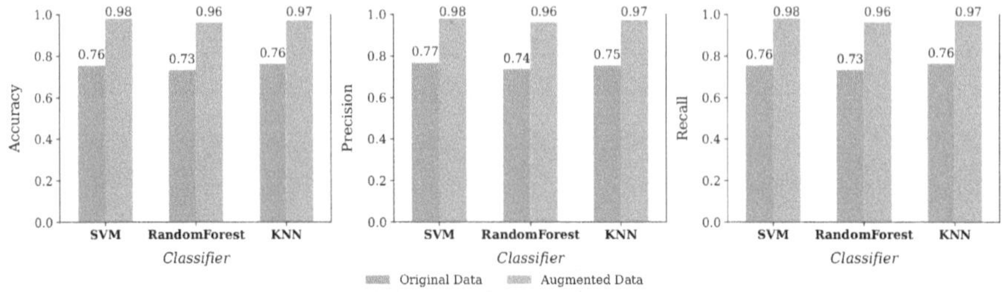

Fig. 7. Comparison of the results obtained in the accuracy, precision and recall metrics of the original and augmented data in each classifier.

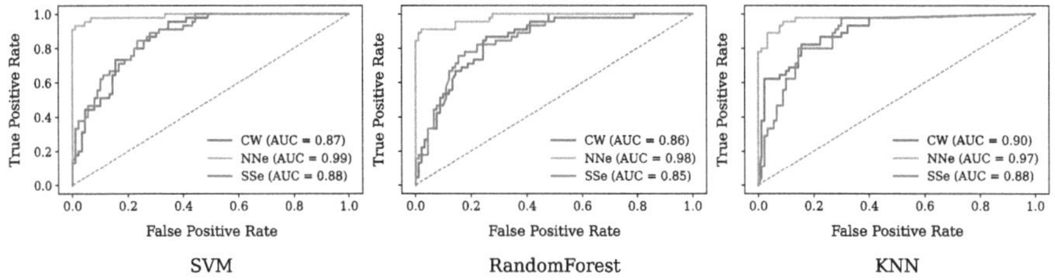

Fig. 8. One vs Rest ROC curves using the original normalized and balanced instances.

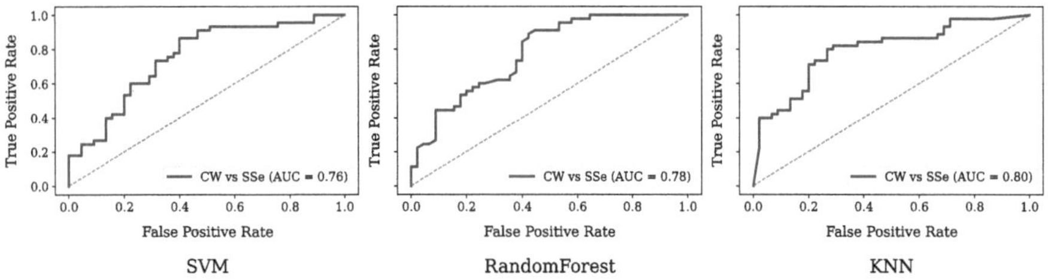

Fig. 9. Binary ROC curve (CW vs SSe) with CW being the positive class. All three classifiers achieved AUC> 0.7.

4 Discussion

Some of the limitations of this study concerned the acquisition of the samples, as they were obtained from a repository containing recordings made with different equipment and at varying sampling frequencies. The quality of the desired signal depended on ambient noise and the applied filtering, which limited the duration of the audio samples and affected the parameter values. A larger sample size is also recommended in order to obtain more representative and robust results. Window size is a parameter that can be further explored with different lengths to assess its impact on the results. Additionally, some acoustic parameters are particularly sensitive to background noise. As noted in [12], the parameters f_0

and alpha ratio yield reliable values up to a background noise level of 47.7 dB(A), whereas jitter and shimmer remain reliable only up to a background noise level of 43.8 dB(A).

5 Conclusions and Future Work

In the 2D and 3D combinations with the greatest centroid distances, there is greater distinction between the street cries of the CW and NNe regions (2D *dist*: 3.02, 3D *dist*: 3.06), whereas between the CW and SSe regions (2D *dist*: 0.61, 3D *dist*: 0.90) a lower degree of differentiation is observed. However, when comparing each region against the rest, the NNe cries exhibit the highest degree of distinction, in contrast with those from SSe, which show the least differentiation.

In the context of street cries, the presence of principal components PC_1 and PC_2 (associated with the vocal tract) in all combinations with the greatest centroid distances suggests that differences exist among the prototypes of regional street cries, linked both to the formants and to their variability. This may represent an intrinsic strategy in the production of street cries, arising from differences in vowel content that generate dynamic changes in the vocal tract. However, PC_5, associated with the vocal folds due to the weight of pitch (f_0), plays a relevant role in the differentiation among classes, including in the distinction of the SSe region from the others, whose prototype is the least distinguishable within the set.

As a result of the performance metrics and ROC curves across the different classification models, it can be inferred that the acoustic parameters jitter, shimmer, f_0, f_0stdv, F_i ($i = 1, \ldots, 4$), and $Disp(F_j - F_i)$ can be used to characterize street cries from different regions of Mexico, particularly those defined in this study as NNe, CW, and SSe.

Future work should implement a more robust experimental design for sample acquisition and investigate additional acoustic parameters, together with strategies employing alternative clustering or classification tools, to achieve a more comprehensive characterization. Moreover, exploring a different distribution of Mexican regions may allow the identification of new classes.

Acknowledgments. This study was supported by a scholarship granted to the first author through SECIHTI, representing the people of Mexico.

References

1. Anikin, A., Barreda, S., Reby, D.: A practical guide to calculating vocal tract length and scale-invariant formant patterns. Behav. Res. Methods **56**(6), 5588–5604 (2024)
2. Borja-Robalino, R., Monleón-Getino, A., Rodellar, J.: Estandarización de métricas de rendimiento para clasificadores machine y deep learning. Revista Ibérica de Sistemas e Tecnologias de Informação **E30**, 184–196 (2020)

3. Essam, F., El, H., Ali, S.R.H.: A comparison of the Pearson, Spearman rank and kendall tau correlation coefficients using quantitative variables. Asian J. Probab. Stat. **20**(3), 36–48 (2022)
4. Freitas, B., D'Amelio, P.B., Milá, B., Thébaud, C., Janicke, T.: Meta-analysis of the acoustic adaptation hypothesis reveals no support for the effect of vegetation structure on acoustic signalling across terrestrial vertebrates. Biol. Rev. **100**(2), 815–833 (2025)
5. Hardt, B., Benedict, L.: Can you hear me now? A review of signal transmission and experimental evidence for the acoustic adaptation hypothesis. Bioacoustics **30**(6), 716–742 (2021)
6. Lalinde, J.D.H., et al.: Sobre el uso adecuado del coeficiente de correlación de pearson: definición, propiedades y suposiciones (2018)
7. Hoffmann, M., Malburg, L., Bergmann, R.: Augmentation of semantic processes for deep learning applications. Appl. Artif. Intell. **39**(1), 2506788 (2025)
8. Hoo, Z.H., Candlish, J., Teare, D.: What is an roc curve? (2017)
9. Jin, X., et al.: A survey on mixup augmentations and beyond. arXiv preprint arXiv:2409.05202 (2024)
10. Johnstone, I.M., Paul, D.: PCA in high dimensions: an orientation. Proc. IEEE **106**(8), 1277–1292 (2018)
11. Li, G., Hou, Q., Zhang, C., Jiang, Z., Gong, S.: Acoustic parameters for the evaluation of voice quality in patients with voice disorders. Ann. Palliative Med. **10**(1), 13036–13136 (2021)
12. Marsano-Cornejo, M.J., Roco-Videla, Á.: Variation of the acoustic parameters: f0, jitter, shimmer and alpha ratio in relation with different background noise levels. Acta Otorrinolaringologica (English Edition) **74**(4), 219–225 (2023)
13. Mendes, A.P., Nunes, N., Ibrahim, S., Coelho, A.C., Francisco, M.A.: Cepstral measures in the fado voice: gender, age and phonatory tasks. J. Voice **37**(1), 9–16 (2023)
14. Mendes, S., Colino-Rabanal, V.J., Peris, S.J.: Adaptación acústica del canto de turdus leucomelas (passeriformes: Turdidae) a diferentes niveles de ruido antrópico, en el área metropolitana de belém, pará, brasil. Rev. Biol. Trop. **65**(2), 633–642 (2017)
15. Morales, P., Rodríguez, L.: Aplicación de los coeficientes correlación de kendall y spearman. Agrollanía **13** (2016)
16. Nacional, F.: Mapa sonoro de méxico (2025). https://mapasonoro.cultura.gob.mx/
17. Paul, L.C., Al Suman, A., Sultan, N.: Methodological analysis of principal component analysis (PCA) method. Int. J. Comput. Eng. Manage. **16**(2), 32–38 (2013)
18. de Santi, N.S., Abramo, L.R.: Obtaining cosmological covariance matrices with machine learning. Boletim da Sociedade Astrônomica Brasileira. Proceedings da XLV Reunião Anual da SAB **34**, 193–197 (2023)
19. Sathyanarayanan, S., Tantri, B.R.: Confusion matrix-based performance evaluation metrics. Afr. J. Biomed. Res. **27**(4S), 4023–4031 (2024)
20. Sharma, J.: Food cries, historical city sounds, and the twentieth century silencing of street vendors. Food Cult. Soc. **24**(1), 16–30 (2021)
21. Silva, L.J.: Todos los escuchan pero poco se sabe de ellos: pregones y expresiones sonoras empleadas en la venta de productos. Cuicuilco **23**(66), 131–148 (2016)
22. Silva, W.J., Lopes, L., Galdino, M.K.C., Almeida, A.A.: Voice acoustic parameters as predictors of depression. J. Voice **38**(1), 77–85 (2024)
23. Smucny, J., Shi, G., Lesh, T.A., Carter, C.S., Davidson, I.: Data augmentation with mixup: enhancing performance of a functional neuroimaging-based prognostic deep learning classifier in recent onset psychosis. NeuroImage Clin. **36**, 103214 (2022)

24. Taverner, C.: Street food: hawkers and the history of London. Oxford University Press (2023)
25. Teixeira, J.P., Gonçalves, A.: Accuracy of jitter and shimmer measurements. Procedia Technol. **16**, 1190–1199 (2014)
26. Tharwat, A.: Principal component analysis-a tutorial. Int. J. Appl. Pattern Recogn. **3**(3), 197–240 (2016)
27. Van Dyk, D.A., Meng, X.L.: The art of data augmentation. J. Comput. Graph. Stat. **10**(1), 1–50 (2001)

Balancing Local and Global Structure Preservation in Dimensionality Reduction

Corinne Orton(✉), Connor R. Shrader, and Anna Little

University of Utah, Salt Lake City, UT 84102, USA
corton@wisc.edu, connor.shrader@utah.edu, little@math.utah.edu

Abstract. Dimension reduction (DR) techniques are widely used to visualize high-dimensional data by embedding it into two dimensions. Existing methods typically emphasize either global or local structure, leading to embeddings that highlight some features while distorting others. We propose a simple hybrid approach that combines PCA and UMAP to balance global and local preservation, together with a new unsupervised metric, Overall Geometric Fidelity (OGF), to quantify this trade-off. Across a range of benchmark datasets, our hybrid method (UPCA) achieves higher overall structure preservation than PCA, UMAP, t-SNE, and PaCMAP. We also find that a fixed 50/50 weighting of PCA and UMAP performs nearly as well as the optimally tuned hybrid, making the method both effective and easy to use. These results demonstrate that hybrid strategies provide a practical and powerful alternative to existing DR methods.

Keywords: Dimension reduction · Geometric structure preservation · Hybrid method

1 Introduction

From cell biology to consumer habits, many modern datasets are high dimensional. To visualize such data, dimension reduction (DR) techniques are frequently used to embed the original data into a lower-dimensional space. Well-known methods such as PCA [12], t-SNE [6], and UMAP [8] each emphasize preserving either local or global structure, and thus exhibit both strengths and limitations. For example, PCA favors global structure at the expense of local clustering, while t-SNE and UMAP emphasize local neighborhoods but may distort the overall layout of the data [5].

Among existing DR techniques, PCA and UMAP represent two of the most widely used yet contrasting approaches. They highlight the trade-off between global and local structure preservation particularly clearly: PCA is linear and emphasizes global variance, while UMAP is nonlinear and emphasizes neighborhood structure. Before introducing our hybrid method, we briefly review the principles of each.

H. R. Arabnia et al. (Eds.): CSCI 2025, CCIS 2969, pp. 48–57, 2027.
https://doi.org/10.1007/978-3-032-28460-0_4

PCA (Principal Component Analysis) is a linear method that identifies the directions of greatest variance in the data (the principal components) and uses them to form the embedding space [12]. Because it is directly tied to the global variance structure of the dataset, PCA excels at preserving large-scale relationships and overall geometry. However, it often fails to capture local clusters, which can hinder tasks such as visualizing group separation.

UMAP (Uniform Manifold Approximation and Projection) is a nonlinear method that assumes the data lie on a low-dimensional manifold within the high-dimensional space [8]. It constructs a fuzzy simplicial complex based on nearest-neighbor relationships and embeds this structure in lower dimensions by minimizing a cross-entropy objective. In practice, UMAP relies on negative sampling to approximate non-neighbor relationships efficiently [4]. This focus on local neighborhoods makes UMAP highly effective at revealing clusters, though sometimes at the cost of distorting global geometry.

Other DR methods include t-SNE [6], PaCMAP [13], PHATE [9], Diffusion Maps [3], Metric MDS, and Laplacian Eigenmaps [2]. We focus here on PCA and UMAP because they are widely used, computationally efficient, and represent complementary approaches to structure preservation. Our hybrid framework could, in principle, combine any two DR methods.

The contrasting strengths and weaknesses of PCA and UMAP can be clearly observed in their embeddings of the Muraro dataset, which contains single-cell transcriptomic data from human pancreatic cells [10]. Figure 1a shows that PCA fails to produce well-separated clusters: the brown cluster is too diffused, and the green, red, orange, and turquoise clusters overlap. Although suboptimal for visualization, this representation conveys meaningful information—namely, that the latter four cell types are more similar to each other than to those represented by more distant clusters, such as brown or light green. In contrast, Fig. 1b shows that UMAP produces distinct clusters, but at the cost of distorting global relationships: for instance, the orange and green clusters, which PCA placed adjacent, now appear on opposite sides of the embedding. Additionally, several points are confidently misclassified.

To address these trade-offs, we propose a hybrid embedding that combines input from both PCA and UMAP. As shown in Fig. 2, the hybrid embedding blends the strengths of PCA and UMAP. Cluster placement resembles the PCA embedding (Fig. 1a) but is more distinct: the green, red, orange, and turquoise clusters remain close yet non-overlapping. At the same time, the separation of clusters resembles UMAP (Fig. 1b), but without fragmenting the red cluster. The hybrid method thus preserves both global and local structure, yielding distinct clusters arranged in a coherent overall layout.

In the following sections, we formalize our hybrid framework and describe how embeddings are constructed by combining PCA and UMAP. We then introduce unsupervised metrics designed to evaluate global, local, and overall structure preservation, and use these to compare our hybrid approach with established DR methods, including PCA, UMAP, t-SNE, and PaCMAP. Finally, we present

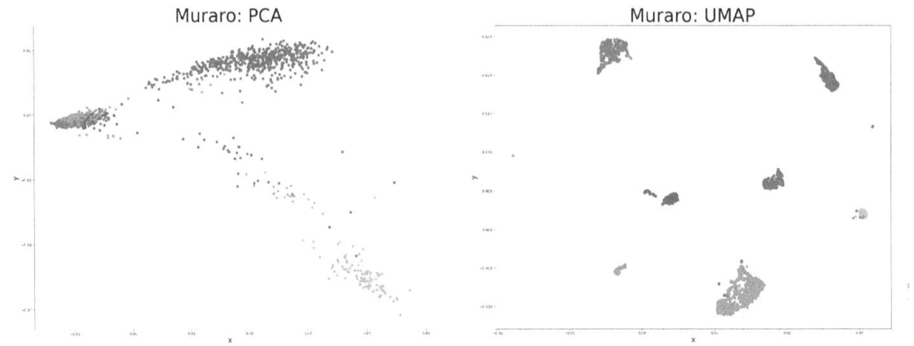

(a) PCA on Muraro dataset

(b) UMAP on Muraro dataset

Fig. 1. Comparison of PCA and UMAP embeddings of the Muraro dataset.

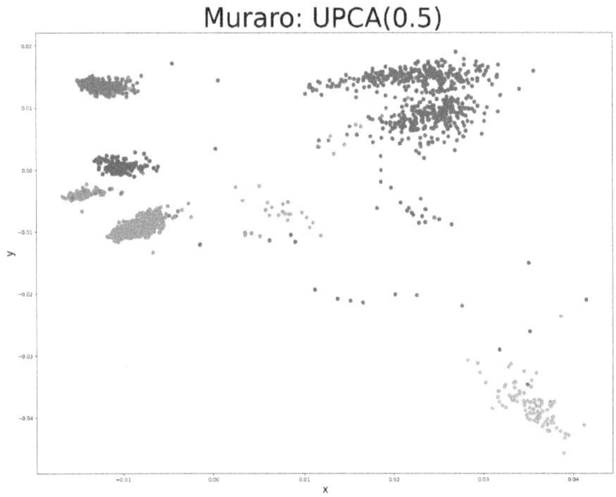

Fig. 2. Hybrid embedding of the Muraro dataset.

results across several benchmark datasets and discuss the strengths, limitations, and future directions of the hybrid method.

2 Methods

2.1 UPCA(α)

We construct our hybrid embedding, denoted UPCA, using the procedure summarized in Algorithm 1. The method blends information from PCA and UMAP according to a weighting parameter $\alpha \in [0, 1]$, with $\alpha = 0$ corresponding to pure PCA and $\alpha = 1$ to pure UMAP. This procedure ensures that both global (PCA) and local (UMAP) structural information are incorporated into the final two-dimensional embedding.

Algorithm 1. UPCA(α)

Input: High-dimensional data $D \in \mathbb{R}^{n \times d}, \alpha \in [0, 1]$
Output: Hybrid embedding $y \in \mathbb{R}^{n \times 2}$
Compute $y_{\text{PCA}} \leftarrow \text{PCA}(D, 2)$; normalize: $y_{\text{PCA}} \leftarrow \frac{y_{\text{PCA}}}{\|y_{\text{PCA}}\|_{\text{Fro}}}$
Compute $y_{\text{UMAP}} \leftarrow \text{UMAP}(D, 2)$; normalize: $y_{\text{UMAP}} \leftarrow \frac{y_{\text{UMAP}}}{\|y_{\text{UMAP}}\|_{\text{Fro}}}$
Scale: $y_{\text{PCA}} \leftarrow (1 - \alpha) \cdot y_{\text{PCA}}, \quad y_{\text{UMAP}} \leftarrow \alpha \cdot y_{\text{UMAP}}$
Concatenate: $M \leftarrow \left[y_{\text{PCA}} \mid y_{\text{UMAP}}\right] \in \mathbb{R}^{n \times 4}$
Compute $y \leftarrow \text{PCA}(M, 2)$
Normalize: $y \leftarrow \frac{y}{\|y\|_{\text{Fro}}}$
return y

2.2 Comparison to Other DR Methods

To evaluate the effectiveness of the hybrid embedding procedure UPCA described in Algorithm 1, we assess its ability to preserve local, global, and overall structure. We test UPCA at $\alpha = 0.25, 0.5, 0.75$, and at an optimally chosen α^* (see Sect. 2.4). Performance is compared to UMAP, PCA, and two additional methods: t-SNE (T-distributed Stochastic Neighbor Embedding), a nonlinear approach that minimizes the divergence between high- and low-dimensional pairwise similarities [6], and PaCMAP (Pairwise Controlled Manifold Approximation), introduced in 2021, which balances local and global structure by preserving both nearest neighbors and "mid-near" pairs [13]. For implementation, we use `scikit-learn` [11] for PCA and t-SNE, and the official libraries for UMAP [8] and PaCMAP [13].

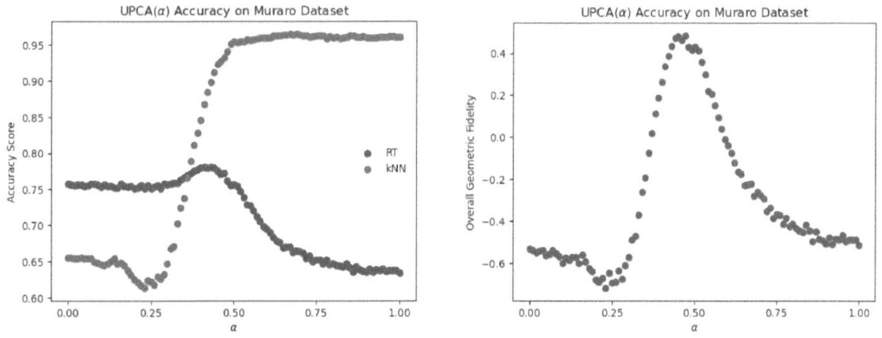

(a) RT and kNN scores on the Muraro dataset across $\alpha \in [0, 1]$.

(b) OGF scores on the Muraro dataset across $\alpha \in [0, 1]$.

Fig. 3. Accuracy scores for UPCA on the Muraro dataset across different values of α.

2.3 Evaluation Metrics

To provide a fully unsupervised evaluation framework that does not rely on labels, we measure structure preservation using only unsupervised metrics. For global structure, we use Random Triplet Accuracy (RT), which generates five triplets for each of the n points in the dataset and records the proportion whose ordered relative distances are preserved in the embedding [13]. For local structure, we use k-Nearest Neighbors Accuracy (kNN), which measures how well a size-k neighborhood is preserved using the `scikit-learn` kNN classifier [13]. Following [5], we set $k = 5$.

Since PCA emphasizes global structure and UMAP emphasizes local neighborhoods, we expect RT to decrease and kNN to increase as α grows. Figure 3a shows that on the Muraro dataset, RT peaks around $\alpha = 0.4$ before declining, while kNN is highest for $\alpha \in [0.5, 1]$. The most informative embedding will lie at a balancing point where kNN is high but RT has not yet dropped substantially; Fig. 3a suggests a value near $\alpha \approx 0.45$ for the Muraro dataset.

To formalize this trade-off, we introduce an additional metric: Overall Geometric Fidelity (OGF). OGF captures how well a method preserves both global and local structure relative to the average performance of baseline methods (PCA, UMAP, t-SNE, PaCMAP) across all datasets. It is defined as a weighted combination of standardized scores:

1. Compute the mean (μ) and standard deviation (σ) of RT accuracy across all four baseline methods and all datasets.
2. For each method (including hybrids) and dataset, compute the z-score of RT: $z\text{RT} = (RT - \mu)/\sigma$.
3. Repeat steps 1–2 for kNN accuracy to obtain zkNN.
4. For each method and dataset, define $\text{OGF} = \frac{1}{2}(z\text{RT} + z\text{kNN})$.

2.4 Automatic Parameter Selection

We define α^* as the value of α that maximizes OGF. To approximate α^* for each dataset, we evaluate α on a discrete grid $\{0, 0.05, 0.1, 0.15, \ldots, 0.95, 1\}$. Table 2 reports the selected α^* values, showing that the optimal choice depends on the dataset. Importantly, the procedure requires no label information, making the hybrid embedding with α^* fully unsupervised.

3 Results

3.1 Global and Local Structure Preservation

As shown in Fig. 4a, UPCA's RT scores generally decrease as α increases, consistent with PCA's emphasis on global structure. PCA, UPCA(0.25), and UPCA(α^*) tend to score highest on RT; t-SNE, UMAP, and UPCA(0.75) score lowest, with UPCA(0.5) and PaCMAP in between. *Note:* When $\alpha^* = \alpha'$ for some fixed $\alpha' \in \{0.25, 0.5, 0.75\}$, any small discrepancies between UPCA(α^*) and UPCA(α') are due to randomness in the RT computation.

In contrast, Fig. 4b shows that kNN accuracy generally increases with α, reflecting UMAP's focus on local neighborhoods. UMAP, UPCA(0.75), t-SNE, and PaCMAP achieve the highest kNN scores; PCA and UPCA(0.25) are lowest, with UPCA(0.5) and UPCA(α^*) in between.

3.2 Overall Performance

Table 1 reports the median z-scores of RT and kNN, as well as median OGF, across all datasets. As expected, PCA has the highest median RT score, reflecting its emphasis on global structure, while t-SNE achieves the highest median kNN score, consistent with its strong local focus. UMAP has high kNN but weak RT, while PaCMAP performs moderately well on both metrics.

Most importantly, UPCA(α^*) achieves the highest median OGF of all methods, confirming that the hybrid approach most effectively balances local and global preservation. UPCA(0.5) is close behind and strongly outperforms other methods, suggesting that $\alpha = 0.5$ is a good general parameter choice without requiring dataset-specific tuning. This is notable because α^* often lies near 0.5 (see Table 2), and because UPCA(0.5) requires no additional parameter search yet achieves nearly the same performance as UPCA(α^*).

Table 1. Median z-scores of RT, kNN, and OGF for each algorithm across all datasets.

Median Values	Algorithm	RT	kNN	OGF
	PCA	0.8586	−1.1689	−0.0823
	t-SNE	−0.1132	0.6588	0.1547
	UMAP	−0.6614	0.6258	−0.2045
	PaCMAP	0.4407	0.6384	0.2396
	UPCA(0.25)	0.8123	−1.2407	−0.0610
	UPCA(0.5)	0.6935	0.1407	0.4239
	UPCA(0.75)	−0.1724	0.6038	−0.0817
	UPCA(α^*)	0.8033	0.2817	0.4508

3.3 Dataset-Specific Variation

While the median results establish the overall advantage of the hybrid approach, performance still varies across datasets. Table 2 shows the α^* value selected for each dataset and the method with the highest OGF, while Fig. 4c illustrates overall trends.

We highlight the following observations:

1. Some datasets' structures are inherently easier to preserve (e.g., CAFs), while others are more challenging (e.g., MNIST).

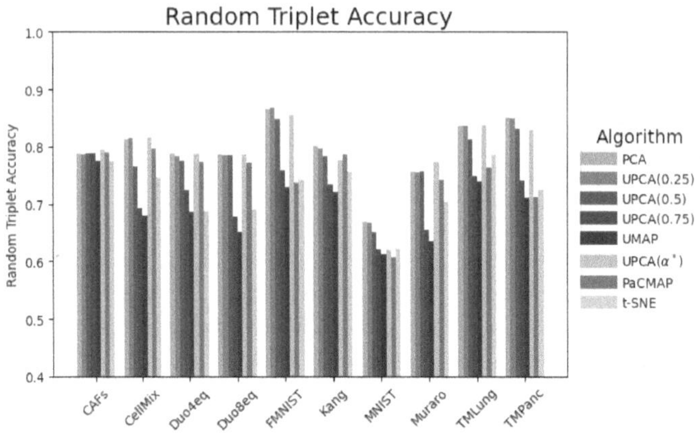

(a) Random Triplet (RT) scores of each method.

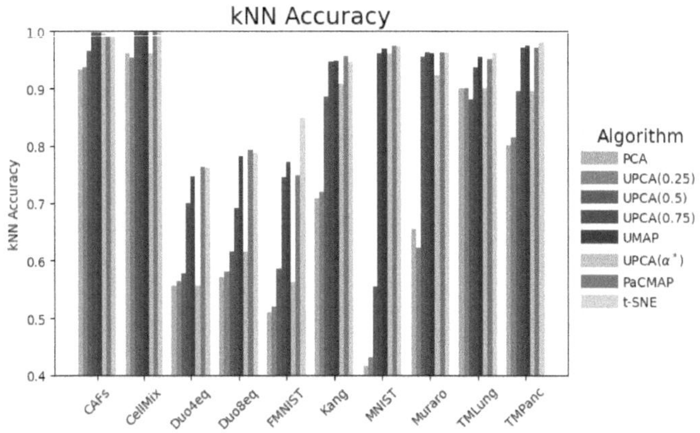

(b) k-Nearest Neighbors (kNN) accuracy of each method.

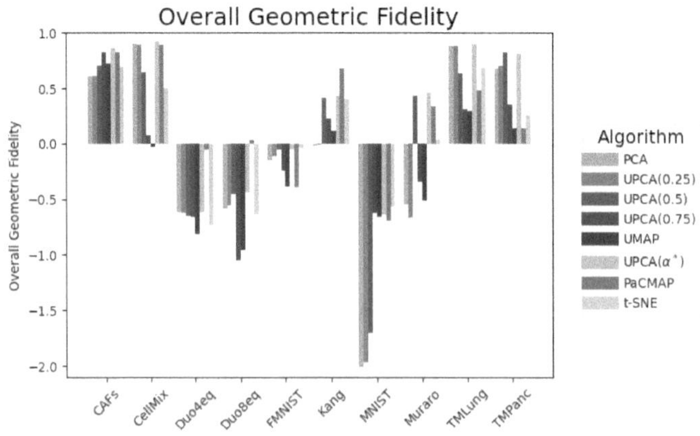

(c) Overall Geometric Fidelity (OGF) of each method.

Fig. 4. Comparison across methods: (a) global structure (RT), (b) local structure (kNN), and (c) overall structure (OGF).

2. The range of suitable α values differs by dataset: on TM Lung, strong PCA influence dominates, while on Muraro only UPCA(0.5) and UPCA($\alpha^* = 0.45$) achieve positive OGF.
3. Most datasets show a clear preference for either t-SNE, PaCMAP, or UPCA. PaCMAP excels on Kang and CellMix, while UPCA and PaCMAP dominate 8/10 datasets overall. On MNIST and FMNIST, t-SNE ranks highest, though all methods yield negative OGF.

Table 2. α^* for each dataset, and method achieving the highest OGF.

Highest OGF Scores		
Dataset	$\alpha^* =$	Highest OGF
CAFs	0.7	UPCA(α^*)
Cell Mix	0.05	UPCA(α^*)
Duo4eq	0.0	PaCMAP
Duo8eq	0.45	PaCMAP
FMNIST	0.45	t-SNE
Kang	0.55	PaCMAP
MNIST	0.75	t-SNE
Muraro	0.45	UPCA(α^*)
TM Lung	0.1	UPCA(α^*)
TM Panc	0.5	UPCA(α^*)

4 Discussion

We have shown that the best choice of dimension reduction technique depends strongly on the dataset, but that hybrid methods can substantially improve robustness across tasks. In particular, UPCA(α^*) generally preserves overall structure more effectively than all other techniques tested, and UPCA(0.5) achieves nearly the same performance without requiring any parameter tuning. This demonstrates that a simple hybrid of PCA and UMAP can provide a practical and effective balance between global and local structure preservation.

We highlight the introduction of our metric Overall Geometric Fidelity (OGF), created specifically to measure a DR transform's geometric robustness. By equally weighting small neighborhood and random ordered triplet preservation, it effectively quantifies the accuracy of an embedding's representation of the data's overall structure. Furthermore, it is completely unsupervised, so it can be used to measure performance on non-benchmarking datasets as well.

Our next steps include developing an efficient algorithm to select α^* automatically, and comparing the total runtime of UPCA(α^*) to other methods. We emphasize that, apart from parameter selection, UPCA's algorithmic complexity cannot exceed that of its components. Because it requires one UMAP computation and two PCA computations (with PCA being very fast), its runtime is comparable to UMAP alone. More broadly, the hybrid framework inherits the scalability of whatever algorithms it combines.

Conclusion. This work introduces and systematically evaluates a simple hybrid approach for dimension reduction that blends the complementary strengths of PCA and UMAP. By providing a method that balances global and local preservation in a fully unsupervised way, we highlight both the limitations of relying on a single embedding technique and the promise of hybrid strategies. Our results show that UPCA, particularly with $\alpha = 0.5$, achieves consistently strong performance across diverse datasets. This suggests hybrid DR methods are a powerful and generalizable alternative to existing techniques, with the potential to improve data exploration and visualization in a wide range of scientific and applied domains.

5 Data and Availability

The CAFs dataset is from [1]. Cell Mix, TM Lung, and TM Panc are from [7]. Duo 4, Duo 8, Kang, Kazer, and Muraro are from [5]. MNIST accessed through `scikit-learn` [11]. FMNIST accessed through umap-learn [8].

Our code is available at https://github.com/corinne-ort/drpaper25.

Acknowledgments. Corinne Orton thanks the University of Utah Office of Undergraduate Research for funding through the Undergraduate Research Opportunity Program. Connor R. Shrader is supported by the National Science Foundation (Grant No. 2139322). Anna Little thanks the National Science Foundation for support from NSF DMS CAREER 2441153, NSF DMS 2309570, and NSF RTG DMS 2136198.

References

1. Bartoschek, M., Oskolkov, N., Bocci, M., et al.: Spatially and functionally distinct subclasses of breast cancer-associated fibroblasts revealed by single cell RNA sequencing. Nat. Commun. **9**(5150) (2018). https://doi.org/10.1038/s41467-018-07582-3
2. Belkin, M., Niyogi, P.: Laplacian eigenmaps and spectral techniques for embedding and clustering. In: Dietterich, T., Becker, S., Ghahramani, Z. (eds.) Advances in Neural Information Processing Systems, vol. 14. MIT Press (2001). https://proceedings.neurips.cc/paper_files/paper/2001/file/f106b7f99d2cb30c3db1c3cc0fde9ccb-Paper.pdf
3. Coifman, R.R., Lafon, S.: Diffusion maps. Appl. Comput. Harmonic Anal. **21**(1), 5–30 (2006). https://doi.org/10.1016/j.acha.2006.04.006. https://www.sciencedirect.com/science/article/pii/S1063520306000546

4. Damrich, S., Hamprecht, F.A.: On umap's true loss function. In: Ranzato, M., Beygelzimer, A., Dauphin, Y., Liang, P., Vaughan, J.W. (eds.) Advances in Neural Information Processing Systems, vol. 34, pp. 5798–5809. Curran Associates, Inc. (2021). https://proceedings.neurips.cc/paper_files/paper/2021/file/2de5d16682c3c35007e4e92982f1a2ba-Paper.pdf
5. Huang, H., Wang, Y., Rudin, C., Browne, E.P.: Towards a comprehensive evaluation of dimension reduction methods for transcriptomic data visualization. Nature (2022)
6. van der Maaten, L., Hinton, G.: Visualizing data using t-sne. J. Mach. Learn. Res. **9**(86), 2579–2605 (2008). http://jmlr.org/papers/v9/vandermaaten08a.html
7. Manousidaki, A., Little, A., Xie, Y.: Clustering and visualization of single-cell RNA-seq data using path metrics. PLOS Comput. Biol. **20**(5), 1–19 (2024). https://doi.org/10.1371/journal.pcbi.1012014
8. McInnes, L., Healy, J., Melville, J.: Umap: uniform manifold approximation and projection for dimension reduction (2020). https://arxiv.org/abs/1802.03426
9. Moon, K.R., van Dijk, D., Wang, Z., Gigante, S., Burkhardt, D.B., Chen, W.S.: Visualizing structure and transitions in high-dimensional biological data. Nat. Biotechnol. **37**, 1482–1492 (2019). https://doi.org/10.1038/s41587-019-0336-3
10. Muraro, M.J., Dharmadhikari, G., Grün, D., Groen, N.: A single-cell transcriptome atlas of the human pancreas. Cell Syst. **3**(4), 385–394.e3 (2016). https://doi.org/10.1016/j.cels.2016.09.002. https://www.sciencedirect.com/science/article/pii/S2405471216302927
11. Pedregosa, F., et al.: Scikit-learn: machine learning in python. J. Mach. Learn. Res. **12**, 2825–2830 (2011)
12. Reris, R., Brooks, J.P.: Principal component analysis and optimization: a tutorial. In: 14th INFORMS Computing Society Conference, pp. 212–225 (2015). https://doi.org/10.1287/ics.2015.0016
13. Wang, Y., Huang, H., Rudin, C., Shaposhnik, Y.: Understanding how dimension reduction tools work: an empirical approach to deciphering t-sne, umap, trimap, and pacmap for data visualization. CoRR arxiv:2012.04456 (2020)

PISTIS-DM: A Semantically Enriched Data Model for Federated and Trusted Data Sharing, Discovery and Utilization of Proprietary Data Assets

Muhammad Ahtisham Aslam(✉), Yury Glikman, Sangeetha Reji, Rosanny Sihombing, and Ahmad Al Asadi

Fraunhofer FOKUS, Kaiserin-Augusta-Allee 31, 10589 Berlin, Germany
{muhammad.ahtisham.aslam,yury.glikman,sangeetha.reji, rosanny.sihombing,ahmad.al.asadi}@fokus.fraunhofer.de

Abstract. Data is potentially becoming the most valuable asset for any organization, and dataspaces are emerging as potential platforms for seamless and trusted sharing and trading of data assets. In EU funded project, PISTIS (Promoting and incentivising Federated, Trusted, and fair sharing and trading of interoperable data assets), we are aiming at developing a reference dataspace (named as PISTIS Platform) for secure, controlled and trusted usage, exchange and monetisation of proprietary data assets. *PISTIS Data Model (PISTIS-DM) is one of the main components* which creates the ground to support above mentioned various tasks of PISTIS dataspace. In this paper, we present the PISTIS-DM and its implementation as knowledge schema for PISTIS dataspace. We also present how the PISTIS-DM is used for semantic enrichment of data assets before making them available for sharing and trading. We also describe the PISTIS Data Modeling Process (PISTIS-DMP) that we developed and adopted for the PISTIS Data Model (PISTIS-DM) development, to overcome the multi-domains and multi-disciplinary nature of demonstrators. Finally, as an evaluation, we present how the PISTIS-DM is integrated into the PISTIS dataspace and is utilized for semantic annotation, discovery, and trading of proprietary data assets.

Keywords: Dataspaces · Ontologies · Data Modeling · Reasoning

1 Introduction

An organizational growth depends heavily on its data assets and how they are used. Keeping in view the value of data, different mechanisms for data storage, representation, utilization and analysis have been proposed [6], out of which ontologies [8] are one of the most effective mechanism for data modeling and management. Additionally, ontologies are being adopted as a mechanism to resolve data interoperability, integration, and heterogeneity issues for the purpose of data sharing and trading through dataspaces.

H. R. Arabnia et al. (Eds.): CSCI 2025, CCIS 2969, pp. 58–73, 2027.
https://doi.org/10.1007/978-3-032-28460-0_5

Even though ontologies provide very effective mechanism for data modeling, querying and representing knowledge in dataspaces, both academia and industry have limitized themselves to domain specific ontology design and development [9] which ultimately limitizes scope of dataspaces for data assets sharing and trading. Also, these domain specific ontologies build the first wall that reduces the multi-disciplinary and multi-sectors sharing, trading and utilization of proprietary data assets. Especially, at such a time when knowledge engineers are working on developing "Global Dataspaces", modeling the multi-disciplinary and multi-sectors ontologies can boost up data globalization through dataspaces. Another important perspective is to harness the availability of multi-domain and multi-disciplinary open data [1,3] according to the Findable, Accessible, Interoperable, and Reusable (FAIR) data principles [10].

To address above-mentioned challenges, in EU funded project, PISTIS[1], we have developed a reference data model (named as PISTIS Data Model (PISTIS-DM)) as knowledge sharing schema for PISTIS Dataspace that aims at supporting multi-disciplinary and multi-domains data discovery, sharing, trading and utilization of proprietary data assets. PISTIS-DM is developed as a joint venture of knowledge engineers and domain experts from three different hubs i.e., Mobility & Urban Planning, Energy and Automotive (with ongoing growth of PISTIS-DM to more sectors and demonstrators). Additionally, depending on the diversity of stakeholders and their background knowledge, a generalized process (named as PISTIS Data Modeling Process (PISTIS-DMP)) has been developed to identify the role and contribution of different stakeholders from various domains. The ultimate goal is to support the multi-disciplinary and multi-domains data spaces that can support controlled utilization of proprietary data assets.

Overall this research paper has following contributions:

- Technical and motivational aspects of PISTIS dataspace
- PISTIS Data Modeling Process (PISTIS-DMP): a data modeling process that is specifically designed for a joint data modeling task with experts from various domains and backgrounds
- PISTIS Data Model (PISTIS-DM): a multi-disciplinary and multi-domains integrated semantic data model
- PISTIS Ontology (PISTIS-Onto): an OWL based implementation of PISTIS-DM
- Evaluation as integration of PISTIS-DM in to the PISTIS dataspace and its utilization in semantic mapping, discovery, sharing and trading of proprietary data assets.

The rest of the paper is organized as follows: we present the motivational background and related work in Sect. 2. Details about PISTIS Data Modeling Process (PISTIS-DMP) are described in Sect. 3 where as PISTIS Data Model (PISTIS-DM) and its OWL-based implementation are discussed in Sect. 4. In Sect. 5, we evaluate our work by describing the integration and utilization of

[1] https://www.pistis-project.eu/.

PISTIS-DM in to the PISTIS dataspace. Finally we conclude our work in the Sect. 6 and highlight some future directions.

2 Background and Related Work

The PISTIS project aims to develop a federated data sharing, trading, and monetization platform designed for the secure, trusted, and controlled exchange and use of proprietary data assets. It builds on advancing technologies such as federated data discovery and sharing, distributed ledger technologies (DLT), data NFTs, and AI-driven data quality assessment and monetization. By leveraging these technologies, PISTIS unlocks the latent potential of data, fostering trust and addressing the concerns of diverse stakeholders, including data providers, owners, and consumers. Through PISTIS, these stakeholders can form a distributed network of interconnected data spaces with built-in governance, breaking down data silos, maximizing data value, and creating new value through derivative assets (e.g., analytics and insights) in a fair and transparent way. It provides a suite of tools to reshape data and intelligence sharing by establishing them as valuable, well-defined assets with clear ownership stakes, balancing the benefits and risks of data sharing while promoting sustainable business models within a thriving data market. One of the key differentiators of the PISTIS project from other data space initiatives is its provision of robust tools for data providers, enabling comprehensive data and metadata quality assurance as well as systematic data valuation. The PISTIS Data Model(PISTIS-DM) forms the foundation for these tools, facilitating semantic interoperability with existing domain standards and other data space platforms. This ensures cross-sector compatibility while remaining adaptable to evolving data needs. Data curated within PISTIS becomes aligned with domain-specific models and is meticulously prepared for seamless consumption by potential users.

Working in these research directions, several research groups are focusing on developing tools, technologies and architectures for seamless data sharing and trading. One key initiative is to develop integrated data models for semantic enrichment of data assets making the meaning of the data clearly understandable both by humans and computer systems. Similarly researchers are working on developing various solutions such as [5], that can be used for semantic mapping between relational databases, RDF, and CSV on the Web (CSVW). These tools provide a robust framework for data transformation. Additionally, a review article [7] highlighted that not only an ontology provides a structure that defines concepts and their relationships on a formal logic, but it also can be seen as the primary tool to enable data integration, sharing, and discovery. The majority of studies often develop a domain specific ontologies [2], although it is also feasible to integrate multi-domain models. For example, a multi-domain model integration architecture based on Knowledge-Graph is discussed in [12] to assist multiple business scenarios. Similarly, in [4], authors presented multi-domain semantic modeling that enhances the ability of the system being studied to interpret and handle complex requirements. Another study [9], describes the importance of semantic integration for data spaces and how establishing a semantic layer is essential for effectively realizing the idea of data spaces. The study also concludes that a more extensive implementation of semantic technologies in enterprises is still necessary in order to fully utilize its potential in dataspaces. Furthermore, the role of semantics in dataspaces, particularly in ensuring data complies with the FAIR principles, has been presented in [11]. Authors describe, how the semantic technologies and standards, such as ontologies, RDF, JSON-LD etc. play an important role in dataspaces by facilitating the integration, linking, mapping, querying different data formats and data sources through the use of Persistent Identifiers (PIDs) and rich metadata. PIDs ensure longevity and accessibility of data by addressing the issue of unsolvable URIs in Linked Data. This eventually supports the FAIR data principles within dataspaces and enhances data interoperability. As an accumulative solution to these issues, in the next sections, we present our PISTIS Data Modeling Process (PISTIS-DMP) and corresponding data model (i.e., PISTIS-DM).

3 PISTIS Data Modeling Process

Integrating semantic technologies into the dataspaces architecture is a key challenge. In this section we describe that how we addressed this challenge by designing, developing and utilizing PISTIS Data Modeling Process (PISTIS-DMP) for developing integrated multi-disciplinary and multi-domain data model (i.e., PISTIS-DM). In PISTIS project, initially we have three demonstrator hubs i.e., Mobility & Urban Planning, Energy and Automotive contributing in the PISTIS-DMP in different capacities such as source data provider, metadata explanatory experts, data semantics formulators, data quality analyzers and data interoperability monitors. The Fig. 1 shows the task oriented PISTIS-DMP that we developed and adopted for PISTIS data modeling task, keeping in view the capacity of different stakeholders, their expertise, background and data assets occupancy. In the following sub-sections we describe how the PISTI-DMP is used for data sharing, cleaning, revising, updating and finally in modeling and implementing the data model.

3.1 Analyzing the Source Datasets

In PISTIS project, we have three demonstrator hubs and several partners, where each partner is a data provider as well. As an initial exercise, around hundred datasets (all together from three demonstrator hubs) are collected and analyzed before they can be used as source for data modeling task. Analyzing the source datasets provided by domain experts resulted in to several issues in using these datasets such as varying level of terminological complexities, multi formats, meshed up, no semantics, noisy, and having no explanatory documentation that can be used to dig further in to the terms and entities that are defined in these datasets. Each of these issues actually resulted in to a task that need to be performed over these datasets before they can be used as sources for data modeling.

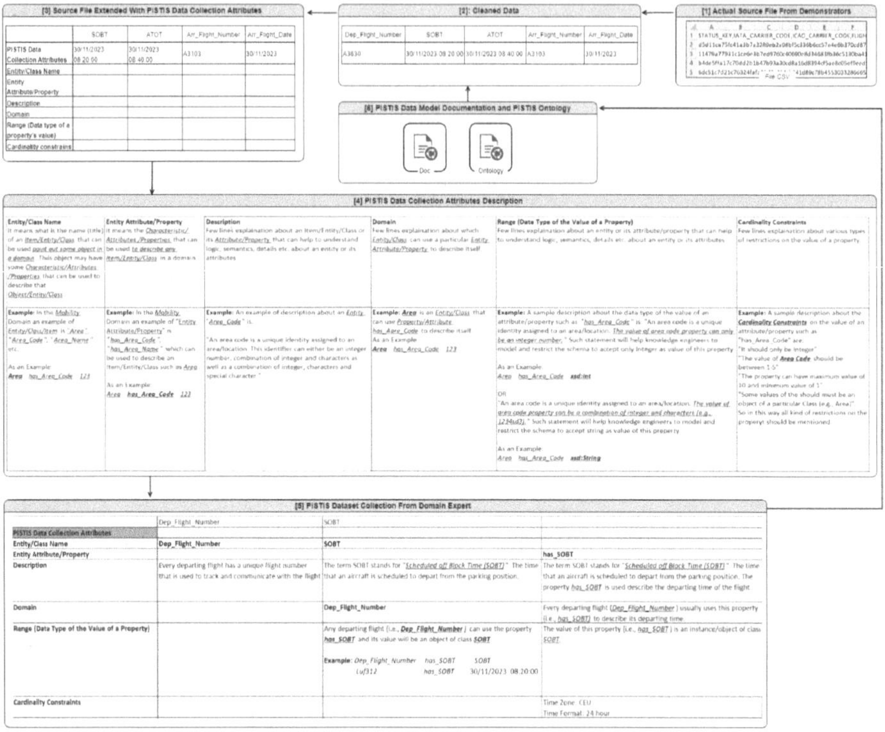

Fig. 1. PISTIS Data Modeling Process - a back and fourth data modeling approach.

The Fig. 2 (a) shows one sample dataset having most of the above mentioned issues in it.

3.2 Data Cleaning

After analyzing and highlighting the issues, the first task/activity is to clean the source datasets (where ever needed). Cleaning the sources datasets involved various sub tasks such as removing the noise, formatting the dataset to a use able format, making differentiation between entities and values, taking care of data lost, and structuring the datasets as much as possible. The Fig. 2 (b) shows an example of producing clean and structured document from a meshed up and noisy dataset. After the cleaning task is completed, knowledge engineers are able to get better picture and understanding of the provided datasets in addition to identifying issues, deficiencies, discrepancies, missing data aspect, as well as identifying the missing metadata and data descriptions. Based on all this missing information in source datasets, a self-explanatory PISTIS Metadata Collection Template is designed to educate the demonstrators with what is missing in provided datasets and what is needed more so that these datasets can be used in data modeling task (which is further explained in the next section).

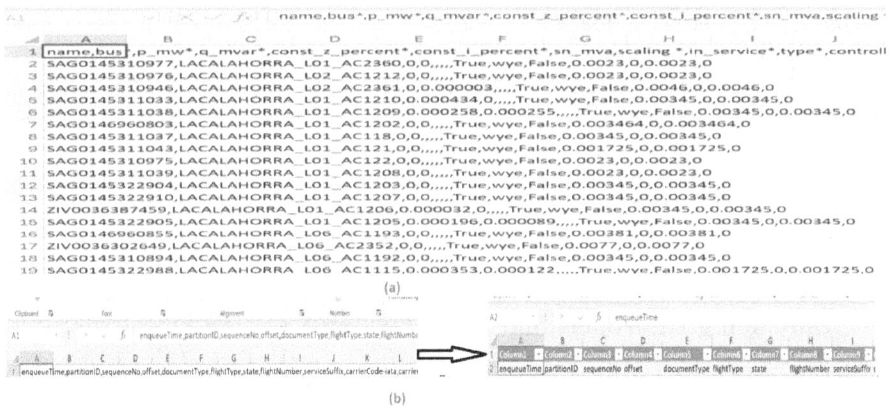

Fig. 2. (a) Sample source meshed up dataset, (b) From mashed up to clean dataset.

3.3 PISTIS Metadata Collection Template

PISTIS Metadata Collection Template (PISTIS-MCT) is a self-explanatory mechanism in the form of structured sheet to collect missing data, metadata and information from domain experts who are well educated in their domains but are not familiar or less familiar with technical aspects and features of data that can be helpful in data modeling tasks. The PISTIS-MCT was shared and explained to project demonstrators in a (PISTIS data modeling workshop), to be filled for each data entity in every provided dataset. The PISTIS-MCT is designed in such a way that it includes almost every missing data entity, its categorization, description, type, semantics, cardinality constraints and reasoning perspectives to maximize the benefit by using OWL and reasoning. The PISTIS-MCT has two perspectives i.e., Class/Entity Perspective and Property/Attribute Perspective. Each of these perspectives are further described below.

Class/Entity Perspective. The Class/Entity Perspective of the PISTIS-MCT describes most of the common aspects required to classify an entity in source datasets as a Class for PISTIS-DM. How and based on what characteristics a data item can be classified as a Class, are well documented in PISTIS-MCT with examples, descriptions and sample code (RDF Triple view) from corresponding domain. Classifying an entity as a Class, for sure, is not enough to decide the modeling role and perspective of a data item, that's why its parent & child class tree, characteristics, equivalency level, disjointedness and other modeling constraints are collected from domain experts by using PISTIS-MCT.

Property/Attribute Perspective. The Property/Attribute Perspective of the PISTIS-MCT describes various aspects (including Cardinality Constraints) required to model a property. The most important perspective of a property i.e., reasoning perspective, is extracted and modeled by understanding the semantics and logic from the description of the property provided by domain experts. Modeling a property as inverse, transitive, reflexive, symmetric etc., is based on semantic understanding of the description of a property. Schematic categorization of properties was helpful in extracting the knowledge which explicitly is not mentioned in the related datasets. This reasoning based data and knowledge extraction was very supportive in semantic mapping, discovery, and sharing of data assets through PISTIS platform.

3.4 Enhancing Interoperability By Reusing Existing Ontologies

One of the key tasks in data modeling is to maximize the interoperability of PISTIS dataspace by reusing existing ontologies (where ever possible). Since, the PISTIS-DM is developed for multi-domains and multi-disciplines therefore, during the modeling task, reuse of existing ontolgoies is taken care to its maximum. After the metadata collection of PISTIS domain datasets from related stakeholders, each data entity is analyzed from its metadata perspective keeping in view data semantics extracted from metadata descriptions. These entity descriptions, definitions, comments and labels are used to justify the semantic interoperability between PISTIS-DM and existing ontolgoies and a mapping is documented based on matching semantics (as shown in the Fig. 3).

Ontology Mapping Example from Various Demonstrators Hubs. In this section we describe different examples of mapping between various datasets items/entities from different sectors to various domain specific existing ontologies. The mapping is done based on the matching semantics that are extracted from the metadata of data sets items/entities collected by using PISTIS-MCT template and existing ontologies. The Fig. 3 shows a sample traffic dataset in Mobility & Urban Planning domain containing a data item "transversalAcceleration" which refers to "travel acceleration of gravity" and is measured with unit "g". The Fig. 3 also shows the cardinality constraints on the values of this entity i.e., value is ranging between -131,075 To 131,072. Also, the description of the data item in the source dataset and "label" and "comments" properties in the existing ontology show the semantic matching between the "transversalAcceleration" item in the source dataset and "acceleration" class of the existing ontology. Based on this matching semantics a mapping is documented between these items and PISTIS-DM is designed accordingly.

Similarly, datasets such as departure bag, flight data, passenger dataset etc. are collected from corresponding demonstrator hubs and their data entities are mapped with various ontology facts such as Air Traffic Management (ATM) ontology and NASA Air Traffic ontology etc. An example of entity to ontology mapping is about the term "SOBT" (i.e., Scheduled off Block Time) with its definition (i.e., "The time that an aircraft is scheduled to depart from the parking position") and nature of possible values (i.e., dateTime). The metadata collected about the "SOBT" entity by using the PISTIS-MCT and the documentation of

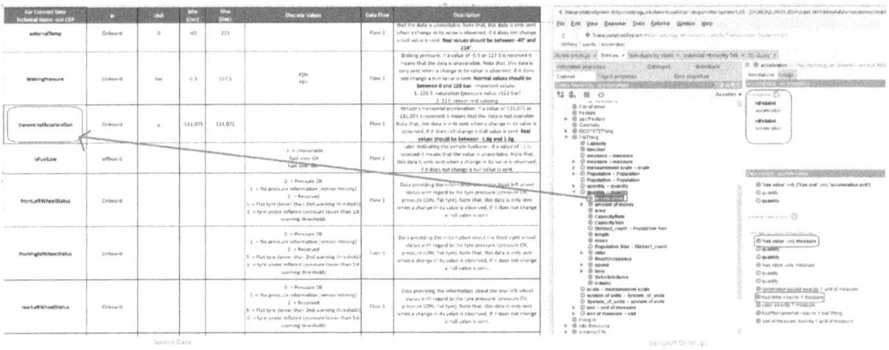

Fig. 3. An example of mapping PISTIS traffic dataset entity to existing Urban System Ontology (Mobility & Urban Planning Domain).

the property "atm:plannedDepartureTime" in the ATM ontology shows a clear semantic matching between them and that's why in PISTIS-DM implementation, this ontology is reused as one of existing ontologies. Similarly, a huge number of datasets are collected from corresponding demonstrator hubs and mapping of various entities in these datasets with most suitable classes/properties of existing ontologies is identified. Various attributes such as label, comments, data types, constraints etc. of Classes and/or Properties of existing ontologies are considered for semantic matching and reuse of existing ontologies.

3.5 PISTIS Data Model Generation

The final step/result of the PISTIS-DMP is the generation of the PISTIS-DM. The finally generated PISTIS-DM consists of its implementation as OWL ontology as well as complete documentation describing the ontology semantics, reuse of existing ontologies and integrated schema from three demonstrator hubs (as of now with potential growth to other sectors and disciplines). The next section (i.e., Sect. 4 describes the finally generated PISTIS-DM in detail).

4 PISTIS Data Model

PISTIS Data Model (PISTIS-DM) is implemented as OWL ontology, accumulating knowledge graph from three different demonstrator hubs (i.e., Mobility & Urban Planning, Energy and Automotive). Datasets in all these three domains are collected from more than fifteen data partners and are made available for browsing and querying locally in N-Triple format by using PISTIS-DM as schema for these datasets. These datasets produce knowledge graphs for user queries that can be used for data assets searching, sharing and trading in trusted environment. The Table 1 shows some sample RDF statements from PISTIS datasets generated according to PISTIS-DM as data schema for several data sources collected from various demonstrators.

Table 1. Sample RDF statements of PISTIS datasets from various demonstrator hubs.

Subject	Predicate	Object
pistis:Dept_Flight_Number	nas:departureRouteAirport	nas:Airport
oeo:Event	pistis:marker_Position	pistis:event_Coordinate
pistis:predicted_Event	pistis:has_Category	pistis:over_Voltage

4.1 PISTIS Data Model Classes

PISTIS-DM is developed and implemented as accumulative set of classes from multiple domains and sectors showing the involvement of domain experts from various PISTIS demonstrator hubs. This is the reason that the parent/child hierarchy contains classes as an integrated graph model from these three demonstrators hubs. Each class is well documented with possibly all of its characteristics. Classes in the PISTIS-DM are modeled by following the best practices and as per the semantic needs of the PISTIS domain datasets. An attribute is properly classified as a class with all its possible characteristics such as parent/child hierarchy, equivalency and disjointedness with other classes. Setting the characteristics of a class such as owl:equivalentClass, owl:disjointWith and/or owl:sameAs adds the reasoning factor to the implementation of the PISTIS-DM classes as well as enhances the mapping and reuse of existing ontologies.

4.2 PISTIS Data Model Properties

Following the standard ontology development practices, PISTIS-DM properties are also categorized as object and data type properties. The PISTIS ontology is enriched with reasoning by maximizing the usage and categorization of its object properties as Transitive, Inverse, Functional, Symmetric etc. Such categorization enhanced the reasoning factor of datasets which ultimately was very helpful in identification and searching of required data assets through PISTIS dataspace, which otherwise might be difficult. The Table 2 shows that how making the property has_Departure_Airport as Inverse of the property has_Arrival_Airport link the departing and arriving airports with each other in a bidirectional way as attributes of a Flight. Here, an interesting factor is that categorization of properties as object or data type properties and all the characteristics of these properties are based on the metadata collected from domain experts by using the PISTIS-MCT as part of PISTIS-DMP.

4.3 How PISTIS-DM is an Integrated Data Model?

As, PISTIS-DM supports semantic mapping of datasets from three different domains and disciplines, many classes and properties have been integrated into one depending on their similar nature\description and matching semantics. Such

Table 2. Machine reasoning example with semantically enriched modeling of *has_Departure_Airport* Property.

<table>
<tr><th>Object Property Name</th><th>Domain</th><th>Range</th><th>Property Type</th></tr>
<tr><td>has_Departure_Airport</td><td>Dep_Flight_Number</td><td>Airport
Stand</td><td>Inverse Of</td></tr>
<tr><td colspan="4">Logical Description</td></tr>
<tr><td colspan="4">Departure Airport property represents as $A = \{$Airport Names as Instances of Airport Class$\}$ $\forall$ Departing Flight $\in$ Airport $\subseteq$ Country $\exists$ $X \in A$ where every instance of Departing Flight is connected by using has_Departure_Airport property with the Airport in specific A</td></tr>
<tr><td colspan="4">Text Description</td></tr>
<tr><td colspan="4">Where various characteristics such as Airport and Stand altogether identifies different attributes of Departing Flight</td></tr>
<tr><td colspan="4">Reasoning</td></tr>
<tr><td colspan="4">If the property (P) links an individual "A" to individual "B" then it's inverse property (e.g., has_Arrival_Airport) will link individual "B" to individual "A". Making these two properties as inverse of each other will help to infer the arrival airport of a flight from departure airport and vice versa</td></tr>
</table>

normalization not only reduced the model complexity but also the size of the PISTIS ontology to great extent, and also made modeling easier between multidisciplinary data entities. PISTIS documentation lists down the properties and classes that have been integrated into one to simplify and standardize the integrated modeling process and to maximize the interoperability in dataspaces between multiple domains and disciplines. The *Date* datatype property is one example which has been expressed with different names (e.g., hasDate, Date etc.) in different ontologies having different domains but common range i.e., xsd:dateTime. This property has been modeled in PISTIS-DM in an integrated way such that it can be used by multi domain properties.

4.4 PISTIS Onto: OWL Based Implementation of PISTIS Data Model

PISTIS-DM is implemented as an OWL ontology by using Protégé as an ontology editor. Initial level of reasoning is implemented and tested by using the Pellet as ontology reasoner. Protégé SPARQL Tab as well semantic as web browser Gruff are used for loading the PISTIS ontology and querying the datasets by using SPARQL as query language. Several Competency Questions are identified and executed against the SPARQL Tab and results are evaluated. The Fig. 4 (a) shows a visual presentation of the ontology model in the semantic web browser Gruff, and Fig. 4 (b) shows the sample SPARQL query and its results. Basically, the Fig. 4 (b) shows a SPARQL query for a competence questions (CQ) "Find a flight arriving to Berlin at 2023-11-30T06:16:00 whereas flight aircraft of type should be B777". Finally, the Fig. 4 (c) shows the ontology metrics with all statistics of ontology components.

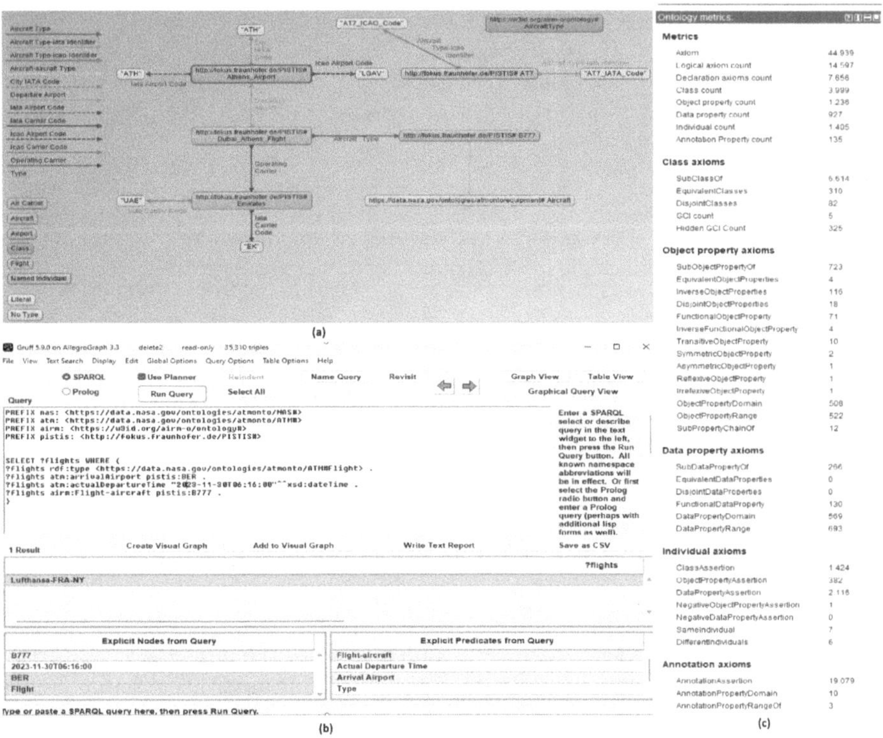

Fig. 4. (a) Visual representation of PISTIS ontology, (b) SPARQL query view, and (c) ontology metrics.

5 Utilizing PISTIS-DM in PISTIS Dataspace for Trading of Data Assets

In this section, first we discuss the whole operational use case about how the PISTIS-DM is integrated into the PISTIS dataspace and how the data enrichment process, before data assets searching and trading takes place. Then we describe the collected real datasets from three different demonstrator hubs and how these datasets[2] are enriched with semantic mapping and made available for trading.

5.1 PISTIS-DM Implementation Use Case in PISTIS Dataspace

The primary goal of the PISTIS-DM is to use it as a knowledge schema to semantically enhance datasets and then make them available for sharing and trading through PISTIS Data Sharing and Trading Platform. The PISTIS platform constitutes of a Data Space where registered users of the platform operate a "PISTIS Data Factory" in their premises which is used to store and pre-process their data for trading. These factories are all connected through a "PISTIS Platform" which is the central marketplace for data monetisation and transaction. Once a data asset is sold, it is transferred via a peer to peer communication channel setup between the PISTIS Data Factories of the Data Provider and the Data Consumer. In the practical application of this platform, the PISTIS-DM plays an integral part in the interoperability of data assets.

Data Enrichment service of the PISTIS platform is a core component which makes use of PISTIS-DM for mapping and semantic enhancement of datasets. This service is built as a comprehensive frontend-backend software solution by combining an intuitive Graphical User Interface (GUI) with a secure and scalable backend and PISTIS-DM is integrated at the backend of this service to serve as a knowledge schema. The GUI allows the user to select a dataset from the Catalogue and visualize it in the Data Enrichment service. Once, user confirms the selected dataset, he can then proceed to choose suitable column names from the classes and datatype properties of the PISTIS-DM. Further datasets specific use case of utilizing the Data Enrichment tool to enrich datasets with PISTIS-DM is discussed in the next section.

5.2 Evaluation

As a part of evaluation process, three demonstrator hubs are involved with real life datasets from respective domains. These datasets are loaded into the PISTIS dataspace (as shown in Fig. 5) for the purpose of data enrichment before they are made available for searching and trading. In total, we have 15 organizations belonging to these three demonstrator hubs which collected and contributed their corresponding domain specific datasets. For the purpose of evaluation and as a use case study, a raw dataset from Athens International Airport, that is part of the first demonstrator hub, Mobility & Urban Planning, is selected for semantic mapping with PISTIS-DM by using the Data Enrichment service.

As a first step, the Data Factory is registered and datasets in CSV format are loaded in to the dataspace. The uploaded datasets are stored in the Factory Data Storage and a metadata of the dataset is recorded and made available to the user through the Factory Data Catalogue. A metadata record displays properties such

[2] These datastes are part of ongoing research project and can be made available on demand.

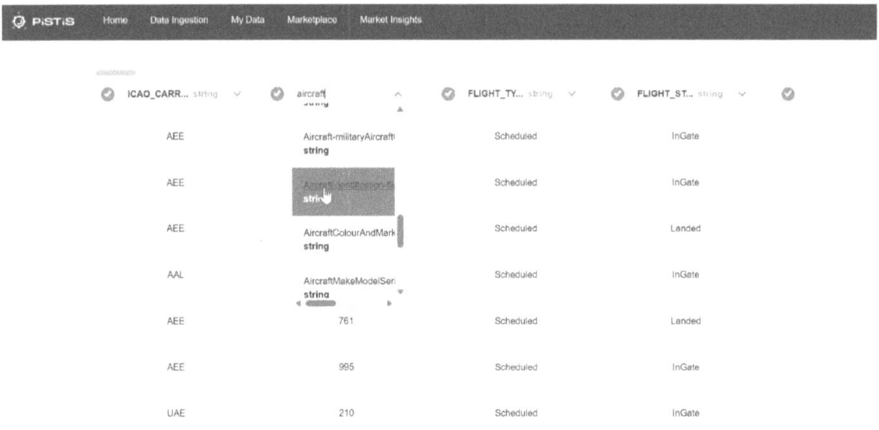

Fig. 5. Visualization and semantic mapping of datasets with PISTIS-DM before trading.

as title, description and distributions which are references of datasets in the Factory Data Storage. From this catalogue page, a user can perform **Data Enrichment** by clicking on the corresponding button on the right side of the distribution. This triggers the Data Enrichment service and displays the dataset along with options to change each of the column names to properties from the PISTIS-DM, as shown in Fig. 5. Each of the drop down menu displays these properties along with their data types. These properties and data types are also linked with their URI Ref. Users can click on these and visualize the corresponding ontology behind these properties to gather more information about it. For example, the ontology behind the property that maps to the column name shown in Fig. 5 is https://airm-o.github.io/airm-o/#AircraftIdentification-flightNumber.

Once a user adds correct mapping to columns of the dataset, the data model mapping process is completed. In case, if a property is mapped to wrong datatype, this process cannot be completed. The user needs to ensure that the property selected has the same datatype as the data in the rows. If this is not the case, the GUI will show an error. Once this process is completed, a new distribution is created in the Factory Data Catalogue which has the new data schema, as shown in Fig. 6. The dataset which was originally in CSV format is also converted to a SQL table and stored in the Factory Data Storage. This dataset can also be queried using the Factory Data Storage API. The user can now chose to offer this dataset up in the Marketplace for trading. In order to do this, the user can click on the **Publish Data** button and chose a distribution to publish to the PISTIS Marketplace. Datasets from all organizations from the three demonstrator hubs can be mapped to their respective domain specific properties in this same manner and can be offered in the PISTIS Marketplace for trading. Figures 5 & 6 jointly show one use case example to load, map, enrich and make a dataset available for searching and trading through PISTIS dataspace. PISTIS-DM is enriched enough to support the mapping of all available datasets from all three different demonstrator hubs.

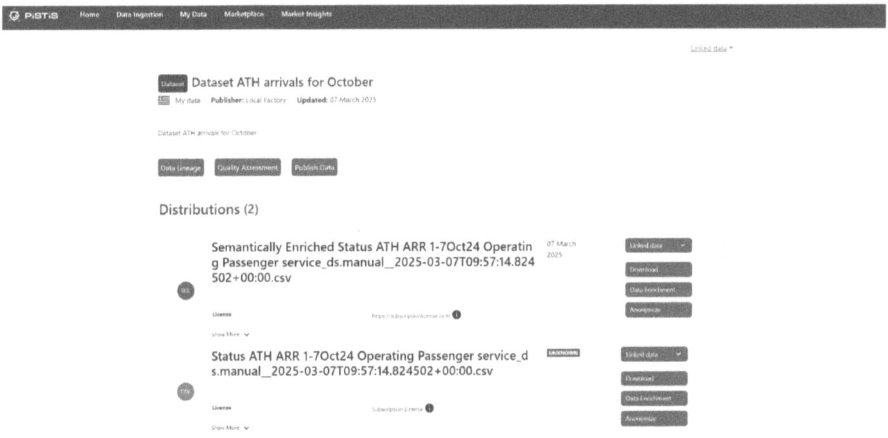

Fig. 6. Semantically enriched dataset stored in the Data Catalogue.

6 Conclusion and Future Work

Dataspaces are rapidly emerging as potential frameworks/platforms for seamless data sharing and trading of data assets. One of the main issues that is limiting the adoption and utilization of dataspaces, is the availability of semantically enriched data models that can be used for mapping and semantic enhancements of data assets before the can be made available for searching, sharing and trading through dataspaces. In this paper we presented an integrated multi-domain and multi-disciplinary data model (i.e., PISTIS Data Model (PISTIS-DM)) that is used as reference schema for mapping and semantic enhancements of data assets before they are made available for smart discovery, sharing and trading through PISTIS dataspace. The PISTIS-DM is implemented as an OWL ontology and integrated into the PISTIS dataspace as a knowledge sharing schema. We also presented the process of utilizing and mapping data assets to PISTIS-DM for their semantically enriched representation. A step before, as system requirements, we presented the PISTIS Data Modeling Process (PISTIS-DMP) that we designed and adopted for the PISTIS-DM development and implementation. We also presented the use case evaluation of our model implementation in the PISTIS dataspace to ensure the practical demonstration and usage of the data model.

Currently, the PISTIS Data Model (PISTIS-DM) is designed for data sharing and utilization among three demonstrators hubs i.e., Mobility & Urban Planning, Energy and Automotive through PISTIS dataspace, with future plans to grow as more demonstrators join the dataspace. In addition to that we are also planning the future work from three perspectives: a) extending the current model such that it support more domain specific data assets entities; b) improving the reasoning model by implementing more complicated property constraints and rules; c) integrating and extending the data model to more demonstrator hubs.

Acknowledgments. This work is funded by the European Union under Grant Agreement 101093016. Views and opinions expressed are however those of the author(s) only and do not necessarily reflect those of the European Union or the European Commission. Neither the European Union nor the granting authority can be held responsible for them.

References

1. Alaoui, K., Bahaj, M.: Semantic oriented data modeling based on RDF, RDFS and OWL. In: Ezziyyani, M. (ed.) AI2SD 2019. AISC, vol. 1105, pp. 411–421. Springer, Cham (2020). https://doi.org/10.1007/978-3-030-36674-2_42
2. Aminu, E.F., Oyefolahan, I.O., Abdullahi, M.B., Salaudeen, M.T.: A review on ontology development methodologies for developing ontological knowledge representation systems for various domains. Int. J. Inf. Eng. Electron. Bus. (2020). https://api.semanticscholar.org/CorpusID:216651445
3. Bizer, C., Heath, T., Berners-Lee, T.: Linked data: the story so far. In: Semantic Services, Interoperability and Web Applications: Emerging Concepts, pp. 205–227. IGI global (2011)
4. Borjigin, S.G., Austin, M.A., Zontek-Carney, E.J.: Semiautomated development of textual requirements: combined NLP and multidomain semantic modeling approach. J. Manag. Eng. **38**(5), 04022053 (2022). https://doi.org/10.1061/(ASCE)ME.1943-5479.0001076
5. Forkel, R., Swanson, D.G., Moran, S.: Converting legacy data to CLDF: A FAIR exit strategy for linguistic web apps. In: Calzolari, N., Kan, M.Y., Hoste, V., Lenci, A., Sakti, S., Xue, N. (eds.) Proceedings of the 2024 Joint International Conference on Computational Linguistics, Language Resources and Evaluation (LREC-COLING 2024), pp. 3978–3982. ELRA and ICCL, Torino (2024). https://aclanthology.org/2024.lrec-main.353/
6. Hauff, M., Comet, L.M., Moosmann, P., Lange, C., Chrysakis, I., Theissen-Lipp, J.: Fairness in dataspaces: the role of semantics for data management. In: SDS@ESWC (2024). https://api.semanticscholar.org/CorpusID:270591688
7. Hitzler, P.: A review of the semantic web field. Commun. ACM **64**(2), 76–83 (2021). https://doi.org/10.1145/3397512
8. Sheth, A.: Panel: data semantics: what, where and how? Presented at the (1997). https://doi.org/10.1007/978-0-387-34913-8_26
9. Sheveleva, T., Koepler, O., Mozgova, I., Lachmayer, R., Auer, S.: Development of a domain-specific ontology to support research data management for the tailored forming technology. Procedia Manuf. **52**, 107–112 (2020). https://doi.org/10.1016/j.promfg.2020.11.020
10. Stocker, M., et al.: Fair scientific information with the open research knowledge graph. FAIR Connect **1**, 19–21 (2023). https://doi.org/10.3233/FC-221513

11. Theissen-Lipp, J., et al.: Semantics in dataspaces: origin and future directions. In: Companion Proceedings of the ACM Web Conference 2023, WWW '23 Companion, pp. 1504–1507. Association for Computing Machinery, New York (2023). https://doi.org/10.1145/3543873.3587689
12. Wang, X., Hu, X., Ren, Z., Tian, T., Wan, J.: Knowledge-graph-based multi-domain model integration method for digital-twin workshops. Int. J. Adv. Manuf. Technol. **128**(1), 405–421 (2023). https://doi.org/10.1007/s00170-023-11874-4

Containerized Architecture for Time Series Data Analysis and Machine Learning Using EdgeX Foundry IoT Platform

Io Masuyama[1(✉)] and Katsumi Wasaki[2]

[1] Graduate School of Science and Technology, Shinshu University, Nagano, Japan
25w6087c@shinshu-u.ac.jp

[2] Department of Electrical and Computer Engineering, Faculty of Engineering, Shinshu University, Nagano, Japan
wasaki@cs.shinshu-u.ac.jp

Abstract. This paper presents a containerized architecture for time-series analytics on the EdgeX Foundry IoT platform. Using open weather data from the Japan Meteorological Agency for two climatically distinct sites—Tokyo (metropolitan) and Shinano-machi in Nagano Prefecture (cold, snowy inland)—we build ARIMA/SARIMA models and automatically select parameters using the Akaike Information Criterion (AIC). We implement a reusable pipeline composed of Preprocessor, Predictor, and Postprocessor services orchestrated with Docker Compose, together with an Auto-Tuning Engine (ATE) that monitors rolling MAE/RMSE and residual periodicity to update model configurations and trigger controlled re-training. In 24-h-ahead temperature forecasting, SARIMA achieves lower errors in the stable, small-amplitude cycles of Tokyo and larger errors under abrupt overnight changes in Shinano-machi. The design emphasizes portability, recoverability after power loss, and edge-friendly operation without notebooks. We also discuss integration with the EdgeX rule engine to drive alerts and actuation from forecasts and anomaly signals.

Keywords: EdgeX Foundry · Internet of Things · Time-Series Forecasting · Auto-Tuning · Edge Computing

1 Introduction

With the widespread adoption of the Internet of Things (IoT), real-time processing and advanced analytics of sensor data have become increasingly important. In smart cities, smart homes, and environmental monitoring, systems that tightly couple sensing with information processing are required. For example, automatically starting a heater when indoor temperature drops rapidly can improve comfort and support elderly residents. To realize such functions, an integrated platform is needed that can collect, store, and analyze data in a unified way. In this study, we ingest open weather data from the Japan Meteorological

H. R. Arabnia et al. (Eds.): CSCI 2025, CCIS 2969, pp. 74–80, 2027.
https://doi.org/10.1007/978-3-032-28460-0_6

Agency (JMA; Japan's national weather service) for two locations with contrasting climates—Tokyo (metropolitan) and Shinano-machi in Nagano Prefecture (a cold inland, snowy region in central Japan)—via EdgeX Foundry, and compare the applicability and forecast accuracy of time-series prediction models. We also containerize the prediction pipeline with embedded machine-learning models and propose a reusable, general-purpose architecture.

2 Time-Series Forecasting Models and Parameter Selection

To analyze time-series data collected per region, we employ ARIMA (Autoregressive Integrated Moving Average) and SARIMA (Seasonal ARIMA) models. SARIMA augments ARIMA with seasonal components and is suitable for real-world time series with periodicity—especially effective for handling diurnal ranges and cyclical variations commonly observed in cold regions. The model is denoted as SARIMA $(p, d, q)(P, D, Q)[m]$, where (p, d, q) are the non-seasonal autoregressive, differencing, and moving-average orders; (P, D, Q) are the seasonal counterparts; and m is the seasonal period. Parameters are selected automatically using the Akaike Information Criterion (AIC), which balances goodness-of-fit and model complexity (smaller is better). Concretely, we perform a parallel search over all combinations of p, d, q, P, D, Q and choose the configuration with the minimum AIC, enabling model construction that adapts to region- and period-specific data characteristics without manual tuning.

3 Implementation of the Prediction System

This study designed a data analysis system based on EdgeX Foundry to efficiently manage and analyze data collected from IoT devices. This system integrates data collection, storage, visualization, and time series analysis, as shown in Fig. 1. The IoT data analysis system designed in this study is based on EdgeX Foundry and can process, store, visualize, and predict data in one process. As shown in Fig. 1, this system consists of multiple modules and services, each with a specific role.

Historical data is sent to EdgeX Foundry via Historical Data Sender. This module enables the reuse of past data for model verification and reanalysis. The utilization of past data is crucial for improving model accuracy and ensuring operational reliability. Sensor data is generated by Sensor Data Simulator. Module functions as a virtual sensor and generates data in real time. The generated data is sent to the EdgeX Foundry's Device Service layer and converted into a format that is easy for the entire system to handle (mainly JSON format). Using a Senser Data Simulator enables system testing without requiring physical sensors. As a result, the overall operation can be verified the system's overall operation without using actual physical sensors.

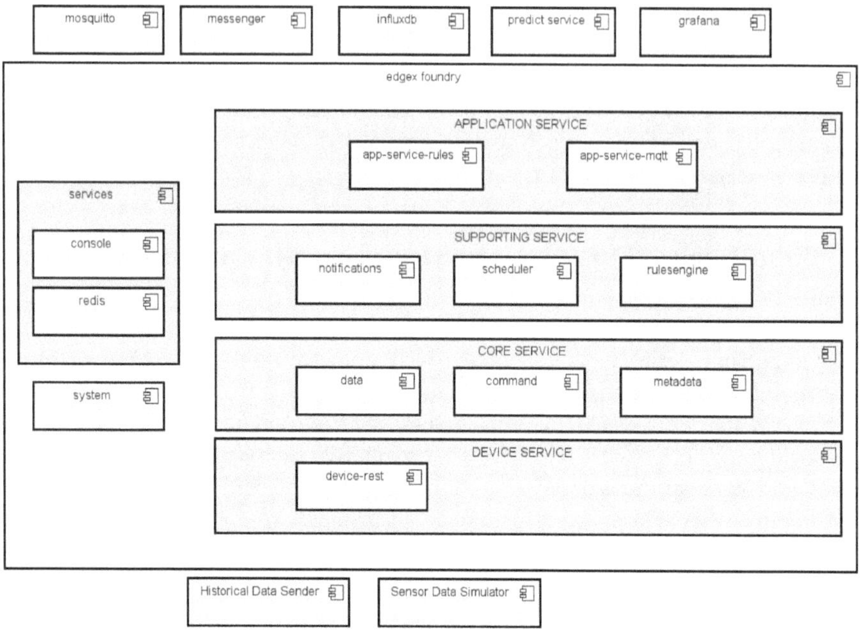

Fig. 1. Overall architecture of the EdgeX-centric IoT data analytics platform used in this study. Time series from the Historical Data Sender and the Sensor Data Simulator are ingested and normalized at the Device Service layer, then routed via Core/Supporting/Application services to storage and visualization (InfluxDB, Grafana) as well as to the forecasting pipeline. Forecasts and anomaly signals are forwarded to the EdgeX rule engine to drive alerts and actuation.

4 Experiments and Evaluation: Regional Temperature Forecasting

This study used open data from the Japan Meteorological Agency in Tokyo and Shinano-machi, Nagano Prefecture, to predict temperatures using SARIMA models and compare the predictive accuracy of each region. For each region, we predicted the temperature 24 h ahead based on 24 h of observations and compared the predicted and actual values to evaluate the model's performance. Figure 2 shows the predicted temperatures in Tokyo. The Tokyo dataset covers early May 2024; a SARIMA model forecasts temperatures up to 24 h ahead. In urban areas, there is relatively little diurnal variation, and the model smoothly follows the actual values. The timing of peaks and troughs is also consistent, indicating good predictive performance. The average absolute error (MAE) is approximately 1.08 °C, and the root mean square error (RMSE)The result was a prediction variance of approximately 1.26 °C, indicating a small margin of error.

Next, Fig. 3 shows the predicted results for Shinano-machi in January 2024. Shinano-machi is a cold region, especially during the winter, when there is a sharp drop in temperature from night to morning. As can be seen in the figure, the model follows the overall trend, but the prediction error increases when the temperature suddenly changes. In particular, the model tends to underestimate the temperature during the nighttime hours. For this period, the MAE is approximately 1.80 °C, while the RMSE is about 2.16 °C. These values indicate that the error is larger than in Tokyo. These results suggest that the SARIMA model performs well in urban areas with stable cycles and small daily variations. However, it tends to struggle in regions with severe climate fluctuations, such as cold regions, where it may not accurately capture sudden changes. This trend is clearly observed in the results. Additionally, since RMSE is strongly affected by outliers, it functions as an indicator that shows the limits of the model when there is a significant difference between the actual and predicted values. On the other hand, since MAE captures average error, it is effective in quantifying the "discrepancy" of the prediction in daily operations. In this study, using these two indicators together made it possible to evaluate the adaptation of the model and its limitations more clearly.

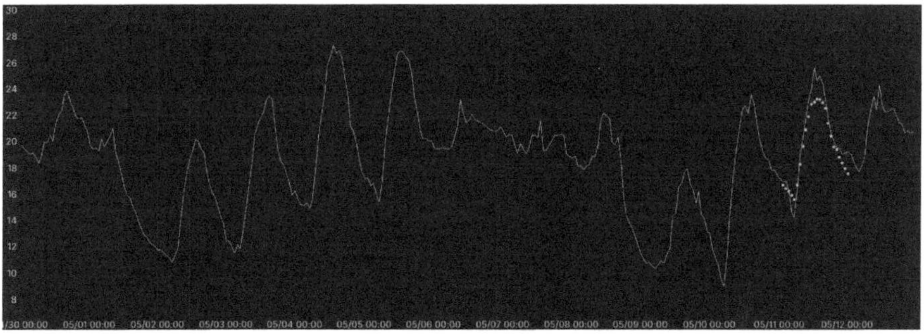

Fig. 2. Temperature Forecasting Results in Tokyo (SARIMA model): The prediction aligns well with the diurnal cycle (MAE $\approx$ 1.08 °C; RMSE $\approx$ 1.26 °C). The Tokyo case serves as a stable baseline; error characteristics differ from the colder inland site (Fig. 3).

5 Container Configuration Design

The predictive pipeline constructed in this study is composed of multiple container services managed by Docker Compose. The main processes are preprocessing, prediction, and postprocessing in three stages, and each process is implemented as an independent container. This allows for the separation of processes and improved maintainability, as well as flexible reconfiguration when switching models. Furthermore, to achieve automatic optimization of the predictive model, an Auto-Tuning Engine (ATE) was introduced. The Predictor can switch between multiple time series models such as ARIMA and SARIMA within a

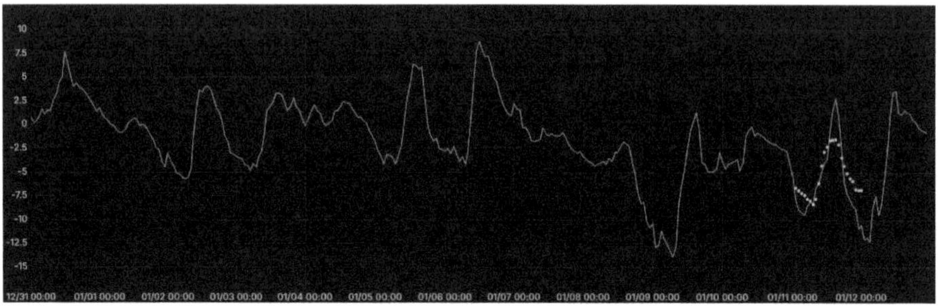

Fig. 3. Temperature Forecasting Results in Shinano-machi (SARIMA model): Abrupt nighttime cooling leads to trough underestimation (MAE $\approx$ 1.80 °C; RMSE $\approx$ 2.16 °C). Conditions are harsher than Tokyo, requiring tighter configuration and operations.

single container, and the ATE uses the prediction accuracy evaluation metrics (RMSE, MAE, etc.) obtained from the Postprocessor and the information about periodicity in the input series to dynamically update the model configuration file (config.yaml). After updating, the Predictor automatically reflects the changes. Switch to the learning or algorithm. Note that the ATE and other service definitions are described in Docker Compose, which ensures reproducibility and portability of the overall configuration (Fig. 4).

This system was designed with the following main requirements: predictability, availability, reliability, maintainability, and portability. Given the constraints of edge computing (Raspberry Pi-level resources, intermittent power, and unstable networks), the system is designed to run continuously without depending on notebook environments. In the event of a power outage or reboot, the system is designed to automatically restore the most recent verified model and settings. In addition to the Docker Compose restart policy and health check, the system uses persistent volumes for models, settings, and logs to ensure that learning results are retained after a reboot. The predictive service can continue without interruption. After recovery, it starts in a conservative maintenance mode (predictive only, with re-learning disabled) and transitions to normal operation once external integrity is confirmed.

The re-learning trigger design avoids reflexive re-learning by adopting the "wait and see + double key" approach. First, a waiting period of a certain length (e.g., 7 days) is established to prevent misfires due to short-term spikes. Then, only when both (i) sustained degradation (e.g., 7-day rolling RMSE $\geq$ 1.15) and (ii) structural drift (significant periodicity or level shift in residuals) are met, the ATE will update the config.yaml and perform re-learning/parameter adjustment. This approach minimizes unnecessary model replacements while enabling a planned response to environmental changes. The postprocessor uses 24-h, 72-h, and 7-day rolling RMSE/MAE and residual autocorrelation and periodicity tests to distinguish between "random spikes" and "structural biases due to day of the week, time of day, etc." If the latter is detected, the ATE proposes remedial measures such as revising the seasonal step and adding dummy features.

After confirming the effectiveness through off-path verification, the decision on whether to promote is made. This configuration enables the system to adapt to changes in the operating environment and sensor data through autonomous re-learning and configuration changes. In the future, we plan to confirm the effectiveness of the proposed method through operational tests that include power outage tests and simulated drift injection, as well as operational metrics such as downtime and mean time to recovery (MTTR).

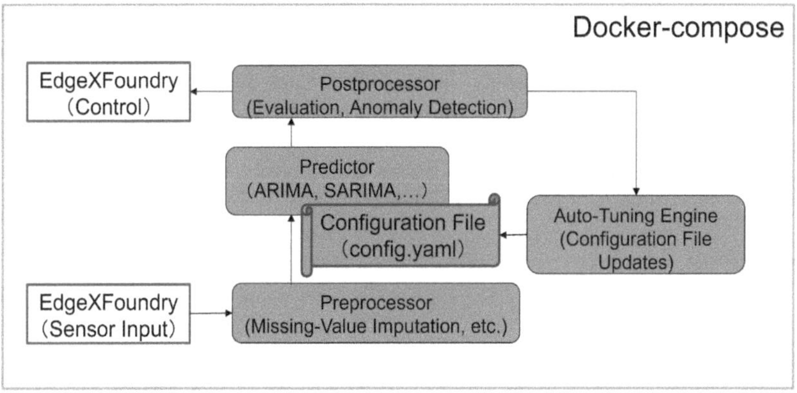

Fig. 4. Containerized forecasting pipeline orchestrated by Docker Compose. The Preprocessor, Predictor, Postprocessor, and Auto-Tuning Engine (ATE) run as independent services. The ATE monitors MAE/RMSE and residual periodicity reported by the Postprocessor, updates `config.yaml`, switches the Predictor's model/parameters, and invokes retraining only when warranted. Models, configurations, and logs are persisted to enable recovery of the last validated state after power loss or reboot. The EdgeXFoundry (Sensor Input/Control) paths indicate how the pipeline interfaces with the platform to realize the "wait-and-see + double-key" retraining trigger and the operational requirements on availability and maintainability (Sect. 5).

6 Conclusion and Future Work

This study aimed to enhance the utilization of sensor data in IoT environments by applying ARIMA and SARIMA models and using AIC to automatically select parameters, thereby eliminating the need for human adjustment in the construction of predictive models. Additionally, a data foundation based on EdgeX Foundry and a containerized predictive pipeline consisting of Preprocessor, Predictor, and Postprocessor was designed, incorporating an Auto-Tuning Engine (ATE) to enable flexible configuration changes. In the temperature prediction experiment in Tokyo and Shinano-machi, it was found that temperature changes were more abrupt in cold regions than in urban areas, resulting in increased prediction errors using SARIMA models. These findings suggest the necessity of adjusting the configuration based on environmental characteristics. The architecture utilizes Docker Compose for high reproducibility and portability, and can autonomously perform retraining and reconfiguration, making it a flexible predictive platform for practical applications.

In the future, in addition to verifying the operation in more complex environments, it will be necessary to diversify the evaluation indicators (e.g., introducing MAPE and abnormal scores) and to organize and accumulate learning models. Furthermore, we plan to strengthen the connection with EdgeX Foundry by connecting the prediction results and evaluation indicators with the rule engine inside EdgeX to apply it to the automatic control system. In particular, it is important to implement a configuration that can send alerts and control commands in real time based on the results of detecting outliers. Real-time alerting and control via the EdgeX Rule Engine will enable smart on-site responses and rapid control in the event of an anomaly, thereby increasing the practicality and value of the IoT platform as a whole.

References

1. Japan Meteorological Agency (JMA). Meteorological Data (2025). https://www.jma.go.jp/jma/index.html
2. EdgeX Foundry. Project homepage (2025). https://www.edgex.org/
3. Jonas Werner. EdgeX Foundry Tutorial v1.0 (2020). https://docs.edgexfoundry.org
4. Grafana Labs. Grafana (project homepage) (2025). https://grafana.com/
5. Cao, K., Liu, Y., Meng, G., Sun, Q.: An overview on edge computing research. Proc. IEEE Access **8**, 85714–85728 (2020)

Evacuation Route Planning Using a GPT Model Pretrained on Geographical Features and Hazard Map Data

Togo Horiuchi[1(✉)] and Katsumi Wasaki[2]

[1] Graduate School of Science and Technology, Shinshu University, Nagano, Japan
25w6084j@shinshu-u.ac.jp

[2] Department of Electrical and Computer Engineering, Faculty of Engineering, Shinshu University, Nagano, Japan
wasaki@cs.shinshu-u.ac.jp

Abstract. The objective of this study is to develop a system that presents pedestrian evacuation routes during disasters using a GPT model fine-tuned with geospatial and facility data derived from hazard maps. To this end, we employed hazard maps from the Geospatial Information Authority of Japan (GSI) and the Google Maps API, which facilitated the acquisition of coordinate data, road names, cadastral information, and surrounding facilities within designated disaster risk areas. In light of the data presented, supplementary training datasets were constructed in JSON Lines format. These datasets encompass natural language inputs, which were subsequently utilized for the purpose of fine-tuning ChatGPT. Through the refinement of GPT with these datasets, the model generated evacuation-related information such as potential shelters and supply points, while considering the disaster risks specific to the target region.

Keywords: Evacuation Route Information · Fine Tuning · Hazard Map

1 Introduction

Japan is frequently subjected to natural disasters, including earthquakes, torrential rains, and typhoons. It is essential for residents to exhibit appropriate evacuation behavior during such events in order to mitigate damage. However, in the context of disaster sites, assessing geospatial and road conditions presents significant challenges. In the initial phase, local residents may encounter difficulties in determining the extent of road damage, which can result in the necessity to retrace their steps or undertake lengthy detours, thereby prolonging the evacuation process.

The objective of this study is to develop a system for presenting evacuation-route information using a GPT model that has been pretrained on geographical requirements and hazard-map peripheral information. The data utilized in this

H. R. Arabnia et al. (Eds.): CSCI 2025, CCIS 2969, pp. 81–91, 2027.
https://doi.org/10.1007/978-3-032-28460-0_7

study consist of hazard maps (GeoJSON format) provided by the Geospatial Information Authority of Japan (GSI) and surrounding facility/road information obtained via the Google Maps API. Utilizing these hazard-map peripheral data, we implement supplementary training to enable the GPT model to present information regarding shelters and locations for acquiring supplies during disasters.

2 GPT and Evacuation Support

2.1 Overview of GPT

GPT (Generative Pre-trained Transformer) [1] is a large-scale language model that demonstrates robust performance in natural language processing. The utilization of knowledge acquired through pretraining on extensive text corpora ensures its versatility across a range of tasks. In this study, we conducted fine-tuning using GPT-4o-mini, a lightweight version of OpenAI's GPT-4o. The model has been engineered to facilitate rapid responses with optimal cost efficiency, while also exhibiting robust performance in the processing of Japanese text.

2.2 Current GPT Model Responses

Current GPT models are trained on general knowledge and have limitations in responses that consider the geography and disaster risks of specific regions. The acquisition of pertinent information, including but not limited to the locations of concrete shelters, supply points, areas at risk, and road conditions, poses a significant challenge due to the absence of readily available data. This limitation hinders the ability to provide precise responses to inquiries. Consequently, they are inadequate for the provision of evacuation support during disasters, and their utilization as is for evacuation route guidance and information presentation poses significant challenges. Consequently, precise calibration concerning regionally distinct geographical characteristics and disaster risks is imperative.

3 Acquiring Data from Hazard Maps

In this study, we conducted fine-tuning of GPT to provide region-specific responses using road segments and surrounding facility information. The processing flow is illustrated in Fig. 1. For the polygon data in the hazard maps described subsequently, we obtain road and facility information with the Google Maps API and save it in JSON format. Subsequently, an auxiliary training dataset is derived from the collected data, and the GPT model is refined to enable the provision of information regarding shelters, medical institutions, and analogous entities in response to user inquiries.

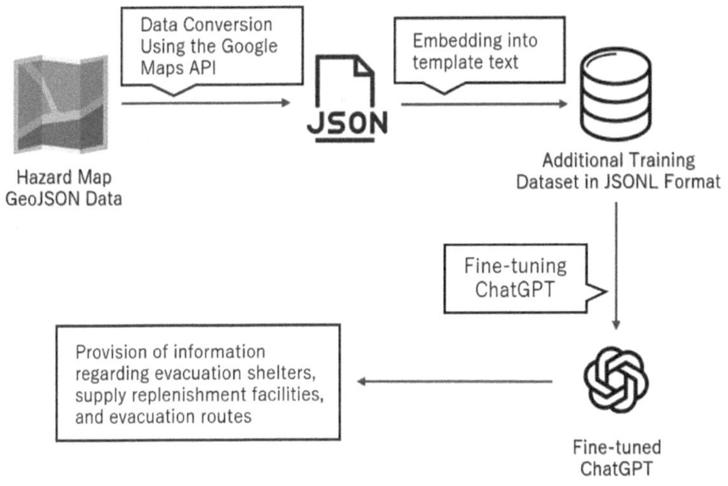

Fig. 1. Overall processing flow of the proposed system

3.1 GeoJSON Structure

The hazard maps utilized in this study are available as open data in GeoJSON format, provided by the GSI [3]. As demonstrated in Figs. 2 and 3, the data are structured as a FeatureCollection object, which stores multiple features. Each feature is associated with attribute information (properties) pertaining to disaster warning areas and polygon topographical data (geometry). As delineated in Table 1, the keys and values are stored in properties, while the polygon shapes are stored in geometry, where the type is designated as "Polygon" and the coordinates are recorded in the coordinates array. The coordinates are expressed in a longitude-latitude sequence, delineating enclosed polygons.

Table 1. Attribute information included in hazard map data

Key	Attribute Information
A33_001	Type of disaster
A33_002	Zone classification
A33_003	Prefecture code
A33_004	Zone number
A33_005	Zone name
A33_006	Location
A33_007	Date of public notice
A33_008	Presence/absence of survey or public notice regarding Special Warning Zone

```
{
"type": "FeatureCollection",
"features": [
{"type": "Feature", "properties": {"A33_001": "2", "A33_002": "2", "A33_003": "20", "A33_004": "D15-201-100", "A33_005": "寺沢2", "A33_006": "長野市", "A33_007": "9999/1/1", "A33_008": "0"},
  "geometry": {"type": "Polygon", "coordinates": [[[138.165325385, 36.643113083], [138.16527011, 36.643075255], [138.165400819, 36.642028342], [138.165535928, 36.642808842], [138.165632022, 36
{"type": "Feature", "properties": {"A33_001": "2", "A33_002": "2", "A33_003": "20", "A33_004": "D15-201-099", "A33_005": "寺沢1", "A33_006": "長野市", "A33_007": "9999/1/1", "A33_008": "0"},
  "geometry": {"type": "Polygon", "coordinates": [[[138.165017774, 36.630097587], [138.165004547, 36.639063931], [138.165168576, 36.638973986], [138.165213553, 36.639087947], [138.165030182, 3
{"type": "Feature", "properties": {"A33_001": "2", "A33_002": "2", "A33_003": "20", "A33_004": "D15-201-101", "A33_005": "金山沢", "A33_006": "長野市", "A33_007": "9999/1/1", "A33_008": "0"},
  "geometry": {"type": "Polygon", "coordinates": [[[138.161057262, 36.648569165], [138.161060606, 36.648522306], [138.16128385, 36.648545463], [138.161481888, 36.648510503], [138.16168874, 36.
{"type": "Feature", "properties": {"A33_001": "2", "A33_002": "2", "A33_003": "20", "A33_004": "D15-201-102", "A33_005": "勝手沢1", "A33_006": "長野市", "A33_007": "9999/1/1", "A33_008": "0"},
  "geometry": {"type": "Polygon", "coordinates": [[[138.162755872, 36.650889946], [138.162726916, 36.650886982], [138.162744562, 36.65081844], [138.162829216, 36.650847968], [138.162795709, 36
{"type": "Feature", "properties": {"A33_001": "2", "A33_002": "2", "A33_003": "20", "A33_004": "D15-201-103", "A33_005": "勝手沢2", "A33_006": "長野市", "A33_007": "9999/1/1", "A33_008": "0"},
  "geometry": {"type": "Polygon", "coordinates": [[[138.164463978, 36.651801852], [138.164434748, 36.651777345], [138.164505483, 36.651699836], [138.164584681, 36.651731882], [138.164478961, 3
{"type": "Feature", "properties": {"A33_001": "2", "A33_002": "2", "A33_003": "20", "A33_004": "D15-201-089", "A33_005": "龍沢", "A33_006": "長野市", "A33_007": "9999/1/1", "A33_008": "0"},
  "geometry": {"type": "Polygon", "coordinates": [[[138.139722864, 36.629617768], [138.138697471, 36.629632284], [138.138526641, 36.629490203], [138.138508711, 36.629210663], [138.138460116, 3
{"type": "Feature", "properties": {"A33_001": "2", "A33_002": "2", "A33_003": "20", "A33_004": "D15-201-092", "A33_005": "滝坂沢", "A33_006": "長野市", "A33_007": "9999/1/1", "A33_008": "0"},
  "geometry": {"type": "Polygon", "coordinates": [[[138.144677297, 36.62731046], [138.144649369, 36.627306502], [138.144643923, 36.627287469], [138.144738583, 36.627300726], [138.144733521, 36
{"type": "Feature", "properties": {"A33_001": "2", "A33_002": "2", "A33_003": "20", "A33_004": "D15-201-093", "A33_005": "畔沢", "A33_006": "長野市", "A33_007": "9999/1/1", "A33_008": "0"},
  "geometry": {"type": "Polygon", "coordinates": [[[138.148390264, 36.629517284], [138.148342929, 36.629521832], [138.148275662, 36.629348757], [138.148239296, 36.629174961], [138.148195072, 3
{"type": "Feature", "properties": {"A33_001": "2", "A33_002": "2", "A33_003": "20", "A33_004": "D15-201-094", "A33_005": "芝沢", "A33_006": "長野市", "A33_007": "9999/1/1", "A33_008": "0"},
  "geometry": {"type": "Polygon", "coordinates": [[[138.150045805, 36.628787192], [138.150019645, 36.628779995], [138.150032627, 36.628674101], [138.150027072, 36.628467703], [138.150100934, 3
{"type": "Feature", "properties": {"A33_001": "2", "A33_002": "2", "A33_003": "20", "A33_004": "D15-201-096", "A33_005": "まむし沢", "A33_006": "長野市", "A33_007": "9999/1/1", "A33_008": "0"},
  "geometry": {"type": "Polygon", "coordinates": [[[138.159625496, 36.632574088], [138.159588682, 36.632556400], [138.159593377, 36.632392301], [138.159686451, 36.632223829], [138.159800101, 3
{"type": "Feature", "properties": {"A33_001": "2", "A33_002": "2", "A33_003": "20", "A33_004": "D15-201-097", "A33_005": "黄金沢", "A33_006": "長野市", "A33_007": "9999/1/1", "A33_008": "0"},
  "geometry": {"type": "Polygon", "coordinates": [[[138.159947954, 36.634211796], [138.159899415, 36.634196336], [138.160048747, 36.634044415], [138.160149609, 36.633976434], [138.160242676, 3
{"type": "Feature", "properties": {"A33_001": "2", "A33_002": "2", "A33_003": "20", "A33_004": "D15-201-098", "A33_005": "太田沢", "A33_006": "長野市", "A33_007": "9999/1/1", "A33_008": "0"},
  "geometry": {"type": "Polygon", "coordinates": [[[138.157389934, 36.640460009], [138.157394497, 36.640429737], [138.157576276, 36.640375723], [138.157761169, 36.640264483], [138.157984655, 3
{"type": "Feature", "properties": {"A33_001": "2", "A33_002": "1", "A33_003": "20", "A33_004": "D15-201-100", "A33_005": "寺沢2", "A33_006": "長野市", "A33_007": "9999/1/1", "A33_008": "0"},
  "geometry": {"type": "Polygon", "coordinates": [[[138.168533437, 36.634766911], [138.16874019, 36.634818677], [138.16900293, 36.634892047], [138.169221428, 36.634958175], [138.169429504, 36.
{"type": "Feature", "properties": {"A33_001": "2", "A33_002": "1", "A33_003": "20", "A33_004": "D15-201-091", "A33_005": "権現沢2", "A33_006": "長野市", "A33_007": "9999/1/1", "A33_008": "0"},
  "geometry": {"type": "Polygon", "coordinates": [[[138.142221098, 36.623213436], [138.142409821, 36.623187227], [138.142610975, 36.623163958], [138.142883309, 36.623146902], [138.143153226, 3
{"type": "Feature", "properties": {"A33_001": "2", "A33_002": "1", "A33_003": "20", "A33_004": "D15-201-099", "A33_005": "寺沢1", "A33_006": "長野市", "A33_007": "9999/1/1", "A33_008": "0"},
  "geometry": {"type": "Polygon", "coordinates": [[[138.168582134, 36.635409223], [138.168834782, 36.635470308], [138.169059251, 36.635542672], [138.169283731, 36.635633872], [138.169503381, 3
```


Fig. 2. Hazard map open data provided by the Geospatial Information Authority of Japan (GSI)


```
{"type": "Feature",
   "properties": {"A33_001": "2", "A33_002": "1", "A33_003": "20", "A33_004": "D15-201-100",
                  "A33_005": "寺沢2", "A33_006": "長野市", "A33_007": "9999/1/1", "A33_008": "0"},
   "geometry": {"type": "Polygon", "coordinates": [[[138.168533437, 36.634766911], [138.16874019, 36.634818677],
                  [138.16900293, 36.634892047], [138.169221428, 36.634958175], [138.169429504, 36.635024274],
                  [138.169640512, 36.635095157], [138.169841267, 36.635169527], [138.170030144, 36.635239176],
                  [138.170214579, 36.635312329], [138.17043287, 36.635405853], [138.17064388, 36.635498095],
                  [138.17085919, 36.63559882], [138.171052395, 36.635692184], [138.171257448, 36.635797567],
                  [138.171484475, 36.635919684], [138.171682105, 36.636035771], [138.171844233, 36.636127967],
                  [138.172007874, 36.636233237], [138.17221273, 36.636363587], [138.172427938, 36.636510729],
                  [138.1726239, 36.636650608], [138.17279629, 36.636780959], [138.168904958, 36.641160811],
                  [138.168211835, 36.641523484], [138.168131252, 36.641642325], [138.16809433, 36.641775797],
```


Fig. 3. Excerpt of hazard map open data (Koichi 1-chome, Amori, Nagano City, Nagano Prefecture, Japan)

3.2 Acquiring Road Information and Evacuation Points

The geometry coordinates contained in the GeoJSON hazard-map data are processed to obtain polygon vertices. Each edge is then divided at approximately 100-meter intervals (0.001 units). From each division point, straight lines are extended in the 30°, 60°, 90°, 120°, and 150° directions. The intersection points of these lines with the other polygon edges are then computed. For each pair of an original point and its intersection point, the Directions API [4] is employed to obtain pedestrian routes, which are then treated as road segments. Segments that intersect the polygon boundary are extracted as exit points, and segments that lie within the polygon are categorized as internal roads. For each endpoint, a placeId is obtained using the Roads API [5], and addresses and names are acquired using the Places API [6]. All information is stored in JSON format. As illustrated in Fig. 4, the warning area and internal roads obtained for Koichi 1-chome, Amori, Nagano City were obtained using this method.

Fig. 4. Aggregated pedestrian route information (blue lines) enabling evacuation outside the debris-flow warning zone (red area). (Color figure online)

3.3 Acquisition of Facility Information

Step 1: Facility Retrieval. To obtain information regarding evacuation-related facilities and locations where supplies can be procured around the polygons indicated by the hazard map, the Google Maps API was utilized. Specifically, the Places API was employed to obtain basic facility information, and the Elevation API [7] was utilized to obtain the elevation of each facility. For each facility, the following data was obtained: the facility name, the primary facility category (PrimaryType) [8], the address, the latitude/longitude coordinates, and the elevation data. The search targets were limited to schools and gymnasiums that functioned as shelters; city halls, fire departments, and police stations that had the potential to serve as disaster response bases; and convenience stores, supermarkets, pharmacies, and hospitals where food and daily necessities can be procured. It was determined that categories with minimal pertinence to evacuation support, such as spas and beauty salons, would be excluded from further consideration.

Step 2: In/Out Classification. For each facility retrieved, the determination is made as to whether its location falls within or outside the polygon delineated on the hazard map. A flag is then assigned to indicate the facility's positional relationship with the disaster warning area, with "in" denoting within the polygon and "out" denoting outside the polygon. For instance, the Amori Post Office, situated in proximity to the center of Fig. 4, is within the designated area and thus designated as "in," whereas Susobana Junior High School, located to the southeast, is outside the designated area and designated as "out."

Step 3: Addition of Elevation Data. The process of obtaining elevation data entails the input of each facility's latitude and longitude to the Elevation API, thereby ensuring the integration of this crucial information as a component of the facility details. This elevation data will be used in future data expansion as material for evaluating elevation differences among facilities and the direction in which disaster impacts may spread.

Step 4: Conversion to JSON Structure. The aforementioned information has been organized in JSON format, including facility name, address, category, latitude/longitude, elevation, and in/out flag, as illustrated in Fig. 5 (e.g., "formattedAddress": "Japan, 1155 Amori, Nagano City, Nagano Prefecture, 380-0941"; "displayName": "Nagano City Amori Elementary School"; "primaryTypeDisplayName": "Elementary school"). These data are then used as variables in the fine-tuning template, as depicted in Fig. 6.


```
{
    "places": [
        {
            "types": [
                "primary_school",
                "point_of_interest",
                "school",
                "establishment"
            ],
            "formattedAddress": "日本、〒380-0941 長野県長野市安茂里１１５５",
            "location": {
                "latitude": 36.639075,
                "longitude": 138.171384
            },
            "displayName": {
                "text": "長野市立安茂里小学校",
                "languageCode": "ja"
            },
            "primaryTypeDisplayName": {
                "text": "小学校",
                "languageCode": "ja"
            },
            "primaryType": "primary_school",
            "in_out": "out",
            "elevation": 378.9743347167969
        },
```


Fig. 5. Facility information data in the vicinity of the hazard area

4 Generating Fine-Tuning Data

4.1 Training Data Format

The generation of question-answer pairs in JSONL format is conducted for the purpose of facilitating supplementary training of GPT. In this format, the conversation with GPT is represented by the speaker's role (system, user, or assistant) and the corresponding message content. The system role constitutes the preliminary configuration message that delineates the comprehensive behavior and constraints of the model. User signifies user input, while assistant represents the GPT response. In addition to the establishment of "question-answer" pairs, the system role is configured to "You are an evacuation guide for XX (cadastral district name)," thereby explicitly delineating the area and situation to the model, thus facilitating the generation of responses within the confines of the established constraints.

4.2 Building Dialogue-Style Data

The generation of training data begins with the preparation of a template text, as shown in Fig. 6 (shown here as an example of the input_data_inout template: system: "You are an evacuation guide for the ${hazard area label}$."; user: "${hazard cause}$ has occurred. Please tell me the nearest safe evacuation site."; assistant: "${facility name}$ is ${inside/outside the warning area}$."). This template is then populated with actual facility and location information to produce the fine-tuning data illustrated in Fig. 7 (e.g., system: "You are an evacuation guide for Nagano Prefecture, Nagano City, Koichi 1-chome."; user: "A landslide has occurred. Please tell me the nearest safe evacuation site."; assistant: "Nagano City Amori Elementary School is located outside the designated landslide disaster warning area and can be used as a safe evacuation site.").

```
input_data_inout = {
    "messages": [{"role": "system", "content": "あなたは${地籍(ハザードエリアラベル)}$の避難ガイドです。"},
                {"role": "user", "content": "${危険要因}$が発生しました。この付近の安全な避難場所を教えてください。"},
                {"role": "assistant", "content": "${施設名}$は${区域区分}$"}]}
input_data_food = {
    "messages": [{"role": "system", "content": "あなたは${地籍(ハザードエリアラベル)}$の避難ガイドです。"},
                {"role": "user", "content": "避難中に食料を調達できる場所はどこですか？"},
                {"role": "assistant", "content": "${施設名}$は${区域区分}$"}]}
input_data_doctor = {
    "messages": [{"role": "system", "content": "あなたは${地籍(ハザードエリアラベル)}$の避難ガイドです。"},
                {"role": "user", "content": "緊急時に治療を受けるための場所はどこですか？"},
                {"role": "assistant", "content": "${施設名}$は${区域区分}$"}]}
input_data_drugstore = {
    "messages": [{"role": "system", "content": "あなたは${地籍(ハザードエリアラベル)}$の避難ガイドです。"},
                {"role": "user", "content": "薬が必要になったときはどこに行けばいいですか？"},
                {"role": "assistant", "content": "${施設名}$は${区域区分}$"}]}
```

Fig. 6. Template text for generating training datasets

```
{
    "messages": [
        {
            "role": "system",
            "content": "あなたは長野県長野市小市1丁目の避難ガイドです。"
        },
        {
            "role": "user",
            "content": "土石流が発生しました。この付近の安全な避難場所を教えてください。"
        },
        {
            "role": "assistant",
            "content": "長野市立安茂里小学校は土砂災害警戒区域(指定済)外に位置しており、安全な避難場所として活用できます。"
        }
    ]
},
```

Fig. 7. Generated supplementary training dataset

5 GPT Response Results with Fine-Tuning

The GPT-4o-mini model was meticulously calibrated for disaster-time evacuation support, leveraging 47 conversational data items generated with hazard maps and the Google Maps API for Koichi 1-chome, Nagano City. The total number of tokens was 12,906, the number of epochs was 3, the batch size was 1, the learning-rate multiplier was 1.8, and the training loss was 0.111.

Inquiries pertaining to shelters were addressed by the pre-training model (Fig. 8-a), which provided generic shelter examples. Following the training intervention (Fig. 8-b), the model showed a significant improvement in its ability to provide concrete answers. These responses were characterized by the specification of specific facility names and the explicit assertion that these facilities were situated beyond the disaster warning area. Inquiries pertaining to the acquisition of supplies were addressed by the pre-training model (Fig. 9-a) through the provision of generic counsel. Conversely, the post-training model (Fig. 9-b) furnished specific, pragmatic responses, encompassing store names and zone-specific information. In the context of medical support, the pre-training model (Fig. 10-a) yielded inaccurate responses, including fictitious facilities. Following the training phase (Fig. 10-b), however, the model demonstrated an improvement in its capacity to present actual facilities.

The study identified several challenges. Following the training phase (Fig. 11-b), the model occasionally generated fictitious hospital names for unseen facilities. In contrast, the more generic responses of the pre-training model reduced the risk of misinformation (Fig. 11-a). Furthermore, due to the template-based nature of the training data, the range of expressions was constrained, resulting in diminished accuracy when the same content was queried with varying phrasings (Fig. 12). Subsequent studies should prioritize the augmentation of linguistic diversity and the provision of design training, thereby empowering the model to opt for non-response (abstention) in circumstances where a response is not feasible.

System message
You are the evacuation guide for Koichi 1-chome, Nagano City, Nagano Prefecture.

User
I am currently at my home in Koichi 1-chome, Nagano City, Nagano Prefecture. A debris flow has occurred, so I need to evacuate. Where should I go for shelter?

Assistant
If you live in Koichi 1-chome, Nagano City, Nagano Prefecture, the following locations are potential evacuation shelters for debris flows:
Nagano City Hall
Koichi Community Center
Nearby schools (e.g., Nagano Municipal Koichi Elementary School)
Please use official government websites or disaster prevention apps to confirm the latest information. It is also important to coordinate with neighbors and evacuate to a safe location.

(a) before fine-tuning

System message
You are the evacuation guide for Koichi 1-chome, Nagano City, Nagano Prefecture.

User
I am currently at my home in Koichi 1-chome, Nagano City, Nagano Prefecture. A debris flow has occurred, so I need to evacuate. Which evacuation shelter should I go to?

Assistant
Nagano City Susobana Elementary School is located outside the designated landslide hazard warning zone and can be used as an evacuation shelter.

(b) after fine-tuning

Fig. 8. Comparison of evacuation shelter responses (before and after fine-tuning)

System message
You are the evacuation guide for Koichi 1-chome, Nagano City, Nagano Prefecture.

User
I am evacuating from Koichi 1-chome, Nagano City, Nagano Prefecture due to a debris flow. I need to buy food.

Assistant
We understand your concerns about evacuating during a debris flow. When purchasing food during evacuation, please note the following:
1.**Choosing a safe location**: First, select a store close to your evacuation area that is deemed safe.
2. **Confirm open stores**: Check in advance which supermarkets or convenience stores are open. It's also helpful to ask volunteers or staff familiar with the evacuation center and surrounding area.
3. **Purchase essential food items**: Focus on buying emergency food and long-lasting items (canned goods, retort pouches, instant foods, water, etc.). Choosing items with longer expiration dates is especially reassuring.
4. **Utilize local information**: Check the latest local disaster prevention information and evacuation orders via social media, radio, TV, etc.
5. **Stay aware of your surroundings**: Pay attention to safety while shopping, and similarly, check for safety when returning to the evacuation shelter.
While evacuating, always be mindful of your surroundings and prioritize safety in all actions.

(a) before fine-tuning

System message
You are the evacuation guide for Koichi 1-chome, Nagano City, Nagano Prefecture.

User
A debris flow has occurred, so I am evacuating from Koichi 1-chome, Nagano City, Nagano Prefecture. I need to buy food.

Assistant
The Seven-Eleven Nagano Technical High School Front Store is located outside the designated landslide hazard warning zone. You can purchase food and daily necessities there during your evacuation.

(b) after fine-tuning

Fig. 9. Comparison of supply acquisition responses (before and after fine-tuning)

System message
You are the evacuation guide for Koichi 1-chome, Nagano City, Nagano Prefecture.

User
A debris flow has occurred in Koichi 1-chome, Nagano City, Nagano Prefecture, so I am evacuating. Please tell me where I can buy medicine and water.

Assistant
Here are several stores near 1-chome, Koichi, Nagano City, Nagano Prefecture where you can purchase medicine and water. However, during disasters like debris flows, business operations and access may be affected, so we recommend checking the latest information beforehand.
1.**Drugstores**
- **Welcia Pharmacy**: There are several locations within Nagano City. Please find the nearest one.
- **Matsumoto Kiyoshi**: This chain also has multiple locations in Nagano City.
2. **Supermarkets**
- **AEON Mall Nagano**: Carries groceries, drinking water, and pharmaceuticals.
- **Tsuruya**: A supermarket offering food and daily necessities.
3.**Convenience Stores**
- Convenience stores like **Seven-Eleven**, **FamilyMart**, and **Lawson** also carry water and some medicines.
Evacuation is the top priority. If moving is difficult or you are in a dangerous situation, stay in a safe place and follow instructions from local authorities and relevant agencies.

(a) before fine-tuning

System message
You are the evacuation guide for Koichi 1-chome, Nagano City, Nagano Prefecture.

User
I'm evacuating due to a debris flow in Koichi 1-chome, Nagano City, Nagano Prefecture. Please tell me where I can buy medicine and water.

Assistant
Aoki Pharmacy Nakagosho Branch is located outside the designated landslide hazard zone and can provide prescription and over-the-counter medications.

(b) after fine-tuning

Fig. 10. Comparison of medical support responses (before and after fine-tuning)

System message
You are the evacuation guide for Koichi 1-chome, Nagano City, Nagano Prefecture.

User
I injured my leg during evacuation and need to go to a hospital.

Assistant
I'm sorry to hear you injured your leg. It's crucial to move to a safe location immediately and receive necessary care. Please follow these steps:
1.**Ask nearby people for help**: Seek assistance from those around you. If you're en route to a shelter or crowded area, you can ask other evacuees or passersby for help.
2.**Locate the nearest hospital**: Depending on the severity of your leg injury, find a nearby hospital and consider how you'll get there. Hospitals or clinics may be available around the Amori district.
3. **Call an ambulance**: If transportation is difficult or the injury seems severe, call 118 (or 119) to request an ambulance. When calling, provide detailed information about your location and condition.
4. **If you can move on your own**: If you can move independently, seek help from people nearby or at the evacuation center to get to a nearby hospital.
Regardless of the injury's severity, we recommend seeking medical attention as soon as possible. Please prioritize safe travel.

(a) before fine-tuning

System message
You are the evacuation guide for Koichi 1-chome, Nagano City, Nagano Prefecture.

User
I injured my leg during evacuation and need to go to the hospital.

Assistant
Saiseikai Nakajo Hospital is located outside the designated landslide hazard warning area and can treat your injury.

(b) after fine-tuning

Fig. 11. Comparison of responses to facilities not included in the training data (before and after fine-tuning)

System message

You are the evacuation guide for Koichi 1-chome, Nagano City, Nagano Prefecture.

User

A debris flow has occurred in Koichi 1-chome, Nagano City, Nagano Prefecture. Where should I go to be safe?

Assistant

ARCS Shinshu Nakaya Amori Store is located outside the designated debris flow hazard zone and can be utilized during evacuation.

Fig. 12. Variation in responses due to differences in phrasing (after fine-tuning)

6 Conclusion and Future Issues

A fine-tuning dataset was generated using hazard maps in GeoJSON format and geographical/road information obtained from the Google Maps API. The GPT-4o-mini model was then fine-tuned. The findings indicated that the model possesses the capacity to generate concrete responses based on facility information pertinent to disaster warning areas. However, limitations such as insufficient training data volume and limited linguistic diversity were also observed.

Future research should focus on improving the model's ability to assess the safety of facilities and the geographical relationships in cases where multiple sediment-disaster polygons are adjacent. A methodology for delineating the area to be refined will be established. This approach will entail the incorporation of a system that enables users to input specific facility names, cadastral districts, or GPS coordinates. The system will then retrieve and process the relevant facility and road information for the designated area in advance. Furthermore, the application of evacuation drills in normal times and the provision of support for diverse evacuation scenarios tailored to geographical conditions-such as the determination of whether escape is possible from regions surrounded by multiple hazard polygons-are also challenges to be addressed.

References

1. OpenAI. ChatGPT (2025). https://chatgpt.com
2. OpenAI. Fine-tuning - OpenAI API (2025). https://platform.openai.com/docs/guides/fine-tuning
3. Ministry of Land, Infrastructure, Transport and Tourism. MLIT Hazard Map Portal Site (2025). https://disaportal.gsi.go.jp/hazardmapportal/hazardmap/copyright/opendata.html
4. Google. Directions API overview - Google for Developers (2025). https://developers.google.com/maps/documentation/directions/overview

5. Google. Roads API overview - Google for Developers (2025). https://developers.google.com/maps/documentation/roads/overview
6. Google. About the Places API (New) (2025). https://developers.google.com/maps/documentation/places/web-service/op-overview
7. Google. Elevation API overview - Google for Developers (2025). https://developers.google.com/maps/documentation/elevation/overview
8. Google. Place Types (New) - Places API - Google for Developers (2025). https://developers.google.com/maps/documentation/places/web-service/place-types

Analyzing Keyword Variations and Trends in COVID-19 Twitter Data

Dan Li, Ching-yu Huang, and Daehan Kwak(✉)

Department of Computer Science and Technology, Kean University, Union, NJ 07083, USA
{lida,chuang,dkwak}@kean.edu

Abstract. Understanding public sentiment during crises such as the COVID-19 epidemic is crucial for researchers and policymakers. This study analyzes the time dynamics of COVID-19-related discussions on Twitter by tracking the life cycle of high-frequency keywords. Tweets were collected using Selenium and Docker, stored in MySQL, and pre-processed through cleaning, stemming, and stop word removal. Three keyword extraction methods, Normal Word Count, RAKE, and YAKE, are applied to construct monthly keyword sets. To address keyword variants (e.g., "vaccine", "vax", "Pfizer"), we manually develop a synonym dictionary and a standardized normalization process. Each keyword is analyzed across three temporal dimensions: frequency, span, and continuity, enabling classification into evergreen words, outbreak words, and other types that reflect persistence or volatility in public concern. Time series visualizations and heat maps reveal clear trends, including sustained attention to "COVID-19" and "vaccines" and short-lived peaks for terms such as "lockdown" and "Trump". Results indicate that more than half of the keywords exhibit explosive growth patterns, while only a small subset demonstrates long-term significance. The combined use of RAKE and YAKE highlights their complementary strengths in capturing semantic breadth and statistical relevance. Overall, the framework offers a reusable approach for tracking public discourse during health crises and supports more accurate public opinion analysis through semantic normalization and temporal modeling.

Keywords: COVID-19 · Temporal Dynamics of Keywords · Social Media Analytics · Public Opinion Evolution · Time Series Analysis

1 Introduction

During the COVID-19 pandemic, social media platforms, especially Twitter (now rebranded as X), became an important source for the public to access breaking news, express their opinions, and share their views. This aligns with prior findings [7], which highlight Twitter as a critical medium for both disseminating pandemic-related measures and capturing public sentiment during COVID-19. Not only did the pandemic limit face-to-face conversations, making online platforms the primary venue for public discourse, but the rapid development of digital media also enabled tweets and posts to spread more quickly and widely

H. R. Arabnia et al. (Eds.): CSCI 2025, CCIS 2969, pp. 92–108, 2027.
https://doi.org/10.1007/978-3-032-28460-0_8

than traditional outlets such as newspapers, exerting a profound influence on public perception.

By tracking the trend of keyword popularity on social media, we can dynamically perceive the flow direction of public sentiment and the rhythm of information dissemination, providing intuitive evidence for understanding the evolution of public opinion in major public events. Compared with high-dimensional methods such as topic modeling, keyword trend visualization reveals key issues such as "what the public is concerned about", "when it will explode", and "whether it will continue to ferment" with a lower threshold. The performance of different keywords in the time dimension reflects how information is initiated, spreads, and weakens on the Internet, forming the "rhythm map" of epidemic public opinion.

Although a large number of studies have focused on the theme evolution and emotional changes of public opinion related to the epidemic, the research on the evolution law of the "life cycle" of keywords in the time dimension is still very limited. Most studies focus on macro themes (such as topic modeling and mood swings), lacking quantitative analysis of the emergence and fading process of keywords themselves. To fill this gap, this paper proposes a systematic approach to extract and normalize the high-frequency keywords related to COVID-19, construct their temporal heat trends, and thereby identify the keywords of "explosive topics" and "continuous attention".

This study will clearly establish a reusable keyword extraction and normalization process, propose keyword life cycle indicators (span, continuity, frequency), and establish classification rules to distinguish evergreen keywords from explosive keywords.

2 Related Work

Text mining for large-scale public health emergencies, such as COVID-19, has established itself as a critical tool for understanding public opinion. [7] enhanced tweet classification by combining Large Language Models (LLMs), specifically Mistral, with word2vec augmentation. Their framework significantly improved semantic capture and robustness regarding complex topics, highlighting the necessity of deep feature extraction for high-quality analysis [7]. Complementing these methodological advancements, recent studies have applied sentiment analysis and topic modeling to specific pandemic measures. Landaverde et al. [6] and Ojeda et al. [9] examined global discourse on face masks and vaccine hesitancy, respectively, linking public sentiment to fluctuating case numbers. Park et al. [10] further expanded this scope by leveraging Reddit data to correlate vaccine perception spikes with major sociopolitical events. Additionally, [8] analyzed the demographic and economic variables driving disproportionate viral burdens, emphasizing the multifaceted nature of the crisis.

Unlike research that focuses primarily on text classification and model performance optimization, this study prioritizes the evolution cycle of public discourse and the time dynamics of keywords. Although Landaverde et al. demonstrated that LLMs possess advantages in handling the semantic complexity of social media slang [7], challenges remain in achieving semantic standardization through structured processes, particularly when addressing morphologically diverse keywords (e.g., 'vaccine' versus 'vax') over time. Building on the prerequisite of maintaining semantic richness [7], this study aims to explore the keyword life cycle, from 'explosive' to 'evergreen', by constructing a specialized thesaurus and integrating multiple extraction algorithms, such as RAKE and YAKE. This approach transforms logic to provide more time-continuous decision-making support for crisis communication.

3 Dataset

The data used in this study originated from the public Dataset "Coronavirus (COVID-19) Tweets Dataset" [4] published by IEEE Dataport, which covers the tweet IDs of COVID-19-related tweets. The data time range is from March 2020 to March 2023, totaling 37 months. Get millions of English tweets every month through the Hydration operation [5] (retrieving the original Tweet based on the Tweet ID). The data content covers related topics such as the spread of the epidemic, public reactions, and epidemic prevention policies.

To ensure the scalability and reproducibility of data scraping, we have built an automated scraping and processing flow based on Docker. The crawling process is built based on Docker containerization technology, with a unified deployment environment to ensure repeatable operation. Implement Tweet scraping using a Selenium Headless Browser and visit the web link corresponding to the Tweet ID one by one. The Tweet ID is derived from a list of published Tweet IDs by a third party (such as the above-mentioned dataset), and is batch-captured on a monthly basis. Capture the data and store it in the structured MySQL master database (master database), and establish the index and keyword fields to accelerate subsequent retrieval.

After the data capture is completed, all tweet texts go through standardized preprocessing and structured transformation before entering the analysis process. This includes text cleaning (removing HTML tags, urls, emojis, and duplicate content), language filtering (retaining English), and stop word removal, etc. The initial steps such as tokenization, sentiment analysis, and stem extraction (stemming) have been achieved. To conduct keyword time trend analysis, we count the occurrence frequency of keywords on a monthly basis and build a Top-50 high-frequency keyword table based on monthly tweets. Ultimately, structured analysis input files are generated for subsequent lifecycle modeling and visual analysis.

4 Methodology

To clearly show the complete data processing pipeline of this study, Fig. 1 presents an overview of the workflow from tweet scraping to final keyword trend analysis and visualization. The process consists of eight major stages: distributed data collection, centralized database storage, large-scale preprocessing, monthly keyword extraction using three methods (WordCount, RAKE, and YAKE), synonym normalization, and finally, keyword lifespan and frequency-based trend modeling. Each step is further detailed in the subsequent subsections.

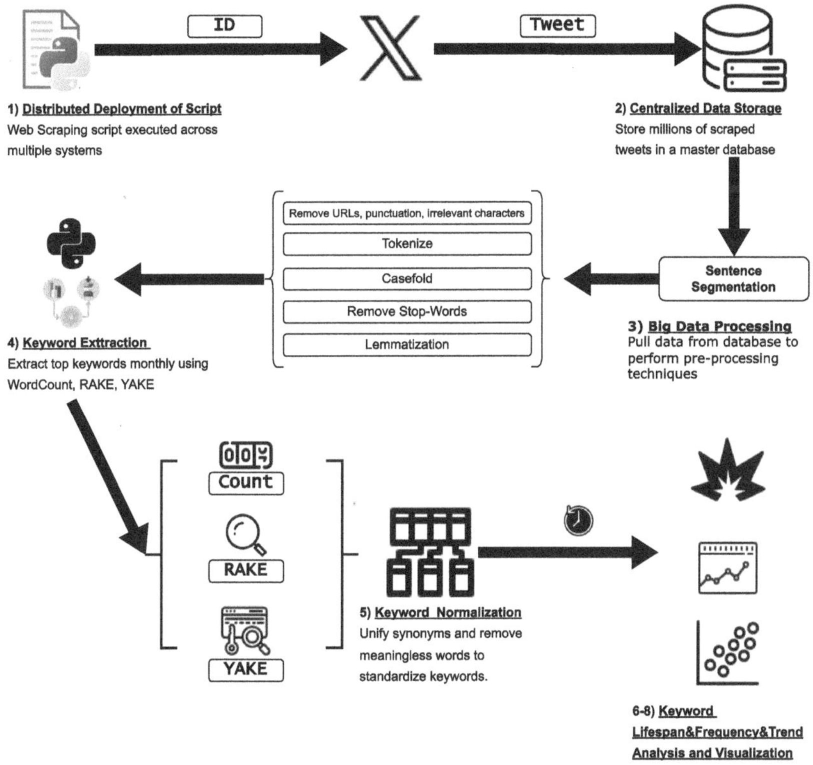

Fig. 1. Overall Workflow of COVID-19 Keyword Analysis

4.1 Preprocessing Data

To enhance the accuracy of keyword extraction and time trend modeling, we first carried out standardized cleaning and format unification processing on the original tweet text. The main cleaning steps include: removing URL links, HTML tags, special characters, emojis, and other content without actual semantic meaning. All text should be uniformly converted to lowercase to avoid word frequency errors caused by differences in upper and lower case. Delete non-English tweets and only retain samples in English to ensure the consistency of keyword semantics.

In addition to the regular stop words, we have also constructed a specific "semantic noise vocabulary list" (Table 2) to eliminate common words with high frequency but no practical analytical significance.

After the cleaning was completed, all tweets were converted into standard word segmentation structures and constructed into the monthly corpus pool required for the subsequent keyword extraction module. All tweets are aggregated by time field and constructed as the monthly tweet corpus. To enhance the comprehensiveness of Keyword Extraction, we employed two mainstream algorithms: RAKE (Rapid Automatic Keyword Extraction) [11] and YAKE (Yet Another Keyword Extractor) [1–3]. After extracting the Top 50 keywords from COVID-19-related tweets, respectively, we uniformly processed the results in lowercase and normalized them using a custom synonym dictionary. Meanwhile, we have established a stopword filtering mechanism to enhance semantic consistency. When comparing the keyword sets, we use Jaccard similarity to evaluate the Overlap degree of terms and measure the ranking consistency through RBO [12] (Rank-Biased Overlap). During the cleaning process, the original tweet ID, timestamp, and other meta-information were still retained, facilitating subsequent traceability and analysis.

Overall, this preprocessing procedure not only enhances the semantic validity of keyword extraction but also lays a clean and unified corpus foundation for subsequent time trend modeling, life cycle calculation, and classification analysis.

4.2 Extract Monthly Keywords

To extract representative keywords, we used three methods to extract the Top 50 high-frequency words from monthly tweets, respectively, and constructed a cross-month keyword set. After each method runs independently, it is uniformly saved as the corresponding folder for convenient subsequent analysis and comparison.

Normal Word Counts. As a baseline method, Normal Word Counts statistically extracts the top-50 high-frequency keywords per month using regular expressions and standard stopword filtering. To ensure scalability, the process employs SQL stream reading with checkpointing to avoid memory overload, enabling efficient, resumable batch processing. The pipeline is fully automated, supports skipping of processed months, and integrates `tqdm` for real-time progress tracking and error recovery.

RAKE [11] short for Rapid Automatic Keyword Extraction algorithm, is a domain-independent keyword extraction algorithm that tries to determine key phrases in a body of text by analyzing the frequency of word appearance and its co-occurrence with other words in the text. Instead of drawing on massive corpora or pre-trained models, RAKE focuses on the text at hand and asks a straightforward question: "which words and word sequences carry weight in this particular document?" The algorithm proceeds by stripping away stop words and punctuation, leaving behind what are called "content words", and then examining how these words co-occur in candidate phrases. Based on the degree of co-occurrence relative to frequency, each word receives a score, and phrases are ranked by the sum of their constituent scores.

YAKE [1–3] stands for Yet Another Keyword Extractor, and it is an unsupervised approach for automatic keyword extraction by leveraging text features. Unlike supervised methods, which presuppose the existence of large annotated corpora that are often unavailable or impractical across languages and domains, YAKE takes a very different route. It relies entirely on statistical features extracted from a single document itself. This design choice may sound deceptively simple, but it directly addresses a long-standing problem in natural language processing: "how to identify terms that are locally important without leaning on dictionaries, external resources, or background knowledge".

4.3 Standardized Keywords

Since different forms of expression can lead to separate statistics of synonyms, we have manually constructed a keyword normalization mapping table to unify words with similar semantics. For example, classify "COVID-19", "corona virus", and "covid-19" as "covid". This need for normalization is especially important in tweets, which frequently contain abbreviations and colloquial variations, making them challenging for traditional NLP tools [7]. The table of keywords used for normalization is in Table 1.

In addition to normalizing synonyms, we further constructed a bad keywords list (Table 2) with no semantic contribution to filter out words that are not representative but have a high frequency in semantic analysis. This list covers words such as "amp", "get", "want", "thing". Common verb pronouns or interjections such as "like", "know", and "really", although they occur frequently, contribute very little in reflecting the semantics related to the epidemic. If they are not excluded, it will dilute the meaningful keyword recognition results.

4.4 Keyword Life Cycle

Based on the monthly keyword data extracted in the method section mentioned in the previous text, this study extracted three core indicators for describing the temporal distribution characteristics of keywords: occurrence frequency ("appear count"), time span ("span months"), and the longest consecutive occurrence months ("max consecutive months"). Among them, the occurrence frequency represents the total number of times a certain keyword appears throughout the 37-month research period, used to measure its overall activity level. The time span refers to the range of months covered by a keyword from its first appearance to its last appearance, reflecting its continuous distribution in the time dimension. The longest consecutive months of occurrence indicates the longest period during which a keyword appears continuously without interruption, used to describe its concentration and stability. Through the comprehensive analysis of these three indicators, we can identify which keywords have continuously received attention throughout the entire epidemic, demonstrating strong long-term and stable characteristics. At the same time, it can also distinguish those short-term topics that only explode in a concentrated manner within a specific period, thereby providing a quantitative basis for the subsequent classification of keyword life cycles (such as "evergreen words" and "burst words").

4.5 Keyword Type Classification Rules

To deeply depict the time lifetime patterns of different keywords, we construct a two-dimensional space based on two indicators: the time span (span months) and the longest consecutive occurrence months (max consecutive months) of the keywords, and classify the keywords according to the rules accordingly. They are classified into three categories: "Evergreen", "Burst", and "Other".

Table 1. Keyword Synonyms and Variants Used for Normalization

Keyword	Synonyms/Variants
covid	covid19, covid-19, coronavirus, corona, corona virus, covid pandemic, coronavirus pandemic, long covid, virus
vaccine	vaccines, vaccination, vaccinated, vax, covid vaccine, unvaccinated
lockdown	quarantine
mask	masks, masking, face covering
death	deaths, died, die, covid deaths
trump	realdonaldtrump
get	got, getting, going
good	well
children	kids, child
year	years
work	working
say	said
go	going
try	trying
use	used
woman	women
fauci	dr fauci
origin	origins
lie	lied
tell	told
day	days
worker	workers
time	times
state	states
month	months
make	made
government	govt
mandate	mandates
test	tested, testing
shot	shots
hancock	matt

Table 2. List of Filtered Low-Relevance Keywords

amp	get	go	way	could	first
say	thing	want	like	right	since
know	need	still	also	two	via
even	try	use	really	may	please
make	tell	good	much	let	sure
day	see	one	due	we	yes
many	take	never	u	new	would
lot					

Evergreen is widely distributed throughout the entire epidemic cycle, characterized by a wide range and a long continuous occurrence time. The classification conditions are: span months ≥ 18 and max consecutive months ≥ 10.

Burst is concentrated explosion that occurs within a certain period of time and then rapidly subsides. The classification conditions are: span months ≤ 5 and max consecutive months ≤ 3. Other words do not have obvious continuity or concentration and are in an intermediate state.

4.6 Time Series of Keywords

To explore the trend of keywords changing over time, we constructed a standardized keyword time series and conducted visual analysis through trend line graphs and heat maps. First, we group and summarize the keyword frequencies extracted each month by month and keyword, and combine the results extracted by different methods (normal word count, RAKE, YAKE) to obtain a unified monthly keyword frequency table. To facilitate the comparison of the relative popularity of different keywords in different periods, we performed Z-score standardization processing on the frequency data by keyword dimension (row) to eliminate the influence of absolute differences in word frequency. We focused on analyzing several representative keywords (such as "covid", "vaccine", "mask", "lockdown", etc.), which were widely used during the epidemic and could represent different types of information dissemination hotspots.

In the visualization section, we utilized `matplotlib` and `seaborn` to generate a line plot (Fig. 2) that illustrates the standardized frequency variation of representative keywords (e.g., "covid", "vaccine") across different months. Heatmap (Fig. 3) displays the monthly activity level of each keyword in terms of color depth, and draws heatmaps from four perspectives: merge method, normal, RAKE, and YAKE, respectively. All images are saved as high-resolution files for subsequent chart display and trend comparison analysis.

5 Results

5.1 Keyword Trend Curves (Z-Score over Time)

To explore the heat change trajectory of high-frequency keywords during the epidemic, we drew the time trend chart Fig. 2 of multiple keywords based on standardized frequency (Z-score) to visualize their dynamic characteristics evolving over time.

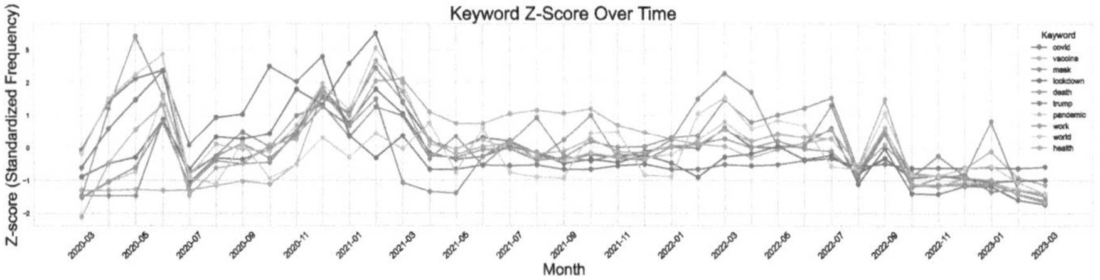

Fig. 2. Trends in the standardized frequency (Z-score) of 10 evergreen keywords from March 2020 to March 2023.

This chart Fig. 2 selects the 10 most representative keywords from Evergreen words It covers the epidemic itself (such as "covid", "pandemic"), intervention measures (such as "mask", "lockdown"), people/events (such as "trump"), social structure (such as "work", "world"), emotions and health (such as "death", "health"). Each keyword has a long span of appearance and a stable number of consecutive months of appearance, representing the core topics that the public has been continuously concerned about during the epidemic. All frequencies have been Z-score standardized before calculation to eliminate the influence of absolute frequency differences of different keywords on the trend chart. The trend chart Fig. 2 was plotted using lineplot, with the horizontal axis representing the months (March 2020 to March 2023) and the vertical axis representing the monthly standardized frequency (Z-score).

It can be observed that "covid" and "vaccine" have maintained a high level of popularity throughout the entire cycle, while "lockdown" and "trump" have shown the characteristics of phased outbreaks. For instance, "covid", as a core thematic term, maintained high popularity throughout the entire cycle and almost ran through the entire epidemic period. It was the "dominant term" in the context of the epidemic, demonstrating its representativeness as an evergreen keyword. May to June 2020 and the beginning of 2021 were two distinct peaks. The popularity of the keyword "vaccine" began to rise rapidly from the end of 2020 and reached its peak in February and March 2021, reflecting that vaccine-related information sparked a lot of discussions at that time. The period of rising popularity corresponds to the stage of vaccine research and development and promotion. It still maintained a high level of popularity in the later stage. And "lockdown", after a short period of concentrated outburst from March to

May 2020, dropped rapidly, showing a typical pattern of explosive topics. The term "trump" gained significant popularity in the second half of 2020, reached its peak in early 2021 (after the US presidential election), and then dropped rapidly, demonstrating a strong association with specific political events. "health" has experienced multiple phased fluctuations, especially before and after the promotion of vaccines, showing a cyclical strengthening of the topic of "public health".

5.2 Keyword Heatmap

To visually display the heat changes of different keywords at different stages of the epidemic, we have drawn a time heat map of the standardized Z-score heat of the keywords, as shown in the Fig. 3. The figure shows three extraction Methods Normal Word Count, RAKE, YAKE, and their combined results (All Methods Combined). For each subgraph, the Top-20 high-frequency keywords in their respective methods were selected, and the heat changes were plotted after standardization by Z-score (the redder the color, the higher the heat). The horizontal axis represents the months (March 2020 to March 2023), and the vertical axis represents the keywords. The colors reflect the trend of heat changes. The popularity of all keywords is standardized by Z-score on each line, respectively, for comparison across time.

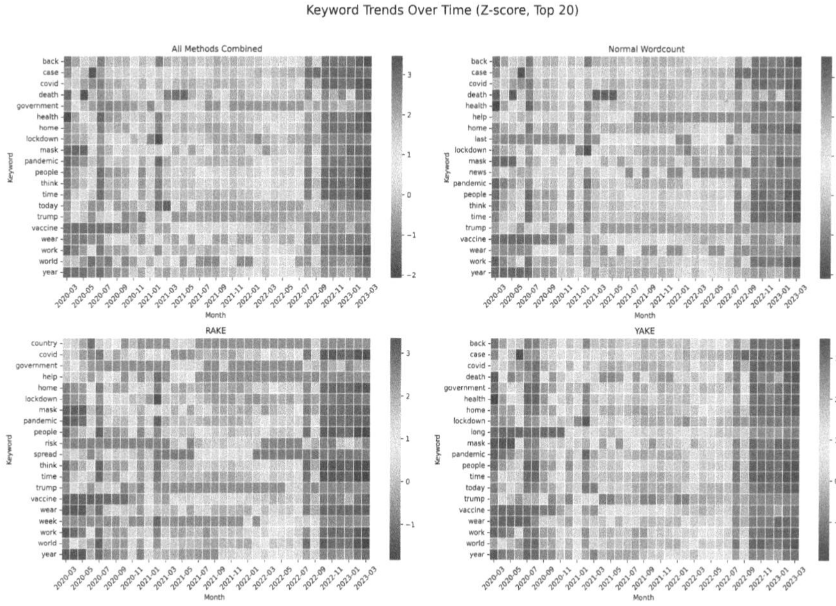

Fig. 3. Keyword Trends Heatmap Over Time (Z-score, Top 20)

From the heatmap Fig. 3 integrating the three methods, it can be seen that keywords such as "covid", "vaccine", and "lockdown" showed a significant peak from the end of 2020 to the beginning of 2021, reflecting the concentrated period of public attention on the epidemic. In the combined graph of all methods, multiple keywords gradually turned to negative Z-scores after 2022, indicating a significant decrease in the public's usage frequency of epidemic-related terms. "lockdown" reached its z-score peak in February 2021, indicating that discussions about lockdown policies were most intense during that period. Death peaked in May 2020, and cases peaked in June 2020, indicating that people were particularly concerned about and discussing topics related to COVID-19 deaths and COVID-19 cases (either confirmed or suspected) at this time.

The Normal Wordcount method prefers to extract high-frequency common words, such as "help", "news", "last", etc., which reflect the social language commonly used in tweets. In the Normal chart, words such as "lockdown", "pandemic", and "people" saw a concentrated increase in popularity at the end of 2020 and the beginning of 2021, showing obvious characteristics of public opinion explosion. The keyword "mask" maintained a moderate level of popularity in multiple periods, indicating that discussions related to masks have persisted relatively continuously. The significant rise of "trump" at the end of 2020 might be related to the US presidential election or policy disputes. Compared with other methods, Normal Wordcount places more emphasis on the usage frequency of the overall tweet corpus rather than semantic uniqueness.

The keywords extracted by the RAKE method are more inclined to semantic salient words, such as "risk", "spread", "country", and other words with the ability to carry context information. "spread" maintained a high level of popularity from May 2020 to the beginning of 2021, indicating that topics related to the spread of the epidemic during this period received widespread attention. The keyword "risk" remained active in the middle of 2021, reflecting the ongoing public discussion on the assessment of epidemic risks at that time. In the RAKE graph, "lockdown" and "pandemic" still show a distinct trend of sudden high fever, verifying their stability in semantic keyword extraction. Compared with Normal, the RAKE method is more capable of unmining structural expression words that are highly relevant to the epidemic situation.

The keywords extracted by YAKE are more diverse and suitable for capturing weak signals in semantic drift and public opinion trends. In the heat map analyzed by the YAK method, the keyword "long" showed a high fever at the beginning of 2021, which might imply the public's emotional expression towards the topic of "long-term epidemic". The fluctuation patterns of the three words "vaccine", "covid", and "trump" in the YAKE graph are highly consistent with the results of the merging method, demonstrating its stability in the extraction of core keywords. Overall, YAKE performed well in capturing ongoing topics, providing data support for subsequent thematic continuity analysis.

5.3 Lifespan Metrics Table (Typical 10 Keywords)

The following Table 3 presents the life cycle indicators of 10 representative Evergreen keywords, including their range of occurrence months, total occurrence times, occurrence time span, and the longest consecutive occurrence months.

The keywords shown in the Table 3 have all persisted during the epidemic, with a long coverage period and strong continuous appearance, reflecting their stability as core topics in the context of the epidemic.

Table 3. Typical 10 Keywords Life Cycle Table

Keyword	Months	Appear count	Span months	Max consecutive months
covid	2020_03–2023_03	37	37	12
vaccine	2020_05, 2020_07–2023_03	34	35	12
mask	2020_06–2023_03	34	34	12
lockdown	2020_03–2023_03	37	37	12
death	2020_03–2023_03	37	37	12
trump	2020_03–2021_03, 2021_06–2021_08, 2021_11–2021_12, 2022_03–2023_03	31	37	12
pandemic	2020_03–2023_03	37	37	12
work	2020_03–2023_03	37	37	12
world	2020_03–2023_03	37	37	12
health	2020_04–2023_03	36	36	12

It can be seen from Table 3 that key words such as "covid", "pandemic", "death", etc. have continuously appeared throughout the 37-month cycle, demonstrating strong temporal stability. Although "vaccine" and "mask" emerged slightly later in the early stage, they have persisted for a long time since then, demonstrating their long-term value during the epidemic prevention period. This table provides a quantitative basis for subsequent keyword classification and trend analysis.

After obtaining the life cycle indicators of all keywords, we further constructed a two-dimensional scatter plot Fig. 4 based on span and the number of consecutive months, and formulated rules to classify the keywords.

5.4 Scatter Plot of Keyword Types

To analyze the temporal persistence of different keywords, we constructed a scatter plot and simultaneously examined the occurrence span months (horizontal axis) and the maximum consecutive occurrence months (vertical axis) of the keywords. Each point in the scatter plot Fig. 4 represents a keyword. The horizontal axis indicates the span from the first occurrence to the last occurrence of the keyword, and the vertical axis represents the number of months with the longest consecutive occurrence. The depth of the color of the dots corresponds to the overall occurrence frequency of the keywords, which helps to distinguish between high-frequency and low-frequency keywords. To avoid label overlap, this study adopts a local divergent labeling strategy. Keywords located at the same or similar coordinates are expanded radially, and guide lines are used to maintain the correspondence with the scattered points, thereby improving readability.

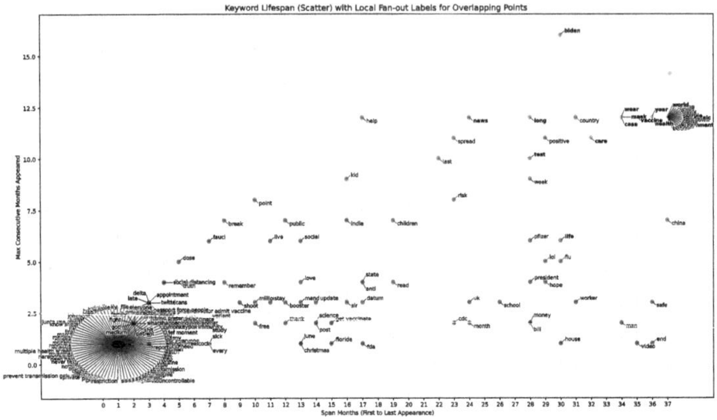

(a) Full scatter plot of keywords

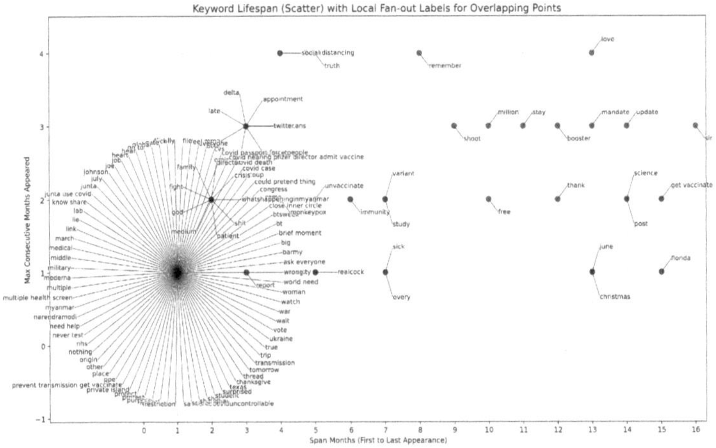

(b) Magnified view of the bottom-left region

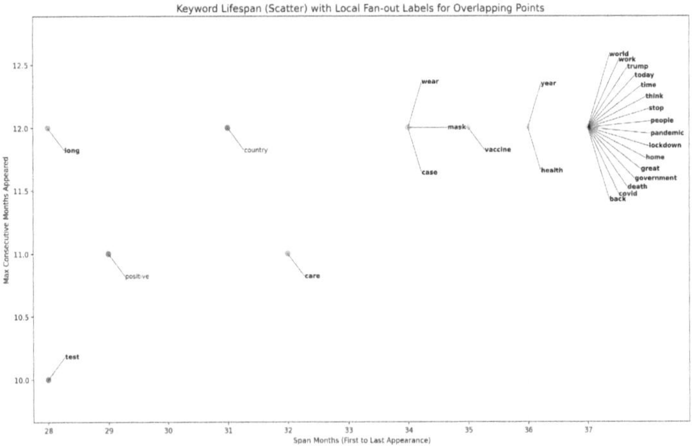

(c) Magnified view of the top-right region

Fig. 4. Analysis of keyword lifespan: (a) overall distribution; (b) magnified view of the bottom left cluster; (c) magnified view of the upper right cluster.

The upper right quadrant (with a large span and long consecutive months) mainly contains "evergreen keywords" such as pandemic, vaccine, and health, which have maintained a high level of attention throughout the entire epidemic process. The lower right area showcases typical "explosive keywords", such as cdc (Centers for Disease Control and Prevention), china, pfizer (pharmaceutical company), safe. Its characteristic is that it appears in a concentrated manner in a short period of time but lacks long-term sustainability. The keywords aggregated in the lower left corner are mostly low frequency or event limiting words. For example, nhs (UK National Health Service) and ppe(Personal Protective Equipment) appear briefly and contribute less to long-term public opinion.

5.5 Count Distribution of Evergreen/Burst/Other

To gain a more systematic understanding of the temporal behavior characteristics of keywords, we classify them into three categories (see Sect. 4 Methodology) based on their occurrence span and the number of consecutive months: evergreen words, explosive words, and other words. The specific classification indicators are in Sect. 4.5. Among all the 200 keywords analyzed, the number of breakout words was the largest, accounting for more than half, while the number of evergreen words was the smallest, only accounting for 15.5%. The classification results show that there are 107 "Burst" words, accounting for 53.5%, 62 "Other" words, accounting for 31.0%, and only 31 "Evergreen" words, accounting for 15.5%. This distribution reveals the high volatility of topics on social media, especially during the peak of the pandemic, when public attention often focused on certain hot events or emotional topics. Relatively speaking, the number of evergreen topics that can continuously attract attention over a long period of time is relatively small, reflecting the scarcity of long-term content in the discourse of the epidemic.

5.6 RAKE and YAKE Keywords Wordclouds

Overall, RAKE demonstrates the advantage of capturing semantic breadth through phrase co-occurrence, while YAKE is better at revealing the trend of statistical significance at the word level. The combination of the two can depict the structural characteristics of Twitter's public discourse during COVID-19 from a complementary perspective.

The keyword extraction results of RAKE and YAKE show partially overlapping but different focuses in COVID-19-related discourses.

As shown in Fig. 5a and Fig. 5b, both methods have identified high-frequency keywords such as "covid", "vaccine", "lockdown", "mask", and "work", reflecting the global public's concern over epidemic prevention measures, vaccination, and changes in work and life during the pandemic. However, in terms of the distribution of supportive words, the two show different tendencies. RAKE is more inclined to capture conceptual and thematic expressions such as "think", "people", "home", "time", and "say", reflecting the subjective emotions and reflective content in social media discussions. YAKE, on the other hand, pays more attention

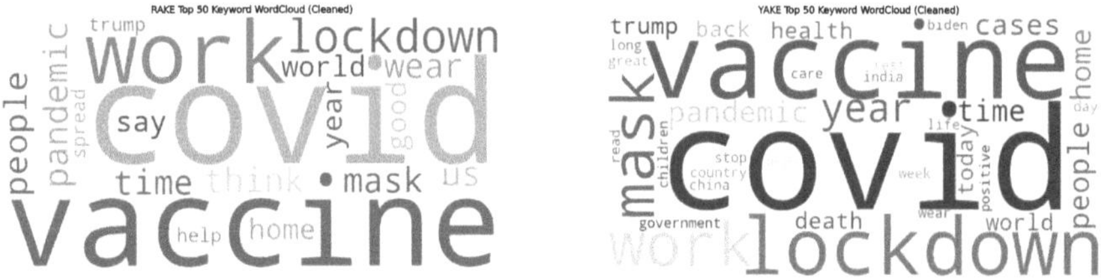

(a) RAKE Top 50 Keywords

(b) YAKE Top 50 Keywords

Fig. 5. Wordclouds of Top 50 Keywords Extracted by RAKE and YAKE

to factual and contextual words such as "cases", "death", "health", "government", "china", "india", and "test", and leans more towards objective descriptions of epidemic events and social issues.

To quantify the overlap degree of the two groups of keywords, the Jaccard similarity (0.266) and the RBO [12] similarity (0.516) were calculated. The results show that the two have certain consistency in the core of the theme, but there are differences in vocabulary and context selection.

6 Discussion

The keyword life cycle analysis highlights the phased nature of public attention during the pandemic, distinguishing long-term core concerns ("evergreen" words like *covid*, *vaccine*, *mask*) from short-lived, event-driven spikes ("burst" words such as *lockdown*, *pfizer*, *china*). This shift reflects an evolution in public psychology from early-stage panic to adaptive normalization. RAKE and YAKE exhibit semantic biases: RAKE favors conceptual and reflective terms, capturing emotional and cognitive layers, while YAKE leans toward factual, event-based expressions, indicating a duality between subjective cognition and objective narrative in discourse. Their moderate overlap (Jaccard 0.266; RBO 0.516) suggests complementary strengths, supporting the use of combined methods for robust coverage. The high proportion of burst keywords (53.5%) underscores the volatility and emotion-driven nature of social media discourse, while the limited share of evergreen topics (15.5%) signals the decline of sustained public engagement. This aligns with the attention cycle in crisis communication: intense but fleeting focus on emerging events, contrasted with waning concern for long-term issues. Sustaining public awareness on enduring health challenges thus calls for renewed communication strategies.

7 Conclusion and Future Work

7.1 Conclusion

This study systematically quantified the temporal evolution characteristics of Twitter discourse during COVID-19 and identified keyword types with different life cycle patterns. Key words such as "covid", "vaccine", and "mask" ran through the entire epidemic and formed the long-term core of public discourse. "Youdaoplaceholder0", "pfizer", "china", etc. show phased characteristics of short-term outbreak. Combining the extraction results of RAKE and YAKE helps to strike a balance between semantic breadth and statistical significance. The keyword lifecycle classification (Evergreen/Burst/Other) provides a quantitative framework for understanding the evolution of public opinion and the diffusion of information.

7.2 Future Work

Future work will focus on modeling the semantic evolution and structural dynamics of COVID-19 discourse. Semantic drift analysis (e.g., Word2Vec/BERT embedding alignment) can be applied to trace meaning shifts in core terms like "vaccine" and "freedom," while multi-word expressions and hashtags (e.g., "herd immunity", #wearamask) may capture richer discourse units. Integrating topic modeling (e.g., LDA, BERTopic) with co-occurrence networks and sentiment analysis enables a multi-dimensional understanding of thematic transitions and emotion-driven diffusion. Additionally, graph-based approaches such as GCNs can uncover high-order semantic relationships within keyword networks. Cross-platform comparisons (e.g., Twitter vs. Reddit) will further reveal convergence and divergence in public discourse. A future direction is to develop a multimodal framework that fuses semantic, emotional, topical, and structural features for interpretable modeling of social media narratives during health crises.

References

1. Campos, R., et al.: YAKE! Keyword extraction from single documents using multiple local features. Inf. Sci. **509**, 257–289 (2020). https://doi.org/10.1016/j.ins.2019.09.013
2. Färber, M., Thiemann, A., Jatowt, A.: To cite, or not to cite? Detecting citation contexts in text. In: Pasi, G., Piwowarski, B., Azzopardi, L., Hanbury, A. (eds.) ECIR 2018. LNCS, vol. 10772, pp. 598–603. Springer, Cham (2018). https://doi.org/10.1007/978-3-319-76941-7_50
3. Campos, R., Mangaravite, V., Pasquali, A., Jorge, A.M., Nunes, C., Jatowt, A.: YAKE! Collection-independent automatic keyword extractor. In: Pasi, G., Piwowarski, B., Azzopardi, L., Hanbury, A. (eds.) ECIR 2018. LNCS, vol. 10772, pp. 806–810. Springer, Cham (2018). https://doi.org/10.1007/978-3-319-76941-7_80
4. Lamsal, R.: Coronavirus (COVID-19) tweets dataset (2020). https://doi.org/10.21227/781w-ef42
5. Lamsal, R., Read, M.R., Karunasekera, S.: BillionCOV: an enriched billion-scale collection of COVID-19 tweets for efficient hydration (2023). https://doi.org/10.21227/871g-yp65
6. Landaverde, E., Ojeda, D., Kwak, D.: An analysis of public sentiment around face mask during COVID-19. In: 2022 International Conference on Computational Science and Computational Intelligence (CSCI). p. 785–789. IEEE (2022). https://doi.org/10.1109/csci58124.2022.00143

7. Landaverde, E., Spencer, A., Kwak, D.: Leveraging data augmentation and large language models for enhanced COVID-19 tweet classification. In: Arabnia, H.R., Deligiannidis, L., Amirian, S., Shenavarmasouleh, F., Ghareh Mohammadi, F., de la Fuente, D. (eds.) CSCE 2024. CCIS, vol. 2252, pp. 174–190. Springer, Cham (2025). https://doi.org/10.1007/978-3-031-86623-4_14
8. Ojeda, D., Champion, A., Huang, C.Y., Kwak, D.: Analyzing covid-19 impact in the us: demographic, economic, and social factors. In: 2023 Congress in Computer Science, Computer Engineering, & Applied Computing (CSCE), pp. 1462–1468. IEEE (2023). https://doi.org/10.1109/csce60160.2023.00242
9. Ojeda, D., Landaverde, E., Huang, C.Y., Kwak, D.: Sentiment analysis and topic modeling on COVID-19 vaccines using twitter data. In: 2022 International Conference on Computational Science and Computational Intelligence (CSCI), pp. 767–771. IEEE (2022). https://doi.org/10.1109/csci58124.2022.00140
10. Park, B., Jang, I.S., Kwak, D.: Sentiment analysis of the COVID-19 vaccine perception. Health Inform. J. **30**(1) (2024). https://doi.org/10.1177/14604582241236131
11. Rose, S., Engel, D., Cramer, N., Cowley, W.: Automatic keyword extraction from individual documents. In: Berry, M.W., Kogan, J. (eds.) Text Mining: Applications and Theory, pp. 1–20. Wiley (2010). https://doi.org/10.1002/9780470689646.ch1
12. Webber, W., Moffat, A., Zobel, J.: A similarity measure for indefinite rankings. ACM Trans. Inf. Syst. **28**(4) (2010). https://doi.org/10.1145/1852102.1852106

Q-Gini: A Multidimensional Inequality Index with Isolation of Its Causal Drivers

Ray Hashemi[1(✉)], Omid M. Ardakani[2], Rushmila Shabneen[1], and Azita G. Bahrami[3]

[1] Department of Computer Science, Georgia Southern University, Savannah, GA, USA
{rayhashemi,rs24222}@georgiasouthern.edu

[2] Department of Economics, Georgia Southern University, Statesboro, GA, USA
oardakani@georgiasouthern.edu

[3] IT Consultation, Savannah, GA, USA

Abstract. The *univariate* Gini (U-Gini) index is the gold standard for quantifying the inequality for a population in reference to a socioeconomic factor such as income, wealth, health, education, etc. However, the U-Gini's focus on a single factor (dimension) does not fully capture inequality in a population. We introduce a new Gini index (Q-Gini) delivering the inequality for multi-socioeconomic factors. The Minimum Squared error approach is used to validate the Q-Gini index by establishing its performance superiority over the U-Gini index. A Modified Neighborhood Clustering is introduced to reduce the multidimensions of the dataset into a set of neighborhoods and then validate the new Q-Gini for the neighborhoods. We also isolate the causal impacts of socioeconomic factors on the Q-Gini. Such isolation is essential because the causal drivers represent targets for intervention by policymakers. The results revealed the performance superiority of Q-Gini over U-Gini in both cases of the non-reduced and reduced dimensions of the dataset. Also, the isolated causal impacts validated the Q-Gini's sensitivity to these fundamental socioeconomic forces.

Keywords: Gini Index · Multidimension Gini Index · Causal impact

1 Introduction

The use of the Gini index (to be precise, univariate Gini—U-Gini) is common for quantifying the inequality for a given attribute (dimension), A, of a population. The U-Gini index for A is calculated using formula (1),

$$\mathrm{U - Gini_A = 1 - L/E} \tag{1}$$

where L is the shaded area under the distribution of the attribute A (Lorenza curve), and E is the area under the equality line for the attribute A (triangle of abc) [1]. This line has a 45-degree angle with both the X-axis and the Y-axis to represent the equality of (x, y) coordinates for every point on this line, Fig. 1.

The U-Gini index is the gold standard for quantifying the inequality within a population in reference to a socioeconomic factor such as income, wealth, health, education,

H. R. Arabnia et al. (Eds.): CSCI 2025, CCIS 2969, pp. 109–121, 2027.
https://doi.org/10.1007/978-3-032-28460-0_9

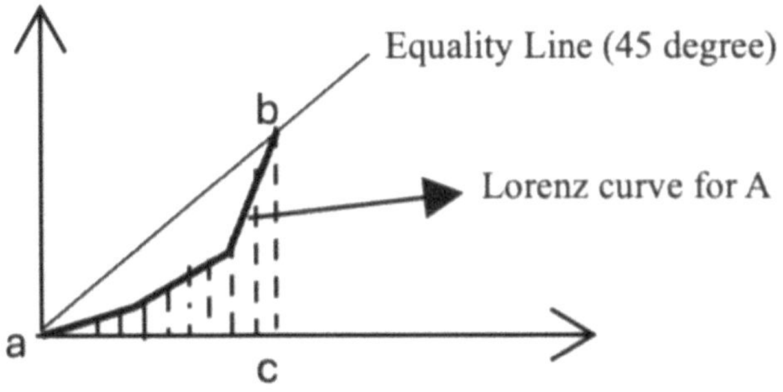

Fig. 1. Equality Income Line and Distribution Curve for Attribute A.

etc. [2, 3]. However, the U-Gini's focus on a single factor (dimension) does not fully capture inequality in a population, because inequality manifests across several interconnected socioeconomic attributes [4, 5]. In this paper, we propose a Gini index capable of delivering the inequality in reference to multi-socioeconomic factors—a comprehensive Gini (Q-Gini) index.

We also investigate the dimensions' reduction effect on Q-Gini. In addition, we isolate the causal impacts of socioeconomic factors on the Q-Gini because (a) these factors exhibit high correlations among themselves, which can lead to spurious inference, and (b) causal drivers represent targets for intervention by policymakers.

The goal of this research effort has four prongs: (i) Introducing the Q-Gini index as a representative of inequality for a multidimensional dataset, (ii) Validating the Q-Gini by comparing its performance with U-Gini, (iii) investigating and validating the effect of the dimension reduction on Q-Gini, and (iv) Delivering the causal drivers of Q-Gini in the dataset.

The rest of the paper is organized as follows: The previous works are the subject of Sect. 2. The Methodology is introduced in Sect. 3. The Empirical results are presented in Sect. 4. The Conclusion and future research are covered in Sect. 5.

2 Previous Works

Inequality measurement is foundational in socioeconomic analysis. It is a known fact that the U-Gini index is used to quantify the inequality for a given socioeconomic index. This is also another known fact that inequality is influenced by more than one socioeconomic index. Therefore, one can claim that the U-Gini index does not fully capture inequality. There is a need for multidimensional extensions of the Gini index. Although such extensions exist [6, 7], their computational validation remains underdeveloped. In particular, challenges remain in aligning composite indices with univariate benchmarks.

The Gini index [1] measures single-attribute Seminal extensions to multiple dimensions include Normative Multidimensional Ginis [6], which generalized the Gini to multiple attributes but omitted computational validation. Distance-Based Approaches [7] proposed Euclidean aggregation yet relied on arbitrary weights. Clustering-Aware Indices [8] advocated attribute grouping but offered no empirical validation protocol. Self-Organizing Map (SOM) for Socioeconomic Clustering [9] established SOMs as topology-preserving tools for high-dimensional data reduction. Applications show efficacy in inequality contexts (e.g., Deboeck for wealth segmentation [10]). Validation of Composite Indices pioneered Monte Carlo sensitivity analysis [11], while Maggino emphasized predictive accuracy as a validation criterion, both of which are underutilized in Gini literature [12].

We propose a comprehensive inequality measure through hierarchical clustering. A novel Mean Squared Error (MSE)-driven comparison of the proposed measure against traditional univariate Gini is employed.

3 Methodology

The details for the four prongs of the goal are presented in the following four subsections.

3.1 Q-Gini Index for a Multidimensional Dataset

To calculate the Q-Gini index for a multidimensional dataset, we build a tree structure out of the dimensions of the dataset in such a way that the root of the tree, as a dimension, be the representative of all dimensions of the dataset. The Gini index for the root dimension is considered the Q-Gini index for the multidimensional dataset.

The tree is a dendrogram [13] that is built using the algorithm Dendrogram, Fig. 2. We explain the algorithm through an example.

Let us assume that we have five attributes of A1,..., A5 in a dataset, and the goal is to calculate the Q-Gini index for the dataset. By applying the algorithm Dendrogram, we generate a tree out of the attributes. A possible instance of such a tree is shown in Fig. 3

Each node of the dendrogram (not leaves) represents a new attribute that is made up of its two children. To decide which two attributes to merged], the total Euclidean distances' square between every pair of attributes is calculated. The pair with the smallest distance is the winner. The reader needs to know that values for the pair of attributes must be of the same scale.

Therefore, all attributes of the dataset D, are normalized to the same scale of 1 to 9, prior to distance calculation using formula (2).

$$Nv_i = 1 + 8\frac{v_i - \min(Aj)}{\max(Aj) - \min(Aj)} \tag{2}$$

where Nv_i is the normalized value of v_i in attribute Aj.

Let us assume that a pair of attributes (A1, A2) is the one with the smallest distance. The new attribute of B is generated, and the values for B are the average of the corresponding values in A1 and A2. Attribute B replaces A1 and A2 in the dataset.

The process of finding a pair of attributes with the minimum distance, averaging them into a new attribute, and replacing the pair with the new attribute continues until the multidimensional dataset is converted into a one-dimensional dataset. This dimension is the root of the tree (E), and it represents all dimensions of the dataset.

```
Algorithm DENDROGRAM
Input : n-dimensional dataset D, with attributes of A1, . . ., An
Output: Dendrogram of D
Method:
 1- Normalize every attribute in D
 2- Trans= Transposition (D);
 3- Min =10;
 4- While (|Trans| >1)
 5-   For i= 1 to |Trans|-1
        For j = i+1 to |Trans|
          If Min > Euclidean distance(row_i, row_j);
          Then Min=Euclidean distance(row_i, row_j);
                Pair = (row_i, row_j)
        End;
      End;
      New_row = Average of corresponding values of rows in the Pair.
      Remove from Trans the rows kept in the Pair;
      Insert New_row into the Trans;
      Min=10;
   End;
 6-End of Method;
```

Fig. 2. Algorithm DENDROGRAM

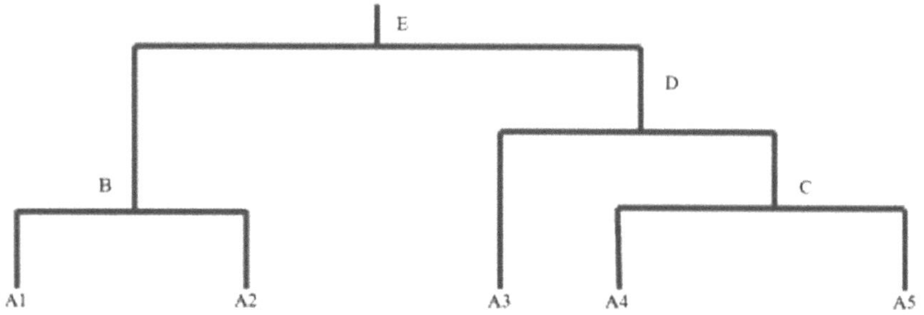

Fig. 3. A Possible Dendrogram for the Five Attributes of A1,…, A5.

3.2 Validation of the Q-Gini

File D is given with n + 1 attributes. One attribute (γ) in D carries the U-Gini value obtained for the population, in reference to one attribute, A, where A is not γ. We remove the attribute γ from the file D and make a new file for γ called D'. Now, D' and D have 1 and n attributes, respectively, and both have the same number of records.

Let TR and TS be a training set and a test set created out of the file (D). The |TR| and |TS| are equal to u1 and u2, respectively such that u1 > u2 and (u1 + u2) = |D|. Both files have n attributes. We calculate the U-Gini index for the attribute Ai (for i = 1 to n) of the TR. As a result, we have n U-Gini indices of ($G^{tr}1,\ldots, G^{tr}n$) –one per attribute. We also calculate Q-Gini for the TR file using the Dendrogram algorithm. The outcome is a Q-Gini index (Q^{tr}) for the entire attributes of TR file. The Information Gain (IG) for each attribute of TR is ($IG^{tr}{}_i = Q^{tr}$- $G^{tr}{}_i$) (for i = 1 to n).

We select the same u1 records from D' and calculate the Q-Gini (Q^{γ}) and information gain for the selected u1 records of D', ($IG^{tr}{}_i = Q^{tr}$- $G^{tr}{}_i$) (for i = 1 to n).

For file TS, we repeat the calculation of U-Gini for each attribute ($G^{ts}{}_1,\ldots, G^{ts}{}_n$), Q-Gini for the entire TS file (Q^{ts})), and information gain for each attribute ($IG^{ts}{}_i = Q^{ts}$-G^{ts}) (for i = 1 to n).

The minimum squared errors in reference to TS are obtained using formulas (3) and (4).

$$MSE_{TR-Q-Gini} = \frac{1}{n}[\sum_{i=1}^{n}\left(IG_i^{tr} - IG_i^{ts})^2\right] \quad (3)$$

$$MSE_{g-Q-Gini} = \frac{1}{n}[\sum_{i=1}^{n}\left(IG_i - IG_i^{ts})^2\right] \quad (4)$$

The Min ($MSE_{TR\text{-}Q\text{-}Gini}$, $MSE_{\gamma\text{-}Q\text{-}Gini}$) determines the superiority of one index over the other one. If the minimum squared error belongs to $MSE_{TR\text{-}Q\text{-}Gini}$, then the validity of Q-Gini is established.

3.3 Investigating and Validating the Effect of the Dimensions' Reduction of Q-Gini

The main question is whether a reduced version of an n-dimensional dataset of TR can be used, and if so, how the Q-Gini of the reduced dataset is comparable with Q^{tr} and Q^{γ}. To answer this question, we reduce the n-dimensional dataset into an m-dimensional dataset, where m < n. The reduction is done by using a selected clustering method that satisfies two conditions of natural clustering and inclusion. The natural clustering condition says that the clustering must take place naturally and not by force.Therefore, those clustering techniques for which the number of final set of clusters, C, must be announced prior to the actual clustering process, perform a forced clustering because each record of the dataset has no choice but to be in one of the C clusters and the natural number of clusters (based on the hidden patterns) within the dataset may be higher, or lower than C.

The inclusion condition says that the centroid of cluster, c_i, must not be used for deciding on inclusion of a new record into c_i because the similarity is not established between two individual records of D, but between a record of D and the centroid of a cluster (a fabricated record).

A neighborhood system [14–17] with some modifications satisfies both conditions of natural clustering and inclusion. The modified version (*Modified-Neighborhood clustering)* This clustering has two parts: *Neighborhood building* and *Neighborhood merging* that are presented in the following two subsections.

Neighborhood Building. The dataset H, which is a copy of the TR file, has n dimensions (A_1,..., A_n) and u1 records (R_1,..., R_{u1}). Each attribute of this file has a representative that is kept separately from H. The representative is a binary string with a length of n, and each bit is a position holder for attributes of A_1 to A_n. Initially, the binary string for attribute A_i contains one "1" value in the bit i of its string. We also need the dataset Shadow-H, which is a duplication of H. Attribute A_i in Shadow-H is considered a shadow of the attribute A_i in H, and it has u1 values of (v_{1i},..., v_{u1i}).

Two attributes of A_i (v_{1i},..., v_{u1i}) and A_j (v_{1j},..., v_{u1j}) of H are considered neighbors, if they satisfy the following conditions:

For k = 1 to n:

1. $0 \leq |(v_{ki} - v_{kj})| \leq \delta$, where $\delta \geq 1$ and
2. $\sum |(v_{ki} - v_{kj})| > 0$

The real variable δ determines the closeness level of two attributes [14, 15]. The neighboring is established based on the closeness of the two attributes and not by closeness of A_i or A_j to a fabricated attribute (centroid of the cluster). However, we replace the shadow of A_i (in Shadow-H) with the averaging of the corresponding attribute values of A_i and A_j. (The shadow of A_i is a centroid of the two neighbors and is kept for later use. This centroid does not play any role in the formation of the final clusters.

We start finding neighbors for A_i of D (for i = 1 to u1). If the attribute A_j (where j = i + 1) of D is the neighbor of A_i, then A_j is in the neighborhood of A_i, and it is denoted as $N(A_i) = \{A_j\}$. The $N(A_i)$ is recorded by setting bit j to one in the bitstring of A_i. The bitstring of attribute A_j does not change. Therefore, each attribute in D has its own neighborhood, and all these neighborhoods are created using the same closeness level. All the neighborhoods are established prior to neighborhood merging.

For $N(A_i)$ the shadow of A_i is the centroid of this neighborhood. However, inclusion of attribute A_b, for example, in $N(A_i)$ is not influenced by the shadow of Ai, but by the closeness of the two attributes of A_i and A_b in H.

Neighborhood Merging. Two neighborhoods of $N(A_i)$ and $N(A_b)$ may merge if the following condition is satisfied:

$$\text{Bitstring}(A_i)\text{Ç}\text{Bitstring}(A_b)^3 d$$

The integer variable d represents a threshold for the number of common attributes between $N(A_i)$ and $N(A_b)$. If a merge happens between $N(A_i)$ and $N(A_b)$, then.

a. $N(A_i) = N(A_i) \cup N(A_b)$,
b. $\text{Bitstring}(A_i) = \text{Bitstring}(A_i) \cup \text{Bitstring}(A_b)$,
c. $\text{Bitstring}(A_b) = 0$, and
d. $\text{Shadow}(A_i)$ = Averaging corresponding values for Shadows of A_i and A_b.

Upon completion of possible merges for all neighborhoods, those neighborhoods with at least one "1" in their Bitstrings are considered a cluster, and there is one shadow attribute for each one of these neighborhoods that can represent the centroid of the neighborhood.

The outcome for applying the Modified Neighborhood clustering on the n-dimensional TR file is an m-dimensional dataset (named file Z). The number of attributes for Z is m and |Z| = |TR|. We repeat the calculation of the U-Gini index for each attribute A_j (for j = 1 to m) in Z, generating (G_1,..., G_m) –one per attribute. One can easily assign the new U-Gini values of (G_1,..., G_m) to the n attributes of the file TR by assigning the same G_i value to those attributes of TR that belong to the cluster, C_i. We also use the Dendrogram algorithm to calculate Q-Gini for Z— Q^z. In addition, the information gain for Z can be calculated and it is ($IG^z{}_i$ = Q^z-G_i) (for i = 1 to n). Furthermore, the MSEz is calculated using formula 5.

$$\mathrm{MSE_z} = \frac{1}{n}[\sum_{i=1}^{n}\left(IG_i^z - IG_i^{ts})^2\right] \tag{5}$$

The Min(MSEtr, MSE_γ, $\mathrm{MSE_z}$) determines the most superior Gini index and its validation.

3.4 Delivering the Causal Drivers of Q-Gini in the Dataset

To determine whether the relationship between dimensions and the Q-Gini is causal or merely correlational, we use the Instrumental Variable (IV). The IV method addresses endogeneity in models, which occurs when an explanatory variable is correlated with the error term. This correlation can bias the estimation of causal relationships. The IV method solves this by introducing an instrument. The IV is a variable that is correlated with the endogenous variable and uncorrelated with the error term.

Instrumental Variables Estimation. To examine potential causal effects of covariates on inequality, we specify a series of exactly identified Instrumental Variables (IV) models. The main challenge is the endogeneity of a regressor X, where $E[X\varepsilon] \neq 0$, which violates the standard linear model assumption and leads to biased Ordinary Least Squares (OLS) estimates. For each endogenous regressor X and a chosen instrument W, we estimate a Two-Stage Least Squares (2SLS) model for each outcome $G \in Q^{TR}$, Q^Z as First Stage: $X = \pi 0 + \pi 1W + v$, Second Stage: $G = \beta 0 + \beta 1X_b + \varepsilon$, where X_b are the fitted values from the first-stage regression. The key identifying assumption is the exclusion restriction: the instrument W affects the outcome G only through its effect on the endogenous variable X [18].

To guard against heteroskedasticity, which is common in cross-sectional economic data, we use heteroskedasticity-robust standard errors using the HC1 estimator [19]. The covariance matrix is given by,

$$\widehat{Var}_{HC1}\left(\hat{\beta}_{2SLS}\right) = \left(\hat{X}^T\hat{X}\right)^{-1}\left(\sum_{i=1}^{n}\hat{\varepsilon}_i^2\hat{x}_i\hat{x}_i^T\right)\left(\hat{X}^T\hat{X}\right)^{-1},$$

where $\hat{X}$ is the matrix of second-stage regressors (including the constant and the fitted values $\hat{X}$), and $\hat{}\varepsilon_i$ are the 2SLS residuals.

For each model, we report several diagnostics. The null hypothesis of a weak instrument is tested using the first-stage F-statistic on the excluded instrument Z. While a common rule-of-thumb suggests $F > 10$ indicates sufficient strength [20], this test is technically undefined under exact identification, as the standard error of the 2SLS estimate is a function of the first-stage error and becomes infinite if the instrument is irrelevant. Results are reported where calculable. A Wu-Hausman test [21, 22] is conducted to test for endogeneity. It tests the null hypothesis that the OLS estimator is consistent and efficient against the alternative that the IV estimator is consistent. A low p-value suggests significant endogeneity, justifying the use of IV.

4 Empirical Results

In our empirical study, we employ an n-dimensional dataset from the U.S. Census Bureau, specifically the 2022 American Community Survey (ACS 5-year estimates). This dataset provides a granular view of socioeconomic variables at the county level across the United States. The variables include Median Household Income (MHI), Per Capita Income (PCI), Total Earnings (TE), Poverty Levels (POV), U-Gini Index of Income Inequality (U-GINI), Unemployment Rate (UR), Educational Attainment (EA), Median Gross Rent (MGR), Median Housing Value (MHV), Internet Access (INT), Health Insurance Coverage (HIC), and demographic breakdowns by race and ethnicity: Percentage of White Population (WP), Percentage of Black or African American Population(BP), and Percentage of Hispanic or Latino Population (HP).

After normalizing the data using formula (2), we calculated the minimum error differences using formulas (3), (4), and (5). The winner was the use of Q-Gini produced for TR produced from the dendrogram. The Q-Gini for the reduced dataset Z obtained by applying the Modified Neighborhood clustering was the second winner.

We employ Instrumental Variable (IV) regression to estimate the causal relationships between socioeconomic factors and the U-Gini. This addresses the endogeneity of key variables. Instead of ordinary linear models, which may provide biased estimates due to endogeneity, IV regression uses external instruments to isolate the exogenous variation in each independent variable. For each socioeconomic variable X_i, we selected a corresponding instrument Y_i that is correlated with X_i but not with the error term $\in_i$, in the Gini model. The IV regression model for each socioeconomic variable is given by

$$G_i = \beta_0 + \beta_1 X_i + \in_i, \tag{6}$$

where G_i is the U-Gini index for population i, β_0 is the intercept, β_1 is the coefficient of X_i, and $\in_i$ is the error term. The coefficients obtained from these models indicate the direction and magnitude of the relationship between each independent variable and income inequality.

Our objective is to determine whether the relationship between regressors and the Comprehensive Gini is causal or merely correlational. The IV method addresses endogeneity in models, which occurs when an explanatory variable is correlated with the error term. This correlation can bias the estimation of causal relationships. The IV method solves this by introducing an instrument. The IV is a variable that is correlated with the endogenous variable and uncorrelated with the error term.

This approach provides a framework to find a causal relationship between variables and the Comprehensive Gini. Each defines an endogenous variable and its corresponding instrument, chosen based on their correlations from the correlation matrix (Table 1). For example, MHI is instrumented by PCI due to their strong correlation ($r = 0.88$) and PCI's relatively low correlation with the Gini coefficient ($r = -0.09$), making PCI a valid instrument that meets the relevance and exogeneity conditions. Similarly, TE is instrumented by UR, given their high correlation ($r = 0.98$), ensuring that UR strongly predicts TE but has a weak direct association with Gini ($r = 0.13$), addressing concerns of endogeneity. POV is instrumented by TE, as their correlation ($r = 0.95$) allows TE to serve as a strong predictor of POV while mitigating the risk of bias in the Gini model. In a similar vein, UR is instrumented by EA, and MGR is instrumented by MHV, both selected for their high correlations with the endogenous variables and minimal direct impact on the Gini coefficient. This method reduces the potential for omitted variable bias and reverse causality. The map of endogenous variables to their instruments is pre-specified based on theoretical considerations and observed correlations.

Table 1. Mapping of Endogenous Regressors to Instrument

Endogenous variable (X)	Instrument (W)
Median household income	Per capita income Total
earnings	Unemployment rate
Poverty rate	Total earnings
Unemployment rate	Educational attainment
Median gross rent	Median home value
Road infrastructure	Educational attainment
Black population %	Hispanic population %
Health insurance coverage	Educational attainment
Hispanic population %	Hispanic population %

The IV regression results indicate the causal impact of various socioeconomic factors on the U-Gini, Table 2. Each coefficient reflects the magnitude and direction of the causal effect of the respective variable on Gini. MHI exhibits a significant negative effect (-0.07, p-value < 0.001), implying that higher median incomes reduce income inequality, confirming the causal relationship between household income and inequality. In contrast, TE and UR have strong positive effects (0.34, p-value < 0.001), suggesting that increased earnings and unemployment rates exacerbate income inequality. Similarly, POV and INT also show significant positive impacts on Gini, indicating their role in widening inequality. BP and HP have large negative coefficients (-1.17 and -1.79, respectively), with highly significant p-values, pointing to the potential disparities in income distribution across different racial demographics.

The following table (Table 3) presents the results of IV regressions for factors, providing estimates of causal relationships between these variables. Each dependent variable (MHI, TE, POV, UR, INT, BP, HIC, HP) is instrumented using other related variables. The rows with zero estimates or not statistically significant estimates are

Table 2. Causal Impacts on Gini

Variable	Estimate	Std. Error	Variable	Estimate	Std. Error
MHI	−0.07	0.01	INT	0.28	0.04
TE	0.34	0.04	BP	−1.17	0.32
POV	0.31	0.04	HIC	0.30	0.04
UR	0.34	0.05	HP	−1.79	0.43

removed. PCI has a strong positive effect on MHI (estimate: 0.89), indicating that higher per capita income leads to a higher median household income. POV, UR, and INT negatively affect MHI, suggesting that as poverty, unemployment, and lack of internet access rise, household income tends to fall. HIC shows a significant positive impact on MHI (estimate: 7.06), indicating that access to health insurance correlates with higher household income, likely reflecting better job quality and benefits in higher-income households.

POV has significant negative impact on TE (estimate: -0.14). Both HIC and INT show strong positive effects on total earnings. MHI and TE have strong negative effects on poverty (estimates: −0.01 and −2.27), confirming the expected causal relationship where increases in income and earnings reduce poverty levels. UR, INT, and HIC all have positive effects on POV, suggesting that higher unemployment and lack of services exacerbate poverty. POV and TE have significant positive effects on UR, with TE indicating that higher earnings can also reflect higher participation in the labor force, thus increasing unemployment in some cases due to labor market friction. INT and HIC reduce unemployment.

TE significantly increases INT (estimate: 1.04). UR has a negative impact on INT (estimate: −0.16), indicating that as unemployment rises, access to internet infrastructure diminishes, potentially due to affordability issues. POV and INT have significant positive impacts on BP, which may indicate higher vulnerability of this demographic to poverty and technological exclusion. MHI has a negative effect on BP (estimate: −0.46). TE and UR both have positive effects on HIC. POV and HIC have large positive effects on HP. UR and INT show negative effects.

Table 3 says that the variables that have a causal effect on UR, for example, are TE, POV, UR, and HIC. The table also says that the variables that have a causal effect on TE are UR, POV, and INT. Clearly, we have a loop between UR and TE. This is a big problem because the graph that will represent the causality among the variables must be an acyclic directed graph.

Table 3. Causal Estimates of Socioeconomic Variables

Dependent Variable	IV	Estimate	Std. Error
MHI	PCI	0.89	0.01
MHI	POV	−1.25	0.17
MHI	UR	−0.76	0.2
MHI	INT	−4.27	0.34
MHI	BP	−0.03	0.01
MHI	HIC	7.06	0.68
MHI	HP	0.01	0.001
TE	POV	−0.14	0.001
TE	UR	0.04	0.01
TE	INT	0.25	0.01
TE	HIC	0.84	0.01
POV	MHI	−0.01	0.001
POV	TE	−2.27	0.06
POV	UR	0.63	0.02
POV	INT	0.82	0.03
POV	HIC	1.8	0.06
UR	TE	0.42	0.06
UR	POV	0.44	0.01
UR	INT	−0.45	0.03
UR	HIC	0.53	0.06
INT	TE	1.04	0.03
INT	POV	0.21	0.01
INT	UR	−0.16	0.01
INT	HIC	−0.03	0.04
BP	MHI	−0.46	0.03
BP	TE	2.67	2.13
BP	POV	0.78	0.53
BP	UR	−0.6	0.63
BP	INT	3.42	1.05
BP	HIC	−5.65	2.11
BP	HP	−0.14	0.01
HIC	TE	0.85	0.01

(*continued*)

Table 3. (*continued*)

Dependent Variable	IV	Estimate	Std. Error
HIC	UR	0.05	0.01
HP	MHI	−0.45	0.03
HP	POV	3.23	0.62
HP	UR	−3.38	0.74
HP	INT	−5.77	1.23
HP	BP	−0.19	0.02
HP	HIC	6.73	2.48

5 Conclusion and Future Research

We introduced the Q-Gini index to quantify inequality across multiple socioeconomic dimensions. The Q-Gini provides a comprehensive evaluation by hierarchically aggregating all dimensions of a dataset into a single inequality dimension. This aspect is missing in the univariate U-Gini. Validation of the Q-Gini used an MSE protocol, demonstrating its superior performance over the conventional U-Gini index.

We also studied effect of dimensionality reduction on inequality measurement. By introducing modified-neighborhood system that satisfies natural clustering and inclusion conditions, we reduced the multidimensional data to representative neighborhoods. Q-Gini calculated on this reduced dataset ranked second among the most accurate measures in our validation framework.

The isolation of causal drivers of the Q-Gini index recognized high correlations among socioeconomic factors, which can lead to spurious inferences. We used IV models to distinguish correlation from causal impact and identify effective policy targets. The IV results confirm that variables such as median household income, total earnings, and poverty levels not only associate with but cause changes in multidimensional inequality. This validates the Q-Gini's sensitivity to these fundamental socioeconomic forces.

For future research, we are considering developing a belief network. This would model the probabilistic relationships between socioeconomic variables and inequality.

Acknowledgments. Our special thanks go to our graduate students, Ms. Noushin Gauhar and Mr. Jeffrey Young for helping us when we needed.

References

1. Gini, C.: Variabilità e mutabilità [Variability and mutability]. Reprinted Memorie di Metodologica Statistica **1**, 211–382 (1912)
2. Atkinson, A.B.: On the measurement of inequality. J. Econ. Theory **2**(3), 244–263 (1970)
3. Ceriani, L., Verme, P.: The origins of the Gini index: extracts from Variabilità e Mutabilità (1912) by Corrado Gini. J. Econ. Inequality **16**(3), 421–443 (2018)

4. Decancq, K., Lugo, M.A.: Weights in multidimensional indices of well-being: an overview. Economet. Rev. **32**(1), 7–34 (2013)
5. UNDP. Human Development Report 2020. United Nations Development Program (2020)
6. Gajdos, T., Weymark, J.A.: Multidimensional generalized Gini indices. Econ. Theor. **26**(3), 471–496 (2005)
7. Andreoli, F., Zoli, C.: From unidimensional to multidimensional inequality: a review. J. Econ. Surv. **34**(4), 754–785 (2020)
8. Lugo, M.A.: Comparing multidimensional indices of inequality: methods and application. In: Inequality and poverty (Vol. 14, pp. 213–236). Emerald Group Publishing Limited (2007)
9. Kohonen, T.: Essential of the self-organizing map. Neural Netw. **37**, 52–65 (2013)
10. Deboeck, G.J.: Financial applications of self-organizing maps. Neural Netw. World **8**(2), 213–241 (1998)
11. Saisana, M., Saltelli, A., Tarantola, S.: Uncertainty and sensitivity analysis techniques as tools for the quality assessment of composite indicators. J. R. Stat. Soc. Ser. A **168**(2), 307–323 (2005)
12. Maggino, F.: Developing indicators and managing the complexity. In: Maggino, F. (ed.) Complexity in society: From indicators construction to their synthesis, pp. 87–114. Springer (2017)
13. Swofford, D.L., Olsen, G.J., Waddell, P.J., Hillis, D.M.: "Phylogenetic inference". In: Hillis, D.M., Moritz, C., Mable, B.K. (eds.). Molecular Systematics, 2nd edition. Sunderland, MA: Sinauer, pp. 407–514 (1996)
14. Yao, Y.Y.: Rough Sets, neighborhood Systems and Granular computing. Proceedings of 1999 IEEE Canadian Conference on Electrical and Computer Engineering, vol. 3, pp. 1553–1558 (1999)
15. Hashemi, R., Danley, J., Tyler, A., Slikker, W., Paule M.: Quality of information granulation: kohonen self-organizing map vs. neighborhood system. Proceedings of the Fourth International Joint Conference on Information Sciences, George Georgiou, Janikow, C., Yao, Y. (eds.). Research Triangle Park, NC, pp 294–297 (1998)
16. Hashemi, R., Bahrami, A., Tyler, N., Antonelli, M., Dahlqvist, B.: Property preservation in reduction of data volume for mining: a neighborhood system approach. The Third International Conference on Data Analytics (DATA ANALYTICS 2014), Laux, F., Pardalos, P.M., Crolotte, A. (eds.), Rome, Italy, pp. 105–111 (2014)
17. Hashemi, R., Bahrami, A., Smith, M., Tyler, N., Antonelli, M., Clapp, S.: Discovery of predictive neighborly rules from neighborhood systems. International Conference on Information and Knowledge Engineering (IKE'13), Las Vegas, Nevada, pp. 119–125 (2013)
18. Angrist, J.D., Kugler, A.D.: Rural windfall or a new resource curse? Coca, income, and civil conflict in Colombia. Rev. Econ. Stat. **90**(2), 191–215 (2008)
19. White, H.: A Heteroskedasticity-Consistent covariance matrix estimator and a direct test for heteroskedasticity. Econometrica: J. Econ. Soc. 817–838 (1980)
20. Staiger, D., Stock, J.: To Econometrica. Econometrica **65**(3), 557–586 (1997)
21. Wu, D.M.: Alternative tests of independence between stochastic regressors and 18isturbances. Econometrica: J. Econ. Soc. 733–750 (1973)
22. Hausman, J.A.: Specification tests in econometrics. Econometrica: J. Econ. Soc. 1251–1271 (1978)

SureSense++: An AI-Optimized Intrusion Detection Framework for IoT Sensor Networks Using Feature Selection and Dimensionality Reduction on IoTNet24 Dataset

Roseline Oluwaseun Ogundokun[1,2(✉)], Rotimi-William Bello[1], Pius Adewale Owolawi[1], and Etienne van Wyk[1]

[1] Department of Computer Systems Engineering, Tshwane University of Technology, Pretoria, South Africa
Ogundokunroseline1@gmail.com, ogundokunRO@tut.ac.za

[2] Department of Computer Science, Redeemer's University Ede, Osun State, Nigeria

Abstract. This study introduces SecureSense++, an AI-optimised intrusion detection framework tailored for IoT sensor networks. Leveraging feature selection and dimensionality reduction techniques, multiple classifiers—including Random Forest, XGBoost, SVM, Extra Trees, LightGBM, KNN, and a Stacked Ensemble—were evaluated on key performance metrics: accuracy, ROC-AUC, and inference time. The Stacked Ensemble model consistently outperformed individual models, achieving a perfect accuracy of 100% and an AUC score of 1.0000, while maintaining efficient inference times suitable for real-time applications. Random Forest, XGBoost, Extra Trees, and LightGBM also demonstrated excellent performance, achieving 100% accuracy and AUC scores of 1.0000, albeit with slightly higher inference latency compared to the ensemble model. KNN achieved an AUC of 0.9987 with minimal classification errors, while SVM, although recording a strong AUC of 0.9932, exhibited significantly lower inference efficiency and a benign class misclassification rate of 11%. These results validate the robustness of SecureSense++ and highlight the effectiveness of ensemble learning combined with dimensionality reduction for securing IoT environments against complex cyber threats.

Keywords: Sensor network · Intrusion detection · Internet of Things · Machine learning

1 Introduction

The proliferation of Internet of Things (IoT) devices has revolutionised modern living, enabling smart environments, innovative healthcare solutions, industrial automation, and intelligent transportation systems. However, this rapid expansion has also introduced significant security vulnerabilities, rendering IoT networks vulnerable to cyberattacks [1, 2]. Traditional Intrusion Detection Systems (IDSs) also lose efficiency in IoT settings due to resource scarcity, network diversity, and the rising sophistication of threats [3,

H. R. Arabnia et al. (Eds.): CSCI 2025, CCIS 2969, pp. 122–128, 2027.
https://doi.org/10.1007/978-3-032-28460-0_10

4]. Thus, there is a requirement for more lightweight, adaptive, and highly accurate IDS solutions that are also easy to implement in dynamic IoT systems.

Several researchers have explored the development of IDS for IoT networks [5, 6]. Doshi et al. [7] proposed a lightweight detection scheme for IoT-oriented DDoS attacks but limited it to a small set of attack classes. Meidan et al. [8] introduced N-BaIoT, a deep autoencoder-based solution, although it suffered from high computational overheads. Diro and Chilamkurti [9] combined deep learning with traditional classifiers but suffered from high training and inference costs. Issa et al. [10] introduce a hybrid feature selection method that combines the Information Gain Ratio and the Black Hole Algorithm, achieving 96.96% accuracy and a 0.89% false positive rate in network intrusion detection by effectively eliminating irrelevant features. Shone et al. [11] and Aloqaily et al. [12] prioritised detection performance but tended to overlook inference speed and resource efficiency, which are essential for real-world IoT deployment.

While these contributions have been valuable to IoT security, key limitations persist. Most frameworks traded inference efficiency for accuracy, without considering it, rendering them less practical for real-world scenarios. More importantly, few attempts unify feature selection, dimensionality reduction, and ensemble learning into a unified, lightweight framework, nor do they rigorously evaluate their models over modern datasets such as IoTNet24.

To bridge these gaps, we present SecureSense++, a lightweight and robust intrusion detection system that combines Random Forest-based feature selection, PCA-based dimensionality reduction, and ensemble learning. The solution provides assured, strong, real-time intrusion detection optimised for future IoT sensor networks.

2 Materials and Methods

This section outlines the dataset utilised, the feature engineering strategies employed for data optimisation, and the modelling techniques adopted to develop the proposed SecureSense++ intrusion detection framework.

2.1 Data Description and Feature Optimisation

The IoTNet24 dataset, a general-purpose benchmark used for conducting IoT intrusion detection studies, was utilised in this study. It encompasses a wide range of attack scenarios simulated in an IoT environment, providing a solid foundation for estimating the performance of intrusion detection models in real-world environments.

Two-stage feature engineering was conducted to improve the detection system. In the first phase, the 84 dataset features were analysed for feature importance using Random Forest, which facilitated the identification and elimination of duplicate and uninformative features. This filtering condensed the feature space, made the model more explainable, and enhanced computational benefits. The pre-cleaned feature set was subjected to Principal Component Analysis (PCA) to reduce dimensionality while maintaining the vital variance required for proper classification. This hybrid feature selection and dimensionality reduction approach significantly reduced model complexity, mitigated overfitting risks, and accelerated training time, making the system suitable for deployment in resource-constrained IoT environments.

2.2 Proposed Model

Figure 1 illustrates the SecureSense++ pipeline, which begins with loading the IoTNet24 dataset, followed by noise elimination using Isolation Forest and data balancing with SMOTE + ENN. Data preparation involves feature selection (using RF-based methods) and dimensionality reduction (utilising UMAP) to feed multiple base models. A stacking ensemble with logistic regression finishes off the framework for testing and deployment.

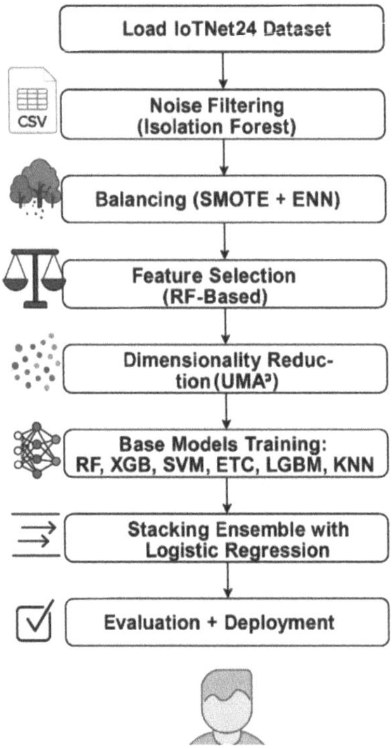

Fig. 1. Proposed Model Architecture

3 Results and Discussion

This section presents a comparison of the SecureSense++ framework across classification accuracy, detection resilience, and computational efficiency for various machine learning models, using confusion matrices, ROC curves, Precision-Recall curves, and radar charts.

The performance evaluation of the proposed SecureSense++ framework is comprehensively illustrated across several figures. As shown in Fig. 3, the confusion matrices indicate that Random Forest, XGBoost, Extra Trees, LightGBM, and the Stacked Ensemble models achieved perfect classification (100% accuracy) for both benign and malicious classes. In comparison, KNN maintained excellent performance with only a 2% misclassification rate, and SVM exhibited slightly lower precision for benign samples (89%). Further validation through the ROC curves in Fig. 2 reveals that Random

Forest, XGBoost, Extra Trees, LightGBM, and the Stacked Ensemble models attained a perfect AUC score of 1.0000. At the same time, KNN and SVM achieved impressive AUC values of 0.9987 and 0.9932, respectively, indicating high true-positive rates across all models. Complementary insights from the Precision-Recall curves in Fig. 4 confirm that the ensemble-based models maintained an AP score of 1.0000, with KNN (0.9998) and SVM (0.9993) trailing closely, emphasising the framework's resilience even under class imbalance. Finally, Fig. 5 presents a radar chart comparing accuracy, ROC-AUC, and inference time, where the Stacked Ensemble model offers the most balanced trade-off between detection performance and computational efficiency, making it highly suitable for real-time IoT deployment scenarios. Collectively, these results validate the superiority and robustness of SecureSense++ in achieving high detection accuracy while maintaining low inference overhead.

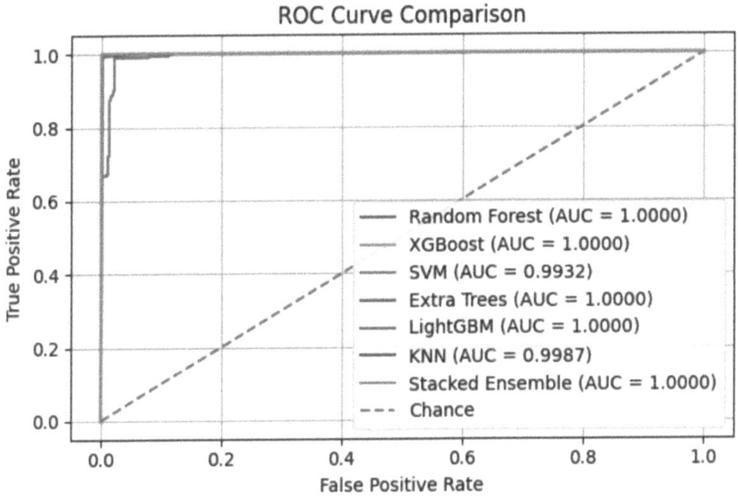

Fig. 2. ROC Curve Comparison Across All Models

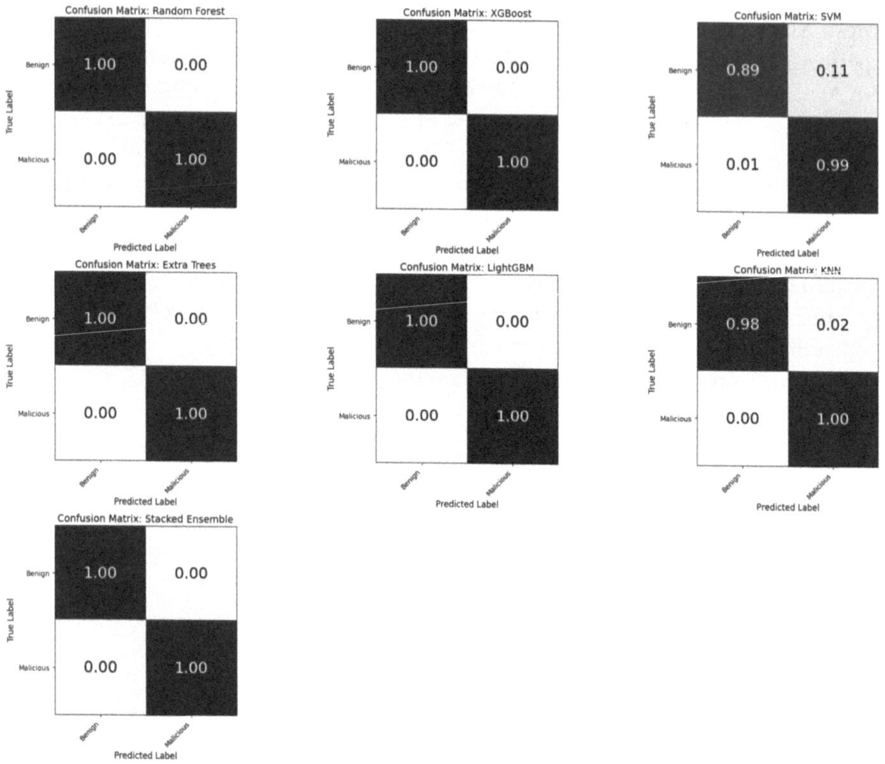

Fig. 3. Confusion Matrices for the Proposed Models

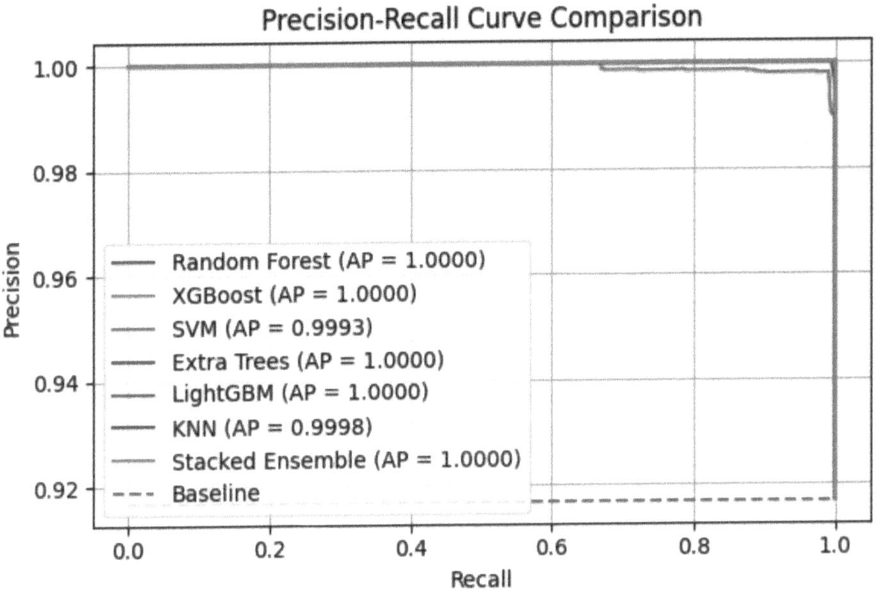

Fig. 4. Precision-Recall Curve Comparison Across All Models

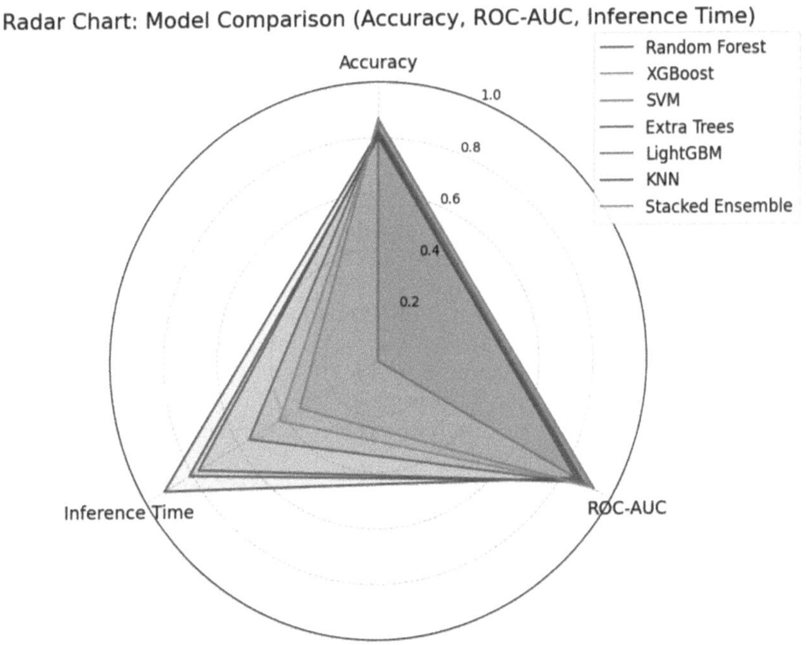

Fig. 5. Radar Chart Showing Accuracy, ROC-AUC and Inference Time of All Models

4 Conclusion

The SecureSense++ model effectively augments intrusion detection in IoT networks by combining optimised feature selection with ensemble learning. Quantitative outcomes reveal that the Stacked Ensemble model achieved optimal accuracy of 100%, an AUC of 1.0000, and better inference-time efficiency than individual models. Similarly, the Random Forest and Extra Trees classifiers achieved 100% classification accuracy and an AUC of 1.0000, albeit with a higher inference delay than the ensemble model, as shown in the radar chart comparison. On the other hand, KNN produced an AUC of 0.9987, and SVM produced 0.9932, even though both models utilised relatively higher inference times and were constrained by classification errors. These findings underscore the importance of maintaining predictive accuracy while ensuring real-time inference capability in IoT security systems. SecureSense++ delivers a scalable, high-performance, and low-computation solution, making a worthwhile contribution to the development of robust, intelligent IoT infrastructures.

References

1. Akande, H.B., Awoniyi, C., Ogundokun, R.O., Oloyede, A.A., Yiamiyu, O.A., Caroline, A.T.: Enhancing network security: Intrusion detection systems with hybridised CNN and DNN algorithms. In: Proc. 2024 International Conference on Science, Engineering and Business for Driving Sustainable Development Goals (SEB4SDG), pp. 1–7 (2024)
2. Oyinloye, T.S., Arowolo, M.O., Ogundokun, R.O., Prasad, R.: Exploring deep learning architectures for enhanced cyber threat detection: a survey. In: Proc. 2024 International Conference on Science, Engineering and Business for Driving Sustainable Development Goals (SEB4SDG), pp. 1–10 (2024)
3. Sicari, S., Rizzardi, A., Grieco, L.A., Coen-Porisini, A.: Security, privacy and trust in Internet of Things: the road ahead. Comput. Netw. **76**, 146–164 (2015)
4. Ogundokun, R.O., Awotunde, J.B., Sadiku, P., Adeniyi, E.A., Abiodun, M., Dauda, O.I.: An enhanced intrusion detection system using particle swarm optimisation feature extraction technique. Procedia Comput. Sci. **193**, 504–512 (2021)
5. Ambusaidi, M.A., He, X., Nanda, P., Tan, Z.: Building an intrusion detection system using a filter-based feature selection algorithm. IEEE Trans. Comput. **65**(10), 2986–2998 (2016)
6. Yin, C., Zhu, Y., Fei, J., He, X.: A deep learning approach for intrusion detection using recurrent neural networks. IEEE Access **5**, 21954–21961 (2017)
7. Doshi, R., Apthorpe, N., Feamster, N.: Machine learning DDoS detection for consumer Internet of Things devices. In: Proc. 2018 IEEE Security and Privacy Workshops (SPW), San Francisco, CA, USA, pp. 29–35 (2018)
8. Meidan, Y., et al.: N-BaIoT—Network-based detection of IoT botnet attacks using deep autoencoders. IEEE Pervasive Comput. **17**(3), 12–22 (2018)
9. Diro, A.A., Chilamkurti, N.: Distributed attack detection scheme using deep learning approach for Internet of Things. Futur. Gener. Comput. Syst. **82**, 761–768 (2018)
10. Issa, S.S., Salih, S.Q., Salman, Y.D., Taha, F.H.: An efficient hybrid filter-wrapper feature selection approach for network intrusion detection system. Int. J. Intell. Eng. Syst. **16**(6) (2023)
11. Shone, N., Ngoc, T.N., Phai, V.D., Shi, Q.: A deep learning approach to network intrusion detection. IEEE Trans. Emerging Top. Comput. Intell. **2**(1), 41–50 (2018)
12. Aloqaily, M., Otoum, S., Al Ridhawi, I., Jararweh, Y.: An intrusion detection system for connected vehicles in smart cities. Ad Hoc Netw. **90**, 101842 (2019)

AI-Enhanced Robustness in a Dementia-Detecting Memory Game

Jiapeng Xu[1], Claire Zou[2], Lawrence Tran[1](✉), and Ruoying Tang[3]

[1] Rancho Cucamonga High School, Rancho Cucamonga, CA, USA
ltkid11792@gmail.com
[2] Cambridge Rindge and Latin School, Cambridge, MA, USA
[3] Los Osos High School, Rancho Cucamonga, CA, USA

Abstract. Dementia is a progressive condition that affects memory and thinking, impacting millions of people worldwide. Early detection is crucial for better treatment, but current methods including PET scans and spinal fluid tests are expensive and invasive. Existing brain training games are often not backed by solid science and cannot be used for reliable screening. We introduce an AI-powered memory game using proven psychological models and intelligent, adaptive gameplay. The game includes five mini-games, each focused on a different type of memory: working, procedural, episodic, semantic, and sensory. Difficulty is adjusted in real time based on player performance. A machine learning model, trained on synthetic medical data, can categorize players into three stages of cognitive health: no decline, mild/moderate decline, or severe dementia, with 86.7% accuracy in empirical testing when no noises in training dataset and 76.6% accuracy with up to 20% of noise data. Our game is a safe, affordable, and easy alternative to monitor brain health over time.

Keywords: Dementia · Psychological Memory Models · AI-assisted Game · Prediction Modeling

1 Introduction

Dementia is a widely known brain disorder of cognitive functions that causes progressive decline in memory and thinking abilities and is common among the older age population. Early detection of the disorder is crucial as it allows patients to access support and treatments to manage their symptoms before they progressively worsen. Traditional medical diagnosis methods such as PET scans of the brain, analysis of cerebrospinal fluid using lumbar puncture and other invasive methods may bring about side effects to the body. Moreover, these procedures are expensive and inconvenient especially during the early stages. This paper aims to create entertaining, harmless, yet effective memory games that achieve dementia detection [8] and provide personalized memory training. AI enabled game engines and machine learning driven prediction models [7] are designed for dementia detection and memory training. With the prediction model trained by a synthetic dataset, preliminary testing results are obtained to illustrate the potential effectiveness of the memory game.

H. R. Arabnia et al. (Eds.): CSCI 2025, CCIS 2969, pp. 129–140, 2027.
https://doi.org/10.1007/978-3-032-28460-0_11

2 Background

2.1 Dementia Diagnosis and Cognitive Training

Current dementia diagnosis methods [5] present several challenges. Medical tests such as PET scans and spinal fluid tests are accurate but expensive, uncomfortable, and hard to access for regular checkups [5]. Similarly, simple memory tests can spot serious disorders but often miss early warning signs [4]. This means many people do not receive medical attention until their condition has progressed.

Many products are available on the market being promoted as brain exercise games, for example, Spot It! matching game or Memory Game for Seniors puzzle activities. However, these have major limitations. They are not designed using proper memory science, do not adapt to different skill levels, and have not been proven to detect dementia. Most importantly, they cannot measure changes in memory over time.

Our project develops a smarter memory game for cognitive testing. It is based on proven scientific models of the human memory system, including the Atkinson-Shiffrin multi store model [1]. The game automatically adjusts its difficulty to match each player's ability while keeping scoring consistent. This allows users to check current memory skills while also tracking changes over weeks or months [8]. The game includes different types of memory challenges. Some test remembrance of recent information, while others check knowledge stored long term. Performance data is analyzed using machine learning algorithms that can detect patterns suggestive of early cognitive decline [7]. This matches new approaches in healthcare that use technology for early detection through continuous monitoring [6]. Our solution combines cognitive science with engaging gameplay to create a dementia screening tool that is both accurate and easy to use. Unlike clinic tests, people can use it at home, while still providing reliable results professionals can trust.

2.2 Levels and Diagnosis of Dementia

The Global Deterioration Scale (GDS) [9] defines seven stages of cognitive decline: no cognitive decline (stage 1), very mild cognitive decline (stage 2), mild cognitive decline (stage 3), moderate cognitive decline (stage 4), moderately severe cognitive decline (stage 5), severe cognitive decline (stage 6) and very severe cognitive decline (stage 7). Patients from stage 5 and up require assistance to survive [9].

Unlike formal diagnosis, our memory game provides simplified categorization of dementia into three stages, corresponding to the seven different clinical stages of dementia. Users diagnosed with Stage A (no impairment or very minor cognitive decline) will be congratulated for a healthy mind, aligning with stages 1 and 2. Stage B (mild to moderate cognitive decline) suggests users seek professional diagnosis and correlates to stages 3 and 4. Stage C (severe cognitive decline) indicates the need for immediate support, aligning with the devastating symptoms of stages 5 through 7.

2.3 Memory Models

Memory models are used in psychology to explain how information is processed, stored, and retrieved. One of the most fundamental models is the Atkinson Schiff model, also known as the multi storage model. This model suggests that memory is composed of three different storage systems: sensory memory, short-term memory, long-term memory, and Baddeley's Working Memory Model [10]. The Atkinson-Schiff model emphasizes the staged storage of memory and the transformation of information from short-term to long-term. It describes sensory registers as low-level sensory systems that temporarily store large amounts of information, only a few items are transferred to short-term memory [1]. Figure 1 illustrates the memory model that our game design is based on, which combines the Atkinson Schiff model and Badley's working memory model [1,10].

1. **Working memory:** a dynamic memory system responsible for temporarily storing and manipulating information to support thinking and problem-solving.
2. **Procedural memory:** a part of long-term memory that involves skills acquired through repeated learning, such as riding a bicycle or typing.
3. **Episodic memory:** stores personal experience and allows recall of past events and contextual details.
4. **Semantic memory:** stores world knowledge, concepts, and facts, such as the principle of gravity.
5. **Sensory memory:** what a person senses and feels with their senses, for example what one smells.

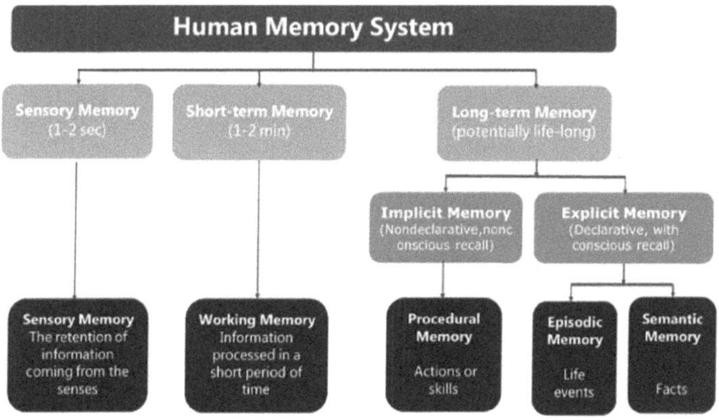

Fig. 1. *The Human Memory Model: Foundation for Game Design*

3 Game Design

3.1 Game Interface and Environment

The game is an AI-assisted memory training tool designed not only to improve cognitive performance but also to help detect early signs of dementia. It offers a personalized and engaging experience by combining interactive subgames with storytelling, trivia, and personal memory recall. To begin, players must create an account and log in via the SignIn/SignUp button as shown in Fig. 2(a). During this process, an AI virtual assistant guides the player through choosing a username and password. Once logged in, players are encouraged to create or update their profiles including personal details including education, career, family, medical history, favorite hobbies, meaningful life events and so on. For memory exercises, players can click on one of the subgame icons shown in Fig. 2(b) to practice a type of memory. For dementia detection, it is recommended that all five subgames are played. Game questions are timed as shown in Fig. 2(c). All game scores and progress will be saved to the player's profile after each round of play, helping build a long-term cognitive profile.

3.2 Game Engine

The game design is rooted in psychological memory models with five subgames aligning with five memory models illustrated in Fig. 1, respectively. Game types include visual-arts-based multiple-choice quizzes, mix & match challenges, intelligently generated personalized puzzles, interactive storytelling or mini-games. Each subgame may begin with a prelude such as a short video, AI conversation, or story fragment to ease players into the experience. For example, the working memory subgame may start with a chat about daily routines before introducing a timed memory task [10].

(a) Game Homepage (b) Subgame Menu (c) Sample Question

Fig. 2. Game Interface Design

Questions are divided into three levels: easy, medium, and hard. Players begin at the easy level and must pass it to move forward. The AI assistant dynamically adjusts question difficulty level based on performance. The game tracks the player's performance and provides real-time feedback, encouragement, and a personalized progress report at the end of each session, detailing cognitive state and tailored recommendations. Figure 3 displays the design framework of the game engine.

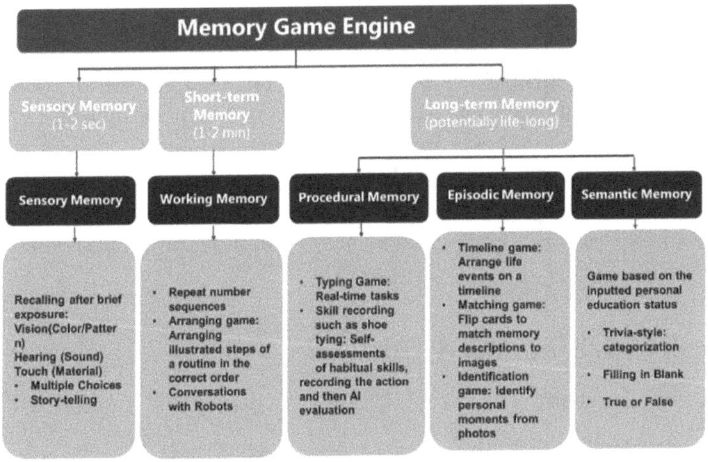

Fig. 3. *Game Engine and Subgames Design Framework*

Working Memory Subgame. The working memory subgame is designed to assess an individual's ability to temporarily hold and manipulate information. It begins with a relatable prompt, such as "Let's chat about your morning routine..." Participants are then asked to complete a sequence-based task, like arranging illustrated steps of that routine in the correct chronological order. Additional variations may involve repeating a short number sequence or recalling a visual pattern after a brief exposure. These tasks target core functions of working memory, including attention, retention, and sequencing.

Procedural Memory Subgame. Procedural memory is a type of long-term memory that stores information about how to perform autonomous actions and skills like typing on a keyboard or riding a bike. The game may contain a real-time activity such as a typing test, a player's self-assessment of a current activity such as the player's performance of a recent biking trip, shoe tying times, or other actions requiring minimal conscious effort. The goal is to tap into skills players learned but do not consciously think about anymore.

Episodic Memory Subgame. The subgame will start with a prelude such as, "Tell me about a day you'll never forget". Then, the player will be shown a set of photos and asked, "Which one is from your wedding day?" Other versions involve placing big life events on a timeline or matching a memory description to an image. This subgame is about reliving meaningful experiences and seeing how clearly they are remembered.

Semantic Memory Subgame). The semantic memory subgame is like a friendly trivia challenge. It looks at how much general knowledge someone has stored away over the years. For example, it starts with "Let's see how much you know about your favorite topics!" From there, the player could be asked to sort items into categories—for example picking out which ones are musical instruments—or match words to their definitions.

Sensory Memory Subgame. This subgame engages players in recalling sensory details from past experiences. It begins with a general question, "What is a smell that instantly takes you back to a specific moment?" Players could then be asked to match that memory with an image, a sound, or a texture that captures the scene. It is about accessing the fleeting impressions stored in sensory memory and expressing how particular sights, sounds, or smells triggered vivid recollections.

4 Empirical/Simulated Results

4.1 Scoring System

The game consists of five subgames as discussed in Sect. 3 and each subgame contains N questions (assume N = 10 in this paper). The player's performance is measured in accuracy (correctness of answers/actions) and performance (time spent in answering questions and/or number of hints used). Score calculations for each game question/action, each subgame and total game score are defined as following:

1. For each game question, a player earns a base score of 3 points for correct answer and 0 point if incorrect, denoted as S[i, j] where i is the subgame number and j is the question number in that subgame. S[i, j] then will be adjusted according to the time spent and hint used in answering the question. Let T[i, j] and H[i, j] be the time and hint adjustments for S[i, j] respectively:

$$S[i,j]_{adjusted} = S[i,j]_{base} + T[i,j] + H[i,j] \mid i \in [0,4], j \in [0, N-1] \quad (1)$$

2. The subgame score SG[i] is the summation of the adjusted scores of all questions in that subgame i. The total game score ST is defined as the weighted average of all subgame scores. Let w[i] be the weight of subgame i. Research work indicates that dementia impacts short-term memory significantly [11]. However, weight assignment will be our future work as discussed in Sect. 6. Given N = 10, the maximum possible total game score after adjustments is 250.

$$SG[i] = \sum_{0}^{N-1} S[i,j]$$

$$ST = \sum_{0}^{4} w[i] * SG[i] \quad (2)$$

Subgame scores as well as the total game score will be rounded to nearest whole numbers. Each subgame also issues a mark, which is the difficulty level (easy, medium or hard) where the player answered the last correct question. Therefore, the game result consists of a total score and five subgame results with each in the form of (subgame name, score, and difficulty level). For example, the result of one game play can be: *[201, (WorkingMem, 40, Medium), (SensoryMem, 45, Hard), (ProceduralMem, 30, Easy), (SemanticMem, 38, Medium), (EpisodicMem, 48, Hard)]*.

4.2 Score Adjustment Strategies

Scores are adjusted based on the player's performance. Assume the baseline time for answering a question is t0 and the player takes t time to answer that question, faster answers (t < t0) may receive bonus points while slower answers (t > t0) may get a penalty. Similarly, a penalty may apply if excess hints are used in answering a question. In the following calculation we scale each adjustment to be in range [−1, 1].

Assume the time baseline for answering a question is t0 and the player's response time for that question is recorded as t, the time adjustment T[i, j] (for question j of subgame i) is calculated as:

$$
\begin{aligned}
T[i,j] &= \frac{t_0 + t}{t_0} \mid \frac{t}{t_0} \leq t_0 \\
T[i,j] &= \frac{-\min(-t_0, t_0)}{t_0} \mid t > t_0
\end{aligned} \tag{3}
$$

For example, if the baseline time is 60 s and the player took 45 s to answer the question, T[i,j] = (60-45)/60 = 0.25. If the player took 75 min to answer the question, then T[i,j] = −0.25 Similarly, hint adjustment H[i,j] is calculated as follows, where h0 is the baseline number of hints allowed, and h is the number of hints used for question j of subgame i.

$$
\begin{aligned}
H[i,j] &= min(1, (h0 - h) * 0.25) \mid h \leq h0 \\
H[i,j] &= -min(1, (h - h0) * 0.25) \mid h > h0
\end{aligned} \tag{4}
$$

For example, if a baseline of 1 hint is allowed, a player using 0 hint will receive 0.25 bonus points, while a player using 4 hints will get −0.75 penalty. The maximum bonus or penalty is 1 point.

4.3 Diagnostic System and Predictive Modeling

As the cognitive functions are very complicated, it is impossible to define the results using mathematical functions. A supervised learning model is built up to predict the player's cognitive stage. It will be very helpful if our model can learn from past experiences. However, as the research is ongoing, we are not able to collect actual clinic data, so a synthetic dataset is established to perform preliminary justification of our design ideas.

The initial synthetic dataset consists of 100 data items with each row containing the player's information saved in the profile (age, gender, education level, physical health condition, prior professional diagnosis result, previous game performance comp) and gameplay result. The gameplay result includes total game score, scores of each subgame, and the difficulty level reached in each subgame. NA is used if a data item is not available for that player. Figure 4 is a snapshot of data format:

Age	Gender	EdLevel	Health	PriorDiagnosis	GameComp	TotalS	WMScore	WMLevel	SMScore	SMLevel	PMScore	PMLevel	EMScore	EMLevel	SMMScore	SMMLevel	Prediction
<50	F	BS	Excellent		Same	201	40	Medium	45	Hard	41	Hard	38	Medium	49	Hard	NoImpairment
70-80	M	HS	Fair	veryMild	Worse	79	10	Easy	8	Easy	16	Medium	10	Medium	35	Easy	Mild-Moderate
50-60	F	ES	Good		Better	235	45	Hard	50	Hard	48	Hard	42	Hard	50	Hard	NoImpairment
60-70	M	HS	Fair	Moderate	Same	110	20	Easy	30	Easy	20	Easy	25	Easy	15	Medium	Mild-Moderate
>80	M	ES	Poor	Mild	Worse	78	20	Easy	18	Easy	10	Medium	20	Easy	10	Medium	Mild-Moderate
60-70	F	BS	Excellent	NoDecline	Better	230	50	Hard	50	Hard	50	Hard	40	Medium	40	Hard	NoImpairment
70-80	F	ES	Good			85	20	Easy	15	Medium	10	Easy	20	Easy	20	Easy	Mild-Moderate
60-70	M	BS	Good		Same	217	50	Hard	45	Hard	40	Medium	40	Hard	42	Hard	NoImpairment
>80	M	HS	Fair		Worse	42	10	Easy	5	Easy	10	Easy	12	Easy	5	Easy	Severe
70-80	M	HS	EMS	Poor	Mild	70	20	Easy	10	Medium	15	Easy	15	Easy	10	Medium	Mild-Moderate
60-70	F	BS	Excellent		Better	207	30	Hard	50	Hard	45	Medium	40	Hard	42	Medium	NoImpairment
60-70	M	ES	Good	NoDecline	Worse	74	20	Easy	12	Medium	10	Easy	20	Easy	12	Easy	Mild-Moderate
50-60	F	HS	Good		Same	215	45	Hard	50	Hard	40	Hard	40	Hard	40	Hard	NoImpairment
>80	M	ES	Poor	Moderate	Worse	35	10	Easy	5	Easy	10	Easy	5	Easy	5	Easy	Severe
70-80	M	ES	Poor	Mild	Worse	106	20	Easy	30	Medium	20	Easy	15	Easy	20	Easy	Mild-Moderate
70-80	F	HS	Excellent		Same	230	50	Hard	50	Hard	45	Hard	40	Medium	45	Hard	NoImpairment
60-70	M	HS	Good	VeryMild	Worse	110	10	Easy	30	Medium	20	Easy	25	Medium	25	Easy	Mild-Moderate

Fig. 4. *Snapshot of Data Format*

The synthetic dataset is generated as follows: m% of data generated by an algorithm using a baseline reference rubric, n% from scenario analysis, and r% with a random generator where n+m+r=100. The r% random data source is used to simulate possible misdiagnosis or just add noises, so we expect r to be a small number. The baseline reference rubric is defined solely based on gameplay result.

Stage A (healthy mind): The player reaches the Hard level for all subgames with a total game score of at least 200.
Stage B (mild/moderate): The player reaches the Hard level in no more than 2 subgames and no more than 2 subgames stay at the Easy level. The total game score is between 100 to 150.
Stage C (severe): The player's total game score is less than 50 and was not able to reach the Medium level in any subgame.

Scenario Analysis with supplement information (to add game scores, comparison with previous game performance, etc.) is illustrated with the following three fictitious cases:

1. **Case 1.** Player 1 is a 35-year-old female. She is a high school history teacher with good physical health. She does not have any record of memory issues. She played five subgames, all reaching the Hard levels. Her total game score is 245.

   Conclusion: Stage A, No Impairment and best wishes for a happy and healthy life.

2. **Case 2.** Player 2 is a retired engineer who supports himself in managing his daily routine and cognitive health. He struggles with short-term memory recall (Working Memory subgame ended at Easy level with a score of 18) but retains strong long-term memory of life events (e.g. Episodic memory reached Hard level), while rest of the subgames stay at Medium level. His total game score is 120.

   Conclusion: Stage B, Mild to Moderate Cognitive Decline. At the end of the game, he is strongly recommended to see his family physician for an official memory test.
3. **Case 3.** Player 3 is an old, retired librarian who spent her life surrounded by books and sharing her passion for literature. According to her medical records, in recent years she has become quieter and more withdrawn, often struggling to follow conversations or recognize familiar faces. She cannot complete the game even with a lot of hints and help. Her total game score is 20 and all subgame stayed at Easy level.

   Conclusion: Stage C, Severe Cognitive Decline. She was immediately connected to the memory support center.

4.4 Results and Discussions

We test the prediction accuracy by varying m, n, and r values. A random forest model is implemented for the classification task. Random forest algorithm has the advantages of high accuracy and robustness, making the model versatile for classification. In data preprocessing, OrdinalEncoder [13] is used to convert categorical variables such as age and education level where order matters, while OneHotEncoder [13] is used to convert categorical variables such as gender where order does not matter. In model building the dataset is split into training dataset (30%) and testing dataset (70%). Table 1 shows the testing results based on the synthetic training dataset and Fig. 5 shows confusion matrices (data size = 1000) for r = 0%, r = 10%, and r = 20% respectively.

Table 1. Prediction of dementia stages based on synthetic training dataset.

m	n	r	Accuracy	
			dataSize = 100	dataSize = 1000
50%	50%	0%	86.7%	88.3%
45%	45%	10%	83.3%	86.4%
40%	40%	20%	76.7%	85.6%

From Table 1 it is observed that the results are improved when data size increases. Also, the smaller the r value the better the prediction accuracy. Figure 5 shows that prediction errors occurred mostly in boundary cases, i.e.

between NonImpairment and Mild-Moderate or between Mild-Moderate and Severe. Though not optimal, the prediction accuracy achieved, ranging from 88.3% (0% random data) to 85.6% (20% random data) for the large dataset is consistent and satisfactory.

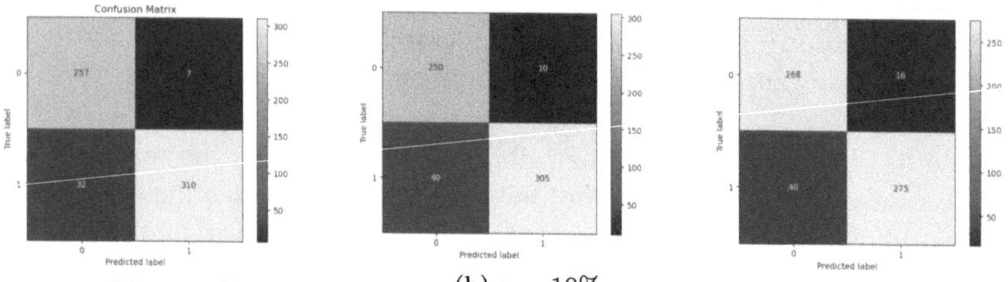

(a) $r = 0\%$ (b) $r = 10\%$ (c) $r = 20\%$

Fig. 5. Confusion matrices from synthetic dataset-based test runs (data size: 1000).

5 Conclusions and Future Work

We have developed a memory-based game designed to support the early detection of dementia in a both engaging and noninvasive way. Compared to traditional screening methods, game-based assessments offer several advantages: they are adaptable, accessible, and less likely to cause stress or frustration for users. Our game prioritizes ease of use while still collecting meaningful cognitive data, particularly related to memory performance.

At this stage, the game is still under development, and several key features are planned for future implementation. One area of focus is enhancing the gameplay mechanics. For example, we plan to introduce Digital Memory Albums, which allow players to scan a QR code and contribute stories, drawings, or reflections to a virtual scrapbook. At the end of the game, a shareable "Memory Lane Album" will compile highlights of the player's journey. The game will also include adaptive difficulty, where the complexity of tasks adjusts based on player age or comfort level, simpler prompts for younger users and more reflective or open-ended questions for older adults. To make gameplay more interactive, we also plan to incorporate AI-driven surprise events. These include scenarios such as a "Time Machine" prompt, which asks players to describe a memory from a specific decade, and a "Robot Interview", where the AI presents rapid-fire, playful questions on unusual topics.

Another area of future work involves improving the game's scoring system. Based on existing research suggesting that short-term memory decline often precedes long-term memory decline in early-stage dementia, we aim to place greater emphasis on performance in working memory tasks. By refining how cognitive performance is assessed, we can better align game outcomes with clinical relevance.

Lastly, we plan to enhance the game's predictive modeling capabilities. While supervised learning currently powers our diagnostic system, we intend to incorporate reinforcement learning to allow the model to learn and improve based on user feedback. This approach enables continuous refinement of the diagnosis process without the need for large, labeled datasets, making it more scalable and adaptive over time.

Acknowledgments. We would like to thank Professor Lan Yang of California State Polytechnic University Pomona for her invaluable guidance and support throughout the development of this project.

Disclosure of Interests. The authors declare that they have no competing interests.

References

1. Atkinson, R.C., Shiffrin, R.M.: Human memory: a proposed system and its control processes. In: Spence, K.W., Spence, J.T. (eds.) The Psychology of Learning and Motivation: Advances in Research and Theory, vol. 2, pp. 89–195. Academic Press, New York (1968). https://doi.org/10.1016/S0079-7421(08)60422-3
2. de Diego, G., et al.: Competencies in the robotics of care for nursing robotics: a scoping review. Healthcare **12**, 617 (2024). https://doi.org/10.3390/healthcare12060617
3. Townsend, D., MajidiRad, A.: Trust in human–robot interaction within healthcare services: a review study. In: Proceedings of the ASME 2022 International Design Engineering Technical Conferences and Computers and Information in Engineering Conference (2022). https://doi.org/10.1115/DETC2022-89607
4. Tsoi, K., et al.: Applications of artificial intelligence in dementia research. Camb. Prisms: Precis. Med. **1** (2022). https://www.cambridge.org/core/journals/cambridge-prisms-precision-medicine
5. Qi, J., Wu, C., Yang, L., et al.: Artificial intelligence (AI) for home support interventions in dementia: a scoping review protocol. BMJ Open **12**(9), e062604 (2022). https://bmjopen.bmj.com/content/12/9/e062604
6. Newby, D., et al.: Artificial intelligence for dementia prevention. Alzheimer's Dementia (2023). https://doi.org/10.1002/alz.13463
7. Javeed, A., et al.: Predicting dementia risk factors based on feature selection and neural networks. Comput. Mater. Continua (CMC) (2023). https://doi.org/10.32604/cmc.2023.033783
8. Lee, L., et al.: Robust and interpretable AI-guided marker for early dementia prediction in real-world clinical settings. eClinicalMedicine (2024). https://doi.org/10.17863/CAM.109598
9. Reisberg, B., Ferris, S.H., de Leon, M.J., Crook, T.: The global deterioration scale for assessment of primary degenerative dementia. Am. J. Psychiatry **139**, 1136–1139 (1982)
10. Baddeley, A.D., Hitch, G.J.: Working memory. In: Bower, G.A. (ed.) Recent Advances in Learning and Motivation, vol. 8, pp. 47–89. Academic Press, New York (1974)
11. Hulme, C., et al.: Short-term memory impairments in Alzheimer-type dementia: evidence for separable impairments of articulatory rehearsal and long-term memory. Neuropsychologia **31**(2), 161–172 (1993). https://www.sciencedirect.com/science/article/abs/pii/002839329390044Z

12. Tulving, E.: Episodic and semantic memory. In: Tulving, E., Donaldson, W. (eds.) Organization of Memory, pp. 381–403. Academic Press, New York (1972)
13. Pedregosa, F., et al.: Scikit-learn: machine learning in Python. J. Mach. Learn. Res. **12**, 2825–2830 (2011)

Detecting Structural Breaks in Latent Feeling and Uncertainty of Expectations of Price Levels Over Time

Carmela Cappelli[1](✉) and Rosaria Simone[2]

[1] Department of Humanities, University of Naples Federico II, Naples, Italy
carcappe@unina.it

[2] Department of Political Science, University of Naples Federico II, Naples, Italy

Abstract. Qualitative data on individuals' perceptions and evaluations are collected through official surveys using rating scales, often administered repeatedly over multiple time waves. In many cases, only aggregated response distributions are made available, with no access to individual-level data. In this context, we combine Atheoretical Regression Trees with CUB models to analyze rating data and identify structural changes in the main characteristics of the response distributions. The chosen modeling framework is well-suited to parameterize both the latent feeling and the uncertainty underlying the ordinal evaluation, with a time-varying structure. Then, the use of Atheoretical Regression Trees involves introducing an artificial covariate that preserves temporal ordering, enabling the segmentation of feeling and uncertainty measures into homogeneous time intervals, which can then be interpreted in light of political and socio-economic events. We illustrate the proposed approach using data from the Consumer Confidence Survey issued monthly by the Italian National Institute of Statistics (ISTAT), assessing whether and to what extent individuals changed their expectations regarding price levels from 1994 to 2019. The proposed method demonstrates satisfactory and interpretable performance similar to the Bai and Perron's method which is considered the gold standard.

Keywords: Survey data · CUB models · Structural Changes · Atheoretical Regression Trees · Price Expectations

1 Motivation

The observed distributions of qualitative data collected through surveys are believed to vary over time in response to socio-economic and political events that influence individuals' perceptions, expectations, and behaviors. Therefore, the problem of change-point detection in key features of ordinal responses must be properly addressed, especially in cases data are released only in aggregated form for public use.

The central narrative of the present research is a case study examining the dynamics of price expectations among the Italian population from 1994 to 2019, based on data from the monthly Consumer Confidence Survey issued by the Italian National Institute

H. R. Arabnia et al. (Eds.): CSCI 2025, CCIS 2969, pp. 141–153, 2027.
https://doi.org/10.1007/978-3-032-28460-0_12

of Statistics (ISTAT): aggregated responses are available for public use as monthly frequency distributions.

Due to the lack of access to individual-level responses, traditional statistical modeling approaches based on regressing the ordinal response on a set of covariates (Agresti, 2015) are not applicable. In this context, modeling on the discrete support provided by the class of mixture models with uncertainty (Piccolo, 2003; Piccolo and Simone, 2019a) offers an effective alternative, as it enables the parameterization of relevant features of the rating distributions without requiring individual-level data. Specifically, the baseline CUB specification assumes a mixture of binomial and uniform distributions to jointly model the latent trait under investigation and the expression of the concomitant framing effects and scale-usage variability, occurring in the form of heterogeneity, inflated frequencies and over-dispersion, among others. In the end, the rating distribution is parameterized using two normalized measures: *feeling*, the probability parameter of the binomial component, and *uncertainty*, the mixture weight associated with the feeling component. Alternative specifications can be introduced such as inflation in frequency for categories conveying a peculiar meaning or wording, and thus being perceived as a *shelter* option (Iannario, 2012). Estimation of parameters for mixture models with uncertainty can be easily performed using the dedicated R package 'CUB' (Iannario et al., 2024).

Assuming a dynamic approach for CUB mixture modeling, the research presented in this paper aims at contributing to the state of the art with the proposal of tree-based method for the supervised identification of the main structural breaks in the average level of the series of *feeling* and *uncertainty* parameters estimated time-wisely.

To this end, Atheoretical Regression Trees (ART) are proposed as an interpretable learning approach based on a recursive partitioning algorithm. This methodology has been successfully applied in various contexts, including time series analysis (see Cappelli et al., 2008; Cappelli et al., 2021a), financial risk evaluation (Cappelli et al., 2021b), and market research (Rea et al., 2010). The procedure consists in a decision tree grown to define homogeneous sub-intervals for the responses retaining the intrinsic time ordering.

Our investigation suggests that ART effectively partitions the series into temporal sub-intervals characterized by different levels of estimated feeling and uncertainty, allowing interpretation in light of political and socio-economic events. Regarding the proposed application, similar results are observed when comparing ART with Bai and Perron's method for change-point detection (Bai, 1997; Bai and Perron, 1998; Zeileis et al., 2002), which is considered the gold standard in the field.

The paper is organized as follows. Section 2 presents the methodological background on mixture models with uncertainty for the analysis of rating data and the ART procedure. Section 3 illustrates the chosen case study Conclusions and comments on future research developments are provided in Sect. 4.

2 Methodological Background

This section provides a brief methodological overview of mixture models with uncertainty and Atheoretical Regression Trees, aimed at motivating the chosen framework and set the notation.

2.1 Mixture Models with Uncertainty for the Analysis of Rating Data

For a response R collected over m ordered categories, the class of mixture models with uncertainty (Piccolo and Simone, 2019a) considers a discrete mixture of a (shifted) binomial distribution with probability parameter $\xi \in [0,1]$ ($b_r(\xi) = \binom{m-1}{r-1}\xi^{m-r}(1-\xi)^{r-1}, r = 1, ..m$), and a discrete distribution u_r for *uncertainty*, resulting in the specification:

$$Pr(R = r) = \pi b_r(\xi) + (1 - \pi)u_r, r = 1, \dots, m \tag{1}$$

The choice of the binomial distribution to represent the latent *feeling* of the response corresponds to the assumption that the final score is the result of a sequence of pairwise comparisons between adjacent categories, in which there is constant probability of rejecting or accepting the lower or upper one due to its suitability to convey the individual sentiment. Therefore, depending on the orientation of the rating scale, either ξ or $1-\xi$ can be referred to as *feeling* parameter.

A relevant feature of the model is that the distribution u_r can be specified in different ways: to account for inflated frequencies at some *shelter* category, say c (by choosing a degenerate distribution concentrated at that category: $D_r^{(c)} = 1$, if $r = c$, $D_r^{(c)} = 0$ if $r \neq c$); for heterogeneity of the distribution (in which case, u_r is the discrete uniform model, yielding the baseline CUB model introduced in Piccolo (2003)); or both:

$$Pr(R = r) = \pi b_r(\xi) + (1 - \pi)\left((1 - \theta)\frac{1}{m} + \theta D_r^{(c)}\right), r = 1, \dots m \tag{2}$$

In the latter case, the model is referred to as CUB with shelter (Iannario, 2012). Given these premises on uncertainty specification, the mixing weight $\pi \in (0, 1]$ quantifies how trustworthy is the expression of the latent sentiment conveyed by the feeling parameter, given the occurrence of diversified sources of discrete noise, embedded in the uncertainty component. Beyond classical Maximum Likelihood methods, these parameters can be estimated using profile likelihood procedures, as done in Simone et al. (2023), or by selecting the best-fitting model within this class, as described in Sect. 3.

The direct parameterization on the discrete support of location, heterogeneity, and framing effects in rating distributions allows for an effective interpretation of the main features of the distribution, useful especially when adopting a comparative perspective to distinguish between response behaviors (uncertainty and shelter effects) and perceptual aspects of the choice (feeling), discriminating groups of respondents or survey items.

Regression modeling can be performed by setting a logit link between feeling and uncertainty parameters and subject-specific covariates $\boldsymbol{x}_i$ and z_i:

$$logit(1 - \xi_i) = \boldsymbol{x}_i\boldsymbol{\gamma}'; \; logit(\pi_i) = z_i\boldsymbol{\beta}'. \tag{3}$$

However, the distinctive characteristic of mixture models with uncertainty with respect to other modeling approaches for rating data is that it does not require the compulsory specification of individual covariates to parameterize a rating distribution according to (2). This circumstance makes the parameters of CUB models particularly suitable for studying the evolution over time of the latent components underlying the rating process in qualitative data collected through official surveys. In fact, such data are typically released publicly only in aggregated form for successive waves.

2.2 ART: Atheoretical Regression Trees for Change Point Analysis

A change point divides time-wise data into segments that differ from each other in terms of a given statistical feature, usually the mean. The automatic detection of change points—when both their number and location are unknown—is useful in several fields of research and applications: it can reveal behaviours in the time-wise data that might otherwise be mistaken for long memory, help identify the presence of outliers, and serves as a preliminary step in forecasting, since a model estimated on a recent segment may provide more accurate predictions.

ART is a procedure that utilizes the recursive partitioning approach of Least Squares Regression Trees (LSRT) (Breiman et al., 1984) to identify multiple change points whose occurrence dates are unknown a priori.

In general, tree growing relies on a data-driven, top-down algorithm known as recursive partitioning. This method generates all possible binary splits from the available information set and then selects the best one based on a goodness-of-split criterion. The selected rule reflects how well the split can identify two mutually exclusive subsets of observations that are as homogeneous as possible with respect to the response variable. In LSRT, the relationship between a response variable and a set of covariates is represented as a binary tree, constructed by recursively splitting the data into two subgroups that become increasingly homogeneous with respect to the response variable. Specifically, the best exhaustive split s of node h into its descendants h_l and h_r is the one that maximizes the reduction in the sum of squares:

$$\Delta SS(h|s) = SS(h) - [SS(h_l) + SS(h_r)],\ SS(h) = \sum_{i=1}^{n(h)} (y_i - \hat{\mu}(h))^2. \tag{4}$$

Since the child nodes h_l and h_r form an exhaustive binary partition of the parent node h, this splitting criterion seeks the split that produces child nodes whose means differ as much as possible in terms of squared differences. In fact, it can be easily shown that this criterion corresponds to the between-group sum of squares:

$$\Delta SS(h|s) = BSS_{y|s}(h) = \frac{n(h_l)n(h_r)}{n(h)^2} (\hat{\mu}(h_l) - \hat{\mu}(h_r))^2. \tag{5}$$

ART is an adaptation of LSRT that aims to partition a response variable Y according to an internal ordering. The term "Atheoretical" refers to the artificial nature of the single covariate used in the partitioning procedure, which is constructed as an arbitrary sequence of strictly increasing numbers $k = 1, 2, \ldots, n$, where $\boldsymbol{n}$ is the number of observations. Based on these premises, ART is a non-parametric approach to analyzing numerical data that follows an intrinsic ordering, which is preserved and conveyed through the artificial numerical covariate generated for the partitioning algorithm. As said in the introduction, ART has proven to be effective in detecting multiple changes occurring at unknown dates in various types of time series.

Indeed, when the target variable is a time series Y_t, the artificial covariate corresponds to time, and the procedure generates sub-intervals with different mean values. As a result, the tree splits can be interpreted as change points in the time series. Hence, ART performs a non-parametric detection of change points in the average value of a response collected at subsequent time observations.

It is worth noting that the procedure mimics Bai and Perron's method (Bai, 1997; Bai and Perron, 1998; Bai and Perron, 2003), but, being based on a recursive algorithm, it is extremely efficient computationally. In contrast, Bai and Perron's method searches for the optimal partition associated with the global minimum of the sum of squares. While ART identifies a local minimum, its results are often very close to—and sometimes coincide with—those of Bai and Perron's method.

Moreover, ART produces a hierarchical structure that provides insight into the relative importance of the detected changes and ultimately yields a more parsimonious partition with a simpler interpretative framework.

In the following, we compare the performance of ART with that of Bai and Perron's method based on the residual sum of squares and the BIC index, assuming an intercept-only regression model with normally distributed errors as the benchmark: $Y_t = \beta + \varepsilon_t, t = t_0, \ldots, T$, against a piece-wise regression model built on the basis of a partition of the observation period $[t_0, T]$ in m adjacent sub-intervals $\{(a_{k-1}, a_k]\}_{k=1}^{m}$, so that:

$$Y_t = \beta_1 + \sum_{k=2}^{m} \beta_k D_{t,k} + \varepsilon_t, \tag{6}$$

with $D_{t,k} = 1$ if $t \in (a_{k-1}, a_k]$, and $D_{t,k} = 0$ otherwise: under this setting, standard ML estimation is performed for regression parameters, error variance and split points $a_0, \ldots, a_m$.

Note that a development of ART, called Theoretical Regression Trees (TRT), has been proposed (Cappelli and Di Iorio, 2013) for the specific case of time series data. In this version, various models are tested within the segments to identify the change point.

3 Detecting Change Points in Price Expectations Collected with Survey Data

The investigation of inflation perceptions and expectations is an important issue because what people think about the price level and how they expect prices to develop in the future affects how they spend, borrow, and invest money today. For example, if someone plans to buy a new car one year from now and believes that prices will not rise, they will likely wait those 12 months before making the purchase. However, if they expect prices to increase, they will buy the car now. Many other consumers may act similarly, and this surge in demand, combined with expectations of wage adjustments, will push businesses to raise the prices of their products and services. This vicious cycle will cause prices to rise.

In other words, prices end up reflecting, in part, what people expect them to be: if everyone expects prices to rise by, say, 2% over the next year, businesses will want to increase prices by at least 2%, driving prices upward. Conversely, if people become accustomed to very low or stable inflation, they may expect prices to remain about the same.

As a consequence, there is a peculiar intersection between global economic forces and people's hopes and fears about the future. From the policymakers' perspective, it is important to analyze perceptions and expectations to understand their behavior over time, enabling more informed decisions and better communication to keep expectations anchored and prices stable. For a discussion in the macroeconomic literature about the role of expectations in inflation dynamics and the rational expectations hypothesis versus survey-based short-term expectations, see, among others, Fuhrer (2011), Pfajfar (2022), and Coibion et al. (2018).

At the official level, national statistical offices are responsible for collecting information about people's perceptions and expectations regarding past and future price levels, respectively. The resulting qualitative data should be appropriately analyzed to investigate the dynamics of people's evaluations using sound statistical methods that provide easy-to-interpret results.

3.1 Data Premise

In the following, we focus on data from the monthly consumer confidence surveys collected and released by the Italian National Statistical Institute from January 1994 to December 2019. This sample survey is conducted monthly, with each wave involving about 2,000 individuals aged 18 and over. The goal is to survey consumers' self-evaluations on economic and financial aspects of life, both at the individual and broader levels.In the sequel, we focus on the question related to price expectations for illustrative purposes:

'How do you think the price level will change over the next 12 months?'

Responses are collected over a scale with $m = 5$ ordered categories, coded as follows: *1 = "fall", 2 = "stay about the same", 3 = "rise slightly", 4 = "rise moderately", 5 = "rise a lot".*

As is common practice for official survey data, individual responses are not available for public use; only the monthly frequency distributions are published. Therefore, subjective characteristics are available at most at the aggregated level. From a statistical point of view, this limitation prevents the application of standard regression models.

Figure 1 displays some instances of the response distributions, at selected time points, whereas Fig. 2 shows the time series of the frequencies for each response category (in %). The exploratory visual inspection of the temporal series of ordinal distributions related to price expectations indicates that the structural features of the distributions (location, dispersion, and possible inflated frequencies) vary greatly over time. Additionally, an excess of frequency in the second category, "*stay about the same*," is observed at several time points. To grasp the temporal evolution of the distribution of price expectations in a whole, see Fig. 3.

Therefore, a model selection within the class of mixture models with uncertainty is performed to identify the best-fitting model and thus optimally parameterize the monthly response distributions in a parsimonious and comprehensive modeling framework.

Subsequently, the ART procedure is applied to the time series of logit-transformed feeling and uncertainty parameters to satisfy the unbounded support requirements of ordinary regression, setting a minimum segment length of 12 months.

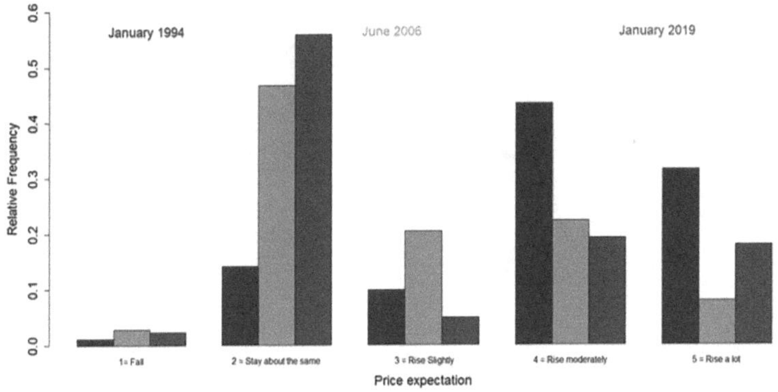

Fig. 1. Distribution of price expectations at selected time points

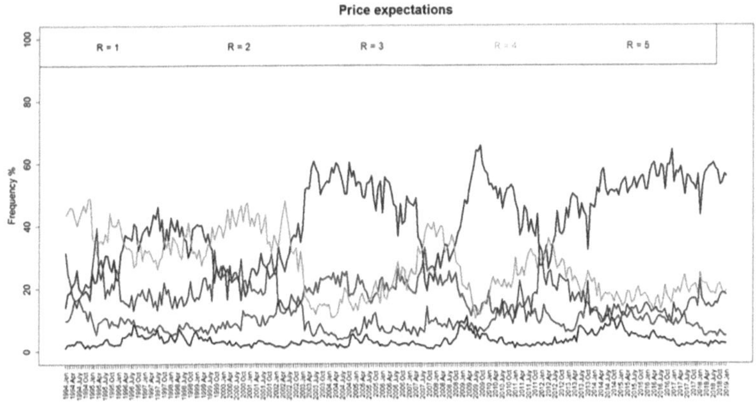

Fig. 2. Time series of the frequencies (in %) of each response options for price expectations

For the data under examination and given the orientation of the scale, it is worth noting that the lower the value of the feeling parameter ξ, the higher the expectation that prices will increase. Thus, 1-ξ is the probability that an individual expects prices to increase, assuming that the rating process is driven by feeling solely, with no uncertainty blurring the response.

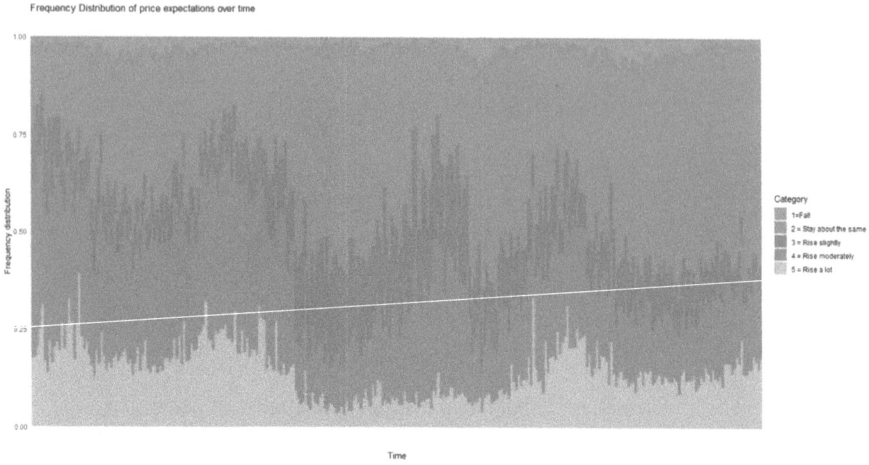

Fig. 3. Evolution over time of the frequency distribution of price expectations

3.2 Detecting Change Points in Feeling of Expectations of Price Levels

Figure 4 shows the selected best tree according to the CART pruning procedure applied to the ART grown for the series of feeling parameters related to price level expectations (cost complexity pruning parameter set to 0.05). The procedure identifies 3 change points and the corresponding four subperiods.

Accordingly, Fig. 5 highlights the change points identified by ART, shown as red vertical lines on the standard time series plot of the feeling parameter for price expectations. Structural changes identified using the best partition from Bai and Perron's method are instead indicated by dashed blue vertical lines to enhance comparison.

Overall, the change points detected by ART appear consistent with the behavior of the series. Moreover, the identified subperiods can be easily interpreted in light of political and macroeconomic events. The first period covers the implementation of policies aimed at reducing inflation to meet the goals of the Maastricht Treaty. It is reasonable to assume that people were aware of the measures required to create the EURO zone and that, among other objectives, keeping inflation low was a priority. Thus, they expected prices to rise in the future.

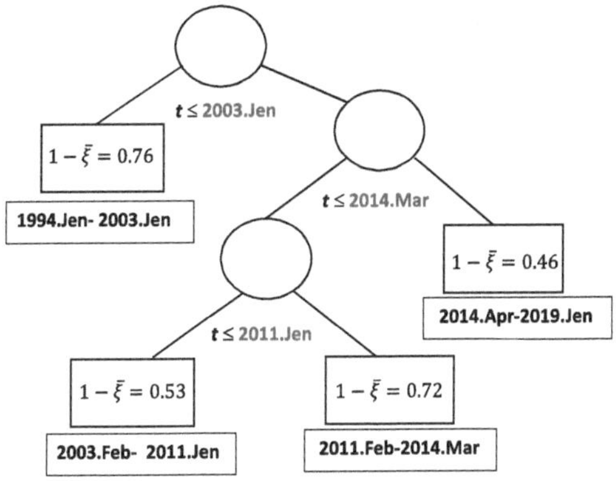

Fig. 4. Atheoretical Regression Tree to identify change points in feeling of expectations and the corresponding subperiods

The second period includes the introduction of the EURO currency, as well as the global financial crisis of 2008, which originated in the United States, followed shortly by the outbreak of the Italian spread crisis. It was a very troubled period marked by recessions and false hopes of recovery, during which people's feelings towards inflation were moderate, around 0.5.

The third period can be interpreted as a reaction to the news that the current government was about to resign and be replaced by a technocratic government composed entirely of unelected professionals, which was perceived as a solution to the Italian sovereign debt crisis. People expected prices to rise following the recession.

The final period is characterized by persistent instability, resulting in widespread fear about future price levels.

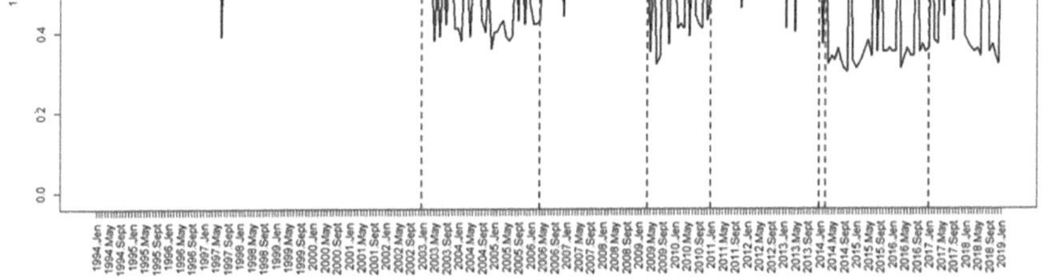

Fig. 5. Time series of the feeling related to price expectations: detected change points with ART and Bai and Perron's procedures are highlighted with vertical lines.

3.3 Detecting Change Points in Uncertainty of Expectations of Price Levels

With respect to the mixing weight of the Binomial component, Fig. 6 shows the final pruned ART, indicating the average level of the uncertainty parameter for each time period, while Fig. 7 displays the corresponding time series with change points highlighted by red vertical lines. The analysis of the uncertainty weight provides additional insights into the dynamics of expectations. Specifically, higher discrete noise is observed from November 2008 to October 2009, from December 2012 to March 2014, and from July 2015 to January 2019. Overall, while the feeling component related to people's expectations remains relatively stable over time, the larger number of subintervals identified in the uncertainty series indicates that price expectations are subject to high uncertainty.

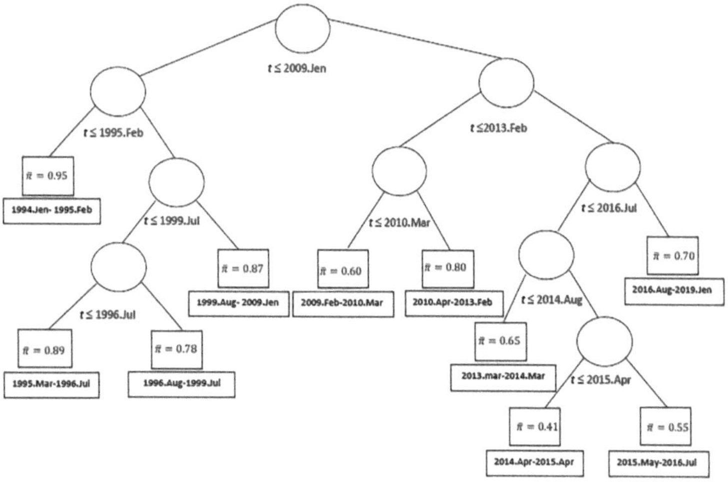

Fig. 6. ART for the series of uncertainty parameters related to price expectations

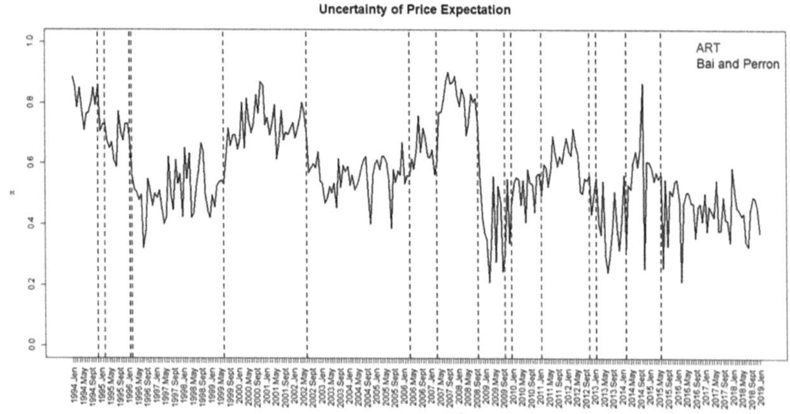

Fig. 7. Time series of the uncertainty parameter related to price expectations: detected change points with ART and Bai and Perron's procedure are highlighted with vertical lines

3.4 Comparisons with Bai and Perron's Optimal Partition

For comparisons of ART performance with Bai and Perron's method, we resorted to the implementation available within the R package 'strucchange' (Zeileis et al., 2002). With respect to the feeling series, the procedure selects a seven-segment partition as the optimal one based on the BIC criterion. The corresponding breakdates are all, but one, coincident with the split points of the maximal, unpruned ART. To foster comparison of performance between the two approaches, Fig. 8 displays the difference in BIC and in Residuals' Sum of Square (RSS) of the intercept model assuming Bai and Perron's and ART partitions for structural changes, for both series. It is seen that pruning ART at cp = 0.05 provides a parsimonious partition with satisfactory performance in comparisons with the optimal one yielded by Bai and Perron's method (see Table 1). Overall, the differences between the two approaches are more pronounced for the uncertainty series than they are for the feeling series.

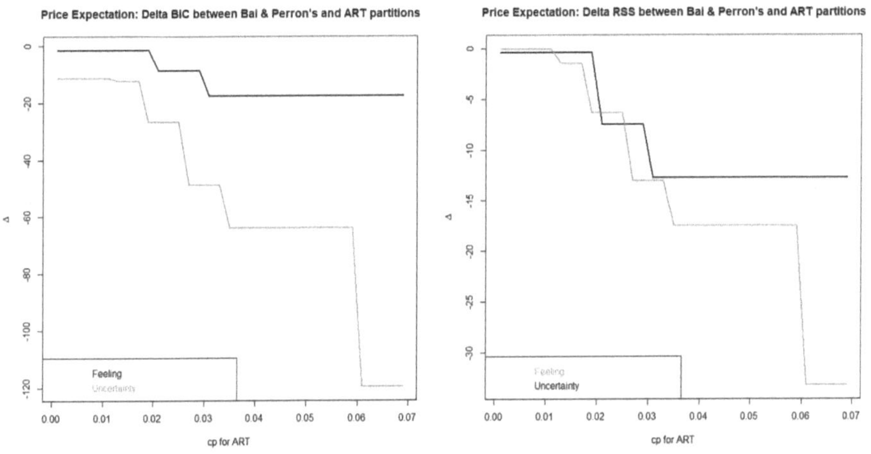

Fig. 8. Delta BIC and Delta RSS to compare Bai and Perron's with ART partitions for varying cp levels to prune ART

Table 1. Synthesis of the goodness of fit of art and bai and perron's partitions

	RSS	loglik	BIC	df	N. periods
	Feeling of price expectation				
Intercept model	176.1473	−346.46	704.34		
ART partition	80.4886	−228.59	502.84	8	4
BP partition	67.7357	−202.63	485.16	14	7
	Uncertainty of price expectation				
Intercept model	130.5315	−301.358	614.1309		

(continued)

Table 1. *(continued)*

	RSS	loglik	BIC	df	N. periods
ART partition	52.8418	−165.259	399.0039	12	6
BP partition	35.3572	−104.789	335.1341	22	11

For the uncertainty series, Bai and Perron's method chooses an 11-segment partition. Of the corresponding breakdates, eight out of eleven are present in the unpruned ART. The other are, at most, 3 months before or after the one identified by ART, confirming the method provides an excellent approximation of the global minimum partition.

4 Concluding Remarks

The proposal put forth by the paper with respect to the state of the art is a new approach for the identification of structural breaks in temporal series of ordinal data distributions, assuming that parsimony is important for the sake of interpretation of time sub-intervals in light of real-data events. The proposal hinges on the methodological combination of Atheoretical Regression Trees on the temporal series of synthetic measures of feeling and uncertainty characterizing time-wise the rating data distributions. These measures are estimated by resorting to the setting of CUB mixture models for the analysis of rating data, useful to characterize the most relevant features of the distribution also in case individual data are not of public use, as customarily for several official survey. Results on the chosen case study provide evidence that ART is effective in deriving a parsimonious partition of the two main latent components of the rating generating process and their dynamics over time, with satisfactory behavior when compared in terms of goodness of fit with the optimal partition entailed by Bai and Perron's method. Even if the proposed approach has offered a two-fold view of the dynamics undergoing the self-evaluation of the temporal dynamics of price expectations, future research developments will be addressed to obtained a unified analysis of the co-breaking of feeling and uncertainty series for the methodological aspects of the proposal, as well as the co-breaking with the series of judgements of price levels from the applicative point of view.

Acknowledgments. This work was supported by grant SI-WCWB from University of Naples Federico II (FRA 2022), DR n 3429, 07/09/2023 (CUP: E65F22000050001).

Disclosure of Interests. The authors declare no potential conflict of interest.

References

Agresti, A.: Foundations of linear and generalized linear models. Wiley Series in Probability and Statistics. ISBN: 1118730038 (2015)

Bai, J.: Estimating multiple breaks one at a time. Economet. Theor. **13**, 315–352 (1997)

Bai, J., Perron, P.: Estimating and testing linear models with multiple structural changes. Econometrica 47–78 (1998)

Bai, J., Perron, P.: Computation and analysis of multiple structural change models. J. Appl. Economet. **18**, 1–22 (2003)
Cappelli, C., Cerqueti, R., D'Urso, P., Di Iorio, F.: Multiple breaks detection in financial interval-valued time series. Expert Systems with Applications, pp 113775–11377 (2021a)
Cappelli, C., Di Iorio, F.: Theoretical regression trees: a tool for multiple structural-change models analysis. In: Grigoletto, M., Lisi, F., Petrone, S. (eds.) Complex Models and Computational Methods in Statistics. Contributions to Statistics. Springer, Milano (2013)
Cappelli, C., D'Urso, P., Di Iorio, F.: Change point analysis of imprecise time series. Fuzzy Sets Syst. **225**, 23–38 (2013)
Cappelli, C., Maddaloni, A., Di Iorio, F., D'Urso, P.: Atheoretical Regression Tres for classifying risky financial Institution. Ann. Oper. Res. **299**, 1357–1377 (2021)
Cappelli, C., Reale, M.: Detecting changes in mean levels with atheoretical regression trees. In: Modelli Complessi e Metodi Computazionali Intensivi per la Stima e la Previsione, pp. 479–484. Cleup, Padova (2005)
Cappelli, C., Simone, R., Di Iorio, F.: CUBREMOT: a tool for building model-based trees for ordinal responses. Expert Syst. Appl. **124**, 39–49 (2019)
Coibion, O., Gorodnichenko, Y., Kadmar, R.: The formation of expectations, inflation and the Phillips curve. J. Econ. Lit. **54**(4), 1447–1491 (2018)
Fuhrer, J.: The Role of Expectations in Inflation Dynamics. Int. J. Cent. Bank. **7**(1), 167–207 (2011)
Iannario, M.: Modelling shelter choices in a class of mixture models for ordinal responses. Stat. Methods Appl. **21**, 1–22 (2012)
Iannario, M., Piccolo, D., Simone, R.: CUB: A Class of Mixture Models for Ordinal Data. R package version 1.1.5 (2024). https://CRAN.R-project.org/package=CUB
Agresti, A., Kateri, M.: The class of CUB models: statistical foundations, inferential issues and empirical evidence. Stat. Methods Appl. **28**(3), 445–449 (2019). https://doi.org/10.1007/s10260-019-00468-8
Pfajfar, D.: The role of expectations in changed inflation dynamics. Int. J. Cent. Bank. **18**(1), 199–238 (2022)
Proietti, T., Frale, C.: New proposals for the quantification of qualitative survey data. J. Forecast. **30**(4), 393–408 (2011)
Simone, R., Corduas, M., Piccolo, D.: Dynamic modelling of price expectations and judgments. METRON **81**, 323–342 (2023)
Simone, R., Piccolo, D., Corduas, M.: Dynamic modeling of price expectations, In: Porzio, G.C., Greselin, F., Balzano, S. (eds.), CLADAG 2019 Book of Short Papers, pp. 449–452. ISBN 978-88-8317-108-6 (2019a)
Zeileis, A., Leisch, F., Hornik, K., Kleiber, C.: Strucchange: an R package for testing for structural change in linear regression models. J. Stat. Softw. **7**(2), 1–38 (2002)

A Fuzzy Approach to Record Linkages

Pratik K. Biswas(✉)

Artificial Intelligence and Data Science, Verizon Communications, Basking Ridge, NJ 07920, USA
pkbiswas@acm.org

Abstract. Record Linkage is the process of identifying and unifying records from various independent data sources. Existing strategies, which can be either *deterministic* or *probabilistic*, often fail to link records satisfactorily under uncertainty. This paper describes an indigenously developed fuzzy linkage method, based on *fuzzy set* techniques, which can effectively account for this *uncertainty* prevalent in the disparate data sources and address the shortcomings of the existing approaches. Extensive testing, evaluation, and comparisons have demonstrated the efficacy of this fuzzy approach for record linkages.

Keywords: Record Linkage · Deterministic Linkage · Probabilistic Linkage · Fuzzy Linkage · Fuzzy Matching · Fuzzy Set Theory · Fuzzy Logic · Fuzzy Weights · Fuzzy System · Fuzzy Clustering

1 Introduction

Data comes in different shapes, sizes, and quality from multiple different sources with many idiosyncrasies. Personal information, encapsulated as strings, often contains typographical and data entry errors, misspellings, over time changes, false, or missing information. Spouses and/or children of a family can use the primary member's information. Households may not have been identified. Therefore, there is a pressing need in every organization to unify, align and merge data gathered from multiple disparate data sources.

Record linkage [1–4] is the task of linking or finding data from multiple data sources that refer to the same entity. There are mainly two types of linkage algorithms, namely, *deterministic* and *probabilistic*. Choosing the best algorithm to use in a given situation depends on many interacting factors, such as time, resources, the research question, and the quantity and quality of the variables or columns (attributes) available to link, including the degree to which they individually and collectively are able to identify an individual uniquely. The key is to develop algorithms to extract and make use of enough meaningful information to arrive at sound decisions. Here, we consider record linkages based only on variables (columns/attributes) whose values are *strings (textual)*.

Deterministic algorithms determine if record pairs agree or disagree on a given set of identifiers (variables/columns/attributes) through *exact* pattern matching, where agreement on a given identifier is assessed as a discrete—*all-or-nothing*—outcome. Match status is determined in a single or in multiple steps. Accordingly, there are two

H. R. Arabnia et al. (Eds.): CSCI 2025, CCIS 2969, pp. 154–173, 2027.
https://doi.org/10.1007/978-3-032-28460-0_13

types, namely, *exact* and *approximate* or *iterative* record matching. Exact deterministic matching is a single step strategy that requires an exact match on the full set of identifiers all at once. Approximate or iterative matching is a multi-step strategy that requires an exact match on one of the several rounds of matching but not on all possible identifiers. In this strategy, records are matched in a series of progressively less restrictive rounds (steps) where pairs that do not meet a first round of match criteria are passed to a next round. If a record pair meets the criteria at any round (step), it is classified as a match. Otherwise, it is a non-match. The iterative deterministic strategy may consist of a sequence of deterministic matches using different match criteria in each successive round.

Probabilistic algorithms determine the *likelihood* (*probability*) of agreement between record pairs on a given set of identifiers through *inexact* (*approximate or fuzzy)* string/pattern matching (on attribute values). Comparison space can be reduced by searching only those matched pairs that meet certain basic criteria – this is known as *blocking*. Probabilistic strategies look at the discriminatory power of each identifier. The weight assigned to agreement or disagreement on each identifier is assessed as a *likelihood ratio*, comparing the *m- and u- probabilities* [3], where *m-probability estimates* the probability that two records *agree* on a particular identifier when they are *true matches* while the *u-probability* estimates the probability that two records *agree* on a particular identifier merely *by chance*. When two records agree on an identifier, an *agreement weight* is calculated by dividing the *m-probability* by the *u-probability* and taking the $\log_2$ of the quotient. When two records disagree on an identifier, a *disagreement weight* is calculated by dividing (1 - *m-probability*) by (1 - *u-probability*) and taking the $\log_2$ of the quotient. The *total linkage weight* for each record pair is obtained by summing the individual linkage weights for all linked pair of identifiers (columns/attributes). This total linkage weight of each record pair is then compared with a pre-defined *probability threshold* and those record pairs whose weights are above this threshold (standard = 0.5) are considered as *linkages*.

In information-poor yet business critical scenarios, where the collected data is of inferior quality, deterministic linkages may result in too few *True Positives*, while probabilistic linkages may result in very high *False Positives* or *False Negatives* depending on the chosen *cut-off* probability threshold.

The quantification of uncertainties can be based on either *probability theories*, using *random* variables, or *fuzzy set theories*, using *intervals* and *fuzzy* variables. *Fuzzy set*-based techniques are increasingly being used for design problems where there is random uncertainty, but information on the statistics of the uncertainty is sketchy. Fuzzy models are for *uncertainties* due to *vagueness* and can often perform better if little information is available.

In this paper, we present a novel and flexible *fuzzy record linkage* solution that combines *fuzzy matching*, *fuzzy set theory*, *fuzzy logic*, *fuzzy constraint*, *fuzzy weights*, *fuzzy rule-based inferencing* and *fuzzy clustering* to provide record linkages that are more purposeful than either deterministic or probabilistic solutions in uncertain scenarios with imprecise and inadequate information.

2 Related Work

Related research falls into two categories: 1) record linkages and 2) fuzzy concepts and capabilities that have been utilized in the proposed method/approach.

Herzog et al. and Winkler [1, 2] reviewed the different record linkage techniques. Fellegi and Sunter [3] provided the theory of probabilistic linkages. Christen [4] represented the record linkage procedure as a workflow, whose steps were cleaning, indexing, comparing, classifying and evaluation. The classified record pairs could be fed back to improve the previous step. Abriel et al. [5] introduced a distance-based record linkage method which used Choquet integral to compute the distance between the linked records as well as supervised learning to determine the optimal fuzzy measure for the linkage. Ouyang et al. [6] considered record linkage as a two-class recognition problem and proposed a three-way classification, which worked in two phases, to identify uncertain points (regions) for further clerical investigation in decision-making. Mierzejewski et al. [7] presented a framework for assessing similarity measure impact on classification confidence associated with probabilistic record linkages. Kaufman et al. [8] proposed an adaptive learning based algorithmic solution for record linkages, by combining fuzzy string matching with machine learning using only a single, error-prone identifier. Ornstein [9] showed how pretrained text embedding based semantic similarity could be incorporated into a probabilistic record linkage procedure yielding considerable improvements in both precision and recall over existing methods.

Zadeh [10] proposed the fuzzy set theory in 1965 to model the uncertain and vague situations more precisely. Dubois and Prade [11] defined the concept of a fuzzy number and Novak [12] alluded to the triangular fuzzy number as the most used type of fuzzy number. Ali et al. [13] described many arithmetic operations on the triangular fuzzy number. Yue et al. [15] extended the traditional definition of neighborhood coverings to fuzzy ones, consisting of membership functions that formed an approximate distribution of neighborhood belongingness. Ruttkay [16] considered fuzzy constraint as a fuzzy relation.

Zadeh [10] also introduced the notion of fuzzy logic along with his proposal of fuzzy set theory. Furthermore, he [17–19] developed the concept of linguistic variables, which was subsequently formalized by Zimmerman [20]. Borovička [21] described the transformation process of linguistic terms to fuzzy numbers. Chen and Hwang [22] discussed the conversion scale. Wang et al. [23] also showed how to calculate fuzzy weights from linguistic terms. Bellman [24] described the fuzzy decision process as a multi-objective programming problem involving fuzzy goals and fuzzy constraints. Mamdani [25] developed a new method of fuzzy inference systems, known as Mamdani-Type Fuzzy Inference. Beginning in 1987, various solutions to the Fuzzy Weighted Average (FWA) problem have been provided. Dong and Wong [26] presented the first FWA algorithm; Liou and Wang [27] presented an improved FWA (IFWA) algorithm; and Lee and Park [28] presented an EFWA. All three algorithms were based on α-cuts. Other solutions included those in [29–32]. Ahmed et al. [33] compared and analyzed the performances of the different algorithms for Fuzzy Analytic Hierarchy Process (FAHP). Bezdek [34] introduced Fuzzy c-means clustering, while Li and Lewis [35] reviewed the application of various fuzzy clustering algorithms.

3 Preliminaries – Concepts and Terminologies

In this section, we clarify key concepts and terminologies and explain certain technologies, which provide the foundation for our work.

3.1 Fuzzy Set Theory

In this subsection, we will review only those concepts of fuzzy set theory [10, 11, 13] that are relevant for us.

Let X denotes the classical set of objects called the *universe*, the generic elements of which are marked as x. A **fuzzy set** A is defined as: $A = \{(x, \mu_A(x))|x \in X, \mu_A(x) \in [0,1]\}$. In the pair $(x, \mu_A(x))$, the first element belongs to the classical set and the second element $\mu_A(x)$, called *membership function*, belongs to the interval [0,1]. We can say that A is a subset of X that has no sharp boundaries.

The **kernel** of a fuzzy set A is defined as $Ker(A) = \{x|\mu_A(x) = 1\}$, while the **support** of fuzzy set A is defined as: $Sup(A) = \{x|\mu_A(x) > 0\}$.

Given $\alpha \in [0, 1]$, the **α-cut** of a fuzzy set A is defined as a crisp set $A_\alpha(orA[\alpha]) = \{x|\mu_A(x) \geq \alpha]\}$.

A **fuzzy number** [11] is a fuzzy set A on $\mathbb{R}$ that must possess at least the following three properties, (i) A must be a normal fuzzy set; (ii) A_α must be closed interval for every $\alpha \in [0,1]$, , (iii) the support of A, A_{0+}, must be bounded. A fuzzy number intuitively represents a value, which is inaccurate.

A **triangular fuzzy number** [12] is represented with three points as follows: $A = (a, b, c), where\, a \leq b \leq c$. Its *membership function* has the shape of a triangle, formalized as:

$$\mu_A(x) = \begin{cases} 0, x < a, x > c \\ \frac{x-a}{b-a}, a \leq x \leq b \\ \frac{c-x}{c-b}, b \leq x \leq c \\ 1, x = b \end{cases} \tag{3.1}$$

The **left** and **right triangular fuzzy number** can be formally written as $A_l = (a, b, b)$, and $A_r = (b, b, c)$, with membership functions [14] as follows:

$$\mu_{A_l}(x) = \begin{cases} 0, x < a, x > b \\ \frac{x-a}{b-a}, a \leq x \leq b \\ 1, x = b \end{cases} \tag{3.2}$$

$$\mu_{A_r}(x) = \begin{cases} 0, x < b, x > c \\ \frac{c-x}{c-b}, b \leq x \leq c \\ 1, x = b \end{cases} \tag{3.3}$$

The **α-cut** of a triangular fuzzy number A = (a, b, c) is defined as:

$$A_a(or\,A[\alpha]) = [a + (b - a)\alpha, c - (c - b)\alpha] \tag{3.4}$$

A **fuzzy neighborhood** [15] of $x \in X$ is defined as:

$FN(x) = \{(y, \mu_{FN(x)}(y)) | y \in X, \forall x, x \in X, d(x, y) < \lambda, \mu_{FN(x)}(y) \in [0, 1] \ \& \ \mu_{FN(x)}(y) = f(d(x, y))\}$, where $d(\cdot)$ is distance function and λ is the threshold. $f(d(x, y))$ can be $(1 - d(x, y)/\max(d(x, y)))$.

Further, $FN_X = \{FN(x_1), FN(x_2), .., FN(x_n)\}$, i.e., the set of fuzzy neighborhoods of data samples forms the **fuzzy neighborhood covering** of the data space X.

A binary **fuzzy relation** R between crisp sets X and Y is a fuzzy subset of $X \times Y$, i.e.,$\{((x, y), \mu_R(x, y)) | \ x \in X, y \in Y, R \subseteq X \times Y, \mu_R(x, y) \in [0, 1]\}$. The membership function indicates the intensity of the relation between x and y.

A **fuzzy constraint** [16] C is a n-ary fuzzy relational mapping from product of finite domain of variables to the [0,1] interval and is defined as:

$$C = \left\{ \begin{array}{l} (\langle x_1, x_2, \cdot\cdot, x_n \rangle, \mu_C(\langle x_1, x_2, \cdot\cdot, x_n \rangle)) \mid \langle x_1, x_2, \cdot\cdot, x_n \rangle \in D_1 \times D_2 \times \cdot\cdot \times D_n, \\ \mu_C(\langle x_1, x_2, \cdot\cdot, x_n \rangle)| \in [0, 1] \end{array} \right\} \tag{3.5}$$

We call the membership function $\mu_C(\langle x_1, x_2, \cdot\cdot, x_n \rangle)$ the *degree of satisfaction* of the constraint C. If $\mu_C(\langle x_1, x_2, \cdot\cdot, x_n \rangle) = 1$, then we say that $\langle x_1, x_2, \cdot\cdot, x_n \rangle$ *fully satisfies* C (crisp constraint). If $\mu_C(\langle x_1, x_2, \cdot\cdot, x_n \rangle) = 0$, then we say that $\langle x_1, x_2, \cdot\cdot, x_n \rangle$ *fully violates* C. We define the *support set* of C as $Sup(C) = \{\langle x_1, x_2, \cdot\cdot, x_n \rangle | \ \mu_C(\langle x_1, x_2, \cdot\cdot, x_n \rangle) > 0\}$. For elements which satisfy the constraint better than a given threshold α, where $1 \geq \alpha > 0$, the $\alpha - support$ set is defined as follows: $Sup(C)_\alpha (or\ Sup(C)[\alpha]) = \{\langle x_1, x_2, \cdot\cdot, x_n \rangle | \ \mu_C(\langle x_1, x_2, \cdot\cdot, x_n \rangle) \geq \alpha\}$.

3.2 Fuzzy Logic, Linguistic Variables and Fuzzy System

Fuzzy logic [10] can be defined as *a many-valued logic in which the truth value of variables can be expressed through linguistic terms and can be* any real number between 0 and 1. It is employed to handle the concept of *partial truth*, where the truth value may range between *entirely true* and *completely false*. Fuzzy logic is an approach to computing based on *degrees of truth* rather than on *true* or *false* (1 or 0) as in Boolean Logic. Fuzzy logic is based on the observation that people make decisions based on imprecise and non-numeric information.

A **fuzzy or linguistic variable** [10, 17–19, 21, 23] is a variable that can accept non-numeric values, i.e., linguistic terms to facilitate the expression of facts and rules. Because natural languages seldom contain enough value terms to express a fuzzy value scale [22], it is common practice to modify linguistic values with adjectives or adverbs.

In general, a linguistic variable has values that are words, and the meanings of these words are fuzzy sets in a certain universe. The values of a linguistic variable are called *terms*. Typical examples of this type of variable could be age, height, intelligence, etc. Their values could be terms like young, high, clever, etc. Each term can be represented by a triangular (or trapezoidal) fuzzy number.

A **fuzzy system** [25] is a repository of fuzzy expert knowledge that can reason with data in vague terms instead of precise Boolean logic. The expert knowledge is a collection of fuzzy membership functions and **fuzzy rules,** used to infer an output based on input variables. Unlike in crisp logic, the antecedent and the consequence of a fuzzy rule can be true to a degree, instead of being entirely true or entirely false. **Fuzzy inference** can be viewed as decision-making with a fuzzy system employing fuzzy implications and fuzzy rules of inference. **Defuzzification** converts fuzzy values into crisp quantities, i.e., it maps a fuzzy set to a single point, given that the point belongs to the support of the fuzzy set.

3.3 Fuzzy Analytic Hierarchy Process

Fuzzy Analytic Hierarchy Process (FAHP) [33, 36] is a method that integrates Analytic Hierarchy Process (AHP) with fuzzy logic theory. FAHP is a multi-criterion technique that works similarly to the AHP method, but scales into a triangular fuzzy number scale for representing the relevance criteria. FAHP can be used to infer fuzzy weights from linguistic terms for fuzzy (linguistic) variables using either the *geometric mean* method [14], the *extent analysis* method, or the *entropy-based* method. These fuzzy weights can be further defuzzified.

3.4 Fuzzy Weighted Average

We have defined the **Fuzzy Weighted Average** as in [29] by applying Zadeh's extension principle with a α-cut to the non-fuzzy weighted average (*wa*). Let $X_1, X_2, \cdot\cdot, X_n$ (variables), and $W_1, W_2, \cdot\cdot, W_n$ (weights) be the fuzzy numbers, then the **fuzzy weighted average** (*fwa*) is an *interval* defined as:

$$fwa(X_1, X_2, \cdot\cdot, X_n, W, W_2, \cdot\cdot, W_n)[\alpha],\ 1 \leq i \leq n$$

$$= \{\ wa(x_1, x_2, \cdot\cdot, x_n, w_1, w_2, \cdot\cdot, w_n) |\, x_i \in X_i[\alpha], w_i \in W_i[\alpha]\}$$

$$= \{\sum_{i=1}^{n} x_i * w_i |\, x_i \in X_i[\alpha], w_i \in W_i[\alpha]; \sum_{i=1}^{n} w_i = 1\} \tag{3.6}$$

We will consider weights which are triangular fuzzy numbers, i.e., $W_i = (l_i, m_i, r_i)$. The condition of fwa-normality is $\sum_{i=1}^{n} m_i = 1$, i.e., their modal values add up to unity. When a set of fuzzy weights is not fwa-normal, they can be made normal by normalizing it. We have used the algorithm in [29] to compute the fuzzy weighted average.

3.5 Fuzzy C-means Clustering

Fuzzy c-means clustering (FCM) [34, 37] is a soft clustering [35] approach, where each data point is assigned a membership score that indicates the degree to which it belongs to a cluster. The primary advantage of fuzzy c-means clustering is that it allows gradual memberships of data points to clusters, measured as degrees in [0, 1]. This provides the flexibility to express overlapping membership, i.e., allowing data points to belong to more than one cluster.

3.6 Fuzzy Matching

Fuzzy matching, also known as Approximate String Matching, is a technique that helps identify two elements of text, strings, or entries that are approximately similar but not exactly the same. It is called fuzzy because of the inexact nature of the matching involved. The closeness of a match, known as *edit distance*, is measured in terms of the number of primitive operations necessary to convert the string into an exact match.

4 Fuzzy Record Linkage

This section describes a fuzzy linkage technique that we are proposing for linking database records. This record linkage technique **uniquely** integrates approximate (fuzzy) string/pattern matching, fuzzy logic, search space reduction through fuzzy constraints, fuzzy weighted linkage estimation using fuzzy (linguistic) weights, fuzzy rule-based inferences, and fuzzy clustering, all through a 5-step sequential procedure/method.

The procedure is highly parameterized. The list of parameters to the proposed procedure includes: *Constraint Type, Field & Value* (default: "None/False"), *Link Columns, Match Algorithm, Logic Type* (default: "Fuzzy"), *Thresholds* (default: 0.9), *Linkage Type* (default: "crisp"), *Crisp Weight Vector* (default: $< 1,1,1 >$), *Linguistic Terms* (default = ("low", "medium", "high"), *Fuzzy Number Scale, FAHP Method* (default: "geometric mean"), *Fuzzy Rule Base, Number of Fuzzy Clusters* (default: 3). Parameters will be italicized in text.

The procedure supports "crisp" or "fuzzy" *Constraint Type* "Boolean" or "Fuzzy" *Logic Type* and "crisp" or "fuzzy" *Linkage Type* for record linkages. "Boolean" *Logic Type* is associated only with "crisp" *Linkage Type* and *Crisp Weight Vector* (or default weights of 1 for the *Link Columns*) to establish record linkages. "Fuzzy" *Logic Type* is associated with either "crisp" or "fuzzy" *Linkage Type* and FAHP generated crisp (ignoring the *Crisp Weight Vector*) or fuzzy weights for the *Link Columns* to establish record linkages. FAHP [33, 36] is used to generate fuzzy weights from *Linguistic Terms* describing the relevance of the *Link Columns*, i.e., the columns to be linked in a record pair. Fuzzy rules from the *Fuzzy Rule Base* connect match scores of the linked columns to the total linkage score for the record pair by treating them all as fuzzy variables. "Fuzzy" *Logic Type* and "fuzzy" *Linkage Type* employ Mamdani type [25] fuzzy inference triggering fuzzy rules. The objective of the procedure is two-fold. First, to associate a total linkage score to each linked record pair, indicating the quality of the linkage so that it can be ranked individually if needed, and second, to cluster the linked record pairs into a pre-specified number of fuzzy clusters based on their linkage qualities so that the different clusters can be interpreted and treated differently as per user requirements. Figure 1 shows the steps of this procedure.

Next, we examine the key steps of the Fuzzy Record Linkage (FRL) procedure and discuss their implementations.

4.1 Linkage Model

This is the first step of the procedure, which sets up the linkage query using the input parameter values. It consists of the following two activities:

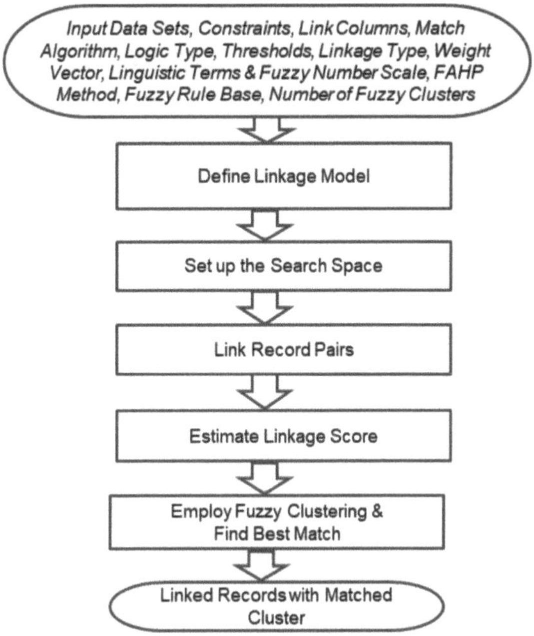

Fig. 1. Proposed Fuzzy Record Linkage Solution.

1. Define the linkage model with the parameters to the procedure, i.e., *Link Columns*, *Match Algorithm*, *Logic Type*, *Thresholds*, *Linkage Type*, etc.
2. Construct a linkage query with the linkage model using the input values to the parameters mentioned above.

The result provides definition and invocation of the linkage model. A linkage query can support either "Boolean" or "Fuzzy" *Logic Type* and "crisp" or "fuzzy" *Linkage Type*.

4.2 Search Space

This step sets up the constrained search space using the input data sets and the constraints (if any). The constraint can be crisp (exact) or fuzzy. This step consists of the following activities:

1. Merge the input data sets. With 2 data sets, e.g., $A \& B$ of sizes:$m \& n$; the search space is $A \times B$ and has $m * n$ (size) record pairs
2. Constrain (*Constraint Type*) the search space $\{(x, y) | x \in A \& y \in B\}$ using blocking with either crisp ($\mu_C(\langle x, y \rangle) = 1$) or fuzzy constraint, *i.e.*, its $\alpha - support$ set with $\mu_C(x, y) \geq \alpha$, defined through either crisp (exact) or fuzzy neighborhoods (i.e., fuzzy neighborhood covering), whose memberships ($\mu_{FN(x)}(y)$) can be determined using either the edit distances or the physical distances (in case of locations)

The result is a merged data set for a constrained search space consisting of paired records.

4.3 Record Pair Linkage

Our focus here is on linking record pairs through the match scores on the linked columns, whose original values are strings. So, this step links each record pair in the constrained search space by applying the selected fuzzy (approximate) string *Match Algorithm* on the *Link Columns* (columns to be linked) to generate the corresponding match scores in [0, 1] and then transforming them if needed, depending on the chosen *Logic Type*. It consists of the following two activities:

1. Execute the linkage query by applying fuzzy (approximate) string matching on the *Link Columns* to generate the respective match scores
2. If *Logic Type* is "Boolean" then
   a. For every column in *Link Columns* of a record pair, set the match score to 1, if match score $\geq$ match *Threshold*; otherwise set it to 0
   b. Ensure that *Linkage Type* is set to "crisp"

The result is the constrained data set from Sect. 4.2, with each linked record pair now additionally including either replaced (1 or 0) or raw match scores in [0, 1] for *Link Columns*.

4.4 Linkage Score Estimation

Linkage score estimation is the most critical step of this procedure. *Total linkage score* (*TS*) reflects the quality, i.e., the strength of record linkage. This step computes the total linkage score for each linked record pair from the match scores and the weights associated with the *Link Columns*, based on *Logic Type* and *Linkage Type*. If "Boolean" *Logic Type* is chosen, then *Linkage Type* can only be "crisp", and the normalized *Crisp Weight Vector* (or default weights of 1 for *Link Columns*) is used to compute the total linkage score of each record pair. FAHP [33, 36] is used only with the "Fuzzy" *Logic Type* for record linkages. It is used to estimate fuzzy weights (triangular fuzzy numbers) from the specified *Linguistic Terms*, reflecting the relative importance of the *Link Columns*, and the *Fuzzy Number Scale* (both parameters to the procedure). This relative importance of the *Link Columns* can come from either the customers (users) or the owner of the data sources, based on their experience and understanding of the data usage. Furthermore, if the chosen *Linkage Type* is "crisp," then the fuzzy weights for the *Link Columns* are defuzzified and normalized (ignoring the *Crisp weight Vector*) for computing the weighted average of the match scores reflecting the total linkage score for the record pair. For example, if two data sets are to be linked on three columns: Name, Address, and City and their deemed relevancies are "low", "medium" and "high", then FAHP with a 3-point number scale can yield their fuzzy weights as (0.1, 0.16, 0.33), (0.15, 0.3, 0.6), (0.28, 0.54, 0.96), which can be further defuzzified, if needed, into 0.17, 0.31, and 0.52 to serve as their normalized crisp weights. However, if the *Linkage Type* is "fuzzy," then fuzzy weights (from FAHP) and match score intervals of the *Link Columns*, represented as triangular fuzzy numbers, are used to estimate the fuzzy weighted average to define the interval of the total linkage score of the linked record pair, also as a triangular fuzzy number. The match scores of the *Link Columns* and the total linkage score of the linked record pair become fuzzy variables whose values can be *Linguistic Terms*, which are fuzzy sets represented by triangular fuzzy numbers based on the intervals and distributions of their original values. Fuzzy rules are used to connect the fuzzy variables, i.e., fuzzy match scores of the *Link Columns* in the antecedent to the fuzzy weighted average (reflecting the total linkage score) in the consequence. A Mamdani type [25] fuzzy controller applying fuzzy inference on these fuzzy rules can then infer a defuzzified total linkage score (*DTS*) for each linked record pair from the raw match scores of the *Link Columns*. In this context, we establish a formal framework for the two types of record linkages supported by the proposed procedure.

Definition 4.1: *A crisp linkage is a record linkage whose quality is estimated as a weighted average (wa) of match scores of the linked columns, where the match scores and weights are both crisp numbers.*

Definition 4.2: *A fuzzy linkage is a record linkage whose quality is estimated as a fuzzy weighted average (fwa) of match scores of the linked columns, where the match scores and weights are both fuzzy numbers.*

Let $S = \langle s_1, s_2, \cdot\cdot, s_n \rangle$ and $W = \langle w_1, w_2, \cdot\cdot, w_n \rangle$ be the match scores (in [0, 1]) and normalized crisp weights of the *Link Columns* $\langle x_1, x_2, \cdot\cdot, x_n \rangle$ and $\langle y_1, y_2, \cdot\cdot, y_n \rangle$ of the linked record pair (x, y) of the input sets A and B, where $MatchAlgorithm(x_i, y_i) = s_i, 1 \leq i \leq n$.

Proposition 4.1: *Crisp linkage is a mapping from $A \times B$ to $\mathbb{R}$, which results in a crisp relation defined as:*

$$CL = \{((x, y), TS) | \, x \in A, y \in B, TS \in \mathbb{R}; TS = wa(s_1, s_2, \cdot\cdot, s_n, w_1, w_2, \cdot\cdot, w_n) = S \cdot W\}$$

Let $\langle FS_1, FS_2, \cdot\cdot, FS_n \rangle$ and $\langle FW_1, FW_2, \cdot\cdot, FW_n \rangle$ be the fuzzy match scores & fuzzy weights of the *Link Columns* $\langle x_1, x_2, \cdot\cdot, x_n \rangle$ and $\langle y_1, y_2, \cdot\cdot, y_n \rangle$ of a linked record pair (x, y) of the input sets A and B.

Proposition 4.2: *Fuzzy linkage is a mapping from $A \times B$ to the* [0, 1] *interval, which results in a fuzzy relation defined as:*

$$\begin{aligned} FL = \{((x, y), \mu_{FL}(x, y)) | \, x \in A, y \in B, \mu_{FL}(x, y) \in [0, 1], \\ \mu_{FL}(x, y) = \mu_{TS}(s), s \in \mathbb{R}; TS \\ = fwa(FS_1, FS_2, \cdot, FS_n, FW_1, FW_2, \cdot\cdot, FW_n)[\alpha], \alpha \in [0, 1]\} \end{aligned}$$

Let S_{CL} & S_{FL} represent the intervals of total linkage scores (TS) for crisp and fuzzy linkages, and let $\subset$ denotes a sub-interval, then the following can be remarked about the intervals in our context.

Remark 4.1: $S_{CL} \subseteq [0, 1]$

Proof: The total linkage score for *crisp linkage* is estimated as the weighted average of match scores (s_i) and normalized crisp weights (w_i) of the *Link Columns*, i.e., $\sum_{i=1}^{n} s_i * w_i$. Since $0 \leq s_i \leq 1$, and $\sum_{i=1}^{n} w_i = 1$, we have $0 \leq \sum_{i=1}^{n} s_i * w_i \leq 1$. This implies that $S_{CL} \subseteq [0, 1]$.

Remark 4.2: $S_{FL} \subseteq [0, 1]$

Proof: The total linkage score for *fuzzy linkage* is estimated as the fuzzy weighted average (*fwa*) of fuzzy match scores (FS_i) and fuzzy weights (FW_i) of the *Link Columns*. The α-cuts of $FS_i \& FW_i$ are intervals and let them be denoted by $[FS_{i\alpha-}, FS_{i\alpha+}]$ and $[FW_{i\alpha-}, FW_{i\alpha+}]$. From Eq. (3.6), for $0 \le \alpha \le 1$, $fwa(FS_1, FS_2, \cdot\cdot, FS_n, FW_1, FW_2, \cdot\cdot, FW_n)[\alpha]=\{\sum_{i=1}^{n} fs_i * fw_i \mid fs_i \in [FS_{i\alpha-}, FS_{i\alpha+}], fw_i \in [FW_{i\alpha-}, FW_{i\alpha+}]; \sum_{i=1}^{n} fw_i = 1\}$ is also an interval. Now, it has been shown in [27] that this interval for *fwa* can be given by $[\sum_{i=1}^{n} FS_{i\alpha-} * FW'_{i\alpha-}, \sum_{i=1}^{n} FS_{i\alpha+} * FW'_{i\alpha+}]$ where $\forall i, FW_{i\alpha-} \le FW'_{i\alpha-} \le FW_{i\alpha+}, FW_{i\alpha-} \le FW'_{i\alpha+} \le FW_{i\alpha+}$ and $FW'_{i\alpha-} \le FW'_{i\alpha+}$. Now, $FS_{i\alpha-}$ and $FS_{i\alpha+}$ are elements of α-cuts of fuzzy set FS_i, defined over the interval [0, 1] of s_i. So, $\forall i, 0 \le FS_{i\alpha-} \le 1$ and $0 \le FS_{i\alpha+} \le 1$. Also, by definition of *fwa*, we have $\sum_{i=1}^{n} FW'_{i\alpha-} = 1$ and $\sum_{i=1}^{n} FW'_{i\alpha+} = 1$. This yields $0 \le \sum_{i=1}^{n} FS_{i\alpha-} * FW'_{i\alpha-} \le 1$ and $0 \le \sum_{i=1}^{n} FS_{i\alpha+} * FW'_{i\alpha+} \le 1$. So, the interval of *fwa* is bounded by 0 and 1, which are its endpoints, i.e., $[\sum_{i=1}^{n} FS_{i\alpha-} * FW'_{i\alpha-}, \sum_{i=1}^{n} FS_{i\alpha+} * FW'_{i\alpha+}] \subseteq [0, 1]$. This implies $S_{FL} \subseteq [0, 1]$.

From Remarks 4.1 & 4.2, we conclude that S_{CL} and S_{FL} are both bounded by [0, 1]. Finally, this step consists of the following activities:

1. If *Logic Type* is "Fuzzy", use *FAHP* [33, 36] to infer fuzzy weights (triangular fuzzy numbers) associated with the *Linguistic Terms* (for the column-relevance) using the *Fuzzy Number Scale* and the specified *FAHP Method*
2. If *Linkage Type* is "crisp" then
   a. If *Logic Type* is "Boolean", then either use the normalized *Crisp Weight Vector* (or default weights of 1 for *Link Columns*); else if *Logic Type* is "Fuzzy", then defuzzify the fuzzy weights (step 1) to generate normalized crisp weights for the *Linguistic Terms*, describing each linked column's relevance/importance
   b. Calculate total linkage score for each linked record pair as $TS = S \cdot W$, along with the weighted score ($s_i * w_i$) of each linked column i, where $S =< s_1, s_2, \cdot\cdot, s_n >$ be score vector and $W =< w_1, w_2, \ldots, w_n >$ be the normalized *Crisp Weight Vector* for the *Link Columns* (assuming identical names for both records)
3. If *Linkage Type* is "fuzzy" then
   a. Extract the α*-cuts* from the triangular fuzzy numbers representing score intervals (with modes) and the fuzzy weights of the *Link Columns* using Eq. (3.4)
   b. Compute total linkage score (*TS*) as a fuzzy weighted average of the α*-cuts* of the fuzzy scores and fuzzy weights using Eq. (3.6) and represent it as a triangular fuzzy number (interval) [e.g., fuzzy match scores of [(0, 0.25, 1), (0.23, 0.55, 1), (0, 0, 1)] and fuzzy weights of [(0.1, 0.16, 0.33), (0.15, 0.3, 0.6), (0.28, 0.54, 0.96)] for the 3 *Link Columns*: Name, Address, and City of a linked record pair produce a fuzzy weighted average of (0.04, 0.21, 1) using Eq. (3.6)]
   c. Define fuzzy sets for the *Linguistic Terms* using sub-intervals from the intervals and distributions of the match scores of the *Link Columns*, as well as from the computed interval for *TS* (from 3. b.)
   [e.g., Fig. 2 shows 3 fuzzy sets for 3 *Linguistic Terms*: "low", "medium" and "high" covering the full intervals of match scores of the 3 *Link Columns*: Name,

Address, City, and *TS* (Score) of the linked record pairs; for the 3 *Link Columns*, the triangular fuzzy numbers representing the 3 fuzzy sets for the 3 *Linguistic Terms* appear as identical as the match scores of *Link Columns* have the same range and similar distributions]

d. Define a Mamdani type [25] fuzzy controller [38] using the fuzzy sets for the match scores, *TS* and the fuzzy rules (from *Fuzzy Rule Base*) connecting the linked column scores to the total linkage score [e.g., a fuzzy rule based on Fig. 2, using the 3 *Link Columns* may have the form:
rule1 ← Rule(Name['high'] and Address['high'] and City['low'], Score['medium']), where Name, Address, City and Score are fuzzy variables, taking values from *Linguistic Terms*: "low", "medium" and "high". Fig. 3a shows the structure of a fuzzy rule, where the nodes represent antecedent, consequence, and terms]
e. Simulate the controller and employ fuzzy inference to estimate defuzzified total linkage score (*DTS*) for each linked record pair from the match scores of *Link Columns* [Fig. 3b shows the defuzzified total linkage score, i.e., *DTS* (Score) for a linked record pair]

The result is the data set from Sect. 4.3, with each linked record pair now additionally including weighted match scores for the *Link Columns* (crisp linkage) and the estimated total linkage score (*TS*/*DTS*).

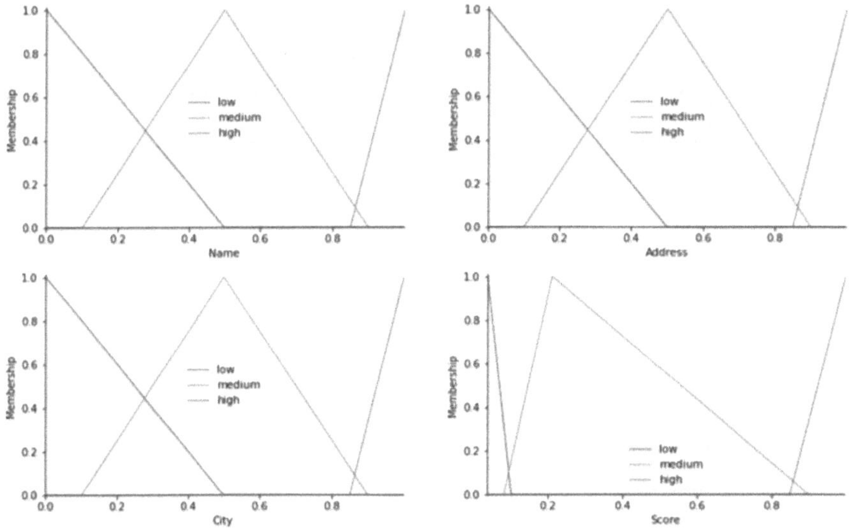

Fig. 2. Fuzzy Sets for Linguistic Terms describing the match scores of *Link Columns* and Total Linkage Score/*TS* (fuzzy variables: City, Address, Name, and Score).

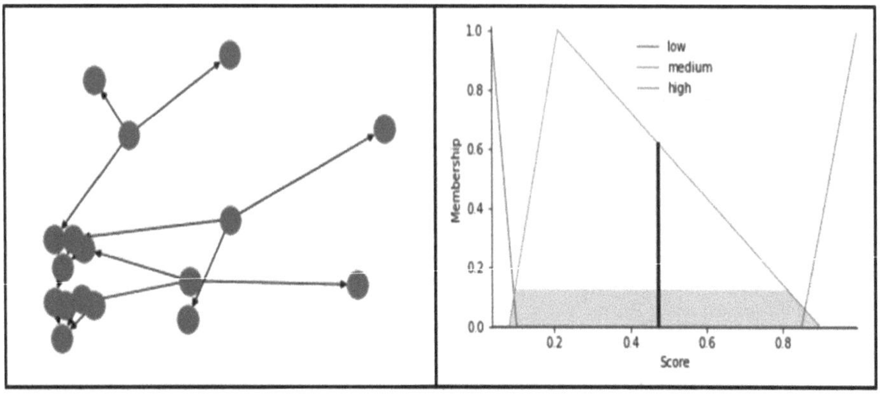

Fig. 3a Fig. 3b

Fig. 3. A Fuzzy Rule from *Fuzzy Rule Base* (Fig. 3a) and an estimated Defuzzified (Black Line from fuzzy set "medium" to raw score on "Score" axis) Total Linkage Score (*DTS*) for a record pair (Fig. 3b).

4.5 Fuzzy Clustering

This is the last step of the main procedure and consists of the following two activities:

1. Use fuzzy c-means clustering on the estimated total linkage scores (*TSs/DTSs*) from the previous step (Sect. 4.4) and generate specified number (*Number of Fuzzy Clusters*) of overlapping clusters
2. For each linked record pair, identify the cluster with the highest membership value as the best match

The result is the data set from Sect. 4.4, with each linked record pair now including the *best linguistic cluster* to which it belongs. Algorithm 1 below formalizes the entire above-mentioned procedure.

Algorithm 1

Algorithm for Fuzzy Record Linkages

Inputs: : *Input Sets (A, B), Constraint (C, λ), Link Columns (linkColumns), Match Algorithm (MA), Logic Type (logicType), threshold (th), Linkage Type (linkageType), Linguistic Terms (terms: {"low", "medium", "high"}, Crisp Weight Vector (CW), Fuzzy Number Scale (ns), FAHP-method (FAHP), FWA-α (α), Fuzzy Rule Base (FRB), Number of Fuzzy Clusters (k)*

Output: Linked record pairs with designated fuzzy cluster

FRL($C, linkColumns, MA, logicType, linkageType, \lambda, th, \alpha, terms, CW, ns, FAHP, FRB, k$)

$n \leftarrow |\, linkColumns\,|$
$x \leftarrow \langle x_1, x_2, \cdots, x_n \rangle$, where $x_i \; \varepsilon \; linkCcolumns$ of Input Set *A*
$y \leftarrow \langle y_1, y_2, \cdots, y_n \rangle$, where $y_i \; \varepsilon \; linkCcolumns$ of Input Set *B*
$XY \leftarrow \{(x,y) \mid \mu_C(x,y) \geq \lambda, (x,y) \in A \times B, y \in FN(x),$
$\mu_C(x,y) \in [0,1], \mu_{FN(x)}(y) \in [0,1],$
$\mu_C(x,y) \leftarrow \mu_{FN(x)}(y),\; \mu_{FN(x)}(y) \leftarrow (1 - d(x,y)/d_{max}),\, d_{max}$
$\leftarrow \max(d(x,y)\}$
$m \leftarrow |\, XY\,|$
$S \leftarrow \{\langle s_1, s_2, \cdots, s_n \rangle_j \mid \langle s_1, s_2, \cdots, s_n \rangle \leftarrow MA(x,y),\; s_i \in [0,1], \quad 1 \leq i \leq n, 1 \leq j \leq m\}$
if $logicType$ == "Boolean" **then**
  $S \leftarrow \{\langle s_1, s_2, \cdots, s_n \rangle_j \mid s_i \leftarrow (0, if\ s_i \leq \text{th};\ 1, if\ s_i > th),$
  $1 \leq i \leq n, 1 \leq j \leq m\}$
  if $linkageType \neq$ "crisp" **then**
    $linkageType \leftarrow$ "crisp"
  if *CW* then
    $TSs \leftarrow \{TS_1, TS_2, \cdots, TS_m \mid TS_j \leftarrow S_j \bullet CW',\; S_j \leftarrow \langle s_1, s_2, \cdots, s_n \rangle_j;\ CW' \leftarrow$
    $\langle w_1', w_2', \cdots, w_n' \rangle,\; w_i' \leftarrow w_i / \sum_{i=1}^{n} w_i,\; CW \leftarrow \langle w, w_2, \cdots, w_n \rangle; 1 \leq i \leq n, 1 \leq j \leq m\}$
  else
    $TSs \leftarrow \{TS_1, TS_2, \cdots, TS_m \mid TS_j \leftarrow S_j \bullet W,\; S_j \leftarrow \langle s_1, s_2, \cdots, s_n \rangle_j,\; 1 \leq j \leq m,$
    $W \leftarrow \langle 1/n, 1/n, \cdots, 1/n \rangle\}$
else ($logicType$ == "Fuzzy")
  $w_i \leftarrow (a_i, b_i, c_i) \leftarrow FAHP(terms, ns),\; a_i \leq b_i \leq c_i, 1 \leq i \leq n$
  if $linkageType$ == "crisp" **then**
    $W \leftarrow \langle dw_1, dw_2, \cdots, dw_n \rangle \leftarrow defuzzify(w_i),\; 1 \leq i \leq n$
    $TSs \leftarrow \{TS_1, TS_2, \cdots, TS_m \mid TS_j \leftarrow S_j \bullet W, \quad S_j \leftarrow \langle s_1, s_2, \cdots, s_n \rangle_j\ 1 \leq j \leq m\}$
  else ($linkageType$ == "fuzzy")
    $s_{i'} \leftarrow (p_i, q_i, r_i), \quad p_i \leq q_i \leq r_i,\; p_i \leftarrow min(s_i),$
    $q_i \leftarrow mode(s_i), r_i \leftarrow max(s_i), 1 \leq i \leq n,$
    $TS \leftarrow (u, v, z) \leftarrow fwa(s_{1'}, s_{2'}, \cdots, s_{n'}, w_1, w_2, \cdots, w_n)[\alpha], u \leq v \leq z$

    for $i = 1\ to\ n$ **do**
      x_i["low"] $\leftarrow (d_i, d_i, e_i),\; d_i < e_i,\; p_i \leq d_i$
      x_i[medium] $\leftarrow (e_i, f_i, g_i), e_i < f_i < g_i$
      x_i["high"] $\leftarrow (g_i, h_i, h_i),\; g_i < h_i,\; h_i \leq r_i$

    TS["low"] $\leftarrow (l, l, m_1),\; l < m_1,\; u \leq l$
    TS["medium"] $\leftarrow (m_1, m_2, m_3), \quad m_1 < m_2 < m_3$
    TS["high"] $\leftarrow (m_3, o, o),\; m_3 < o,\; o \leq z$

    $r_i \leftarrow TS[t] \leftarrow x_1[t]\ and\ x_2[t]\ and \cdots and\ x_n[t], t \in terms$
    $FRB \leftarrow \{r_1, r_2, \cdots, r_{3^n} \mid 3^n$: number of rules}
    controller ← *fuzzyController(FRB)*
    linkage ← *simulation(controller)*
    $DTSs \leftarrow \{DTS_1, DTS_2, \cdots, DTS_m \mid DTS_j \leftarrow fuzzyInfer(linkage(S_j)),\; S_j$
    $\leftarrow \langle s_1, s_2, \cdots, s_n \rangle_j, 1 \leq j \leq m\}$

$u \leftarrow FCM(TSs/DTSs, k)$, where k is the number of clusters
$cluster\_memberships \leftarrow argmax(u, axis = 0)$

5 Performance Evaluation

We have evaluated the performance of the proposed fuzzy record linkage method on benchmark US hospital data sets and compared its performance with those of deterministic and probabilistic linkage methods on the same data sets using both crisp and fuzzy constraints as well as Boolean and Fuzzy logic.

5.1 Experimental Setup

We implemented the fuzzy record linkage method in a *python* based *Jupyter Notebook*, using packages like *pandas, numpy, swifter, recordlinkage* and *scikit-fuzzy (skfuzzy)*. For the fuzzy controller and the clustering algorithm, we used the *control* and the *cluster.cmeans* modules from *skfuzzy*. We tested the proposed method on two publicly available benchmark data sets of US medical facilities [39], one consisting of **5339** records and the other consisting of **2697** records. The first was an internal data set that contained basic hospital account number, name and ownership information. The second data set contained hospital information (called provider) as well as the number of discharges and Medicare payments for specific "heart failure" procedures. The constrained search space, comprising either "exact same state" (crisp constraint) based record pairs from the two data sets or the fuzzy neighborhoods (using edit distances) of the "same state" (fuzzy constraint) based record pairs, included 475,830 and 998,860 record pairs, respectively. The records from the two data sets were exactly matched on City and approximately matched (fuzzy string matching) on 2 other columns, namely, Hospital Name and Hospital Address using *Levenshtein* and *Jaro-Winkler* edit distances. We used both Boolean and Fuzzy logic as well as crisp and fuzzy linkages. We considered relevance criteria of "low", "medium" and "high" for Hospital Name, Hospital Address and City respectively, for estimating their fuzzy weights. Accordingly for fuzzy linkages, we defined 3 fuzzy sets for each of the 3 *Link Columns* based on the distribution of their match scores, as shown in Fig. 2. We inputted *Fuzzy Rule Base* associating the three *Link Columns* (fuzzy variables) with a fuzzy score variable and employed a Mamdani type fuzzy controller [25, 38] to infer defuzzified total linkage score (*DTS*) for each record pair (Fig. 3b). We applied fuzzy c-means clustering algorithm on the estimated total linkage scores (*TSs/DTSs*) of the linked record pairs and generated 3 overlapping clusters, namely, *Matches, Possible Matches,* and *Non-matches.* We compared our model's performance with that of open-source, off-the-shelf packages like *recordlinkage* (exact deterministic) [40] and *splink* (probabilistic) [41] only on the crisply constrained search space, as these packages did not support fuzzy constraints. For non-package-specific strategy comparisons, we used the fuzzily constrained search space.

5.2 Evaluation Metrics

We evaluated our method for its *effectiveness*. For evaluation of the *effectiveness* of our method, we have considered the number of *Matches* and *Possible Matches* amongst the linked record pairs in the constrained search spaces. We have compared our method with deterministic and probabilistic linkages on only the number of *Matches.* Note that deterministic linkage results in discrete *all-or-nothing* outcome. Therefore, there are only

Matches or *Non-matches* but no *Possible Matches.* For probabilistic linkage, we have considered 0.5 (model standard) as cut-off probability threshold and a probability ≥ 0.5 as a *Match.* For us, *Non-matches* were those record pairs that neither *matched* nor *possibly matched*, i.e., (*Total number of linked record pairs* −(*Matches* + *PossibleMatches*)) They are not shown in Figs. 4 and 5 as their numbers were much larger compared to the other two clusters and were not the focus of our interest.

5.3 Discussion on Results and Model Comparisons

Figure 4 compares the performances of the proposed method with those of *recordlinkage* (for exact deterministic) and *splink* (for probabilistic) packages on the crisply constrained search space consisting of **475,830** linked record pairs. For the proposed solution, we used: i) "Boolean" logic to evaluate the results of approximate string matching on the *Link Columns,* to be consistent with the results from the other packages (any match score ≥ 0.9 was a *Match* and resulted in a score of 1), and ii) "crisp" linkage with both user provided weights as well as the default weights of 1 (1/3 as normalized) for *Link Columns* to compute the total linkage score (*TS*) for each record pair. For user provided weights, we used the normalized defuzzified weight vector of $< 0.17, 0.31, 0.52 >$ from the *Linguistic Terms* of "low", "medium" and "high" for Hospital Name, Hospital Address and City. Figure 4 shows that the proposed solution, with user provided and default weights, found i) 2669 and 2701 *Matches*, which were in between 2207 and 2740, the number of *Matches* found by *recordlinkage* and *splink,* as well as additionally ii) 6862 & 6882 *Possible Matches* in the two cases.

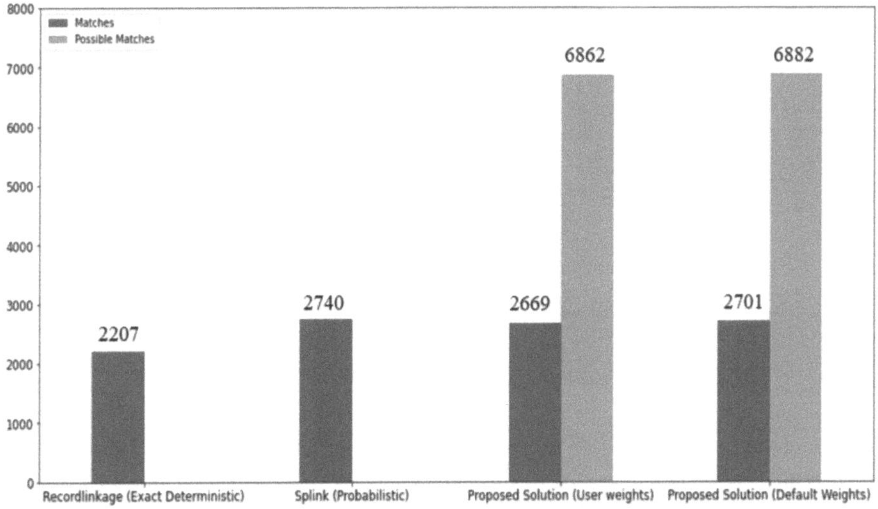

Fig. 4. Comparison with Existing Packages (Crisp Constraint).

Figure 5 compares the performances of the proposed method with those of deterministic and probabilistic linkages on the fuzzily constrained search space consisting of **998,860** linked record pairs. We implemented deterministic linkages by using exact string matching (on all *Link Columns*), "Boolean" *Logic Type*, "crisp" *Linkage Type* with the default weight of 1 for each linked column and estimated the total linkage score (*TS*) for each record pair as the average of the match scores of the *Link Columns*. For probabilistic linkages, we calculated the *m-* and *u-probabilities* from a small sample of the constrained search space and then computed a probabilistic weight vector for the *Link Columns* using these probabilities. Subsequently, we estimated the total linkage score (*TS*) for each record pair as the dot product of the probabilistic weight vector and the raw match scores from the string matches for that pair. For proposed crisp linkages (using "Fuzzy" *Logic Type* and "crisp" *Linkage Type*), we estimated the total linkage score (*TS*) of each record pair using steps 2. a. & 2. b. of the Linkage Score Estimation algorithm (Sect. 4.4); while for proposed fuzzy linkages (using "Fuzzy" *Logic Type* and "fuzzy" *Linkage Type*), we estimated the same and defuzzified (*DTS*) using steps 3. a. – 3. e. of the same algorithm. Figure 5 shows that the proposed solution found 9367 & 2459 *Matches*, which were in between 2207 and 99841, the number of *Matches* found by deterministic and probabilistic linkages respectively. Besides, it also displays that the proposed solution found 392792 & 8460 *Possible Matches* in the two cases.

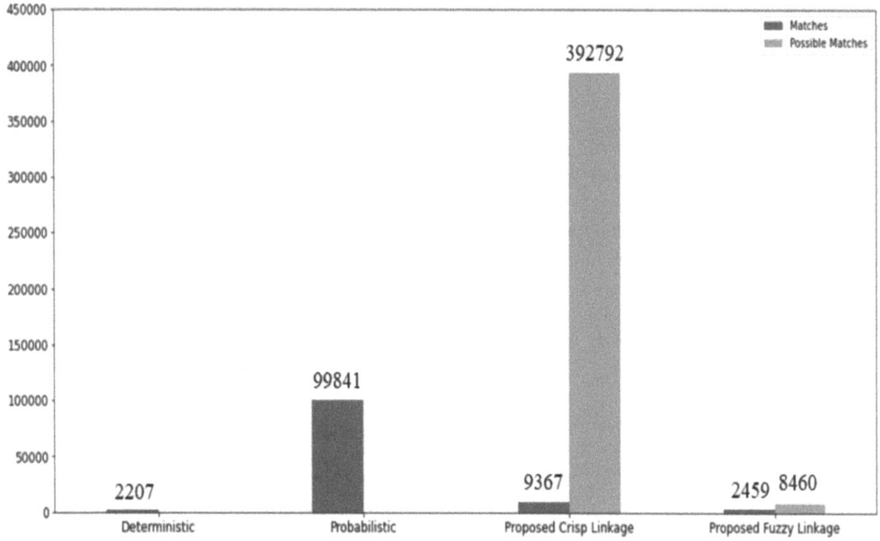

Fig. 5. Comparison with Existing Strategies (Fuzzy Constraint).

Figure 6 shows the results of proposed fuzzy linkages of two record pairs from Hospital Accounts and Hospital Reimbursements (i.e., the two tables used in this experiment) around the boundary between the two clusters, i.e., *Matches* and *Possible Matches*, based on the *Link Columns*: Hospital Name, Hospital Address and City in the order of importance: "low", "medium" and "high". The record pair at the top represents a member assigned to the cluster of *Matches* with the *lowest* defuzzified total linkage score (*DTS*), while the pair at the bottom represents a member assigned to the cluster of *Possible*

Matches with the *highest* defuzzified total linkage score (*DTS*). Note that the record pair at the top matches *fully* on City, Hospital Address and *partially* on Hospital Name. The record pair at the bottom also has *full* matches on City and Hospital Address and only a *partial* match on Hospital Name. However, the record pair at the top matches on more words than the pair at the bottom (3 against 1, with the name on the right being fully included in the name on the left in the top pair) on the Hospital Name, which has the lowest importance. Hence, the two record pairs have eventually ended up in two different clusters at their boundary.

Match			
Facility Name	THE HOSPITALS OF PROVIDENCE MEMORIAL CAMPUS	Provider Name	PROVIDENCE MEMORIAL HOSPITAL
Address	2001 N OREGON ST	Provider Street Address	2001 N OREGON ST
City	EL PASO	Provider City	EL PASO
State	TX	Provider State	TX
ZIP Code	79902	Provider Zip Code	79902
County Name	EL PASO	Total Discharges	39
Phone Number	(915) 577-6011	Average Covered Charges	51544.7
Hospital Type	Acute Care Hospitals	Average Total Payments	6954.26
Hospital Ownership	Proprietary	Average Medicare Payments	6050.74
Possible Match			
Facility Name	ADVENTHEALTH WATERMAN	Provider Name	FLORIDA HOSPITAL WATERMAN
Address	1000 WATERMAN WAY	Provider Street Address	1000 WATERMAN WAY
City	TAVARES	Provider City	TAVARES
State	FL	Provider State	FL
ZIP Code	32778	Provider Zip Code	32778
County Name	LAKE	Total Discharges	152
Phone Number	(352) 253-3300	Average Covered Charges	24835.8
Hospital Type	Acute Care Hospitals	Average Total Payments	5596.34
Hospital Ownership	Voluntary non-profit - Church	Average Medicare Payments	4665.42
Hospital Accounts		**Hospital Reimbursements**	

Fig. 6. Match and Possible Match of record pairs between Hospital Accounts & Hospital Reimbursements.

6 Conclusion

In this paper, we have presented a fuzzy record linkage technique to address some of the challenges associated with deterministic and probabilistic record linkages, with regards to their low true positives or high false positives (false negatives), under uncertainty with imprecise information. We have combined fuzzy string matching, fuzzy set theory, fuzzy logic, fuzzy constraints, fuzzy weights, fuzzy rules and inferences with fuzzy clustering to generate clusters of *Matches* and *Possible Matches* of linked record pairs, to provide a solution that has several advantages over the existing ones. Firstly, it provides a highly parameterized, flexible solution with a wide range of choices. Secondly, it combines multiple strategies to benefit from their complementary advantages. Thirdly, it produces a more balanced, controllable and an objective solution to the desired level of granularity. The number of *Matches* can be expected or controlled to be between those of deterministic and probabilistic linkages to overcome shortcomings associated with either low true positives or high false positives. Lastly, it can serve multiple purposes with results of both *Matches* and *Possible Matches*, thereby better accommodating user interests. For example, a service provider using door-to-door marketing campaigns can target customers linked as *Matches*, whereas the same provider can target the *Possible Matches* for web or email based bulk promotions. It may be noted here that probabilistic linkage can also generate non-overlapping intervals of *Matches* and *Possible Matches*, using multiple (two) crisp cut-off probability thresholds. However, such classification would then become very subjective in nature, based purely on user's choice of the thresholds. Finally, we have established the effectiveness of the proposed record linkage solution through extensive evaluations and performance comparisons.

Currently, efforts are underway to link and unify Verizon customers and prospects, associated with multiple disparate data sources, with global identification using the proposed technique.

Disclosure of Interests. The author states that he was an employee of *Verizon,* and this paper addresses work performed in course of the author's employment.

References

1. Herzog, T.N., Scheuren, F.J., Winkler, W.E.: Data quality and record linkage techniques, vol. 1. Springer (2007)
2. Winkler, W.E.: Record Linkage. Encyclopedia of Environmetrics. J. Wiley (2013)
3. Fellegi, I.P., Sunter, A.B.: A theory for record linkage. J. Am. Stat. Assoc. **64**(328), 1183–1210 (1969)
4. Christen, P.: Data matching: concepts and techniques for record linkage, entity resolution, and duplicate detection. Springer Science & Business Media (2012)
5. Abriel, D., Navarro-Arribas, G., Torra, V.: Choquet integral for record linkage. Ann. Oper. Res. **195**, 97–110 (2012)
6. Ouyang, T., Pedrycz, W., Pizzi, N.J.: Record Linkage Based on a three-way decision with the use of granular descriptors. Expert Syst. Appl. **122**, 16–26 (2019)
7. Mierzejewski, K., Mazurek, M.: A New Framework for Assessing Similarity Measure Impact on Classification Confidence Based on Probabilistic Record Linkage Model. Procedia Manufacturing **44**, 245–252 (2020)
8. Kaufman, A.R., Klevs, A.: Adaptive Fuzzy String Matching: How to Merge Datasets with Only One (Messy) Identifying Field. Political Analysis **30**(4), 590–596 (2022). https://doi.org/10.1017/pan.2021.38
9. Ornstein, J.T.: Probabilistic Record Linkage Using Pretrained Text Embeddings, pp. 1–12. Published online, Political Analysis (2025)
10. Zadeh, L.A.: Fuzzy sets. Inf. Control **8**(3), 338–353 (1965)
11. Dubois, D., Prade, H.: Fuzzy Sets and Systems: Theory and Applications. Academic Press, New York (1980)
12. Novak, V.: Základy fuzzy modelování. BEN, Praha (2000)
13. Ali, M.Y., Sultana, A., Khan, A.F.M.K.: Comparison of Fuzzy Multiplication Operation on Triangular Fuzzy Number. IOSR Journal of Mathematics **12**(4), 35–41
14. Gupta, M., Bhattacharjee, D.: Multi objective problem in fuzzy environment where resources are triangular fuzzy number. Eur. J. Sci. Res. **46**(1), 99–106 (2010)
15. Yue, X.D., Chen, Y.F., Miao, D.Q., Fujita, H.: Fuzzy neighborhood covering for three-way classification. Inf. Sci. **507**, 795–808 (2020)

16. Ruttkay, Z.: Fuzzy constraint satisfaction. Proceedings of 1994 IEEE 3rd International Fuzzy Systems Conference, pp. 1263–1268 vol. 2 (1994)
17. Zadeh, L.A.: The concept of a linguistic variable and its application to approximate reasoning – I. Inf. Sci. **8**(3), 199–249 (1975)
18. Zadeh, L.A.: The concept of a linguistic variable and its application to approximate reasoning – II. Inf. Sci. **8**(4), 301–357 (1975)
19. Zadeh, L.A.: The concept of a linguistic variable and its application to approximate reasoning – III. Inf. Sci. **9**(1), 43–80 (1975)
20. Zimmermann, H.J.: Fuzzy Set Theory and its Applications. Kluwer Academic Publishers, Boston (1991)
21. Borovička, A.: Fuzzy weights estimation method based on the linguistic expression of criteria relevance. Central European Rev. Econ. Issu. **17** (2014)
22. Chen, S.J., Hwang, C.L.: Fuzzy Multiple Attribute Decision Making: Methods and Applications. Springer-Verlag, Berlin (1992)
23. Wang, M., Wang, L.: Fuzzy weight's calculation method. 2010 6th International Conference on Advanced Information Management and Service (IMS), pp. 172–175 (2010)
24. Bellman, R.E., Zadeh, L.A.: Decision making in a fuzzy environment. Manage. Sci. **17**(4), 141–164 (1970)
25. Mamdani, E.H.: Application of fuzzy algorithms for control of simple dynamic plant. Proceedings of the Institution of Electrical Engineers. **121**(12), 1585–1588 (1974)
26. Dong, W.M., Wong, F.S.: Fuzzy weighted averages and implementation of the extension principle. Fuzzy Sets Syst. **21**, 183–199 (1987)
27. Liou, T.S., Wang, M.-J.J.: Fuzzy weighted average: An improved algorithm. Fuzzy Sets Syst. **49**, 307–315 (1992)
28. Lee, D.H., Park, D.: An efficient algorithm for fuzzy weighted average. Fuzzy Sets Syst. **87**, 39–45 (1997)
29. Broek, P.V., Noppen, J.: Fuzzy Weighted Average: Alternative approach. NAFIPS - Annual Meeting of the North American Fuzzy Information Processing Society, pp. 126–130 (2006)
30. Guh, Y.-Y., Hon, C.-C., Wang, K.-M., Lee, E.S.: Fuzzy weighted average: A max-min paired elimination method. Comput. Math. Applicat. **32**, 115–123 (1996)
31. Liu, F., Mandel, J.M.: Aggregation using the Fuzzy Weighted Average as Computed by Kernik-Mandel Algorithms. IEEE Trans. on Fuzzy Sys. **16**(1), 1–12
32. Pavlacka, O., Talasová, J.: Applications of the Fuzzy Weighted Average of Fuzzy Numbers in Decision Making Models. EUSFLAT Conf. (2007)
33. Ahmed, F., Kilic, K.: Fuzzy Analytic Hierarchy Process: A performance analysis of various algorithms. Fuzzy Sets Syst. **362**, 110–128 (2019)
34. Bezdek, J.C.: Pattern Recognition with Fuzzy Objective Function Algorithms. Plenum Press, New York (1981)
35. Li, J., Lewis, H.W.: Fuzzy Clustering Algorithms — Review of the Applications. 2016 IEEE International Conference on Smart Cloud (SmartCloud), pp. 282–288 (2016)
36. Buckley, J.J.: Fuzzy Hierarchical Analysis. Fuzzy Sets Syst. **17**, 233–247 (1985)
37. Fuzzy c-means Clustering. https://pythonhosted.org/scikit-fuzzy/auto_examples/plot_cmeans.html
38. Fuzzy Controller. https://pythonhosted.org/scikit-fuzzy/userguide/fuzzy_control_primer.html
39. Practical Business Python. https://pbpython.com/record-linking.html?hmsr=joyk.com&utm_source=joyk.com&utm_medium=referral
40. Python Record Linkage Toolkit. https://recordlinkage.readthedocs.io/en/latest/
41. Splink: https://www.robinlinacre.com/introducing_splink/

Cyber Warfare, Cyber Defense, Cyber Security, and Post-Quantum Security

Taking Cyber Warfare to the Airwaves: Virtualized Electronic Warfare and Directed Energy Systems Using Sparse Arrays and Software Defined Radios

Jeffrey Wallace[1(✉)], Miroslav Vukovic[1], William Edwards[1], David Simpson[2], Jeffrey Reed[2], Nishith Tripathi[2], and Jason Hilker[3]

[1] Rocket Technology Systems LLC, 1717 K Street NW, Suite 900, Washington, DC 20006, USA
{jeff,miro,bill}@roc-tec.systems
[2] Virginia Polytechnic Institute and State University, 3625 Potomac Avenue, Alexandria, VA 22305, USA
{simpsond,reedjh,nishith}@vt.edu
[3] SMBG Services, LLP, 23035 Ruth Lane, Waynesville, MO 65583, USA
jason@smbgservices.com

Abstract. Software Defined Radios are enabling a new era of cyber warfare where programmable attacks can now be launched via radio frequency (RF) systems. As such, modern electronic warfare (EW) and directed energy weapon (DEW) systems are evolving toward distributed, software-defined architectures that can adapt to contested radio-frequency (RF) environments. This paper explores the theoretical and operational viability of virtualized EW/DEW systems built on sparse antenna arrays and software-defined radios (SDRs).

We present a rigorous theoretical foundation for sparse arrays, deriving key performance relationships that enable large-aperture effects with fewer elements. We model power-modulated jamming techniques for low probability of intercept (LPI), providing signal processing analyses of their interference patterns. We then describe an original prototype implementation of a stealth jammer/DEW node system using multiple NI/Ettus SDR platforms (X410, X310, N310, and B210), referencing internal design documents and operational scenarios in counter-sUAS and space-based interceptor applications. Results from our prototype and simulations indicate that a network of coordinated SDR-based nodes can coherently combine signals as a sparse array to achieve effective jamming and directed energy effects, while offering flexibility and survivability in contested RF environments.

The operational usefulness of distributed, virtualized EW/DEW capabilities is highlighted, showing promise for next-generation defense systems. We conclude that virtualized sparse-array EW/DEW is not only theoretically sound but also practical with current technology, laying a foundation for future research and deployment in military and security domains.

H. R. Arabnia et al. (Eds.): CSCI 2025, CCIS 2969, pp. 177–205, 2027.
https://doi.org/10.1007/978-3-032-28460-0_14

Keywords: Electronic Warfare · Directed Energy Weapon · sparse array algorithm · synthetic RF system/aperture · sub-nanosecond timing

1 Introduction

Electronic warfare (EW) and directed energy weapons (DEWs) are critical components of modern defense strategy, especially as the battlespace becomes increasingly contested in the RF spectrum. Traditional EW/DEW systems have often relied on large, monolithic platforms ? for example, single high-power jammers or dedicated laser systems. These approaches, while powerful, can lack flexibility and may be vulnerable to detection or suppression. In contrast, recent advances in software-defined radio (SDR) technology and networking enable a new paradigm: virtualized, distributed EW/DEW systems composed of many small, coordinated nodes. Such systems leverage concepts of sparse arrays and coherent signal combining to emulate a larger aperture or higher power system from multiple low-cost elements.

This paper investigates the viability of these emerging approaches. In particular, we focus on systems wherein multiple SDR-based nodes act in concert as a sparse phased array for jamming or directed energy effects. By ?virtualized,? we mean that the EW/DEW functionality is realized in software and can be dynamically allocated to different hardware nodes ? much like virtualization in computing ? allowing adaptive reconfiguration in response to threats. This concept aligns with broader defense trends emphasizing network-centric and distributed operations, which promise asymmetric advantages such as resilience, covertness, and scalability. Indeed, militaries worldwide are beginning to explore networked and distributed EW concepts for congested and denied environments, seeking benefits like coherent multi-node operations and disposable, attritable assets that can overwhelm adversaries [1].

Crucially, software-defined radios provide a hardware platform ideally suited for such experimentation and deployment. SDRs can be reprogrammed to transmit or receive across a wide range of frequencies and waveforms on the fly. As noted in a recent industry article, ?Software-defined radio (SDR) will continue to play a key role in c-UAS solutions for the foreseeable future.? SDR-based designs thus future-proof EW/DEW systems against evolving threats by enabling rapid updates to jamming waveforms or communication techniques. In counter-drone (c-UAS) scenarios, where small unmanned aircraft are a fast-growing threat, SDRs already underpin many solutions due to their flexibility and responsiveness [2]. Our work leverages SDRs to push this flexibility further ? into coordinated multi-node operations and even directed energy effects.

This paper makes both theoretical and practical contributions. On the theoretical side, we develop a foundation for sparse array beamforming in EW, including mathematical derivations of array patterns and power gain for non-uniform arrays. We also introduce and analyze power-modulated jamming techniques as a means of achieving LPI/LPD (low probability of intercept/detection) while still disrupting enemy sensors. On the practical side, we describe an original prototype system consisting of several SDR nodes (NI USRP family) that implement a stealthy jammer and DEW platform. We use a counter-sUAS prototype and a Space Force Golden Dome Space-Based Interceptor (SBI) ? to illustrate how such distributed systems can be applied in real operational scenarios. Finally, we discuss the operational usefulness of these distributed, virtualized EW/DEW capabilities in contested RF environments, identifying both the opportunities and the remaining challenges (such as synchronization and coordination overhead).

The remainder of this paper is organized as follows. In Sect. 2, Theoretical Foundations, we review sparse array theory and derive key equations relevant to distributed EW/DEW arrays, and we formulate the signal processing model for power-modulated jamming. Section 3, Implementation of a Virtualized EW/DEW Node System, details the hardware and software architecture of our prototype, including the SDR platforms used and how we achieved coherent operation. Section 4, Operational Scenarios, describes two use cases ? counter-sUAS defense and a space-based missile interceptor ?constellation? ? to demonstrate how the theory and prototype translate into real-world applications. Section 5, Discussion, assesses the results, advantages, and limitations of the approach. Finally, Sect. 6, Conclusion, summarizes our findings and outlines future research directions.

2 Theoretical Foundations

By distributing signal generation and control across multiple coordinated SDR nodes, we effectively synthesize a virtualized EW/DEW system architecture. This structure allows for spatially distributed beamforming and adaptive interference techniques without reliance on fixed infrastructure. Figure 1 illustrates the architecture underlying our distributed EW/DEW concept. A central controller manages coordination and virtualization logic, issuing tasking and synchronization data across a resilient network to remote SDR nodes. Each node acts as a spatial element in a sparse array, collectively synthesizing desired beam patterns or modulation effects. This architecture provides fault tolerance, reconfigurability, and enhanced survivability in contested RF environments.

2.1 Sparse Array Antenna Theory

In a conventional phased array, antenna elements are arranged in a regular grid (e.g., a uniform linear array, ULA) with spacing on the order of half the wavelength ($\lambda/2$) to avoid grating lobes. If an array of N elements has element spacing $d > \lambda/2$, a ULA will produce grating lobes ? essentially duplicate maxima in the radiation pattern ? which are undesirable as they can send energy in unintended directions or allow ambiguities in sensing. However, maintaining half-wavelength spacing for a large aperture requires a very large number of elements (since number of elements $\approx$ aperture size$/(\lambda/2)$). For example, to form an effective aperture dozens of meters across at X-band frequencies, a full ULA would need thousands of elements, which is cost-prohibitive and complex.

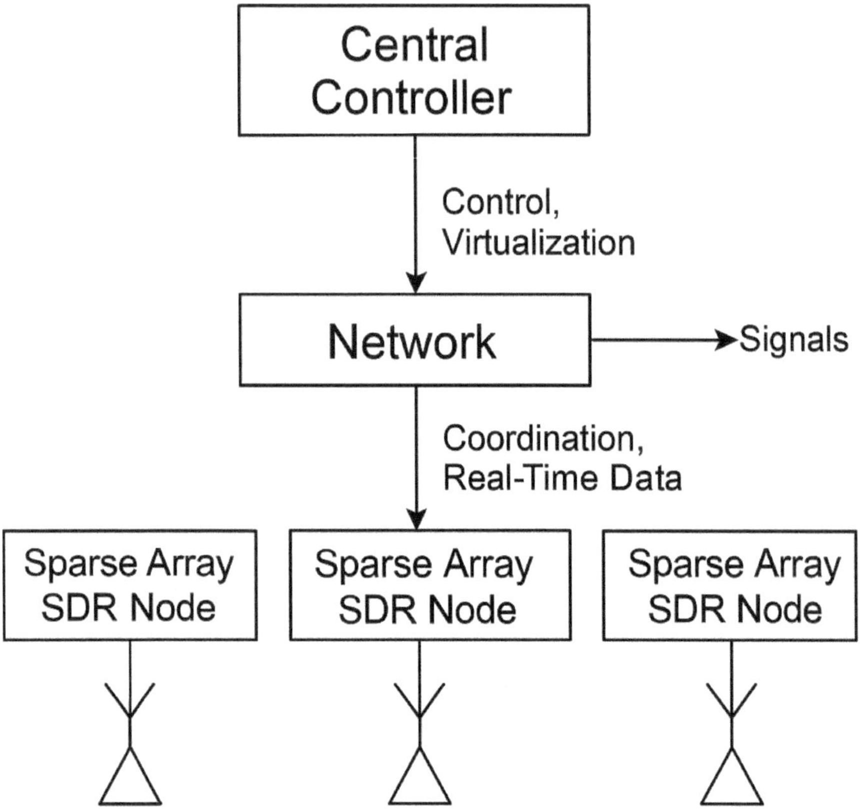

Fig. 1. Virtualized EW/DEW system architecture showing a central controller, networked communication layer, and multiple sparse array SDR nodes. The system enables coordinated signal generation and adaptive EW/DEW operations across distributed platforms.

Sparse arrays address this by using fewer elements spread over a larger aperture, typically with non-uniform spacing. The idea is that by strategically choosing the element positions, one can achieve almost the same resolution and aperture size as a filled array, but with far fewer elements. The trade-off is that the sidelobe pattern of the array becomes irregular and potentially higher in amplitude than a comparable ULA ? in other words, sparse arrays avoid true grating lobes but instead yield a complicated sidelobe structure that must be managed.

As stated in one comprehensive study, sparse arrays allow element separations greater than half a wavelength while still avoiding regular grating lobes due to the non-uniform element placement [3]. The cost of this is an unpredictable sidelobe behavior that necessitates careful optimization of the element layout and weighting [4].

Mathematically, consider an array of N transmitting elements located at positions $x_1, x_2, \ldots, x_N$ along a line (generalization to 2D or 3D arrays is straightforward). For simplicity, assume isotropic elements and equal weights (the extension to complex weights for beamforming is standard). The array factor (far-field radiation pattern gain relative to a single element) as a function of angle θ (measured from broadside) can be written as:

$$AF(\theta) = \sum_{n=1}^{N} e^{jkx_n \sin\theta}, \tag{1}$$

where $k = 2\pi/\lambda$ is the wavenumber. For a ULA with spacing d, $x_n = (n-1)d$ and this sum simplifies to a well-known sinc function form. But for a sparse, non-uniform array, there is no simple closed form ? $AF(\theta)$ will have an irregular pattern of maxima/minima. The peaks of $|AF(\theta)|$ define the main lobe and sidelobes of the radiation (or jamming) pattern. The beamwidth (main lobe width) is largely governed by the total aperture span $L = \max(x_n) - \min(x_n)$; roughly, the narrower the main lobe needs to be (for high angular resolution or narrow jamming focus), the larger L must be. This is the same as a filled array of length L, meaning a sparse array can approach the resolution of a filled array if spread over the same total length. Meanwhile, the sidelobe levels depend on the specific distribution of the x_n. A uniform or periodic spacing yields uniformly spaced grating lobes if $d > \lambda/2$. A non-uniform (aperiodic) spacing breaks the periodicity and thus avoids true grating lobes; instead, sidelobes appear at various angles with generally lower envelope. The goal in sparse array design is often to minimize the highest sidelobe (peak sidelobe level) and avoid large ?holes? in the effective aperture coverage.

Figure 2 illustrates the practical impact of using sparse arrays in virtualized EW/DEW systems. Although sparse configurations exhibit wider main lobes and slightly elevated side lobes, they enable substantial hardware simplification and spatial diversity?crucial advantages in distributed deployments. These trade offs are particularly beneficial in scenarios demanding rapid reconfiguration, resilience, and cost-effective scalability.

One way to analyze a sparse array is via the concept of a difference co-array. The set of all pairwise element separations (differences $x_i - x_j$) is called the co-array. For beamforming or direction-finding, these differences correspond to attainable phase shifts or time delays across the array. A filled ULA of N elements has a co-array that contains every multiple of the base spacing up to $(N-1)d$, with many redundancies. Sparse arrays seek to maximize the number of unique separations (to emulate a larger filled array) while minimizing redundancies (which waste elements). Minimum redundancy arrays (MRA) were one of the earliest formulations of this problem: an ideal MRA minimizes repeated spacings and covers as many distinct spacings as possible. For example, with M elements, the maximum number of unique spacings (degrees of freedom for beamforming) is $M(M-1)/2$ (if we count both positive and negative differences).

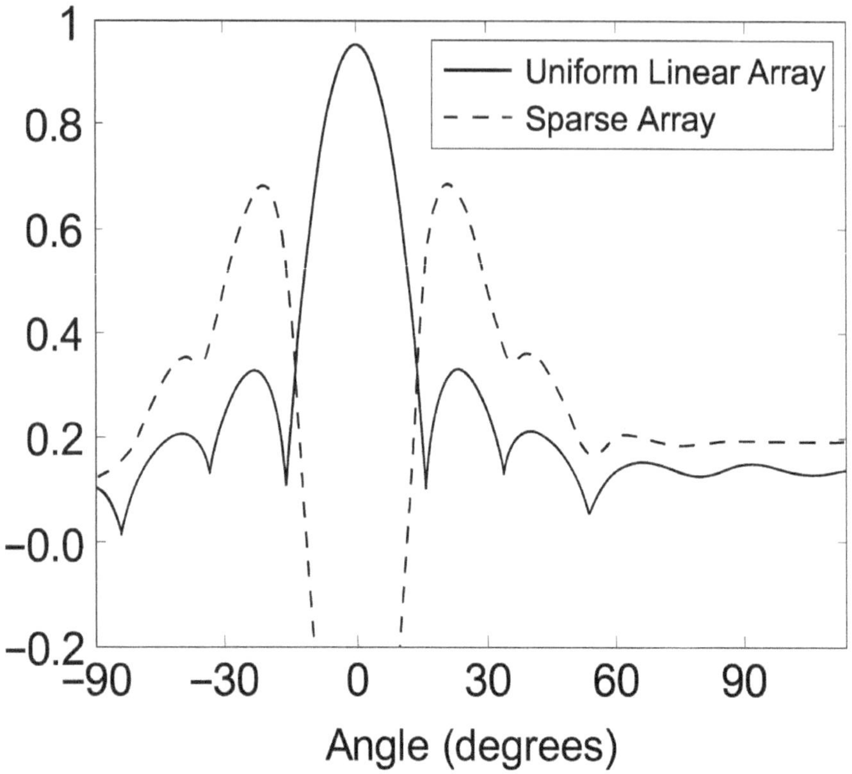

Fig. 2. Comparison of beam patterns between a Uniform Linear Array and a Sparse Array. While the uniform array offers narrower main lobes and tightly packed side lobes, the sparse array trades spatial resolution for lower hardware complexity and broader lobe coverage?demonstrating a viable trade off for distributed, virtualized EW/DEW systems.

An ideal sparse design would achieve close to this number of unique spacings, meaning it performs nearly like a fully populated aperture of M elements.

It is known that perfect MRAs (no redundant spacings, no missing spacings) are only possible for very small M (up to 4 elements in a linear array). Beyond that, any sparse array will have to accept either some redundancy or some ?holes? (missing spacings) in the co-array. Nevertheless, numerous designs ? such as minimum hole arrays, nested arrays, and coprime arrays ? have been devised to systematically construct sparse layouts that offer good performance with far fewer elements than a comparable filled array.

In summary, sparse arrays can drastically reduce hardware complexity for a given aperture size, which is very attractive for distributed EW/DEW. Instead of one platform carrying a dense array, we can have multiple smaller platforms each with a single (or few) elements, distributed over space. When their signals are combined coherently, they act as a sparse array that can achieve high gain (or narrow beams) equivalent to a large antenna. This is the principle that allows a cluster of networked SDR nodes to function as one giant jammer or sensor. However, to realize this in practice, two key requirements must be met: synchronization and calibration. All nodes must share a common timing/frequency reference to maintain phase coherence, and the relative phase offsets due to different propagation paths or hardware must be calibrated out [3]. In Sect. 3 we discuss how our implementation addresses these challenges.

2.2 Power-Modulated Jamming and LPI Considerations

One concern with conventional jamming is that a strong, steady jamming signal is easily detectable by adversaries. A continuously emitting jammer reveals its presence (and location, via direction-finding) due to its high power signature. To mitigate this, we explore power-modulated jamming, where the jammer intentionally varies its transmit power over time in a known pattern. The goal is to create intermittent interference that still disrupts enemy systems, but makes it harder for them to detect and track the jammer. This technique is inspired by LPI/LPD principles commonly used in stealth radars and communications, where transmission is spread out or varied to be less conspicuous.

We define a transmit power modulation function $P(t)$ for the jammer. For instance, $P(t)$ might be a periodic function like:

$$P(t) = P_0 \left[1 + m \cos(2\pi f_m t)\right], \tag{2}$$

where P_0 is the average power, $m \in [0, 1]$ is a modulation index (with $m = 1$ giving 100

From a signal processing perspective, power modulation is effectively amplitude modulation of the jamming signal (if the waveform shape remains the same and only amplitude varies). This will inevitably broaden the spectrum of the jamming signal: for example, if the jammer was transmitting noise centered at some frequency band, turning that noise on and off will introduce sidebands or a spreading of the noise spectrum. The benefit, however, is on the intercept side: an enemy trying to detect the presence of a jammer typically uses energy detection or looks for consistent interference. A power-modulated jammer might fall below detection thresholds intermittently or appear like random fluctuations. Indeed, varying the transmission power can make the jammer more difficult to detect on enemy sensors. The downside, as noted by prior work, is that whenever the jammer is at a low-power state, the enemy?s communications or sensors get a respite ? some of their signals might slip through the cracks. In other words, power modulation inherently reduces jamming effectiveness during the ?low? intervals, potentially allowing the target to communicate briefly. Thus, there is a trade-off between LPI stealthiness and jamming impact.

To quantify this trade-off, consider a simple metric: jammer-to-signal ratio (JSR) at the target receiver. Let the target?s own signal power (from its transmitter or a protected signal) be S. A constant jammer with power P yields a JSR $= P/S$ (ignoring path losses for simplicity). If we apply a duty cycle D

on-off modulation (for example), the target sees jammer power P for fraction D of the time and near 0 for fraction $(1 - D)$. Over a long averaging, the average JSR would be $D \cdot (P/S)$. However, if the target?s receiver or algorithm can exploit the off-times (for instance, by buffering or error-correcting through the outages), they might effectively communicate at those moments. Conversely, for detection of the jammer, an energy detector integrating over a certain window might miss the jammer if the duty cycle or modulation makes the energy sporadic. The detection probability depends on the integration time versus the jammer?s on/off periodicity. A savvy jammer might use randomized modulation so that the enemy cannot simply time their communications in the gaps.

Another form of power modulation is a carefully adaptive power control that reacts to the target?s behavior. For example, one could imagine a jammer that estimates the distance of a drone (via radar reflection or other means) and then modulates its power in a way to mislead the drone?s navigation (as explored in the JAMMING-ME concept by Tedeschi et al.). They propose a jammer that can do ?distance bounding? ? measuring the drone?s distance ? and then adaptive jamming: varying $P(t)$ to spoof the drone into thinking it?s either farther or closer than it really is [5,6]. While our work does not explicitly implement that advanced tactic, it underscores the potential richness of power modulation strategies.

In our prototype system, we implemented a basic form of power-modulated jamming: a pseudorandom pulse jamming waveform. The SDR-based jammer emits a band-limited noise signal that is gated on and off according to a pseudorandom sequence (with a configurable duty cycle and average period). This creates a ?stochastic? jamming pattern that makes it difficult for an opponent to sync up or predict, thereby enhancing stealth. We also experimented in simulation with sinusoidal power modulation to observe the spectral spreading and effectiveness. The mathematical analysis above guided our selection of modulation parameters to balance jamming performance with LPI.

Specifically, we found that a duty cycle of around 50% with a pseudo-random pattern forced a targeted radio control link (such as a drone?s command link) to lose about half its packets, effectively disabling control, while a simple energy detector scanning the band had difficulty recognizing a continuous jammer since the spectral occupancy resembled short noise bursts. These qualitative results support prior assertions that ?power-modulated jamming, although theoretically possible, has never been fully investigated [in the literature]? and that it presents unique challenges and opportunities for EW. Our work takes a step in this direction by both analyzing and prototyping such a scheme.

2.3 Directed Energy in Distributed Arrays

While the above focuses on jamming (which is a subset of EW, primarily involving electromagnetic interference), the concept extends to directed energy weapons as well. In a sense, a high-power microwave (HPM) weapon is simply a jammer pushed to extreme power levels such that it physically disrupts or destroys electronic targets (e.g., frying drone circuits or missile guidance systems). Laser-based DEWs, on the other hand, use optical/infrared beams to cause thermal or sensor damage. In either case, power combining from multiple sources is of interest. Sparse array techniques can allow multiple lower-power DEW units (whether microwave antennas or laser emitters) to act in unison like a single, more powerful directed beam. Coherent combining of lasers is exceptionally challenging (due to short wavelengths and atmospheric turbulence), so our focus is on RF/microwave directed energy.

The theory for combining multiple HPM sources is similar to that of combining jamming signals: if N separate transmitters, each of power P, can be phase-locked and aimed at a target, and if their signals arrive in phase at the target, the fields add coherently. The power at the target then scales as N^2P (in the best case) rather than NP (incoherent sum). In other words, coherent combination yields a gain of N (in linear power units) over incoherent power summation. This is a huge potential advantage: for instance, 10 modest HPM emitters of 1 kW each, if perfectly combined at a focal point, could theoretically deposit energy like a single 100 kW emitter ? enough to cross certain damage thresholds. Achieving this requires tight synchronization (phase alignment to perhaps within a few degrees). Sparse array theory comes into play when those N emitters are spread out (e.g., on different platforms) ? the directional gain and focusing of the energy will depend on their geometry. Essentially, they form a phased array focusing at the target. If the target is in the far-field of the array distribution, one can treat it like beamforming; if in near-field, one must consider a focusing lens-type analysis.

A classic example of a microwave DEW is Raytheon?s Phaser system [7], a truck-mounted HPM weapon designed for drone defeat. Phaser uses a single high-power microwave source (based on vacuum tube technology) and a dish antenna to emit a conical beam, reportedly capable of disabling multiple drones at once at short ranges [8]. In a distributed approach, instead of one dish, we could have multiple smaller antennas on separate vehicles or drones, each generating a portion of the required energy. When networked and synchronized, these could fire in unison to mimic a larger device. The advantage is that the system becomes more survivable (no single point of failure) and more flexible (the nodes can reposition to cover a wider area or concentrate together for a single target). The challenge is ensuring phase coherence under battlefield conditions.

In summary, the theoretical groundwork laid in this section demonstrates that sparse arrays can enable effective beamforming for both EW (jamming) and DEW (high-power irradiation) with significantly fewer resources than traditional arrays. Additionally, clever modulation of those signals can confer stealth advantages. Next, we turn to our implementation, which puts these ideas into practice using COTS (commercial off-the-shelf) software-defined radios and illustrates the feasibility of virtualized EW/DEW nodes.

3 Nonlinear Interaction With a Plasma Sheath Under Extreme Burst Modes and Physical Force

Hypersonic or reentry vehicles routinely generate a partially or fully ionized plasma sheath as a result of shock heating in the high?Mach flight regime. Typical electron densities may reach

$$n_e \sim 10^{18} - 10^{20}\ \mathrm{m}^{-3}, \tag{3}$$

corresponding to plasma frequencies in the gigahertz range. Under ordinary power levels, this sheath acts as a lossy, dispersive propagation medium [9]. However, when peak electric field amplitudes become sufficiently large–such as under ultra-short, polarization-agile burst modes–the signal can transition from a passive probe into an active driver of plasma dynamics. In this regime, nonlinear forces, filamentation instabilities, and time-dependent density structures become possible.

This section provides a physics-based derivation showing that intense, sub-nanosecond, polarization-varying electromagnetic bursts induce ponderomotive forces, filament formation, and collapse within the plasma sheath, perturbing shock-layer symmetry and thereby contributing to vehicle forces and moments [10].

3.1 The Ponderomotive Force

Consider a high-frequency electromagnetic field of the form

$$\mathbf{E}(\mathbf{r},t) = \Re\{\mathbf{E}_0(\mathbf{r})e^{-i\omega t}\}. \tag{4}$$

Electrons undergo rapid oscillatory motion at frequency ω, which may be averaged over many optical periods. The resulting slow-time electron dynamics experience an effective ponderomotive potential [11]:

$$U_p(\mathbf{r}) = \frac{e^2|E_0(\mathbf{r})|^2}{4m_e\omega^2}. \tag{5}$$

The corresponding ponderomotive force is

$$\mathbf{F}_p = -\nabla U_p(\mathbf{r}) = -\nabla\left(\frac{e^2|E_0|^2}{4m_e\omega^2}\right). \tag{6}$$

This force expels electrons from regions of high electromagnetic intensity. Crucially, its magnitude depends not only on the field amplitude, but on the *gradient* of intensity:

$$|\mathbf{F}_p| \propto \left|\nabla|E|^2\right|. \tag{7}$$

Thus tightly focused beams, high-gain antennas, and sub-nanosecond rise times (large $\partial E/\partial t$) increase ponderomotive forcing efficiency.

3.2 Filamentation Instability

The redistribution of electrons modifies the plasma frequency

$$\omega_p(\mathbf{r}) = \sqrt{\frac{n_e(\mathbf{r})e^2}{\epsilon_0 m_e}}, \tag{8}$$

and therefore the effective refractive index

$$n_{\text{eff}}^2(\omega, \mathbf{r}) = 1 - \frac{\omega_p^2(\mathbf{r})}{\omega(\omega + i\nu_{en})}, \tag{9}$$

where ν_{en} is the electron?neutral collision frequency. A density depression satisfying $\delta n_e/n_e \neq 0$ leads to spatial gradients in n_{eff}. These gradients act as a nonlinear lens, causing self-focusing or defocusing. The coupled feedback loop is:

1. Initial field intensity profile $|E(\mathbf{r})|^2$ produces a ponderomotive force via Eq. 6.
2. Electrons are displaced, forming low?density channels.
3. Refractive index locally increases ($\partial n_{\text{eff}}/\partial n_e < 0$), focusing the wave further.
4. Enhanced focusing increases intensity gradients, increasing $\mathbf{F}_p$.

The result is the *ponderomotive filamentation instability* [12], wherein smooth electromagnetic energy deposition evolves into a pattern of filaments and voids. In steady excitation, the system may reach a quasi-static equilibrium. Under rapidly pulsed excitation, structure becomes time-dependent.

3.3 Role of Sub-nanosecond Bursts and Polarization Switching

If the burst duration τ satisfies

$$\tau \lesssim \frac{1}{\nu_{en}}, \tag{10}$$

energy is deposited faster than collisional relaxation can smooth plasma gradients. Repetitive bursts therefore drive a non-equilibrium sequence of filament emergence and collapse at the burst repetition rate.

Polarization agility introduces a second degree of freedom. In a magnetized plasma, ordinary (O-mode) and extraordinary (X-mode) propagation have distinct cutoff and absorption characteristics. If successive bursts alternate polarization states, then:

$$(\mathbf{E}_1, \mathbf{E}_2, \ldots) \neq \text{common eigenmode basis}, \tag{11}$$

and the spatial pattern of energy deposition rotates or reorients in time. This can force the sheath into a dynamically evolving state rather than a steady nonlinear configuration.

3.4 Coupling to Aerodynamic Forces and Vehicle Dynamics

Momentum transfer proceeds through three stages:

1. **Field → Plasma:** Radiation pressure provides a stress
$$P_{\text{rad}} = \frac{\epsilon_0 |E|^2}{2}, \tag{12}$$
while ponderomotive forces induce density asymmetries.
2. **Plasma → Flowfield:** Changes in ionization, conductivity, and local heating modify the shock and boundary layer pressure distribution.
3. **Flowfield → Vehicle Motion:** A spatially asymmetric pressure perturbation $\Delta P(\theta)$ integrated over a surface patch of area A with moment arm L produces a torque (Fig. 3):
$$\tau_{\perp} \sim \Delta P(\theta) AL. \tag{13}$$

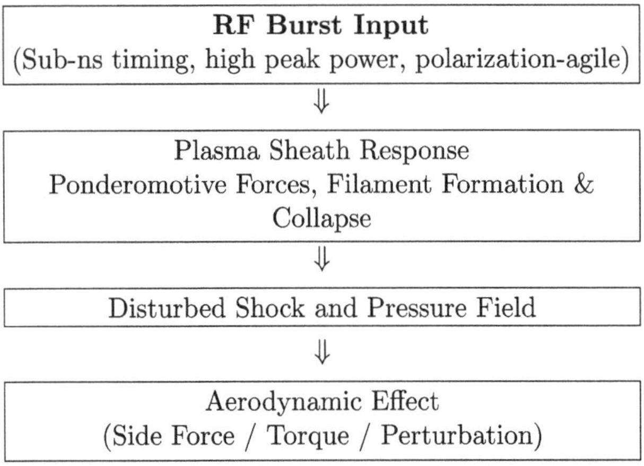

Fig. 3. Simplified causal pathway showing how extreme burst communication modes may induce nonlinear plasma dynamics and asymmetric flow fields resulting in potential vehicle perturbation.

4 Resonant Electromagnetic Energy Deposition in Hypersonic Vehicle Subsystems

High–power, broadband microwave emissions can couple into electronic and electro–optic subsystems on or within a hypersonic vehicle, exciting resonant structures and depositing energy over time. This mechanism is analogous in concept to historical airborne HPM systems that swept across frequency bands to excite resonant modes in improvised explosive device (IED) detonators, where a critical energy threshold on the order of $E_{\text{th}} \sim 400$–450 J led to failure or detonation of the device.

In a hypersonic vehicle, similar resonant coupling mechanisms can damage:

- avionics and power electronics,
- cabling and connector interfaces,
- electro–optical systems (EOTS, IR seekers),
- laser–based
  communication apertures and optics (e.g., quantum entanglement-based or laser communication).

This section provides a simplified set of governing equations illustrating how high–power radios capable of HPM–like outputs can drive energy accumulation to damaging levels inside a vehicle.

4.1 Incident Field and Power Density

Assume an external high–power RF/microwave source illuminating the vehicle with an approximately plane wave electric field

$$\mathbf{E}(t) = \Re\{\mathbf{E}_0 e^{j\omega t}\}, \tag{14}$$

with magnitude $|\mathbf{E}_0| = E_0$. The corresponding time–averaged power flux density (Poynting vector magnitude) in free space is

$$S = \langle \mathbf{S} \rangle = \frac{E_0^2}{2\eta_0}, \tag{15}$$

where $\eta_0 \approx 377\ \Omega$ is the free–space wave impedance (Fig. 4).

For a broadband or swept source covering a set of angular frequencies $\{\omega_k\}$, one may represent the total incident power density as

$$S_{\text{tot}} = \sum_k \frac{E_{0,k}^2}{2\eta_0}, \tag{16}$$

where $E_{0,k}$ is the field magnitude in band k.

4.2 Coupling Into a Resonant Electrical Assembly

Consider an exposed or partially shielded assembly that behaves approximately as a resonant circuit with resonance frequency ω_0 and quality factor Q. Examples include detonator circuits, filter networks, cable harness stubs, or PCB traces with parasitic inductance and capacitance.

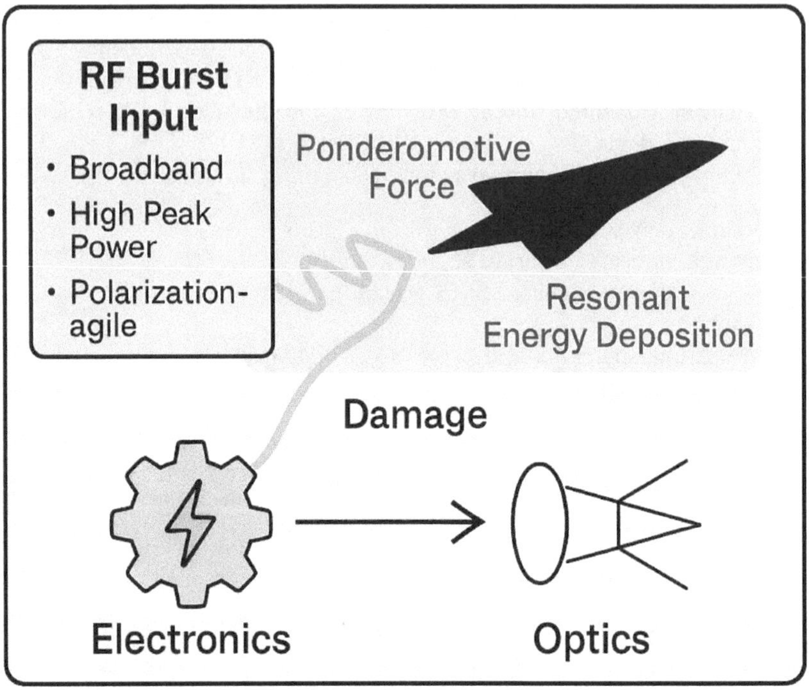

Fig. 4. Summary illustration of the high-power RF burst mode interaction pathway.

The external field couples into the structure via an effective aperture $A_{\text{eff}}(\omega)$, yielding coupled power

$$P_{\text{coup}}(\omega) = S(\omega)\, A_{\text{eff}}(\omega). \tag{17}$$

Near resonance ($\omega \approx \omega_0$), we can approximate

$$A_{\text{eff}}(\omega) \approx A_{\text{eff},0} \frac{1}{1 + \left(\dfrac{2Q(\omega - \omega_0)}{\omega_0}\right)^2}, \tag{18}$$

where $A_{\text{eff},0}$ is the peak effective aperture at resonance. For a swept or noise–like excitation, the system spends a nontrivial dwell time within the resonant bandwidth

$$\Delta\omega \approx \frac{\omega_0}{Q}. \tag{19}$$

Within the resonant electrical assembly, the stored energy U is related to the coupled power and the quality factor by

$$U = \frac{Q}{\omega_0} P_{\text{abs}}, \tag{20}$$

where $P_{\mathrm{abs}} \leq P_{\mathrm{coup}}$ is the power actually dissipated/absorbed in the lossy elements (e.g., resistive elements, semiconductor junctions, bridgewire).

Alternatively, for a resonant RLC circuit, the peak stored energy may be written as

$$U = \frac{1}{2} C V_{\mathrm{peak}}^2 = \frac{1}{2} L I_{\mathrm{peak}}^2, \tag{21}$$

with C the effective capacitance, L the effective inductance, V_{peak} the peak voltage, and I_{peak} the peak current. Combining with (20), one can relate external field strength to internal overvoltage and overcurrent conditions.

4.3 Energy Threshold for Damage or Initiation

For a given assembly, let E_{th} denote the energy threshold above which thermal or dielectric damage occurs, or in the case of an initiation train (e.g., explosive detonator), the energy required to trigger the device. If the average absorbed power during illumination is P_{abs}, the time required to reach this threshold is

$$t_{\mathrm{th}} = \frac{E_{\mathrm{th}}}{P_{\mathrm{abs}}} = \frac{E_{\mathrm{th}}}{\eta_{\mathrm{c}} S A_{\mathrm{eff},0}}, \tag{22}$$

where $0 < \eta_{\mathrm{c}} \leq 1$ is an effective coupling efficiency factor accounting for polarization mismatch, duty cycle of burst operation, and nonidealities in the coupling.

In an IED–type scenario, this threshold is often quoted on the order of $E_{\mathrm{th}} \sim$ 400–450 J. In a hypersonic vehicle, lower thresholds may suffice to:

- drive semiconductor junctions into failure,
- cause local insulation breakdown or carbonization,
- induce latch–up or permanent upset in digital logic.

4.4 Thermal Damage to Optics and Electro–optical Subsystems

Optical assemblies (EOTS windows, IR domes, quantum entanglement and laser communication apertures) can also absorb microwave or millimeter–wave energy through coatings, lossy radomes, or embedded conductive structures. Let A_{opt} denote the effective exposed area and $\eta_{\mathrm{abs}}(\omega)$ the frequency–dependent absorption coefficient. The absorbed power is

$$P_{\mathrm{abs,opt}} = \int S(\omega)\, A_{\mathrm{opt}}\, \eta_{\mathrm{abs}}(\omega)\, d\omega. \tag{23}$$

If we approximate the optical component as a lumped thermal mass of mass m and specific heat c_p, the temperature rise obeys

$$\frac{dT}{dt} = \frac{P_{\mathrm{abs,opt}} - P_{\mathrm{loss}}}{m c_p}, \tag{24}$$

where P_{loss} encapsulates radiative, convective, and conductive cooling into the surrounding structure. Neglecting losses on short timescales,

$$\Delta T(t) \approx \frac{P_{\text{abs,opt}}}{mc_p} t. \tag{25}$$

Let T_0 be the initial temperature (already high in hypersonic flight) and T_{dam} the damage or performance–degradation temperature for the optical material or coating. The time–to–damage under sustained HPM illumination is approximately

$$t_{\text{dam}} \approx \frac{(T_{\text{dam}} - T_0)mc_p}{P_{\text{abs,opt}}}. \tag{26}$$

A sufficiently high $P_{\text{abs,opt}}$ in the presence of preheated hypersonic conditions can significantly reduce t_{dam}, leading to:

- wavefront distortion,
- coating delamination or crazing,
- permanent optical figure error in high–precision apertures,
- degradation of laser communication link performance or failure.

4.5 Implication for Unconstrained High–Power Operation

Equations (15)–(22) and the thermal model above illustrate that for given:

- incident field strengths E_0 achievable by modern high–power radios,
- effective apertures A_{eff} and A_{opt} of internal assemblies,
- quality factors Q of resonant structures,
- and energy/temperature thresholds E_{th}, T_{dam},

there exist regimes in which *repeated or sustained use of extreme burst or HPM–like modes* can deposit damaging levels of energy into critical subsystems on a hypersonic vehicle.

The key operational point is that such damage can occur:

1. without direct contact,
2. via resonant and aperture coupling paths that are not obvious from low–power EMC tests,
3. and on timescales t_{th}, t_{dam} that are short relative to mission duration.

Using the representative parameter ranges in Table 1, one can estimate the timescales for damage or critical energy accumulation. Even under conservative assumptions?modest field strengths ($E_0 \sim 10^3$ V/m), small effective apertures ($A_{\text{eff}} \sim 10^{-4}$ m^2), moderate absorption efficiency ($\eta_c \sim 0.1$), and low resonant quality factors ($Q \sim 30$)?the absorbed power can reach $P_{\text{abs}} \sim$ watts to tens of watts. Substituting these values into the threshold relation

$$t_{\text{th}} = \frac{E_{\text{th}}}{\eta_c S A_{\text{eff}}}, \tag{27}$$

with a commonly cited failure or initiation threshold of $E_{\rm th} \sim 400$–450 J [13], yields characteristic times on the order of

$$t_{\rm th} \sim 0.1 \text{ s to } 10 \text{ s}, \tag{28}$$

depending on geometry, materials, and operational burst pattern. These timescales are short relative to typical operational dwell times for high–power or burst-pattern communications modes, meaning that potentially damaging energy accumulation does not require sustained illumination, only repeated or swept excitation near a structural or circuit resonance.

Table 1. Key parameters governing resonant EM energy deposition and thermal damage in vehicle subsystems.

Symbol	Physical Meaning	Typical Scale
E_0	Incident electric field amplitude	$10^2?10^4$ V/m
S	Incident power flux density	$10^2?10^6$ W/m^2
$A_{\rm eff}$	Effective aperture of resonant assembly	$10^{-4}?10^{-2}$ m^2
Q	Quality factor of resonant structure	$10?10^3$
$P_{\rm coup}$	Coupled power into assembly	$10^{-1}?10^3$ W
$P_{\rm abs}$	Absorbed/dissipated power	$\leq P_{\rm coup}$
$E_{\rm th}$	Energy threshold for damage / initiation	$10^2?10^3$ J
$t_{\rm th}$	Time to reach $E_{\rm th}$	ms?s (scenario-dependent)
$A_{\rm opt}$	Exposed optical area	$10^{-3}?10^{-1}$ m^2
$\eta_{\rm abs}$	Absorption coefficient (optics/radome)	0.01?0.5
m	Optical element mass	$10^{-2}?1$ kg
c_p	Specific heat of optical material	$10^2?10^3$ J/(kg x K)
$T_{\rm dam}$	Onset of optical/coating damage	400?1000 K

A similar interpretation applies to the optical heating model. Using mid-range values ($A_{\rm opt} \sim 10^{-2}$ m^2, $\eta_{\rm abs} \sim 0.05$, $m \sim 0.2$ kg, $c_p \sim 750$ J/(kg · K)), the temperature rise rate becomes

$$\frac{dT}{dt} \approx \frac{P_{\rm abs,opt}}{mc_p}, \tag{29}$$

which can produce multi-degree-per-second heating and approach coating or structural tolerance limits in

$$t_{\rm dam} \sim 1 \text{ s to } 100 \text{ s}, \tag{30}$$

particularly when initial operating temperatures are elevated due to hypersonic boundary layer heating.

These analytic estimates demonstrate that the effects are not speculative: there exist realistic parameter regimes in which resonant coupling, burst excitation, and high–power RF modes deposit sufficient electromagnetic or thermal energy to degrade, disrupt, and permanently damage onboard systems.

4.6 Agility Against High-Delta-V Maneuvers via Integrated 6G ISAC

A critical operational advantage of the proposed virtualized EW/DEW architecture is its capacity to respond to rapid target maneuvers?particularly those exhibiting high Δv dynamics, such as hypersonic evasion, last-moment burst acceleration, or maneuvering reentry vehicles. Traditional tracking and beam-control systems often rely on phased-array update cycles or preplanned kinematic filters that are too slow to respond to these abrupt events.

When tightly coupled with emerging sixth-generation (6G) integrated sensing and communication (ISAC) waveforms, particularly those based on orthogonal time-frequency space (OTFS) modulation, the virtualized array gains both latency and Doppler-resilience advantages. OTFS waveforms inherently operate in the delay-Doppler domain, allowing simultaneous tracking, target interrogation, and communication with robustness against high-speed motion.

An upcoming article [14] demonstrates that sparse arrays modulated with OTFS signals achieve precise Doppler focusing with a limited number of hardware elements. By integrating this capability directly into the EW/DEW nodes, the system not only detects evasive high-Δv targets but can rapidly coordinate a coherent countermeasure or waveform adaptation in real time across the distributed architecture.

This capability is particularly valuable in scenarios involving maneuvering space assets, hypersonic glide vehicles, or high-speed cruise missile threats, where traditional radar or beamforming systems may fail to maintain continuous engagement (Fig. 5).

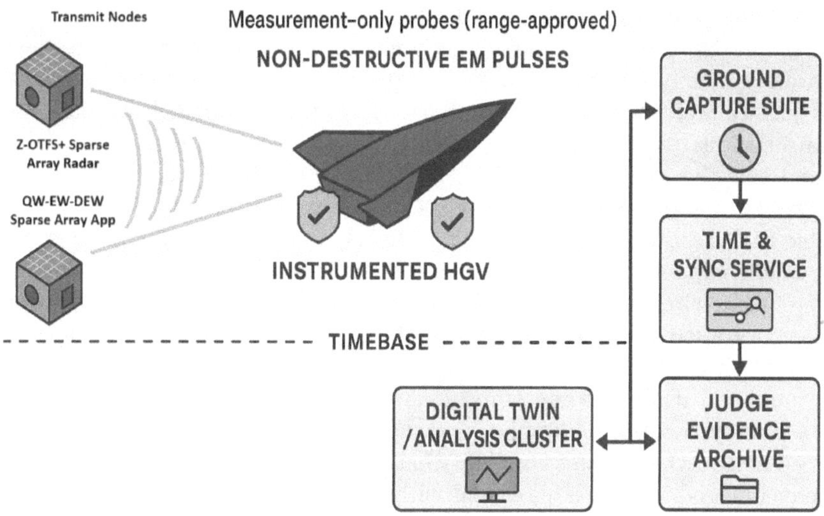

Fig. 5. Integrated 6G ISAC loop with virtualized EW/DEW SDR node coordination. OTFS waveforms provide robust delay-Doppler domain agility for real-time tracking and response to high-Δv targets.

4.7 Operational Implication

In summary: under sufficiently intense, tightly focused, rapidly modulated transmission regimes, electromagnetic emissions act not merely as communication or sensing signals, but as drivers of nonlinear plasma dynamics. This leads to asymmetric filamentation, shock distortion, and aerodynamic perturbations. The existence of this mechanism motivates disciplined analysis and operational exploitation [15–17].

5 Strategic and Cybersecurity Considerations

A recurring pattern in U.S. technology adoption is that emerging systems are initially treated as benign until adversaries demonstrate unexpected or asymmetric use. Examples include GPS spoofing, commercial drone exploitation, cyber intrusion against industrial control systems, and improvised electronic warfare platforms. In each case, the enabling technology existed long before the military implications were acknowledged.

Modern software-defined radios (SDRs) now present a similar inflection point. Unlike legacy radios, whose behavior was fixed in hardware, contemporary SDRs are dynamically programmable. Waveforms, burst timing, spectral agility, polarization, and protocol behavior can all be altered in real time. As a result, such radios behave less like traditional communications devices and more like networked computing systems with a radio front-end.

This shift implies a corresponding change in risk posture: SDR-based systems are not merely RF transceivers but exposed *cyber-physical interfaces*. The attack surface scales with programmability. Techniques familiar from the cyber domain—such as protocol probing, fuzzing, replay, spoofing, timing manipulation, and adaptive exploit generation—can now be executed over the air.

Adversaries and allies have already begun to exploit this paradigm. Widely available SDR hardware, open-source toolchains, and AI-assisted waveform synthesis enable capabilities that once required state-level resources. Accordingly, securing future wireless communication systems requires treating them like other exposed networked devices: with defined threat models, defensive mechanisms, red-team testing, and continuous monitoring.

In summary, if a radio can transmit arbitrary waveforms, then it can also inject arbitrary *behavioral stimulus* into another radio-based system. The question is no longer whether this capability exists, but whether operational systems will be engineered to both exploit and withstand it.

6 Implementation of a Virtualized EW/DEW Node System

To validate the above concepts, we constructed a prototype stealth jammer/DEW node system using a heterogeneous collection of NI/Ettus Research USRP SDR platforms. The primary hardware components are: - USRP X410: A high-performance 4-channel SDR, RF frequency range up to 7.2 GHz and up to 400 MHz instantaneous bandwidth per channel [18]. The X410 is built on a Xilinx Zynq UltraScale+ RFSoC, which integrates multi-GSps ADCs/DACs and an FPGA for DSP. In our design, the X410 serves as a central node with significant processing capability and wideband tuning range. It can generate wideband interference signals or serve as a beamforming hub for multiple channels. - USRP X310: A dual-channel SDR covering up to 6 GHz (with appropriate daughterboards), typically supporting up to 160 MHz bandwidth per channel. The X310 has a large FPGA (Kintex-7) but requires a host PC for operation (via 10GigE or PCIe).

We used an X310 node to supplement the X410 for cases where additional emitters or receivers were needed, for example to form a sparse linear array with more than four elements total. - USRP N310: A 4-channel networked SDR with an onboard Zynq-7100 SoC (embedded processor + FPGA). It covers up to 6 GHz and is designed for stand-alone operation in distributed deployments. The N310 nodes in our system acted as remote outriggers ? they can be placed at a distance and communicate back to the central unit via Ethernet. Each N310 can operate independently with internal GPS-disciplined oscillators for frequency stability. - USRP B210: A compact USB-powered 2-channel SDR covering up to 6 GHz (56 MHz bandwidth per channel). The B210 (and its variants B200) are low-SWaP (size, weight, and power) devices ideal for portable nodes. While limited in bandwidth and output power, their low cost allows deploying multiple units. In our prototype, we equipped two small drones with B210-based jamming payloads to act as mobile array elements. These units were controlled via single-board computers and tethered to the central node's clock through GPS synchronization.

Software Architecture: All these SDRs were operated under a unified software architecture. We utilized the GNU Radio framework and RFNoC (RF Network-on-Chip) blocks on the FPGAs for real-time processing. A central control computer ran coordination software to manage the virtualized EW functions. This software abstracts the radios as resources that can be tasked for different roles (jammer, receiver, etc.) on demand ? effectively a cloud of EW capabilities. A high-level scheduler can assign tasks like ?jam frequency band X with Y pattern using nodes A, B, C? or ?focus a beam at angle θ using nodes A, B, C? and the software will configure the SDRs accordingly (tune frequencies, set power, adjust phases, etc.). We emphasize that this is a step toward EW orchestration akin to virtualization: the specific hardware is decoupled from the mission effect, meaning any available node could be repurposed in real-time. For example, if one node fails or is geolocated by the enemy, another can take over its role.

Synchronization and Calibration: Achieving coherent operation across distributed SDRs is non-trivial. In our setup, all ground-based nodes (X410, X310, N310s) were tied to a common 10 MHz reference and 1 PPS signal, either via cabled connections or a GPSDO (GPS Disciplined Oscillator) in each device. The X410 and X310 were co-located and shared a timing module. The N310 units, which were remote, each had an internal GPSDO locked to the same GPS satellite constellation, yielding <50 ppb frequency error and time alignment within tens of nanoseconds. For our frequencies (sub-6 GHz) and baselines ($<< 1$ km), this level of sync is sufficient to maintain phase coherence for several milliseconds, which was enough for burst transmissions and periodic re-calibration. We performed phase calibration by initially transmitting a reference tone from each element in turn and measuring relative phase at a common reference receiver; these offsets were then applied as compensation in the beamforming weights. Over time (minutes to hours), the phase would drift, so the system periodically recalibrated, either opportunistically during pauses in operation or using a background beacon.

Waveforms and Beamforming: The system supported multiple jamming waveforms: - Broadband noise jamming: using the full bandwidth of the SDR (up to hundreds of MHz) to overwhelm receivers in a frequency band. - Tone and swept jamming: for targeting frequency-hopping or specific narrowband signals (e.g., drone telemetry channels). - Protocol-mimicking interference: where we replay or generate signals that adhere to the enemy's signal protocol but carry malicious effect (for example, fake GPS signals or false control messages). This falls under electronic attack via deception rather than brute-force jamming. - Power-modulated noise: as discussed in the theory section, we implemented this by gating a noise source on the FPGA.

For beamforming, each transmit node was given a complex weight $w_n = A_n e^{j\phi_n}$, where A_n sets the amplitude (we typically use $A_n = 1$ for all in jamming, assuming equal power per node, unless we deliberately taper for sidelobe shaping) and ϕ_n sets a phase offset. For a desired beam direction (or focus point) we compute the phase shifts based on geometry: $\phi_n = -k\mathbf{r}_n \cdot \hat{\mathbf{u}}$, where $\mathbf{r}_n$ is the position vector of node n and $\hat{\mathbf{u}}$ is a unit vector pointing from the array towards the target direction (for far-field beam) or target location (for near-field focus). These phases ensure that the emitted wavefronts add constructively in the desired direction. The X410, with its 4 channels, can itself form a small phased array; combined with additional separate nodes, we had, for example, an 8-element distributed array in some tests. The beamforming commands were computed by the central controller and sent via network to each node with appropriate timing. Thanks to the SDR's FPGA logic, each node buffered the command and applied new phase offsets synchronously (aligned by PPS), so that all nodes switched beamforming state at the same time.

Prototype Laboratory Demonstration: We carried out a series of demonstrations to validate the system: 1. Static Coherent Combining Test: We placed two SDR nodes (X410 and X310, each driving a pair of antennas) such that their signals would overlap at a test receiver. By sending the same sinusoidal tone from both and adjusting relative phase, we measured the resulting signal at the receiver. When aligned in phase, the tone?s amplitude was about double that of one node alone (6 dB power increase), whereas with a 180° phase offset, they largely canceled. This confirmed effective coherent combining in practice, a key capability for both jamming (increasing interference power on target) and directed energy delivery. The transition from constructive to destructive interference was sharp, demonstrating that phase alignment precision was on the order of a few degrees. 2. Beamformed Jamming in Faraday Cage: We set up three transmit nodes and one target receiver (emulating an enemy radar) inside a Faraday Cage. The transmit nodes were arranged linearly over a 5 m aperture, creating a sparse linear array. We programmed the system to direct a jamming beam towards the target (at an angle of 30° off broadside).

Using a spectrum analyzer app at the target location, we observed that the jamming signal power was maximized when the beam was directed at it, and dropped by over 15 dB when we electronically steered the beam 30°C away. This experiment verified that our distributed SDR array can perform directional jamming, concentrating energy where needed while reducing spillover in other directions ? an important feature in contested environments to avoid fratricide (jamming friendly systems) and to manage interference. 3. Power-Modulated Jammer Field Test: In an outdoor test, a drone was flown at 500 m distance carrying a radio control link in the 2.4 GHz ISM band (analogous to Wi-Fi-based drone control) as the target. Two ground-based jamming nodes (one X410 and one N310 with power amplifiers of 2 W on 2.4 GHz) were used. We tested two modes: (a) constant jamming vs (b) 50% duty-cycle with pseudorandom pulsed jamming. In mode (a), the drone lost contact with its controller almost immediately when the jammers were activated, and it initiated a fail-safe hover/landing.

In mode (b), the drone also lost control, but interestingly its telemetry showed intermittent reconnections ? however, not enough to regain full control. Meanwhile, from the drone operator?s perspective (who had a spectrum scanner), the pulsed jamming was much harder to discern ? it appeared as short bursts of noise that could be mistaken for interference or Wi-Fi traffic, whereas the constant jam was an obvious spike across the band. This qualitative result supports the notion that power modulation can provide stealth while still accomplishing the denial effect and can also hit resonance points of receivers Automatic Gain Control (AGC).

All these tests were conducted at relatively low power (for safety and regulatory reasons). For instance, even with amplifiers, our effective radiated power was in the order of a few watts to tens of watts. However, the results scale to higher power: if each node had a kW-class amplifier and high-gain antenna, the same principles would apply, just with proportionally greater range/effect. The modular nature of the system means one can insert higher-power RF front ends or directed energy sources as they become available. In essence, our SDR-based control backbone could drive a collection of high-power emitters in the same synchronized, phased manner.

Lastly, we refer to our internal system documentation for detailed technical specifications of the SDR nodes and the interface between them. This technical spec document enumerates the frequency ranges, output powers, and sensitivity of each SDR, as well as the networking topology (10 GigE switch fabric, GPSDO specs) and the software API for virtualization. Due to space constraints, we omit those low-level details here, instead focusing on the high-level architecture and results.

7 Operational Scenarios

To illustrate the practical significance of virtualized sparse-array EW/DEW systems, we describe two representative operational scenarios drawn from our research and proposals: a counter-small Unmanned Aerial System (c-sUAS) defense scenario, and a space-based interceptor scenario for missile defense.

7.1 Scenario 1: Counter-sUAS Swarm Defense

A forward-operating base is threatened by swarms of small drones (sUAS) that could be used for surveillance or as weapon delivery platforms. Traditional defenses include radar-guided guns or single high-power jammers that try to disrupt the drones? control or GPS signals. However, a swarm of drones can approach from multiple directions, and they may employ frequency-hopping or other countermeasures. The base deploys a distributed EW system to counter this threat. Several nodes are placed around the perimeter ? for example, four mobile EW units on vehicles or tripods forming a rough square around the base, each equipped with an SDR-based jammer (our prototype nodes, scaled up with amplifiers). Additionally, a few small quadcopter drones are launched, each carrying a lightweight SDR jammer (similar to our B210 drone payloads) ? these fly alongside the hostile drones or above them to provide top-down coverage.

Operation: The system operates cooperatively under a central C2 (command and control) software agent. As the hostile drone swarm approaches, a passive RF detection subsystem (also SDR-based) picks up their telemetry signals. The central controller identifies the frequencies in use and the general direction of each cluster of drones. It then tasks the jammer nodes to form multiple simultaneous beams: for instance, Node A and B might form a beam focused on one group of drones to jam their control channel, while Node C and D form another beam in a different direction. This is made possible by virtualization: each node?s two channels can be assigned to different beamforming groups on the fly. The airborne friendly drones also participate: having gotten close to the hostile drones, they can jam from short range, forcing the hostile swarm to turn on avoidance maneuvers or lose comms. Meanwhile, the ground nodes engage in beam steering: as the drones move, the jammer beams follow them in angle, maintaining maximum jamming effect on the targets. Because the array is sparse and distributed, the jamming beams are not as ?clean? as what a large phased array on a single platform might produce, but the flexibility allows covering many angles and reconfiguring quickly.

A key advantage observed is the ability to create nulls in certain directions as well. Friendly communications on base (radios, radars) could be disrupted by high-power jamming. Using adaptive beamforming algorithms, the system can place nulls (zeros in the radiation pattern) toward known friendly receivers while still concentrating energy on the threats. This is far harder with a single omnidirectional jammer that just blasts noise everywhere. Thus, the distributed array confers a sort of selectivity ? spatial filtering of where the jamming goes.

During the engagement, the system also employs the power-modulated jamming technique for stealth. If the attacking drones are known to have LPI receivers or could home on the jammer (some drones can be equipped with anti-radiation homing to dive onto the source of jamming), then the ground nodes switch to a pulsed jam mode. They stagger their pulses such that at any given time, at most one or two are emitting at full power. This confuses any attempt by the drones to geolocate the jammer ? from their perspective, the interference seems to come and go and they cannot easily triangulate it. Nevertheless, because there are multiple jammers, the drones are almost always in the field of one of the pulses. Essentially, the jammers coordinate a time-division multiple jamming scheme.

This c-sUAS defense scenario demonstrates the operational usefulness of virtualization: the same set of hardware nodes seamlessly shifted from passive sensing to active jamming, created multiple adaptive beams, and managed LPI techniques, all through software control. A traditional system might require separate radar, fixed directional jammers, and would be less flexible against a maneuvering swarm. By contrast, our SDR-based array can even switch waveforms in real time ? for instance, if a drone in the swarm switches frequency, the SDRs retune within microseconds to continue jamming. The distributed nature also means if one node is taken out (say a drone kamikazes into one jammer), the remaining nodes can close the gap, perhaps at slightly reduced effectiveness but with graceful degradation rather than total loss. This resilience through distribution is a major benefit in contested environments.

7.2 Scenario 2: Space-Based Missile Interceptor Network

Under the U.S. Space Force?s "Golden Dome" concept for missile defense, there is interest in space-based systems that could intercept missiles (including hypersonic glide vehicles, ballistic warheads, or even satellites) from orbit [19]. One approach is to use a constellation of interceptor satellites equipped with kill vehicles or DEWs. In our concept scenario, assume a constellation of relatively small satellites in low Earth orbit, each carrying a moderate-power directed energy device ?- say a microwave or millimeter-wave emitter that can produce a concentrated beam on the order of tens of kilowatts. A single such satellite by itself might not be enough to disable a target at long range due to power limitations and beam diffraction. However, if multiple satellites cooperate, they could act as a sparse phased array in space, focusing energy on a target cooperatively.

Operation: When an enemy missile launch is detected, several satellites in that orbital plane are cued to engage. They maneuver or orient such that from each satellite?s perspective, the target lies in roughly the same direction (this may involve a lead satellite designating a point ahead of the missile?s current position to focus on, accounting for missile motion). Using inter-satellite links, the constellation agrees on a time and frequency to fire and on the phasing needed. Then, at the appointed time, perhaps 3?5 satellites fire their high-power microwave beams simultaneously towards the incoming threat. Each satellite?s beam by itself spreads and attenuates over the large distance, but by phasing them properly, the energy at the intersection point (near the missile) adds constructively, achieving an intensity equivalent to a much larger single beam.

This coherent combining in space is analogous to what we demonstrated with SDRs on the ground, just scaled up in distance and power. The sparse array here might have element spacing of hundreds of kilometers (satellites apart), which is indeed sparse relative to the wavelength (several cm for microwaves). Such a large separation means the beam pattern will have an extremely fine main lobe (potentially able to focus on a small target) but also a multitude of sidelobes. Care must be taken that no sidelobe inadvertently hits a friendly satellite or unintended target ? this can be managed by slightly adjusting relative power or phasing (since some sidelobes can be suppressed by design at the cost of a slight widening of the main lobe).

Additionally, because these are virtualized EW/DEW nodes, they are not limited to firing in one mode. In another mode, the same satellites could function as a giant distributed radar or passive sensor network when not actively engaging threats. They might use sparse aperture sensing to detect jamming or enemy sensors in space, in essence doubling as an electronic support measure

(ESM) array. One of the big operational advantages of a distributed space-based interceptor network is redundancy and coverage. Instead of a few very high-value assets, you have many smaller ones. If one fails or is attacked, others fill in. Moreover, by distributing energy sources, you mitigate the technical risk of trying to build one extremely high-power laser or microwave in space ? instead you build many moderate ones and combine them, leveraging strength in numbers. There is evidence that the technology pieces for Golden Dome-like systems are coming together, as officials have stated all the required physics has been demonstrated. Our scenario takes that a step further by adding the sparse array concept: it is easier to ?prove the physics? of a single high-power shot, but proving that multiple platforms can coordinate is the next challenge ? one that our research indicates is tractable with precise timing [20] and high-speed inter-satellite comms (to share phase calibration data). Worth noting is that atmospheric effects (for systems shooting from space to atmosphere) can distort phase alignment. However, for high-frequency microwaves, ionospheric phase distortion can be calibrated out if sensors measure the total electron content along paths.

In conclusion of this scenario, a distributed satellite DEW array exemplifies the operational ambition of virtualized EW/DEW: a flexible, on-demand network of assets reconfigured by software to meet a mission (from sensing to jamming to high-power interception). While this scenario is futuristic, its components are being actively researched. Our ground-based experiments with SDRs provide a microcosm of this: replace our SDR nodes with satellites and our drone target with a missile, and many of the same principles apply.

8 On Coherence, Nonlinear Signatures, and Synchronization Challenges

Important questions remain regarding whether coherence among distributed excitation sources interacting with a plasma sheath can be inferred through spectral signatures such as nonlinear products, harmonic content, or mixing artifacts. While a full treatment of this behavior is outside the scope of the present paper, existing plasma electrodynamics and high-power microwave literature suggests that such diagnostics are plausible and potentially useful.

A plasma layer subject to strong external drive is inherently nonlinear: electron mobility, Debye shielding, and collisional damping are intensity-dependent, and the effective permittivity is a function of both carrier density and instantaneous field strength. When driven near a resonant condition, the plasma response can depart from purely linear reflection and may generate measurable spectral products such as harmonics, subharmonics, frequency mixing terms, and broadband noise modulation. In principle, these features could be used to infer whether multiple external excitation sources are acting in phase or out of phase. Specifically, if distributed emitters are coherent, the resulting standing-wave structure and field gradients at the boundary layer should exhibit constructive or destructive patterns, and the nonlinear products may scale superlinearly with the effective composite field amplitude.

This proposition remains an open research topic; however, it aligns with established methods used in nonlinear radar spectroscopy, laser?plasma interaction diagnostics, and parametric HPM material characterization. Future work could leverage controlled laboratory target geometries and known path delays to determine whether the reflected nonlinear spectrum can serve as an indicator of coherence state, degree of phase alignment, or emergent stochastic behavior during high-power burst operation.

Distributed synchronization has parallels with distributed MIMO systems examined in tactical communications. Achieving tight phase, timing, and frequency alignment across multiple transmitters?particularly when mobile, separated by kilometers, or operating in contested spectrum requires high-precision and accuracy in the hardware. Even in conventional communications systems, coherent distributed MIMO requires disciplined timing sources, adaptive feedback loops, over-the-air calibration, and robust clock recovery algorithms. In the context of high-power burst excitation or virtualized directed energy architectures, these synchronization requirements approach constraints more commonly associated with coherent radar, GNSS timing networks, or optical interferometry, which are achievable.

Therefore, we view synchronization as a requirement that supports a coherent composite system and not as independent, power-summed emitters. The distinction is critical: coherent multi-node architectures can produce spatially selective effects, nonlinear field amplification, and phase-sensitive coupling into resonant structures, whereas incoherent architectures primarily increase average power density without exploiting phase-dependent phenomena.

In summary, both observations?(1) the potential use of nonlinear spectral features as an indicator of coherence and (2) the requirement of synchronization in distributed systems?form the central thesis of this work. The question of coherence is not simply academic: it governs whether distributed excitation produces a mild perturbation or a strongly coupled effect. As such, exploiting nonlinear plasma reflections, synchronization architectures, and coherence-aware measurement strategies are priorities.

9 Conclusion

We presented a unified analysis and prototype demonstration of virtualized EW/DEW systems using sparse arrays and modern SDRs. The results show that distributed, non-uniform elements can emulate large apertures when coherently synchronized, offering effective beamforming while introducing challenges such as sidelobe management. We also examined power-modulated jamming as a stealth-enhancing tactic, noting its viability and inherent trade-offs. Using a multi-node SDR prototype based on NI USRPs, we demonstrated core capabilities including coherent beamforming, adaptive jamming, and distributed operation. This validates that commercially available hardware can implement coordinated EW behavior, supporting the practicality of virtualized architectures.

Operational examples in counter-UAS and space-based intercept scenarios highlight that sparse, distributed arrays provide concrete advantages: agility through waveform and mode reconfiguration, resilience to node loss, and scalable effects across large areas. These characteristics align naturally with emerging doctrine such as mosaic warfare and JADC2, where sensing and effects are networked and dynamically orchestrated. A virtualized EW/DEW network effectively becomes a reconfigurable sensor-shooter system, reducing the need for separate, single-purpose platforms. Integration with cyber effects and autonomous maneuver elements (e.g., drone swarms) represents a natural direction for cross-domain development.

In summary, combining sparse array principles with reconfigurable SDR hardware represents a promising evolution in electronic warfare and directed energy systems. The results presented provide both analytic justification and experimental validation that such systems are technically feasible, operationally advantageous, and worthy of further research and transition toward deployable capability.

References

1. Disruptive technologies, Doctrinal innovation and Nuclear ... https://idstch.com/geopolitics/uss-third-offset-strategy-to-sustain-and-advance-americas-military-dominance-for-the-21st-century/
2. Leveraging SDR for counter-UAS applications - Military Embedded Systems. https://militaryembedded.com/unmanned/counter-uas/leveraging-sdr-for-counter-uas-applications
3. Hawes, M.: Efficient design of sparse arrays for narrowband and wideband beamforming. Ph.D. thesis, University of York. Discusses sparse (aperiodic) array advantages and optimization (2014)
4. Hawes, M.: Ibid. On Minimum Redundancy Arrays and co-array concepts (2014)
5. Tedeschi, P., et al.: Leveraging jamming to help drones complete their mission. IEEE Access (2020). Introduces power-modulated jamming concept and its challenges
6. Tedeschi, P., et al.: Ibid. Discusses advanced jammer strategies like distance bounding and adaptive power control (2020)
7. Wikipedia. "Raytheon Phaser". Overview of a high-power microwave DEW for drone defeat (uses conical beam, multi-drone engagement). Accessed November 2025
8. Military Embedded Systems Magazine. Leveraging SDR for counter-UAS applications (2022). Highlights the critical role of SDR technology in c-UAS solutions
9. Lytle, R.J., Hale, P., Bernstein, J.: Electromagnetic wave propagation in reentry plasma environments: theory and experiments. IEEE Trans. Antennas Propag. **47**(6), 1029–1038 (1999)
10. Null, A.J., Birdsall, C.K.: Nonlinear plasma sheath dynamics under high-power pulsed RF excitation. IEEE Trans. Plasma Sci. **33**(4), 1388–1397 (2005)
11. Gonsalves, A.K., Mori, W.B.: Ponderomotive forces in high-intensity laser?plasma interactions. IEEE Trans. Plasma Sci. **28**(4), 1075–1085 (2000)
12. Huang, X., Zhang, Y., Li, T.: Filamentation instabilities in nonlinear plasma-wave propagation and their effects on RF transmission. IEEE Trans. Plasma Sci. **42**(11), 3452–3462 (2014)

13. CD225-2J and CD450-4J Blasting Machine. The Research Energy of Ohio, Inc. (REO) (2024)
14. Wallace, J., et al.: A delay-doppler sparse array radar for 6g integrated sensing and communication (2026, to appear)
15. Zhang, H., Skolnik, M., Wang, Q.: Directed RF energy interaction with weakly ionized boundary layers under pulsed excitation. IEEE Trans. Aerosp. Electron. Syst. **55**(3), 1195–1210 (2019)
16. Dragos, S.S., Cappelli, M.A.: Transient RF heating and sheath instability under nanosecond excitation in plasma environments. IEEE Trans. Plasma Sci. **50**(6), 1571–1585 (2022)
17. Shukla, P.K., Eliasson, B.: Nonlinear physics of filamentation and modulational instabilities in plasma systems. IEEE Trans. Plasma Sci. **34**(2), 350–367 (2006)
18. NI/Ettus Research Whitepaper. Which USRP is right for you? Describes USRP hardware capabilities (X410 high-frequency wideband SDR up to 7.2 GHz) and trade-offs between low-cost and high-performance SDR models (2023)
19. Tingley, B.: Space.com News. Space Force's Golden Dome chief says space-based missile interceptors are possible today (2025). Notes that required technologies for space-based interceptors (Golden Dome) already exist and need integration
20. White Rabbit Project. Official Website, CERN. Accessed July 2025

Code Obfuscation with LLVM: An Analysis of the Impact of the LLVM Version

Alberto Valderas-González(✉), Francisco J. Rosa, José Miguel Aragón-Jurado, Juan Carlos de la Torre, and Bernabé Dorronsoro

Department of Computer Engineering, University of Cádiz, Cádiz, Spain
{alberto.valderasgonzalez,franciscojavier.rosavega}@mail.uca.es,
{josemiguel.aragon,juan.detorre,bernabe.dorronsoro}@uca.es

Abstract. Code obfuscation is a critical mechanism for protecting software against reverse engineering and intellectual property theft. While traditional approaches rely on manual transformations or specialized tools, compiler infrastructures offer an automated and reproducible alternative to achieve large-scale automatic obfuscation. This article explores the potential of LLVM as a systematic framework for code obfuscation, leveraging its code optimization transformations not only for performance improvement but also as a source of structural complexity that hinders program analysis. To this end, we evaluated all LLVM versions from 3.9 to 19.1, considering both the legacy and the new pass managers, and analyze their ability to generate obfuscating effects. A cellular genetic algorithm (cGA) is employed to address the Software Obfuscation Problem (SOP), searching for optimal sequences of passes that maximize obfuscation. The experimental results, obtained using the PolyBench benchmark, show that the evolution of LLVM does not follow a monotonic trend: older versions often achieve higher levels of obfuscation than recent ones, which have been simplified in favor of efficiency and maintainability. These findings highlight the dual role of compiler optimizations and the need to reconsider modern infrastructures when the goal is software protection. Our study provides a reproducible methodology and insights for future research in compiler-assisted obfuscation.

Keywords: Code obfuscation · LLVM · Compiler infrastructure · Software evolution · Genetic algorithms

Acronym: CSCI-RTCW

1 Introduction

In recent years, software development has experienced a paradigm shift, moving from rigid and highly structured systems to simpler and more flexible approaches [19]. This change answers to the need of accelerating development

H. R. Arabnia et al. (Eds.): CSCI 2025, CCIS 2969, pp. 206–220, 2027.
https://doi.org/10.1007/978-3-032-28460-0_15

and deployment cycles, prioritizing rapid implementation in production to maximize return on investment in the shortest time possible.

This situation has made necessary the creation of systems and development platforms adapted to this new paradigm, providing support to increasingly dynamic and accelerated development environments. A key strategy to achieve these objectives is the use of optimization options offered by compilers, which play a fundamental role in improving the performance and code efficiency, facilitating the work of developers and programmers. In addition, the current interest in fast, specific, and short-term developments has increased the necessity of quick and accessible optimizations to make the best possible use of the underlying hardware, without the need of having a deep understanding on it.

However, in recent times these same optimizations have begun to be exploited in ways that go beyond their original purpose [14,21,22] One of the most active research directions in this regard is code obfuscation [27]. Broadly speaking, code obfuscation explores techniques to conceal the logic of a program without altering its functionality, with the goal of hindering reverse engineering and protecting intellectual property. Although manual obfuscation of large and complex projects is highly challenging, compiler-driven transformations provide an automated and systematic way to generate code that is both optimized and harder to analyze.

This emerging line of research highlights an interesting duality: the same techniques originally conceived to enhance execution speed or memory usage can also be redirected towards security objectives. In practice, the application of certain optimizations modifies the resulting machine code in non-trivial ways, which increases the difficulty of reconstructing the original logic of the program. As a result, code that has undergone aggressive optimization may inadvertently incorporate obfuscating effects, and these effects can be deliberately reinforced by researchers seeking to harden software against reverse engineering.

Within this context, the LLVM [23] compilation infrastructure emerges as a highly useful tool. Together with clang, it not only enables automatic application of specific optimizations (also called passes) to improve runtime on different hardware architectures, but also provides a flexible experimentation environment where such optimizations can be reinterpreted for obfuscation purposes.

This work presents a systematic study of the obfuscation capacity achieved through compiler optimizations, focusing specifically on the LLVM infrastructure. To this end, we evaluate all LLVM versions from 3.9 to 19.1, comparing the transformations introduced by both the legacy and the new pass managers. For each version and pass manager, the Software Obfuscation Problem (SOP) is addressed using a cellular genetic algorithm (cGA), designed to identify the optimal compilation sequence that maximizes the obfuscation level of the software. This study enables us to assess whether the progressive evolution of LLVM, in terms of the quantity and quality of available passes, correlates with an enhanced ability to hinder reverse engineering.

The rest of the document is organized as follows: Sect. 2 reviews related work on code optimization and obfuscation. Section 3 then defines SOP, the combinatorial optimization problem aimed at increasing the obfuscation level of a given software. After that, Sect. 4 outlines the experimental methodology, divided into two parts: the first focusing on pass selection and filtering, and the second on code obfuscation. Section 5 describes the experimental setup, while Sect. 6 presents and discusses the results. Finally, Sect. 7 provides the main conclusions and outlines possible directions for future work.

2 Related Work

Compilers have historically been designed with the objective of improving performance and portability. This can be achieved by the use of specific transformations that infrastructures, such as LLVM [23], provide. However, some of the transformations they perform, such as loop unrolling, inlining, or instruction reordering, can also be applied for other purposes such as code obfuscation [16], as they alter the structure of the generated machine code in ways that complicate reverse engineering.

2.1 Compiler Optimization

The search space of possible compiler optimization sequences is vast, and their effects, both on performance and on code structure, are highly non-linear [11]: the benefit of a given optimization often depends on the specific context in which it is applied, as well as on the presence or absence of other transformations. This phenomenon, commonly known as the *phase-ordering problem* [28], implies that the outcome of applying the same set of optimizations in different orders can vary significantly. Moreover, the optimization space is riddled with local optima, where small changes in a sequence may lead to disproportionate shifts in runtime behavior, memory usage, or code structure.

This complexity has motivated the application of heuristic and metaheuristic techniques, such as genetic algorithms (GAs) [8], simulated annealing [30], or reinforcement learning [29], to guide the selection and ordering of optimizations. Within this context, evolutionary approaches are particularly attractive, as they can explore diverse combinations and even adapt to multi-objective criteria [7], an appropriate approach given the existing trade-off between certain objectives such as runtime and energy consumption [5].

2.2 Code Obfuscation

Code obfuscation has emerged as a key mechanism to protect software against reverse engineering and intellectual property theft [26]. Its primary goal is to transform a program into a semantically equivalent version that is harder to understand, without modifying its intended functionality. Traditional approaches to obfuscation can be classified into three main categories [18]:

1. *Data obfuscation* techniques aim to make data representations hard to analyze or reverse-engineer while preserving functionality. Common methods include

value encoding and encryption, splitting and merging values, pointer mangling, and injecting fake data, all of which raise resilience against inspection at the cost of added complexity, size, and runtime overhead [15].
2. *Layout obfuscation* techniques rearrange code and data structures by function or block reordering, struct field permutation, or padding, to obscure program organization and remove all information that could provide semantics to the code design [10].
3. *Control flow obfuscation* techniques aim to modify a program's execution paths to make its logic harder to analyze or reverse-engineer. Techniques include flattening, splitting or merging basic blocks, inserting opaque predicates, and adding bogus or redundant branches, which increase resistance to static and dynamic analysis at the cost of performance and code complexity [9].

These techniques, while effective, often require manual effort or ad-hoc tuning, and can result in significant performance penalties when applied indiscriminately [4]. Recent advances have shifted attention towards more systematic and automated obfuscation strategies, leveraging compiler infrastructures and program transformations to achieve scalable results.

Based on this observation, several research efforts have explored the deliberate use of compiler optimizations as a tool for obfuscation [1,20], exploiting their ability to automatically produce variants of the same program that are functionally identical but structurally distinct. Some approaches rely on GAs to guide the selection and combination of these optimizations in LLVM [23]. For example, the Software Obfuscation Problem (SOP) models this search as a combinatorial optimization task [12,27], later extended in the Two-level Software Obfuscation Problem (TSOP), where the synergy from the combination of compiler optimizations with traditional obfuscation techniques is demonstrated [6].

However, and taking into account the fast evolution of actual compilers, there is no approach in the literature which raises the question of whether this rapid development corresponds to an improvement on their obfuscation capabilities. This article builds on this observation, with the aim of finding an answer to this question by studying multiple versions of the LLVM infrastructure, performing a deep and well-formed comparison between them.

3 Software Obfuscation Problem

This section introduces the Software Obfuscation Problem (SOP) [12], a combinatorial optimization problem aimed at maximizing the obfuscation quality of a program by applying generic code transformations.

Given a program P and a set of transformations $T = \{t_1, \ldots, t_k\}$ that produce semantically equivalent code P', solving SOP consists of finding a sequence $\boldsymbol{S} = [s_1, \ldots, s_N]$, where $s_i \in T$, that maximizes the obfuscation quality of P':

$$\text{Maximize } \ F(P, \boldsymbol{S}) = \text{ObfuscationQuality}(P'), \tag{1}$$

where P' is obtained by applying the sequence of transformations $\boldsymbol{S}$ to P in the same order, which directly affects the resulting code. Transformations may be repeated ($s_i = s_j$, $i \neq j$) since multiple applications often lead to further code changes [13].

4 Methodology

The methodology of this study is divided into two separate segments. On the one hand, Sect. 4.1 describes the approach studied to obtain a usable list of passes for every LLVM version, for both legacy and new pass managers. On the other hand, Sect. 4.2 focuses on the code obfuscation process, introducing the cGAs as the technique employed to maximize obfuscation using the lists of passes obtained in the previous section.

4.1 Pass Selection

Due to the nature of the LLVM compilation infrastructure, and to ease replicability and experimental isolation, Docker environments have been used. This approach allows the management of different versions of the LLVM infrastructure efficiently, isolating each experiment to avoid unintended consequences. Note that both `ubuntu:20.04` and `ubuntu:22.04` were used as base for the images; the former one for every LLVM version prior to 12 while the latter for versions from the already mentioned one.

The list of available passes in each version has been obtained using `--help` option from `opt`. This list is stored and filtered, excluding the passes specific to hardware architectures, thus allowing a more generalized analysis, independent of a particular architecture.

In particular, passes related to x86, ARM, RISC-V, MIPS, PowerPC, SPARC, SystemZ, and other specific architectures have been filtered. Likewise, passes associated with graphics and parallel processing technologies, as well as those linked to specific environments such as WebAssembly and various microcontrollers, have been excluded. Elements related to specific optimizations, such as Objective-C and Windows, have also been filtered.

Once the selected filters were applied to the collection of passes for each version, they were classified according to their compatibility with the structure of the new or old pass manager. For this purpose, PolyBench [25], a benchmark suite widely used and validated in previous studies on LLVM [27], was employed. This test suite, specifically designed to evaluate the performance of compiler optimizations, allowed the generation of binary files on which different passes were applied, thus facilitating their analysis and classification within the context of the transformations carried out by each compiler version.

After that, the categorization of the passes is carried out using the following methodology:

1. Each pass from the previously filtered list is tested using the different semantics of both pass managers on a copy of the original binary. If the application

of the pass is successful, it is added to the list corresponding to the pass manager used.

2. For the case of the new pass manager, additional scripts are executed that allow the passes to be subcategorized according to the available explicit nestings: `function`, `module`, `loop`, and `cgscc`.
3. After categorizing the passes, the task is to find the biggest feasible combination among those that support similar nesting types. This involves working through the list of suitable passes, attempting to iteratively combine them. During each iteration, a pass is added to the combination list and applied to a duplicate of the created binary. If this application and the subsequent execution are error-free, the pass is retained in the list; otherwise, it is removed, and the process continues with the next pass..

It is worth noting that, since the new pass manager was introduced in LLVM version 11 and the old one became deprecated, its usage required the options `-enable-new-pm=0` for versions 13–16 and `--bugpoint-enable-legacy-pm` for version 17 and later. In the most recent releases, its use is discouraged, except in specific research and debugging scenarios.

4.2 Code Obfuscation

Once a list of usable passes has been obtained for each version, a cellular genetic algorithm (cGA) [3] is applied to obfuscate the software program, solving SOP for each LLVM version. cGAs are characterized by their structured population, where individuals are arranged in a toroidal mesh. This structure introduces the notion of distance between members of the population, enabling the establishment of groups of individuals with similar characteristics depending on the zone of the mesh they occupy. This choice is motivated by its ability to sustain population diversity for longer than traditional genetic algorithm designs, such as panmictic ones [2], thereby delaying premature convergence.

Algorithm 1. Pseudocode of a cGA.

```
procedure cGA
    pop ← newPopulation()
    while not terminationCriterion() do
        for all indiv in pop do
            parent ← getNeighborhoodPartner(pop, indiv)
            offspring ← recombination(indiv, parent)
            offspring ← mutation(offspring)
            fitness ← evaluation(offspring)
            replacement(pop, indiv, offspring, fitness)
        end for
    end while
end procedure
```

The pseudocode of the cGA used in this work is shown in Algorithm 1. After generating and evaluating a random initial population, the algorithm enters an evolutionary loop where, in each generation, individuals are updated one by one. For each individual, neighbors are identified, two distinct parents are selected, and an offspring is produced through recombination and mutation. The offspring is then evaluated and may replace the current individual according to a replacement policy. As in classical GAs, different variants of selection, recombination, mutation, and replacement can be employed.

To guide the search, the cGA requires an objective function that quantifies the level of obfuscation. Since no single metric fully reflects software obfuscation quality, three common measures are considered [1,6,12,27]:

- *Assembly lines of code*: total number of lines in the x86 assembly of the program.
- *Cyclomatic complexity* [24]: software complexity based on decision points, computed from the control flow graph (CFG) as $CC = e - n + 2 \cdot b$, where b is the number of basic blocks, e edges, and n nodes.
- *Nesting level complexity* [17]: maximum nesting depth of branch nodes in the CFG, defined as one plus the depth of the parent control structure (if any).

Only the number of assembly lines of code is used as the optimization objective; the other two serve to analyze the achieved obfuscation.

5 Experimental Setup

The problem to be solved is defined as finding the sequence of passes that maximizes the obfuscation quality of a given software program. Since using the LLVM infrastructure allows avoiding any unwanted code transformation from the compiler, it ensures that the final generated code is direct cause of the sequence of passes applied. The number of passes available in each version and pass manager results from the prior selection and filtering process.

As already mentioned, the algorithm used is a cGA, whose configuration parameters are shown in Table 1. Both the algorithm choice and its parameterization are adopted from previous research [27]. The population consists of 25 randomly initialized individuals arranged in a square grid of size 5×5, with a linear-5 (L5) neighborhood that includes the individual itself among its north, south, east and west neighbors. Each individual chromosome is represented as an array of 300 integers representing a sequence of available passes with possible repetitions. The genetic operators employed are: binary tournament for parent selection, two-point crossover for recombination, random mutation, and an elitist strategy for replacement. Recombination is applied 90% of the times, while mutation occurs with a probability of 0.1, consisting of replacing a transformation with another randomly selected valid pass.

As the software to be obfuscated, PolyBench [25] was chosen—a well-known collection of benchmark programs frequently used in obfuscation research [1, 27]. A subset of sequential applications covering a wide range of computational operations was selected to build a tailored benchmark for the experiments.

Table 1. Configuration parameters of the experimentation.

Cellular Genetic Algorithm	
Length of individuals	300
Population size	25
Population shape	5×5
Population initialization	Random
Neighborhood	L5
Parents selection	Current indiv. + Binary Tournament
Recombination	Two Points Crossover
Recombination probability	$\rho_c = 0.9$
Mutation	Random Value
Mutation probability	$\rho_0 = 0.1$
Replacement	Replace if better
Number of iterations	25
Independent runs	10

6 Results and Discussion

6.1 Evolution of Passes Between Versions

Once the pass filtering process was completed, the total number of usable passes, as well as the number of combinable ones, was counted for each version studied. Before analyzing the results, it is important to mention that the 0 values reported for versions 3–10 in the new pass manager responds to the fact that it was not available yet. From version 11 onward, both legacy and new pass managers are represented.

Figure 1 illustrates the number of applicable and combinable passes across LLVM versions and pass managers. The legacy pass manager grows steadily until version 11, then stabilizes, and drops sharply in version 17, with its deprecation. In the particular case of our study, where onli passes compatible with PolyBench software are considered, a similar trend appears with minor variations, as pass applicability depends on factors such as hardware, program characteristics, and ordering; notably, version 10 allows fewer consecutive passes than the adjacent versions.

On the other hand, the new pass manager, introduced in version 11 and gaining relevance from version 12, can also invoke legacy passes; therefore, the total number of available options combines both sets. Unlike its predecessor, its growth in available passes is less pronounced, reflecting its purpose of reducing technical debt by merging functionalities, adjusting behaviors through nestings, and lowering the number of passes required for certain tasks. Thus, while the overall number of passes is smaller, their flexibility and combinability are greater. However, in the most recent versions, even the new pass manager shows a sharp reduction, indicating a clear trend toward simplicity.

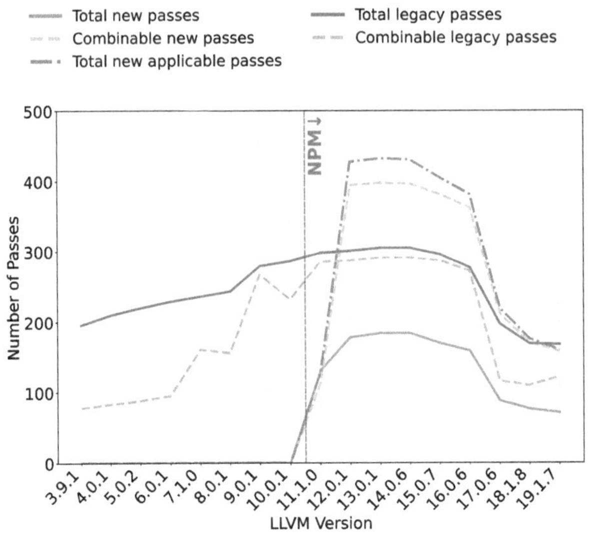

Fig. 1. Evolution of LLVM passes by version and pass manager (3.9.1–19.1.7).

Moreover, an analysis on the categorization of available passes for each version has been carried out, with the results for both legacy and new pass managers described in Fig. 2. The categories considered for this study are: analysis, optimization, inlining and promotion, vectorization and parallelism, coroutines, instrumentation, transformation, cleanup, visualization and debugging, and other specialized passes.

In the legacy pass manager (see Fig. 2b), versions 3–8 maintained a balanced distribution of passes across categories, with a slight increase in specialized options. This was followed by a sharp growth in versions 9–11, driven by the integration of the Polly framework and the addition of passes for advanced optimizations and static code analysis. Versions 12–15 emphasized performance improvements through optimization, analysis, profiling, and security, while eliminating redundant passes. From version 16 onward, the progressive deprecation of the old pass manager led to a significant reduction and restructuring, resulting in a smaller but more sophisticated set of passes focused on floating-point optimizations, parallelization, vectorization, and efficient instrumentation for analysis and debugging.

Finally, the comparative analysis of the new pass manager (see Fig. 2c), covering versions 11–19, highlights its evolution since its introduction in LLVM 11. In versions 11 and 12, several analysis and optimization passes were removed as their functionalities were integrated into new ones, enabling advanced analysis and more effective memory and control flow optimizations such as dead code elimination. Versions 13–15 reinforced this trend, with notable advances in visual analysis, parallel program optimizations, and loop handling with Polly. From version 16 onward, the strategy focused on simplifying and unifying the set of passes, replacing redundancies with safer and more efficient alternatives, while adding new passes for kernel security, exception handling, control flow protection, and optimizations for atomic operations and memory accesses to improve efficiency and adaptability across hardware environments.

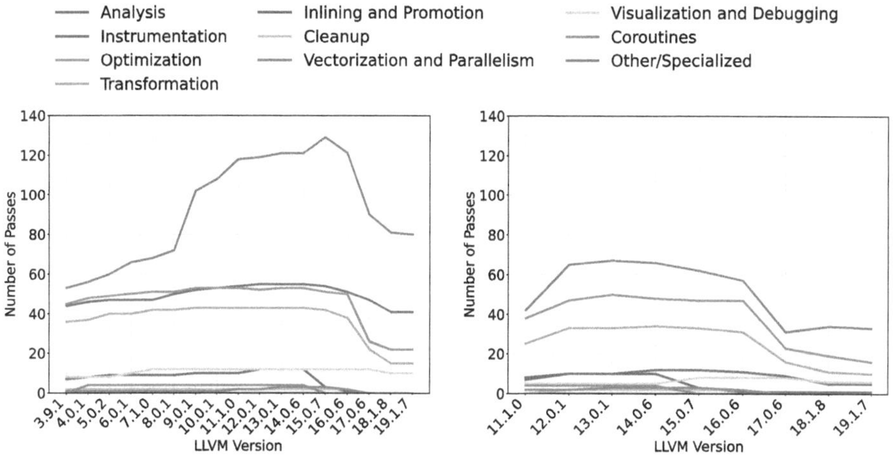

(a) Legacy pass manager (b) New pass manager

Fig. 2. Categorization of LLVM passes by legacy and new pass managers across versions 3.9.1–19.1.7. The legacy manager (a) grows then declines in passes as it was deprecated, while the new manager (b) is more compact and specialized, reflecting LLVM's shift to unified, efficient optimization.

6.2 Code Obfuscation

After obtaining the available combinable passes for each LLVM version, the study of code obfuscation capabilities through optimizations, across versions and pass managers, can finally be carried out. To this end, a total of 10 runs of the cGA are performed for each LLVM version and pass manager, reporting the results of the best solution from each run. This allows verifying how the impact of each studied case evolves in terms of the number of lines of code for Polybench. It is worth noting that the original program contains 6,025 lines of code, which provides a baseline to assess the effectiveness of the obfuscation methodology.

Figure 3 shows the distribution of the best solutions obtained from ten cGA runs for each LLVM version and pass manager. As illustrated in Fig. 3a, the legacy pass manager does not consistently exceed the one-million-line threshold until version 9, where a sudden increase in this metric occurs. Immediately after, the legacy pass manager reaches its peak in version 10, generating over 2 million more lines of code than the median value of the next highest version.

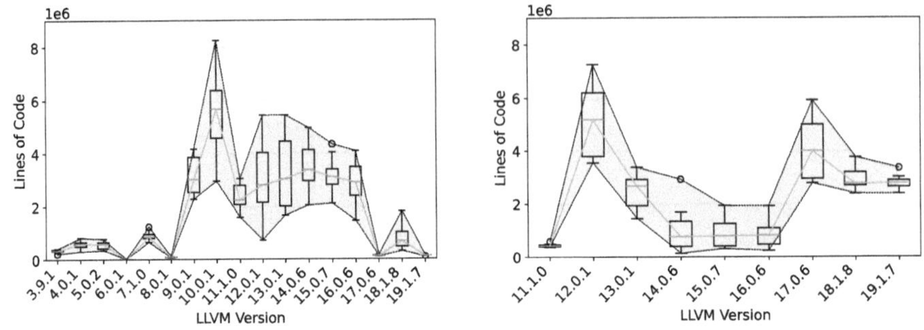

(a) Legacy pass manager (b) New pass manager

Fig. 3. Assembly code line counts of the best solutions obtained in each cGA run, reported across LLVM versions and pass managers.

After that, from versions 11 to 16, the number stabilizes around an average of 3 million lines, though with a fairly wide distribution comparable to that of the two immediately preceding versions. Finally, the fitness value drops sharply from version 17, being versions 17 and 19 the second and third lowest counts overall—only after version 6, which fails to surpass even 100,000 lines.

Moving on to the new pass manager (see Fig. 3b), the initial release, version 11, follows a behavior similar to that of the earliest legacy pass manager versions, prowling around the half-million mark. After that, version 12 shows the opposite trend, with the median number of lines increasing nearly tenfold compared to its predecessor. From there until version 16, a rapid decrement in the fitness value is observed, followed by a big increase in the final three versions, reaching approximately three and a half million lines of assembly code. Nevertheless, none of these versions matches the best results obtained with LLVM version 10.0.1 under the legacy pass manager, which produced the highest number of lines for Polybench—an improvement of 136,960.65% over the non-optimized code.

These results suggest that, while the evolution of the LLVM infrastructure generally reduces the number of passes—by consolidating them and enhancing optimization—this does not necessarily translate into better code obfuscation. Achieving the best obfuscation results therefore requires analyzing each LLVM version individually to determine which is most suitable for the target software and the specific obfuscation quality metrics.

To further deepen this study, the results of the 10 runs of the cGA starting from LLVM version 11 are compared, categorized according to the pass manager used. Figure 4 presents the corresponding boxplots grouped by version and pass manager. A clear trend emerges: the legacy pass manager achieves better performance in increasing the lines of code up to version 17, when it is finally deprecated and its effectiveness in this regard drops sharply. In contrast, the new pass manager shows a substantial improvement in terms of lines of code generation, outperforming the legacy pass manager by a wide margin in this metric.

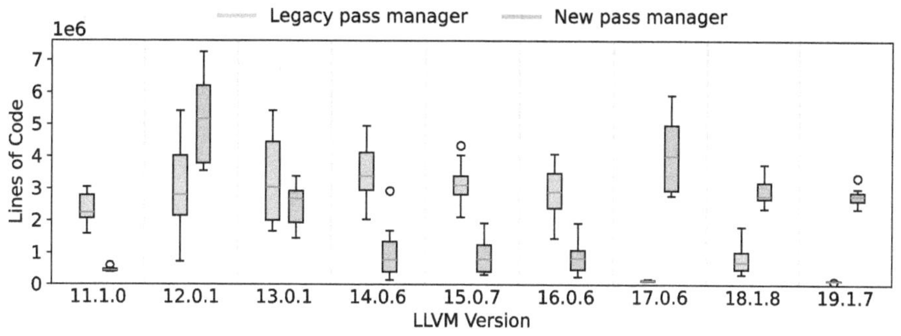

Fig. 4. Number of lines of code obtained by legacy and new pass managers, side by side, for LLVM versions between 11.1.0 and 19.1.7.

However, it can be seen how the new pass manager clearly outperforms the legacy pass manager in LLVM version 12. From the obtained results, we can conclude that the choice of version and pass manager for code obfuscation ultimately depends on the specific software being optimized and the obfuscation metric under consideration. In addition, it was shown that using a newer pass manager or a more recent LLVM version does not necessarily lead to better results in code obfuscation.

As a final contribution of this study, the cyclomatic complexity and nesting level metrics (often considered in software obfuscation [1,27]) were evaluated over the 10 runs of the cGA for a selected compiler version. Version 9.0.1 was chosen as it represents the first release of the legacy pass manager to surpass the one-million–line threshold, positioning itself as one of the most effective versions for code obfuscation while maintaining very low dispersion across runs. The values obtained for these metrics, along with the corresponding lines of code and the baseline measurements from the unobfuscated benchmark, are presented in Fig. 5.

As illustrated in the first graph—and consistent with previous figures—the number of lines of code produced by this methodology increases by nearly three orders of magnitude. Furthermore, cyclomatic complexity also exhibits a substantial rise in most cases, reaching values approaching 13, 000 independent paths. In contrast, nested complexity does not follow the same trend: despite a few high-complexity outliers, its average value remains below that observed in the baseline compilation. This behavior may be attributed to optimization techniques such as loop unrolling, which significantly expand the code size while simultaneously reducing the nesting depth. These findings suggest the need of simultaneously optimizing several metrics, or finding new more meaningful ones.

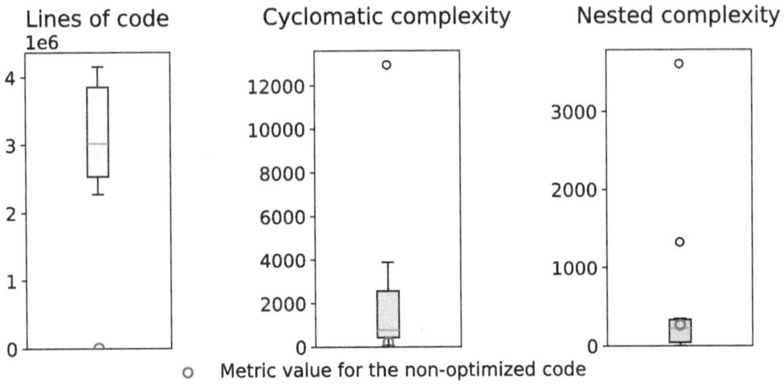

Fig. 5. Number of lines of code, cyclomatic complexity and nested complexity obtained by the cGA using LLVM version 9.0.1.

7 Conclusions and Future Work

This study introduces a novel structured methodology to analyze the evolution and impact of different LLVM versions and pass managers on software obfuscation quality metrics. First, a comprehensive classification of all available passes in each version was performed, separating them according to the two pass managers. Second, for the new pass manager, additional subdivisions were established to account for the various levels of nesting. Finally, a cGA was applied together with the previously classified passes for each LLVM version and pass manager, aiming to maximize the number of assembly code lines in Polybench.

From the results obtained, it can be concluded that the evolution of LLVM does not establish a direct correlation between optimization capabilities and code obfuscation potential. The legacy pass manager achieves the strongest effect in version 10.0.1, with a 136,960.65% increase over the non-optimized Polybench baseline, while later versions—even with reduced and consolidated passes—fail to reproduce this outcome. Generally, better obfuscation results were obtained using the legacy pass manager, although the new pass manager becomes the best one in later releases. However, exceptions were found. Overall, newer LLVM versions or pass managers do not guarantee stronger obfuscation levels; the choice must depend on the specific software and the obfuscation metrics of interest.

These findings have practical implications for software protection workflows. The proposed methodology and analysis pipeline can be integrated into CI/CD environments to automatically evaluate and select LLVM versions and pass configurations that maximize code obfuscation while maintaining performance. This integration would enable continuous adaptation to LLVM updates, providing developers with a reproducible and automated mechanism to strengthen software protection throughout the development lifecycle.

As future work, it would be interesting to study other metrics when evaluating the solutions obtained by each pass manager, like cyclomatic complexity or nesting level. Moreover, a multi-version compilation using both pass managers and/or LLVM versions might provide even better results, offering a potential advantage for practical compiler-based obfuscation techniques.

Acknowledgments. This publication is part of GRASS (FEDER-UCA-2024-A2-23) and gCODE (PIDI_2024_01307) projects, both funded by European Regional Development Funds (ERDF) and the Consejería de Universidad, Investigación e Innovación de la Junta de Andalucía. J.M. Aragón-Jurado would like to acknowledge the Spanish Ministerio de Ciencia, Innovación y Universidades for the support through FPU21/02026 grant. Juan Carlos de la Torre would like to acknowledge the University of Cadiz for its support through "Técnicas avanzadas de Inteligencia Artificial aplicadas al Tratamiento de Imágenes en entornos reales" project (I+i 2024/03).

References

1. Ahmed, H., Hyder, M.F., ul Haque, M.F., Santos, P.C.: Exploring compiler optimization space for control flow obfuscation. Comput. Secur. **139**, 103704 (2024)
2. Alba, E., Dorronsoro, B.: The exploration/exploitation tradeoff in dynamic cellular genetic algorithms. IEEE Trans. Evol. Comput. **9**(2), 126–142 (2005)
3. Alba, E., Dorronsoro, B.: Cellular Genetic Algorithms, vol. 42. Springer (2008). https://doi.org/10.1007/978-0-387-77610-1
4. Aonzo, S., Georgiu, G.C., Verderame, L., Merlo, A.: Obfuscapk: an open-source black-box obfuscation tool for android apps. SoftwareX **11**, 100403 (2020)
5. Aragón-Jurado, J.M., Bangash, A.A., Dorronsoro, B., Ali, K., Hindle, A., Ruiz, P.: Does faster mean greener? Runtime and energy trade-offs in iOS applications with compiler optimizations. Sustain. Comput.: Inform. Syst. 101166 (2025)
6. Aragón-Jurado, J.M., Jareño, J., De la Torre, J.C., Ruiz, P., Dorronsoro, B.: Two-level software obfuscation with cooperative co-evolutionary algorithms. In: 2024 IEEE Congress on Evolutionary Computation (CEC), pp. 1–8. IEEE (2024)
7. Aragón-Jurado, J.M., de la Torre, J.C., Ruiz, P., Dorronsoro, B.: Automatic software tailoring for green internet of things. Internet Things 101521 (2025)
8. Aragón-Jurado, J.M., de la Torre, J.C., Ruiz, P., Galindo, P.L., Zomaya, A.Y., Dorronsoro, B.: Automatic software tailoring for optimal performance. IEEE Trans. Sustain. Comput. **9**(3), 464–481 (2023)
9. Balachandran, V., Tan, D.J., Thing, V.L., et al.: Control flow obfuscation for android applications. Comput. Secur. **61**, 72–93 (2016)
10. Bichsel, B., Raychev, V., Tsankov, P., Vechev, M.: Statistical deobfuscation of android applications. In: Proceedings of the 2016 ACM SIGSAC Conference on Computer and Communications Security, pp. 343–355 (2016)
11. Cooper, K., et al.: Exploring the structure of the space of compilation sequences using randomized search algorithms. J. Supercomput. **36**, 135–151 (2006). https://doi.org/10.1007/s11227-006-7954-5
12. de la Torre, J.C., Aragón-Jurado, J.M., Jareño, J., Varrette, S., Dorronsoro, B.: Obfuscating LLVM intermediate representation source code with NSGA-II. In: 15th International Conference on Computational Intelligence in Security for Information Systems (CISIS 2022), pp. 181–191. Springer, Cham (2022). https://doi.org/10.1007/978-3-031-18409-3_18
13. de la Torre, J.C., Ruiz, P., Dorronsoro, B., Galindo, P.L.: Analyzing the influence of LLVM code optimization passes on software performance. In: Information Processing and Management of Uncertainty in Knowledge-Based Systems. Applications, pp. 272–283. Springer (2018)

14. Franz, M.: E unibus pluram: massive-scale software diversity as a defense mechanism. In: Proceedings of the 2010 New Security Paradigms Workshop, pp. 7–16 (2010)
15. Fukushima, K., Kiyomoto, S., Tanaka, T., Sakurai, K.: Analysis of program obfuscation schemes with variable encoding technique. IEICE Trans. Fundam. Electron. Commun. Comput. Sci. **91**(1), 316–329 (2008)
16. Guelton, S., Guinet, A., Brunet, P., Martinez, J.M., Dagnat, F., Szlifierski, N.: Combining obfuscation and optimizations in the real world. In: IEEE 18th International Working Conference on Source Code Analysis and Manipulation (SCAM), pp. 24–33. IEEE (2018)
17. Harrison, W., Magel, K.: A complexity measure based on nesting level. SIGPLAN Not. **16**(3), 63–74 (1981)
18. Hosseinzadeh, S., et al.: Diversification and obfuscation techniques for software security: a systematic literature review. Inf. Softw. Technol. **104**, 72–93 (2018)
19. Jabbari, R., bin Ali, N., Petersen, K., Tanveer, B.: What is DevOps? A systematic mapping study on definitions and practices. In: Proceedings of the scientific workshop proceedings of XP2016, pp. 1–11 (2016)
20. Junod, P., Rinaldini, J., Wehrli, J., Michielin, J.: Obfuscator-LLVM-software protection for the masses. In: IEEE/ACM 1st International Workshop on Software Protection, pp. 3–9. IEEE (2015)
21. Koo, H., Chen, Y., Lu, L., Kemerlis, V.P., Polychronakis, M.: Compiler-assisted code randomization. In: 2018 IEEE Symposium on Security and Privacy (SP), pp. 461–477. IEEE (2018)
22. Lin, Z., Riley, R.D., Xu, D.: Polymorphing software by randomizing data structure layout. In: Flegel, U., Bruschi, D. (eds.) DIMVA 2009. LNCS, vol. 5587, pp. 107–126. Springer, Heidelberg (2009). https://doi.org/10.1007/978-3-642-02918-9_7
23. LLVM: The LLVM Compiler Infrastructure. https://llvm.org/. Accessed 7 Mar 2025
24. McCabe, T.: A complexity measure. IEEE Trans. Softw. Eng. **SE-2**(4), 308–320 (1976)
25. Pouchet, L.N.: PolyBench/C – Homepage of Louis-Noël Pouchet. http://web.cse.ohio-state.edu/~pouchet.2/software/polybench/. Accessed 7 Mar 2025
26. Schrittwieser, S., Katzenbeisser, S.: Code obfuscation against static and dynamic reverse engineering. In: Filler, T., Pevný, T., Craver, S., Ker, A. (eds.) IH 2011. LNCS, vol. 6958, pp. 270–284. Springer, Heidelberg (2011). https://doi.org/10.1007/978-3-642-24178-9_19
27. de la Torre, J.C., Jareño, J., Aragón-Jurado, J.M., Varrette, S., Dorronsoro, B.: Source code obfuscation with genetic algorithms using LLVM code optimizations. Log. J. IGPL jzae069 (2024)
28. Touati, S.A.A., Barthou, D.: On the decidability of phase ordering problem in optimizing compilation. In: Proceedings of the 3rd Conference on Computing Frontiers, pp. 147–156 (2006)
29. Wang, H., et al.: Automating reinforcement learning architecture design for code optimization. In: Proceedings of the 31st ACM SIGPLAN International Conference on Compiler Construction, pp. 129–143 (2022)
30. Zhong, S., Shen, Y., Hao, F.: Tuning compiler optimization options via simulated annealing. In: Second International Conference on Future Information Technology and Management Engineering, pp. 305–308. IEEE (2009)

STL-QCRYPTO: Strategic Industrial Framework for Quantum-Resistant Cyber Defense

Arit Kumar Bishwas[1(✉)] and Mousumi Sen[2]

[1] PWC USA, San Francisco, CA 94105, USA
aritkumar.official@gmail.com
[2] SETI Institute, Mountain View, CA 94043, USA

Abstract. As quantum computing progresses, its capacity to undermine widely deployed cryptographic systems presents a profound challenge to contemporary cybersecurity. This paper introduces a strategic roadmap designed to help industries anticipate and mitigate risks arising from quantum-enabled attacks. Specifically, we develop a quantum-resistant cryptographic framework that provides a systematic and practical approach for securing critical infrastructures. We propose a novel framework, termed **STL-QCRYPTO**, which delivers tailored, sector-specific methodologies for implementing quantum safe security systems. By offering a structured roadmap, STL-QCRYPTO equips industries with actionable strategies to ensure long-term resilience against the disruptive security implications of quantum computing.

Keywords: Post-Quantum Cryptography (PQC) · Quantum-Safe Security · Quantum Key Distribution (QKD) · Critical Infrastructure Protection · Quantum Cryptography

1 Introduction

We are living in an era of remarkable innovation, where digital advancements are transforming our daily lives and reshaping the way we experience the world. From the cloud [1] to artificial intelligence [2], and now quantum computing [3], technological progress is accelerating at an unprecedented pace. With each breakthrough, new opportunities emerge, offering the potential to revolutionize industries, enhance the human experience, and solve complex global challenges.

Quantum computing is enabling advanced AI, new materials [4], faster drug discovery [5], and improved environmental sensing with quantum-based sensors [6]. These advances mark a leap toward a more intelligent and connected future. Quantum advancements bring both promise and complexity, demanding attention to security and ethics. Fully functional quantum systems with a few hundred error-free qubits could undermine digital cybersecurity, threatening current encryption methods [7], particularly those based on prime factorization [8]. In 1996, Shor's algorithm [9] showed that quantum computers could break classical

H. R. Arabnia et al. (Eds.): CSCI 2025, CCIS 2969, pp. 221–239, 2027.
https://doi.org/10.1007/978-3-032-28460-0_16

cryptography like RSA [10], highlighting the urgency of quantum-safe solutions. This milestone spurred global efforts to develop quantum-resistant cryptography, and as quantum technology advances, businesses must prioritize adopting such methods to secure digital systems.

Despite impressive progress [11,12], large-scale quantum computers may take time to arrive, but their threat is immediate. Data stolen today could be decrypted in the future once quantum systems mature, exposing sensitive information. Since many current algorithms are vulnerable, businesses must adopt quantum-resistant encryption now to safeguard long-term security. This paper introduces a strategic framework for mitigating quantum attacks in industry, structured across three levels—foundational, intermediate, and advanced—based on implementation priority and complexity. The framework outlines adoption processes for each level, explores industrial use cases, and highlights regulatory considerations to ensure compliance while strengthening defenses against quantum threats. Further detail elaboration can be found in our arxiv preprint [13].

2 Security Landscape and Quantum Computing

Quantum computing has advanced rapidly, promising exponential speedups over classical systems [14–17]. It could transform optimization, AI, materials design, and drug discovery, but also threatens classical cryptography, as secure channels may be broken by powerful quantum computers. Quantum cryptography addresses this risk by enabling quantum-secure key distribution, making interception nearly impossible [18].

Quantum cryptography traces back to 1970, when Wiesner [19] proposed quantum money, inspiring QKD research. Building on EPR principles [20], Bennett and Brassard introduced the BB84 protocol in 1984 [21], demonstrating secure key distribution via photon polarization. Ekert's 1991 scheme [22], based on Bell states and entanglement [23], and the BBM92 protocol in 1992 [24] further advanced QKD and remain foundational to modern quantum cryptography.

2.1 Essential Quantum Mechanical Principles

Classical mechanics is intuitive, but quantum mechanics at the atomic scale shows unusual phenomena. Quantum computers exploit these, using qubits that can be both "0" and "1" simultaneously, and operate through key quantum principles.

- Superposition: In quantum mechanics, superposition means a quantum state can exist in multiple states at once, enabling quantum computers to explore many possibilities simultaneously [25].
- Entanglement: In quantum physics, entanglement links particles so changes in one affect the others instantly, a principle behind QKD protocols such as Ekert91 and BBM92, though its cause is still unknown [26].

- Interference: In quantum mechanics, interference arises when wavefunctions overlap, leading to constructive or destructive effects. In quantum computing, this principle controls qubits and optimizes algorithms by manipulating probabilities beyond classical limits [27].
- No-cloning theorem: The no-cloning principle forbids copying quantum states, as any interaction collapses superposition into a classical state [28].
- Decoherence: Although quantum states exist in superposition, measurement collapses them into a single state, preventing simultaneous observation of multiple states [29].

The quantum threat to current security systems demands urgent action. Due to complex infrastructures, a strategic plan is needed. We propose a quantum-safe framework to mitigate risks and prepare organizations for future threats.

2.2 Current Cryptographic Framework

Cryptography is divided into symmetric and asymmetric methods. Symmetric uses one shared key for encryption and decryption, while asymmetric uses a public–private key pair. Both secure data but differ in overhead, key distribution, and use cases.

Asymmetric Cryptography: Asymmetric cryptography [30], notably RSA, secures data using the difficulty of factoring large numbers. It employs a public key for encryption and a private key for decryption but is slower and needs larger keys than symmetric methods. Faster factorization would force even bigger primes, raising costs. Figure 1 shows this process.

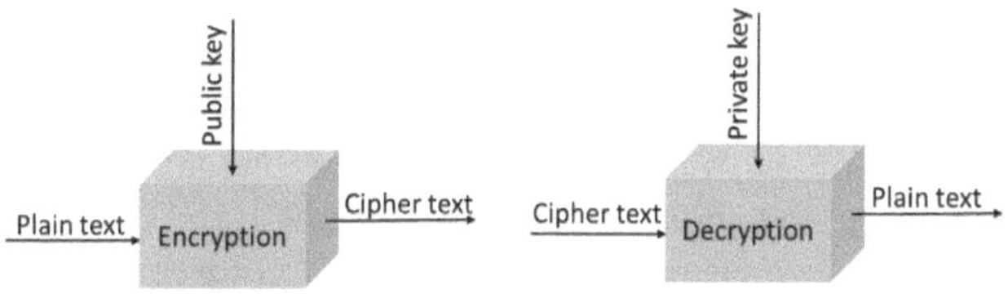

Fig. 1. Encryption and decryption.

Symmetric Cryptography: In symmetric cryptography [31], sender and receiver share the same secret key, usually generated by a random number generator. It is faster and needs shorter keys than asymmetric methods while remaining secure. AES [32], the most common algorithm, supports 128-, 192-, and 256-bit keys (Fig. 2).

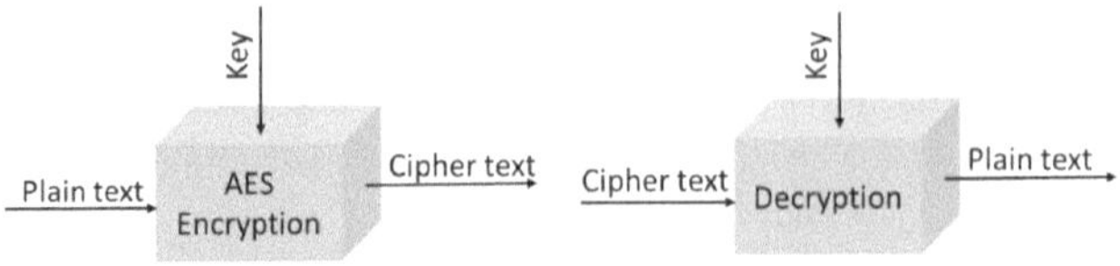

Fig. 2. Encryption and decryption with AES protocol.

3 Proposed Industrial Strategic Framework: STL-QCRYPTO

We propose three strategic levels to guide the shift from classical to quantum-safe security. This phased roadmap addresses immediate risks, mid-term upgrades, and long-term innovations, helping organizations adopt quantum-resistant solutions with minimal disruption and future-proof their infrastructure.

In our strategic roadmap for adopting quantum-safe security, we introduce "strategic transition level" (STL) framework which outlines three key levels of defence. STL-1 (foundational level) focuses on securing systems using asymmetric cryptography in anticipation of quantum threats. STL-2 (intermediate level) and STL-3 (advanced level) address more advanced security measures, emphasizing symmetric cryptography strategies to safeguard against potential quantum attacks. This multi-tiered approach ensures comprehensive protection as the quantum landscape evolves.

3.1 Strategic Transition Level 1 (STL-1): Foundation Classical Way Post-quantum Safety

NIST recently released the first post-quantum cryptography standards [33], urging the industry to act now. Lattice- and hash-based schemes dominate this effort, with three algorithms standardized to resist classical and quantum attacks. While no quantum computer yet threatens current security, Moody [34] warns agencies to prepare in advance. NIST selected CRYSTAL-Kyber [35] as the main key encapsulation mechanism for encryption, and three digital signature schemes—CRYSTAL-Dilithium [36], FALCON [37], and SPHINCS+ [38]—for standardization (Fig. 3). These quantum-resistant algorithms ensure authentication, integrity, and rotection of sensitive data in the digital era.

We recommend STL-1 as the immediate priority for securing sensitive data. It is cost-effective, easy to implement, and requires minimal infrastructure changes, enabling a smooth transition with little disruption. Compared to the more complex STL-2 and STL-3 strategies, STL-1 addresses urgent vulnerabilities while laying the foundation for future security upgrades.

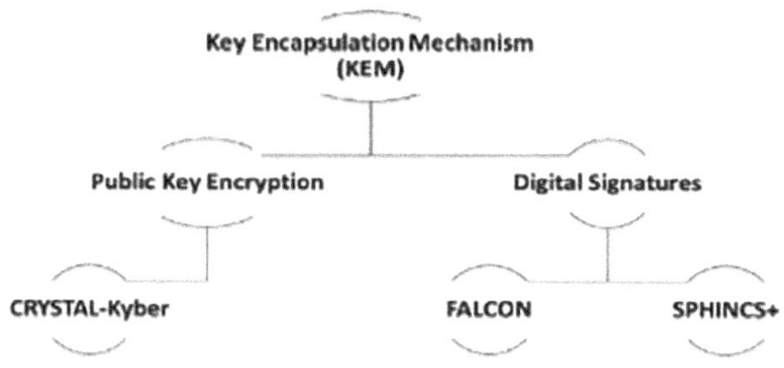

Fig. 3. Primary key encapsulation mechanism recommended by NIST.

3.2 Strategic Transition Level 2 (STL-2): Intermediate Quantum Enhanced Approach

Currently, quantum systems are not mature enough to fully test STL-1. While STL-1 relies on advanced classical algorithms to counter future quantum attacks, it may not provide complete protection. For long-term security, we recommend moving to STL-2 after STL-1.

STL-2 adopts a hybrid quantum–classical model, where Quantum Random Number Generators (QRNGs) [39] provide true randomness beyond classical methods. By combining QRNG keys with classical algorithms like AES [40], STL-2 strengthens defenses against post-quantum attacks. This approach improves resilience over STL-1 and enables gradual integration of quantum technologies without disrupting existing systems.

Implementing QRNGs costs more than STL-1 due to hardware, training, and maintenance, but the security benefits justify the expense. As future quantum attacks could be far costlier, investing in QRNGs is a proactive step to protect critical assets and ensure long-term cybersecurity resilience.

3.3 Strategic Transition Level 3 (STL-3): Advanced Quantum Only Approach

While QRNGs enhance hybrid security, the classical component remains vulnerable, making a pure quantum approach necessary. STL-3 adopts Quantum Key Distribution (QKD) [41], often seen as quantum cryptography, which extends symmetric methods using quantum principles. By leveraging superposition and entanglement, QKD enables secure key exchange, offering strong protection against current and future quantum threats. STL-3 requires complex adoption and high infrastructure costs, making feasibility analysis vital to identify gaps and prioritize the shift to quantum solutions. Careful planning ensures a smooth and cost-efficient transition.

Figure 4 highlights the strategic framework and associated levels for transition from classical to quantum security eco-system.

Table 1 shares important parameters that help in determining the feasible strategic level that we should focus on to consider.

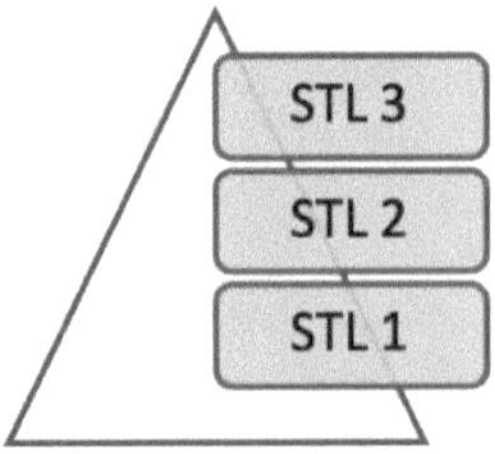

Fig. 4. Strategic transition level framework.

Table 1. Post-Quantum Readiness Across Strategic Transition Levels (STL)

STL	Priorities	Transition Efforts	Implementation Complexity	Transition Cost	Special Skill Requirement	Adoption Timeline	Quantum Safe
1	High	Low	Low	Low	Low	Immediate	Moderate
2	Medium	Medium	Medium	Medium	Medium	0–1 year	Intermediate
3	Medium	High	High	High	High	1–2 years	Advanced

3.4 Structured Industrial Framework for Quantum-Resilient Cryptography Transition

Quantum attacks pose major risks, but the bigger challenge is integrating quantum-resistant methods into classical infrastructures. To address this, we propose "QCRYPTO," a strategic framework for seamless quantum infusion or ransformation of existing cryptographic systems. The QCRYPTO framework offers a structured roadmap for transitioning from classical to quantum-resilient cryptography. It supports sectors like finance, healthcare, and critical infrastructure in mitigating quantum threats through phased adoption aligned with industry needs and global standards, ensuring robust and compliant cybersecurity.

The proposed strategic QCRYPTO framework is a seven-stage, strategic roadmap designed to guide industries in the adoption of quantum-resistant cryptography. Each stage builds upon the last, ensuring a methodical and high-impact transition toward quantum-resilient security. The framework is structured as follows:

1. Quest: This stage initiates the exploration of quantum threats and opportunities, driving industries to investigate the current landscape and emerging quantum technologies.
2. Commence: Here, the organization identifies vulnerabilities within its classical cryptography systems that are susceptible to quantum attacks, laying the groundwork for targeted intervention.
3. Review: At this stage, businesses evaluate their current infrastructure and risk levels, prioritizing areas for immediate improvement based on operational needs and the severity of potential quantum threats.

4. Yield: This involves the gradual integration of quantum-resistant solutions, ensuring compatibility with existing systems and fostering a culture of innovation and adaptability within the organization.
5. Pivot: A strategic overhaul of the organization's cryptographic architecture takes place, incorporating advanced quantum-resistant algorithms and enhancing security protocols in critical areas.
6. Transcend: In the final stage, industries scale their quantum-resilient security measures, amplifying defences across all digital assets, ensuring comprehensive protection, and preparing for future quantum advancements.
7. Observe: After deployment, solutions must be continuously monitored for vulnerabilities, with regular updates and patches applied as new threats arise. This proactive approach keeps security resilient against evolving challenges.

This progression helps industries secure current systems while future-proofing operations in the quantum era. By integrating QCRYPTO and STL, we present the comprehensive "STL-QCRYPTO" framework (Fig. 5).

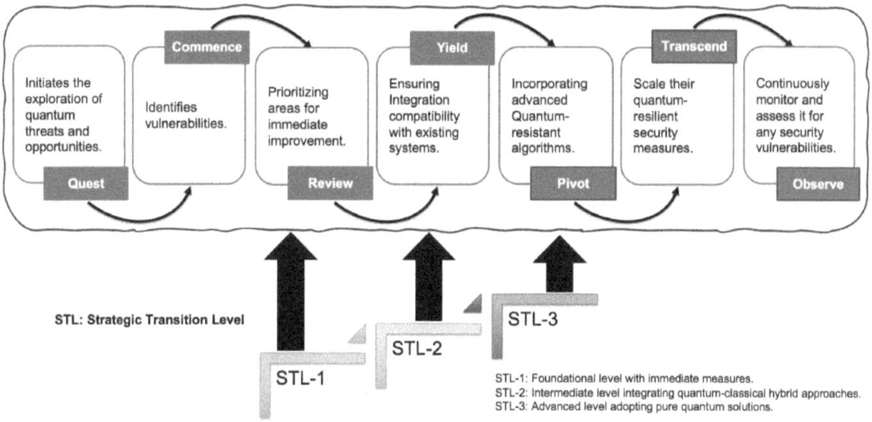

Fig. 5. "STL-QCRYPTO" framework, a comprehensive complete strategic and solutioning framework to handle the quantum attacks.

4 Industrial Adoption Strategy

The transition to post-quantum cryptography requires a proactive and strategic approach, especially in terms of hardware management. Small organizations can leverage cloud-based services, while mid-size and large organizations may need to invest significantly in upgrading data centers, HSMs, and network infrastructure to support the computational demands of PQC. Figure 6 reflects the proposed adoption strategy.

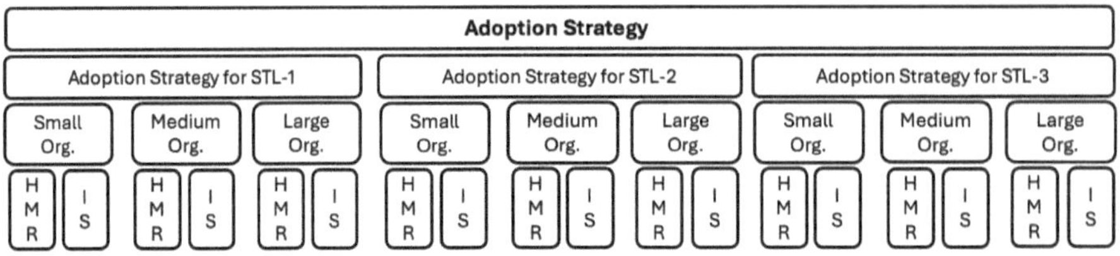

Fig. 6. STL–QCRYPTO framework adoption strategy for small, mid-sized and large organizations. Here, STL–(1/2/3), HMR, and IS represent "Strategic Transition Level (1/2/3) ", "Hardware Management Requirements", and "Implementation Strategy" respectively.

4.1 Adoption Strategy for STL-1

Small Organizations: Small organizations often rely on cloud-based or outsourced IT with limited resources. Quantum-safe algorithms like CRYSTALS-Kyber or FALCON add overhead, so hardware optimization and basic key management via HSMs or enclaves are needed. Implementation should focus on cloud-based PQC services, quantum-safe VPNs/TLS, and software-based end-to-end encryption to avoid major hardware costs. For instance, a small e-commerce firm can adopt cloud HSMs supporting Kyber or Dilithium to secure data without heavy infrastructure investment.

Mid-Size Organizations: Mid-size organizations must upgrade HSMs for PQC key management, enhance processing power, and ensure legacy systems can run both classical and quantum-safe algorithms. Implementation should adopt hybrid cryptography for compatibility, invest in PQC-ready HSMs (e.g., supporting Dilithium, FALCON), and upgrade firewalls, routers, and VPNs for stronger network security. A phased hardware replacement, starting with critical systems, ensures smoother transition.

Large Organizations: Large organizations need scalable HSMs, upgraded data centers, and efficient key management to handle the heavy demands of PQC algorithms like SPHINCS+ and Kyber. Implementation should follow an enterprise-wide migration strategy, starting with critical systems, supported by quantum-safe HSMs, hybrid infrastructures for compatibility, and major data center upgrades. Dedicated quantum security teams must oversee cryptographic operations and coordinate with vendors for readiness.

4.2 Adoption Strategy for STL-2

Implementing STL-2 PQC with QRNGs requires hardware strategies tailored to organizational size. Small firms can rely on cloud-based or plug-and-play QRNGs, while mid-size and large enterprises need dedicated hardware, scalable HSMs, and hybrid systems. Best practices include vendor collaboration, regular security audits, and phased adoption starting with critical systems like financial transactions and customer databases.

Small Organizations: Small organizations with minimal infrastructure can adopt QRNGs through cloud services or low-cost plug-and-play devices, integrating them with classical systems for secure key management. Implementation should focus on cloud-based QRNG offerings, affordable on-premise devices, and hybrid protocols like CRYSTALS-Kyber. For example, a small e-commerce firm can enhance security by using cloud-based QRNG keys for online transactions without heavy hardware investment.

Mid-Size Organizations: Mid-size organizations with mixed cloud and on-premise infrastructure must integrate QRNGs without disrupting operations and adopt advanced key management for hybrid encryption. Implementation should include dedicated QRNG hardware for sensitive data, use of quantum-safe algorithms like CRYSTALS-Dilithium, hybrid cloud/on-premise systems, and data center upgrades. For example, a healthcare provider can deploy QRNG hardware to secure patient data while supporting hybrid cryptographic approaches.

Large Organizations: Large organizations with global infrastructures need high performance QRNGs and HSM integration to manage secure keys at scale. Implementation should focus on enterprise-grade QRNG systems with HSMs, phased hybrid encryption using Kyber for key exchange and Dilithium for signatures, and scalable upgrades to data centers and networks. For example, a multinational bank could deploy QRNG-powered HSMs across global data centers to secure large-scale cross-border transactions.

4.3 Adoption Strategy for STL-3

STL-3 integrates QKD with QRNG to deliver maximum quantum-safe security. Small organizations can rely on cloud-based services, mid-size firms on dedicated QKD hardware, and large enterprises on global QKD networks with enterprise-grade QRNGs. Across all sizes, best practices include vendor collaboration, phased rollout beginning with critical systems, and regular audits to ensure compatibility and security.

Small Organizations: Small organizations with minimal infrastructure can adopt quantum security using cloud-based QKD and low-cost plug-and-play QRNG devices. Implementation should focus on managed QKD services for sensitive systems, QRNG tokens to enhance key randomness, and classical PQC for less critical areas. For example, a small online business could secure payments through QKD cloud services while using QRNG keys for encrypted communications.

Mid-Size Organizations: Mid-size organizations with hybrid infrastructures should adopt dedicated QKD nodes and robust QRNG devices to protect sensitive data like finance or healthcare records. Implementation includes deploying QKD in critical systems, upgrading data centers for secure key distribution, and combining PQC with QKD in hybrid models. For example, a healthcare provider could secure patient records with QKD while using QRNG-enhanced encryption for communications and storage.

Large Organizations: Large organizations with global infrastructures need enterprise level QKD and QRNG integrated with HSMs for secure key management across data centers and offices. Implementation should focus on building global QKD networks, deploying enterprise-grade QRNG systems, and ensuring compliance in regulated sectors like finance or defense. For example, a global bank could secure inter-office communications with QKD while using QRNG-based HSMs to manage keys across data centers.

5 Impact Assessment: Industry Sectors at Risk

Post-quantum threats demand phased mitigation across various industries. STL-1 introduces PQC algorithms like CRYSTALS-Kyber, STL-2 adds QRNG-based hybrid security, and STL-3 deploys QKD for unbreakable key exchange. Together, these stages secure critical infrastructures against quantum-era risks (Fig. 7). We now turn to a discussion of the following key industry use cases.

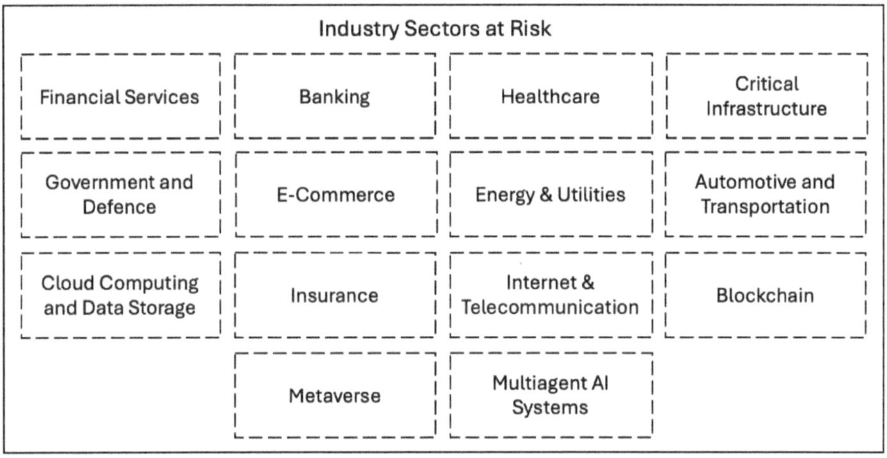

Fig. 7. Some of the key industries identified as being at significant high risk in a post-quantum world.

5.1 Financial Services

Financial transactions, particularly international fund transfers and high-value payments, are highly vulnerable to advanced cyber threats, including quantum attacks. To mitigate these risks, *CRYSTALS-Kyber* can secure cross-border payment networks like SWIFT by encrypting transaction data with quantum-resistant algorithms, while *CRYSTALS-Dilithium* ensures authentication and integrity through quantum-safe digital signatures, providing long-term protection of financial records.

5.2 Banking

Authentication in online and mobile banking requires robust, unpredictable mechanisms to verify user identities and protect transactions. Quantum Random Number Generators (QRNGs) can strengthen this process by generating truly random One-Time Passwords (OTPs) for two-factor authentication [42] and securing Public Key Infrastructure (PKI), ensuring that keys, signatures, and certificates remain resistant to quantum attacks.

5.3 Healthcare

Protecting patient data, electronic health records (EHRs), and telemedicine communications is critical in healthcare. Hospitals can address this by using Quantum Key Distribution (QKD) to secure EHRs and patient data, enabling confidential communication between facilities and medical devices such as MRI scanners and central databases, thereby ensuring quantum-safe protection of sensitive information [43].

5.4 Critical Infrastructure

Smart grids and critical infrastructure systems depend on secure communication between distributed assets and control centers to maintain reliable service. *CRYSTALS-Kyber* can encrypt data exchanges between sensors, meters, and control systems, while *FALCON*'s compact signatures and fast verification secure grid commands with minimal latency, providing quantum-resistant, real-time protection.

5.5 Government and Defence

Governments and defense systems must secure sensitive and classified communications from quantum threats to maintain confidentiality and authenticity. *SPHINCS+* can protect classified documents with long-term quantum-safe signatures, while *CRYSTALS-Dilithium* efficiently authenticates directives and communications. QRNGs provide truly random keys for military systems such as satellites and tactical radios, and QKD safeguards diplomatic and defense channels by enabling secure key exchange against quantum-enabled adversaries [44].

5.6 E-Commerce

E-commerce platforms must protect online payments and sensitive customer data from quantum threats by strengthening encryption and ensuring unpredictable security. *CRYSTALS-Kyber* can safeguard payment data with quantum-resistant encryption, while *CRYSTALS-Dilithium* provides digital signatures to ensure transaction integrity. QRNGs generate truly random keys for secure sessions, and QKD enables safe key distribution between servers and financial institutions, securing high-value transactions and sensitive communications.

5.7 Energy and Utilities

Critical infrastructure, especially grid communication systems, must be protected from quantum-enabled cyberattacks. *CRYSTALS-Kyber* can encrypt smart grid data, securing energy usage transmissions between consumers, utility providers, and grid controllers against future quantum threats.

5.8 Automotive and Transportation

Autonomous driving systems must be protected from quantum-enabled adversaries attempting to intercept or manipulate vehicle communications. *CRYSTALS-Kyber* can secure vehicle-to infrastructure (V2I) and vehicle-to-vehicle (V2V) communications by encrypting control commands and data transmissions between autonomous vehicles and traffic management systems, ensuring safety and trust.

5.9 Cloud Computing and Data Storage

Cloud systems must be protected from quantum threats by securing sensitive data, safeguarding encryption keys, and ensuring transmission confidentiality against eavesdropping. In this context, *CRYSTALS-Kyber* can encrypt stored and transmitted data, *QRNGs* generate unpredictable keys for storage and communication security, and *QKD* ensures safe key distribution between data centers and user devices, preventing interception and maintaining quantum-safe confidentiality.

5.10 Insurance

Insurance records and customer data must be protected from future quantum-based cyberattacks. *CRYSTALS-Kyber* can encrypt sensitive information, including policyholder details, claims data, and financial transactions, ensuring resilience against quantum-enabled breaches.

5.11 Internet and Telecommunications

Telecom providers must secure VoIP, VPN, and encrypted channels, safeguard satellite-based communications, and ensure quantum-safe transmission across 5G networks, data centers, and undersea cables. *FALCON* can secure certificates and session initiations with efficient quantum-safe authentication, while *SPHINCS+* provides long-term hash-based signatures. *QRNGs* generate unpredictable keys for satellite links, protecting sensitive communications, and *QKD* secures fiber, data centers, and undersea cables, with industry trials such as BT exploring its use for 5G networks [45].

5.12 Blockchain Applications

Blockchain networks rely on cryptography to secure nodes, validate transactions, and protect smart contracts, but quantum computing threatens their integrity and tamper resistance. *CRYSTALS-Kyber* can secure on-chain data with quantum-safe encryption, while *SPHINCS+* safeguards smart contract signatures from forgery. *QRNGs* enhance consensus by generating unpredictable randomness, and *QKD* secures key distribution for validating transactions and contracts, ensuring long-term quantum resistance [45].

5.13 Metaverse Applications

Metaverse platforms must secure digital identities, assets, and transactions to ensure user privacy and protect virtual economies from quantum threats. *CRYSTALS-Dilithium* enables quantum-safe identity verification, while *SPHINCS+* protects digital assets such as NFTs from forgery. *QRNGs* generate strong keys for avatars, assets, and user data, and *QKD* [46] secures identities and transactions, ensuring the integrity and resilience of virtual economies.

5.14 Complex LLM-Based Agentic AI Systems

In large multiagent systems [47] powered by LLM-based agents [48,49], secure communication is critical for coordinating decisions and safeguarding sensitive data. *FALCON* provides fast quantum-safe signatures for interagent communication, while *CRYSTALS-Kyber* encrypts exchanged data. *QRNGs* generate unpredictable keys for communication channels, and *QKD* enables real-time, quantum-resistant coordination in domains such as autonomous fleets, smart cities, and supply chains.

6 Policy and Regulatory Considerations for Quantum-Resistant Security

The regulatory landscape for quantum security is evolving. Governments and industries must adapt to advancing technologies, with bodies like NIST driving standards. Sector-specific guidelines and cross-border compliance are essential for a unified global framework.

Standardization and Certification: NIST is finalizing PQC standards, including CRYSTALS-Kyber for encryption and CRYSTALS-Dilithium for signatures, with deployment planned for critical infrastructure [33]. Certification frameworks like FIPS 140-3 are also being updated by NIST and ENISA to include quantum-resistant algorithms [50].

Data Protection Regulations: GDPR may require quantum-safe encryption to protect sensitive data in sectors like finance, healthcare, and government [51]. In the U.S., CISA is preparing guidance for critical areas such as energy, finance, and defense to meet emerging quantum-safe standards [52].

Transition and Interoperability: NIST advises a hybrid model that combines classical and quantum-safe algorithms to protect data during migration [52]. Both NIST and ENISA stress interoperability so classical and quantum-resistant systems can operate together throughout the transition [33].

Critical Infrastructure and Supply Chain Security: ENISA stresses that sectors like energy, telecom, healthcare, and defense must adopt quantum safe solutions [33,52]. Similarly, the NSA is pushing quantum-resistant cryptography into defense and public sector supply chains [53].

Mandates and Compliance Deadlines: The U.S. Quantum Computing Cybersecurity Preparedness Act requires federal agencies to assess vulnerabilities and plan migration to quantum-safe solutions [51,53]. At the global level, bodies like the G7 and OECD are working toward a unified framework for cross border quantum security [52,53].

Research and Innovation Incentives: Governments in the U.S. and EU are promoting public-private partnerships and funding to accelerate quantum-safe cryptography, with NIST and ENISA encouraging collaboration [52]. Financial grants are also being considered to help SMEs adopt quantum-resistant solutions [33,54].

Sector-Specific Regulations: In finance, SWIFT is collaborating with NIST on quantum-safe frameworks [52], while in healthcare, regulations like HIPAA are expected to adopt quantum-resistant requirements to protect sensitive data [51].

7 Implementation Challenges

Both technical and business challenges related to quantum cryptography implementation require strategic planning, incremental adoption, and investment in infrastructure and human capital.

7.1 Technical Challenges

Implementing secure quantum communication faces challenges such as integrating quantum hardware with classical systems, managing scalability and costs, and developing standardized protocols with error correction for reliable cryptography.

Technical Challenges for Implementing STL-1: PQC algorithms such as CRYSTALS Kyber, Dilithium, and FALCON demand higher computational resources, generate larger keys, and create bandwidth and storage overhead, often exceeding the capacity of legacy hardware like HSMs and routers. To address these challenges, organizations should upgrade infrastructure, adopt hybrid and cloud-based solutions, use compression with efficient key management, and phase in hardware upgrades such as HSMs, processors, and accelerators to ensure a smooth transition.

Technical Challenges for Implementing STL-2: QRNG adoption faces challenges with legacy integration, latency in real-time systems, and costly specialized hardware. Organizations can mitigate these by phased library upgrades with hybrid use, strategic device placement supported by HSMs, and leveraging cloud-based QRNG services to reduce cost and improve scalability.

Technical Challenges for Implementing STL-3: QKD faces challenges of limited range beyond 100–150 km, complex integration with classical systems, and vulnerabilities such as side channel attacks. Solutions include quantum repeaters or satellite-based QKD [55], hybrid frameworks with gradual upgrades and cloud services, and regular audits with advanced hardware design supported by security-focused providers.

7.2 Business-Oriented Challenges

Business challenges in quantum cryptography include high adoption costs, complex integration with existing systems, limited skilled talent, and uncertainty from evolving standards that complicates investment and compliance.

Business-Oriented Challenges for Implementing STL-1 Strategy: PQC adoption is hindered by high costs, limited expertise, and regulatory uncertainty. Organizations can mitigate these challenges through staged upgrades supported by grants, open-source and cloud services, staff training and certification, academic or consulting partnerships, and proactive alignment with evolving standards such as NIST guidelines.

Business-Oriented Challenges for Implementing STL-2 Strategy: QRNG adoption is challenged by high costs, regulatory uncertainty, and limited expertise. Organizations can mitigate these through phased deployment and cloud-based services [56], engagement with regulators to shape standards, staff training with academic partnerships, and interim reliance on consultants or service providers.

Business-Oriented Challenges for Implementing STL-3 Strategy: QKD adoption is constrained by high deployment costs, unclear global standards, and talent shortages. Organizations can mitigate these through phased rollouts and QKD-as-a-service to reduce expenses [57], active engagement with NIST, ETSI, and industry groups to shape standards, and investments in training, academic partnerships, or short-term reliance on consultants and quantum firms.

8 Conclusion

In conclusion, quantum computing presents both major opportunities and serious challenges. It promises breakthroughs in AI, materials science, and healthcare, but also threatens cybersecurity by breaking classical encryption like RSA. Though the timeline for large-scale quantum systems is uncertain, the urgency of addressing security risks is clear.

Our strategic framework offers a roadmap with three levels—Foundational, Intermediate, and Advanced—allowing organizations to prioritize based on readiness and resources. By structuring adoption in phases, the approach guarantees baseline resilience from initial implementations and systematically strengthens security through advanced, future-resilient mechanisms.

We outlined industrial use cases from communication to finance, stressing the need to align with regulatory requirements to remain secure and compliant. Proactive adoption of quantum-safe encryption is essential, as waiting for fully functional quantum computers risks exposing critical data. Acting now enables industries to safeguard digital assets and ensure long-term security in the quantum era.

References

1. Kumar, R., Srivastava, A.: Targeted influence maximization based on cloud computing over big data in social networks. IEEE Trans. Cloud Comput. **10**(2), 155–165 (2024)
2. Zhang, Y., Liu, J., Wang, H.: Artificial intelligence in the modern era: a comprehensive survey of recent developments. IEEE Access **12**, 2136–2149 (2024)
3. Gupta, M., Lee, S.: Quantum computing: exploring the paradigm shift in modern computing systems. IEEE Trans. Quantum Eng. **5**(1), 34–49 (2024)
4. Mazza, L., Jakub, P., Booth, T.M.: Quantum simulation for material discovery. IEEE Trans. Quantum Eng. **5**(2), 12–23 (2024)
5. Kumar, A., et al.: Quantum computing in drug discovery: bridging the gap to pharmaceutical development. IEEE J. Emerg. Top. Comput. **12**(1), 80–89 (2024)
6. Gomez, J., Li, C., Gupta, R.K.: Quantum-based environmental sensing technologies. IEEE Sens. J. **24**(3), 230–242 (2024)
7. Martín-López, E., et al.: Experimental realization of shor's quantum factoring algorithm using photonic qubits. Nat. Photonics **6**(12), 773–776 (2012)
8. Rivest, R.L., Shamir, A., Adleman, L.: A method for obtaining digital signatures and public-key cryptosystems. Commun. ACM **21**(2), 120–126 (1978)
9. Shor, P.W.: Algorithms for quantum computation: discrete logarithms and factoring. In: Proceedings of 35th Annual Symposium Foundations of Computer Science (FOCS), Santa Fe, NM, USA, pp. 124–134 (1994)
10. Vandersypen, M.K.L., et al.: Experimental realization of shor's quantum factoring algorithm. Nature **414**(6866), 883–887 (2001)
11. Preskill, J.: Quantum computing in the NISQ era and beyond. Quantum **2**, 79 (2018)
12. Ye, Z., et al.: Quantum topology optimization via quantum annealing. IEEE Trans. Quantum Eng. **4**, 1–12 (2023)
13. Bishwas, A.K., Sen, M.: Strategic roadmap for quantum-resistant security: a framework for preparing industries for the quantum threat. arXiv preprint arXiv:2411.09995 (2024). Submitted November 2024
14. Bishwas, A.K., Pichai, A., Som, A.: Molecular unfolding formulation with enhanced quantum annealing approach. arXiv preprint arXiv:2403.00507 (2024). https://doi.org/10.48550/arXiv.2403.00507
15. Bishwas, A.K., Mani, A., Palade, V.: An investigation on support vector clustering for big data in quantum paradigm. Quantum Inf. Process. **19**(4), 1–17 (2020)

16. Bishwas, A.K., Mani, A., Palade, V.: Gaussian kernel in quantum learning. Int. J. Quantum Inf. **18**(03), 2050006 (2020)
17. Bishwas, A.K., Mani, A., Palade, V.: Big data classification with quantum multiclass SVM and quantum one-against-all approach. In: Proceedings of 2016 2nd International Conference on Contemporary Computing and Informatics (IC3I), Pune, India (2016)
18. Shannon, C.E.: Communication theory of secrecy systems. Bell Syst. Tech. J. **28**(4), 656–715 (1949)
19. Wiesner, S.: Conjugate coding. Sigact News **15**, 78–88 (1983)
20. Einstein, A., Podolsky, B., Rosen, N.: Can quantum-mechanical description of physical reality be considered complete? Phys. Rev. **47**, 777 (1935)
21. Bennett, C.H., Bessette, F., Brassard, G., Salvail, L., Smolin, J.: Experimental quantum cryptography. J. Cryptol. **20**, 78 (1989). See also: C. H. Bennett and G. Brassard, SIGACT News; S. Wiesner, SIGACT News 15, 78 (1983)
22. Ekert, A.K.: Quantum cryptography based on bell's theorem. Phys. Rev. Lett. **67**, 661 (1991)
23. Clauser, J.F., Horne, M.A., Shimony, A., Holt, R.A.: Proposed experiment to test local hidden-variable theories. Phys. Rev. Lett. **23**, 880 (1969)
24. Bennett, C.H., Brassard, G., Mermin, N.D.: Quantum cryptography without bell's theorem. Phys. Rev. Lett. **68**, 557 (1992)
25. Krinner, S., et al.: Quantum communication over long distances with superconducting qubits. Nature **605**, 669–674 (2022)
26. Tang, J.S., Li, C.F., Xu, X.Y., et al.: Realization of high-dimensional quantum entanglement. Nat. Photonics **11**(3), 180–185 (2017)
27. Scully, M.O., Zubairy, M.S.: Quantum interference: from quantum optics to quantum computing. Proc. IEEE **95**(6), 1031–1050 (2007)
28. Scarani, V., Iblisdir, S., Gisin, N., Acin, A.: Quantum cloning. Rev. Mod. Phys. **77**(4), 1225–1256 (2005)
29. Zurek, W.H.: Decoherence, einselection, and the quantum origins of the classical. Rev. Mod. Phys. **75**(3), 715–775 (2003)
30. Diffie, W., Hellman, M.E.: New directions in cryptography. IEEE Trans. Inf. Theory **22**(6), 644–654 (1976)
31. Rijmen, V., Daemen, J.: AES proposal: Rijndael. In: Proceedings of 1st Advanced Encryption Standard (AES) Candidate Conference, pp. 1–18 (1998). https://csrc.nist.gov/publications/detail/fips/197/f
32. Daemen, J., Rijmen, V.: The Design of Rijndael: AES - The Advanced Encryption Standard. Springer (2002)
33. NIST. NIST releases first 3 finalized post-quantum encryption standards. U.S. Department of Commerce, News Release (2024). https://www.nist.gov/news-events/news/2024/08/nist-releases-first-3-finalized-post-quantum-encryption-standards. Accessed 24 Sept 2024
34. ABA Banking Journal. NIST unveils encryption standards to protect against quantum computer cyberattacks (2024). https://bankingjournal.aba.com/2024/08/nist-unveils-encryption-standards-to-protect-against-quantum-computer-cyberattacks/. Accessed 24 Sept 2024
35. Dolmeta, A., Martina, M., Masera, G.: Hardware architecture for crystals-kyber post-quantum cryptographic sha-3 primitives. In: 2023 18th Conference on Ph.D Research in Microelectronics and Electronics (PRIME), Valencia, Spain, pp. 209–212 (2023)

36. Li, X., Lu, J., Liu, D., Li, A., Yang, S., Huang, T.: A high speed post-quantum crypto-processor for crystals-dilithium. IEEE Trans. Circuits Syst. II Express Briefs **71**(1), 435–439 (2024)
37. Seo, S.C., An, S.W., Choi, D.: Accelerating falcon post-quantum digital signature algorithm on graphic processing units. Comput. Mater. Continua **75**(1), 1963–1980 (2023)
38. Sphincs: Sphincs+ (2024). https://sphincs.org/. Accessed 24 Sept 2024
39. Ma, X., Yuan, X., Cao, Z., et al.: Quantum random number generation. NPJ Quantum Inf. **2**(1), 16021 (2016)
40. Mandal, S., Anand, R., Rahman, M., et al.: Implementing Grover's on AES-based AEAD schemes. Sci. Rep. **14**, 21105 (2024). https://doi.org/10.1038/s41598-024-69188-8
41. Bennett, C., Brassard, G.: Quantum cryptography: public key distribution and coin tossing. In: Proceedings of IEEE International Conference on Computers, Systems and Signal Processing, Bangalore, India, pp. 175–179 (1984)
42. Herrero-Collantes, M., Garcia-Escartin, J.C.: Quantum random number generators. Rev. Mod. Phys. **89**(1), 015004 (2017)
43. Toshiba Europe. Ensuring long-term-secure government and medical communications with QKD (2024). https://www.toshiba.eu/quantum/resources/ensuring-long-term-secure-government-and-medical-communications-with-qkd/. Accessed 25 Sept 2024
44. Haw, J.Y., et al.: Maximization of extractable randomness in a quantum random-number generator. Phys. Rev. Appl. **3**(5), 054004 (2015)
45. Savage, N.: Keeping secrets in a quantum world — cryptographers are preparing for new quantum computers that will break their ciphers (2024). https://www.nature.com/articles/d41586-023-03336-4. Accessed 24 Sept 2024
46. Dwivedi, Y.K., et al.: Metaverse beyond the hype: multidisciplinary perspectives on emerging challenges, opportunities, and agenda for research, practice and policy. Int. J. Inf. Manage. **66**, 102542 (2022). https://doi.org/10.1016/j.ijinfomgt.2022.102542
47. Grossi, D., Rosenfeld, A., Talmon, N.: Advances in multi-agent systems research: EUMAS 2021 extended selected papers. SN Comput. Sci. **4**, 587 (2023). https://doi.org/10.1007/s42979-023-02098-0
48. Zhou, L., Schellaert, W., Martínez-Plumed, F., et al.: Larger and more instructable language models become less reliable. Nature (2024). https://doi.org/10.1038/s41586-024-07930-y
49. Hubert, K.F., Awa, K.N., Zabelina, D.L.: The current state of artificial intelligence generative language models is more creative than humans on divergent thinking tasks. Sci. Rep. **14**, 3440 (2024). https://doi.org/10.1038/s41598-024-53303-w
50. Understanding FIPS 140-3 certification: what it is and why it matters (2024). https://abr.com/2024/08/16/understanding-fips-140-3-certification-what-it-is-and-why-it-matters/
51. Thales Group. Data privacy and security concerns in quantum computing (2020). https://cpl.thalesgroup.com/blog/encryption/data-privacy-security-concerns-in-quantum-computing
52. Cybersecurity and Infrastructure Security Agency. CISA's post-quantum cryptography (PQC) initiative (2024). https://www.cisa.gov/quantum
53. National Security Agency. NSA cybersecurity perspectives on quantum key distribution and quantum cryptography (2020). https://www.nsa.gov/Press-Room/News-Highlights/Article/Article/2394053/nsa-cybersecurity-perspectives-on-quantum-key-distribution-and-quantum-cryptogr/

54. National Security Agency. Post-quantum cryptography: CISA, NIST, and NSA recommend how to prepare now (2023). https://www.nsa.gov/Press-Room/Press-Releases-Statements/Press-Release-View/Article/3498776/post-quantum-cryptography-cisa-nist-and-nsa-recommend-how-to-prepare-now/
55. European Telecommunications Standards Institute. Quantum key distribution (2024). https://www.etsi.org/technologies/quantum-key-distribution
56. McKinsey & Company. Quantum technology monitor. McKinsey & Company. Quantum technology monitor. Available online (2023)
57. BitSight. Gartner 2024 strategic roadmap for managing threat exposure (2024). https://www.bitsight.com/resources/gartner-2024-strategic-roadmap-managing-threat-exposure-ppc

A Framework for Improving Anomaly-Based Network Intrusion Detection in IoT Systems Using Hybrid GWO with Jaya

Shaimaa Kouka, Mohammed Azmi Al-Betar, and Qussai M. Yaseen(✉)

Artificial Intelligence Research Center (AIRC), Department of Information Technology, Ajman University, Ajman 346, UAE
{s.kouka,m.albetar,q.yaseen}@ajman.ac.ae

Abstract. The Internet of Things (IoT) field is an active area of research, and its applications extend to healthcare, industrial, and educational sectors. The growing global demand for IoT applications has exposed cybersecurity challenges and issues, which have urged academic researchers in the security and intelligent systems domains to investigate robust intrusion detection systems. This work proposes a powerful anomaly-based NIDS model for IoT systems based on a metaheuristic wrapper-based feature selection model. The feature selection problem of NIDS in IoT systems is modeled as a binary combinatorial optimization problem. The proposed model uses three metaheuristic models, which are Grey Wolf Optimizer (GWO), Jaya, and a hybrid GWO with Jaya. Moreover, to enhance the performance of the proposed metaheuristic models, it is suggested to utilize two key contributions. First, a population initialization mechanism (NSHI) to enhance the quality of the initial population. Second, a novel exploration operator to enhance exploration capabilities and diversity.

Keywords: GWO · Jaya · intrusion detection · optimization · IoT intrusion detection

1 Introduction

IoT technologies serve various sectors such as home automation, healthcare, and agriculture [16]. Along with the advancement, IoT presents cybersecurity threats and vulnerabilities where intruders take advantage of security gaps at each layer of the IoT architecture. Machine learning algorithms are employed in modern IDS for recognizing a system's activities as normal or an attack [2]. Machine learning algorithms allow modern IDS to effectively process in real-time large amounts of traffic data with many algorithms, and with the variety of machine learning algorithms, issues like data linearity and complexity are handled successfully. Moreover, since machine learning IDS extracts patterns

H. R. Arabnia et al. (Eds.): CSCI 2025, CCIS 2969, pp. 240–252, 2027.
https://doi.org/10.1007/978-3-032-28460-0_17

from the data, it provides adaptations to changes and new cyberattack tactics by continuously learning and updating the model [20].

Intrusion detection systems are a crucial element of IoT systems, as they protect users from facing the consequences of trusting vulnerable systems. The communication of IoT devices over the network generates a vast amount of data. Some of the network traffic information is vital for identifying cyberattacks, while some is noisy and irrelevant. Therefore, addressing the feature selection problem in NIDS of IoT systems improves the classifier's ability to identify malicious activities and enhances each individual's security.

Selecting features that boost the classifier performance is challenging due to the large search space estimated to be $(O(2^m))$, where m represents the size of the original feature set. The stochastic metaheuristic search is not sufficient in the case of a large search space, and the quality of the initial population elevates the metaheuristic algorithm's performance.

As stated in the No Free Lunch (NFL) theorem by Eiben and Smith [6], Wolpert and Macready [25], realistically, no existing metaheuristic algorithm outperforms all other metaheuristics in all optimization problems. Considering the NFL theorem, to achieve competitive performance, we have to address the weakness of metaheuristic algorithms in solving the feature selection problem of NIDS of IoT systems.

Existing related work, as shown in the next section, on metaheuristic-based NIDS for IoT revealed a robust foundation. However, we detected an absence of studies solving the NIDS for IoT problem using the Jaya algorithm, and only one GWO implementation was found based on our search and knowledge.

This work presents a methodology for developing an effective wrapper-based feature selection model that optimizes the anomaly-based NIDS of IoT systems performance while eliminating unnecessary features for classification decisions. The main contributions of this work are summarized as follows:

- Design a binary GWO and binary Jaya to solve the feature selection problem in NIDS for IoT systems.
- Propose a novel heuristic population initialization method (NSHI) for improving the algorithm's performance, such that its positive influence on the model's performance is validated using the Friedman statistical test on the five NIDS datasets.
- Propose a novel exploration operator that enhances the algorithm's exploratory capabilities and diversity, such that its positive influence on the model's performance is validated using the Friedman statistical test on the five NIDS datasets.
- Propose a binary hybrid GWO with Jaya for the feature selection problem in the anomaly-based NIDS of IoT systems.

The rest of the paper is organized as follows. Section 2 discusses some related work. Section 3 formulates the feature selection problem mathematically and discusses the methodology of this work. Finally, Sect. 4 summarizes the methodology, discusses the challenges, and presents the future work.

2 Literature Review

This section discusses some related work tackling the feature selection for network intrusion detection in IoT systems using metaheuristic algorithms.

Asadi et al. [4] introduced *BD-PSO-V*, a wrapper-based feature selection process based on particle swarm optimization (PSO). The performance of *BD-PSO-V* was enhanced by handling outliers detected through the X-Mean clustering algorithm. The final set of features was assessed using an ensemble machine learning model composed of DNN, SVM, and C4.5 on IoT-based NIDS datasets and six adversarial attacks. Similarly, Samdekar et al. [21] developed a wrapper-based feature selection comprised of a firefly algorithm (FA) and SVM classifier. The yielded results of the experiments confirmed a superior performance of the FA with SVM compared to the filter-based approaches, chi-square, PCA, and Extra Tree classifier on the BoT-IoT dataset, achieving an accuracy of 99.38%.

Mafarja et al. [14] proposed an augmented Whale Optimization Algorithm (WOA) with the elitist tournament method (*WOA-V-ET*). The V-shaped and S-shaped transfer functions were employed to binarize the continuous search space. *WOA-V-ET*'s performance using a KNN classifier was tested on the N-BaIoT dataset, where it proved its effectiveness in terms of accuracy and number of selected features.

Shitharth et al. [22] developed the *ADC-DBScan-LNB* model that consists of a clustering algorithm (ADC-DBScan) and the PPGO-based feature selection with the *ADC-DBScan-LNB* LNB classifier. The proposed clustering algorithm ADC-DBScan merges the anticipated distance-based clustering technique and the density-based spatial clustering of applications with noise (DBScan). The LNB uses data chunking that is useful for large-scale datasets. The yielded results of the experimental studies conducted using both standard NIDS and NIDS for IoT systems concluded a competitive performance of the *ADC-DBScan-LNB* against a variety of classifiers and metaheuristics.

Halim et al. [11] proposed an innovative genetic algorithm (GA)-based feature selection. The proposed model (*GbFS*incorporated the novel fitness function using SVM, KNN, and XGBoost classifiers. *GbFS*'s performance was tested along with traditional feature selection techniques such as sequential feature selectors (SFS) and state-of-the-art models, such that in most cases *GbFS* outperformed those models.

Alghanam et al. [1] proposed a wrapper-based feature selection that consisted of an improved Pigeon-inspired Optimization (PIO). Leveraging the wisdom of the crowd, an ensemble of machine learning classifiers was used in parallel, namely local outlier factor (LOF), isolation forest (iForest), and one-class SVM (OC-SVM). The (LS-PIO) model also embedded local search algorithms, including the Hill Climbing and Tabu Search algorithms, to enhance the model's exploitation power, and modified both the map and compass operators and the landmark operators. Additionally, data reduction using K-Means clustering boosted the performance by reducing the dataset complexity at the preprocessing stage. The experimental results showed that Tabu Search as a local search performed better than Hill Climbing in terms of FPR, TPR, and the number of selected features. The proposed model is lightweight and achieves high accuracy, making it well-suited for IoT systems.

Two distinct hybrid models were proposed for the wrapper-based feature selection problem. The proposed GTO by Ka- reem et al. [12] was hybridized with BSA to enhance its exploitation capabilities, and developed a control randomization parameter, as well as introduced a non-linear transfer function for controlling and switching between the diversification and unification stages. On the other hand, the hybrid model by Dey et al. [5] used both filter-based and multi-objective wrapper-based feature selection techniques. Initially, the population of Non-dominated Sorting Genetic Algorithm III(NSGA-III) was ranked using the filter-based metrics Chi-square, PCC, and MI, which consequently improved the convergence of the search. Furthermore, both hybrid models selected the same search objectives: reducing the error rate and the number of selected features. However, Kareem et al. [12] selected KNN, and Dey et al. [5] used SVM classifiers.

3 The Proposed Model

This section discusses the general scheme of the proposed model, which is depicted in Fig. 1. The main contribution in this work is part of the wrapper-based feature selection stage, where a novel heuristic population initialization called N Selective Heuristic Initialization (NSHI) process and an exploration operator are proposed. The proposed NIDS for IoT systems model is composed of three stages: data preparation (discussed in Sect. 3.2) and wrapper-based feature selection (discussed in Sect. 3.3).

Figure 1 illustrates that the methodology prepares the IoT traffic for classification by removing null records and duplicate values, if present, and converting categorical attributes to numerical format. Next, the dataset distribution is transformed to the standard normal distribution, where the mean value is zero and the standard deviation is one. Then, the metaheuristic algorithm works on searching for the subset of the feature set that optimizes the search objectives. The multi-objective problem is handled as a weighted sum objective function where the objective's weight indicates the objective's priority among the problem objectives. In addition, a binarization step is embedded for searching in the binary search space. Eventually, after the execution of this stage is terminated, the best-found feature list is passed to the next stage. Finally, the classification algorithm is trained only on the selected features, not the full features of the dataset. Then, the trained model is assessed on the same selected features of unseen data (testing data) to analyze the model's capabilities in terms of accuracy, feature subset size, detection rate, precision, F1-score, FPR, FNR, and execution time.

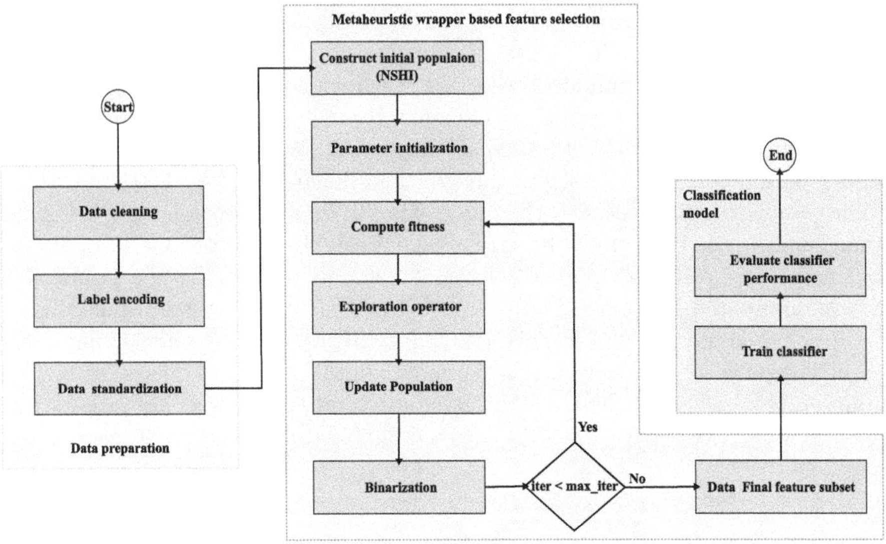

Fig. 1. General scheme of the proposed model for the NIDS in IoT systems

3.1 Datasets Description

A common practice in the reviewed studies on IoT-based NIDS is evaluating the proposed NIDS performance on both IoT-based and standard NIDS datasets to test the proposed model on a diverse range of attack patterns and evaluate the yielded trade-off between generalization ability and IoT-focused performance. We benchmarked the performance using five public datasets, namely NSL-KDD Tavallaee et al. [24], UNSW-NB15 Moustafa and Slay [17], BoT-IoT [13], ToN-IoT Alsaedi et al. [3], CIC-IoT2023 Neto et al. [18]. The first two datasets are the standard NIDS, while the latter three datasets are specialized in IoT-based systems attacks and traffic patterns. The list of datasets includes simulated network attacks of traditional and up-to-date attack tactics, IoT-specific attacks, and standard NIDS attacks.

A common characteristic among all five datasets is the imbalanced normal-to-attack classes ratio, which is logical insight considering the nature of the problem. Typically, benchmark datasets are provided as two separate data files for training and testing, like NSL-KDD and UNSW-NB15. For the remaining three datasets, we applied random stratified data splitting with an 80% training to 20% testing ratio. The main purpose of using stratified-based data splitting is to avoid altering the class distribution when using traditional random splitting and to preserve the dataset complexity and the variance in the attack patterns. Regarding the 80:20 ratio, it is the most effective and common splitting ratio considered by Escorcia-Gutierrez et al. [9], Fatani et al. [12], and Kareem et al. [8].

To overcome the very large sizes of these datasets, we have generated a reduced version of UNSW-NB15, BoT-IoT, ToN-IoT, and CICIoT 2023, taking 60%,5%,40%, and 7% of each dataset, respectively. The purpose of the data sampling is to reduce the training time to a reasonable time while sustaining the data complexity and variance by applying stratified sampling per label. The sampled version of the datasets in Tables 1, 2 are utilized for model comparisons and investigating the impact of our novel operators on the overall performance. It is worth noting that NSL-KDD was not sampled since it already has a reasonable training time.

Table 1. Sampled Datasets: UNSW-NB15, BoT-IoT

(a) UNSW-NB15 (60%)

	Training	Testing	Total
Normal	33600	406	34006
Analysis	1200	350	1550
Backdoor	1048	2453	3501
DoS	7358	6679	14037
Exploits	20036	3637	23673
Fuzzers	10910	11323	22233
Genetic	24000	22200	46200
Reconnaissance	6295	2098	8393
Shell Code	680	227	907
Worm	78	26	104
Total	105205	49399	154604

(b) BoT-IoT (5%)

	Training	Testing	Total
Normal	19	5	24
DDoS	77066	19265	96331
DoS	66007	16506	82513
Reconnaissance	3646	908	4554
Theft	3	1	4
Total	146741	36685	183426

Table 2. Sampled Datasets: ToN-IoT, CICIoT 2023

(a) ToN-IoT (40%)

	Training	Testing	Total
Normal	334	83	417
Backdoor	6,400	1,600	8,000
DDoS	6,400	1,600	8,000
Injection	6,400	1,600	8,000
MITM	6,400	1,600	8,000
Password	92,458	23,114	115,572
Ransomware	6,400	1,600	8,000
Scanning	6,400	1,600	8,000
XSS	6,400	1,600	8,000
Total	137,592	34,397	171,989

(b) CICIOT 2023 (7%)

	Training	Testing	Total
Normal	3,074.96	768.74	3,843.70
DDoS	95,156.32	23,789.43	118,945.75
DoS	22,653.96	12,632.69	35,286.65
Recon	992.67	248.29	1,240.96
Web-Based	69.30	17.57	86.87
Brute Force	36.54	9.17	45.71
Spoofing	1,362.20	340.62	1,702.82
Mirai	7,375.48	1,843.94	9,219.42
Total	130,721.43	39,650.45	170,371.88

3.2 Data Preparation

In the data preparation step of the proposed model, dataset issues like duplicated entries, null, and missing values are tackled. Although some algorithms such as XGBoost and deep learning can handle missing values, it does not eliminate the need for this process since a complete dataset provides more accurate knowledge Garcıa et al. [10]. The action taken for null, missing, and duplicated records is to drop those records since they represent a small percentage of the whole dataset and their influence on the final output is insignificant.

Network traffic data contain categorical attributes such as communication protocols, flags, transactions, and attack labels. Assuming the attribute has k unique values, using label encoding, each of the unique values is mapped to a number from zero to $(k - 1)$. For example, the attack attribute has two possible values, normal and attack, which are mapped to the numbers zero and one, respectively. This approach is time and memory-efficient and does not increase the feature selection problem complexity because it maintains the original number of attributes, unlike other approaches, i.e., one-hot-encoding Elmasry et al. [7]. Although it introduces an order to the attribute's unique values in the mapping process, it is still preferred over techniques that do not assume order. Encoding techniques such as one-hot encoding are deemed impractical for datasets with numerous attributes where each has a large number of distinct values.

The distribution of each attribute is transformed to the standard normal distribution after transforming the categorical attributes from categorical to numerical form, where the mean value is zero and the standard deviation is equal to one. Data normalization has a significant impact on the overall performance due to normalization's ability to level out the influence of each attribute Garcıa et al. [10]. Moreover, data standardization is preferred over min-max normalization due to the nature of the intrusion detection problem in IoT systems. Often, a high ratio of outliers exists in the dataset, and new attack patterns are continuously evolving, which results in uncertainty of some attributes' maximum and minimum values. The presence of outliers in the dataset and the uncertainty of maximum and minimum values of attributes gives standardization an advantage over min-max normalization Garcıa et al. [10]. Standardization is especially useful since SVM and KNN, which we plan to use, are sensitive to feature scaling. In contrast, XGBoost is a tree-based algorithm that is not significantly affected by feature scaling atsat et al. [23].

3.3 Wrapper Based Feature Selection

Three metaheuristic models are proposed to be used in this work, which are binary Jaya, binary GWO, and binary hybrid GWO with Jaya. GWO was selected since it is a powerful metaheuristic algorithm with a simple and flexible design and balanced exploration and exploitation Makhadmeh et al. [15]. Similarly, Jaya is a simple and effective metaheuristic algorithm with great flexibility Zitar et al. [26]. The absence of Jaya algorithm-specific parameters simplifies parameter settings experiments. However, both GWO and Jaya have limitations and are inefficient in solving large search space problems. Therefore, NSHI and the exploration operator were incorporated. Furthermore, a binary hybrid GWO with Jaya is proposed to solve the feature selection problem in NIDS for IoT systems.

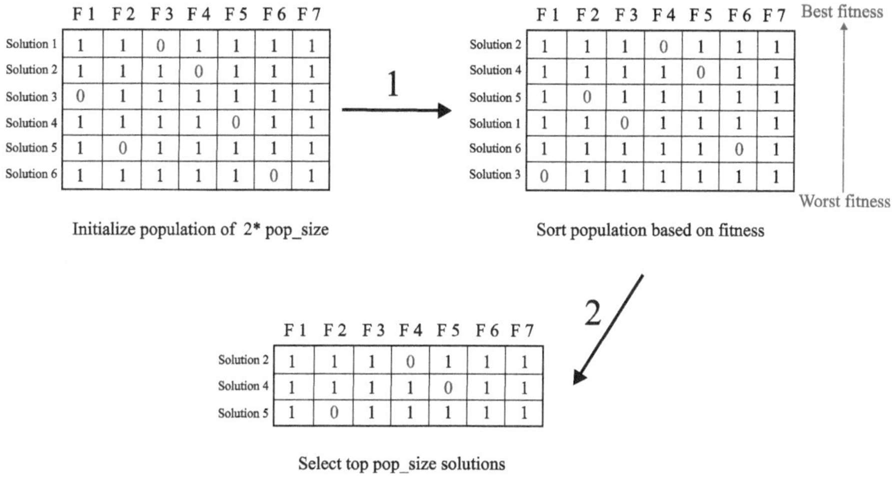

Fig. 2. N-Selective heuristic initialization (NSHI)

3.4 The Proposed N-Selective Heuristic Initialization (NSHI) for Population Initialization

The proposed novel heuristic initialization, inspired by selective initialization Eiben and Smith [6], is summarized in Fig. 2. Like the selective initialization technique, NSHI generates a population of double the actual population size (pop-size), and instead of the traditional random initialization technique, each solution consists of nshi-n randomly selected attributes of the full feature set. After generating the population, it is sorted based on fitness score such that the fittest pop-size solutions are selected to form the initial population.

Figure 3 shows the metaphoric bi-objective space of a feature selection problem of two conflicting minimization objectives, which are the number of features and the error rate. Assuming the full feature set (problem-N) contains 20 features, the search space is $2^m = 2^{20} = 1,048,576$.

The proposed heuristic approach in this paper works on reducing the search space by effectively guiding the search in the multi-objective space. Setting one of the objectives (number of features) to a constant value nshi-n (e.g. 14) in NSHI limits the initial search space $C(m, N) = C(20, 14) = 38,760$. This is highlighted in the blue box, thus intensifying the search with the primary objective of obtaining a subset of features of size nshi-n that achieves a minimal

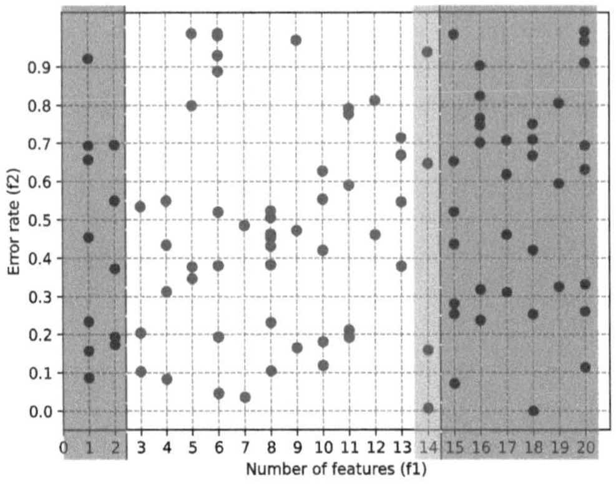

Fig. 3. Bi-objective space of feature selection problem with originally 20 features

error rate within the population. Moving forward in the search, the high-quality solution among the initial population guides the search to promising regions. However, according to the discussed literature review about this problem, the minimum number of features is more than two features. Hence, the search is constrained to at least three features, as depicted inside the left-side red box in Fig. 3. Therefore, the search is directed to reduce the number of features while improving or maintaining the best-found accuracy. A solution might have more than nshi-n features only if it was randomly re-initialized due to reaching fewer than three features, or if the addition of those features yields a higher fitness value.

Figure 4 visualizes the probability of each feature being part of the feature subset when the experiment is repeated 1000 times. The figure shows approximately the same probability of each feature from the original feature set being chosen as part of the solution's feature subset.

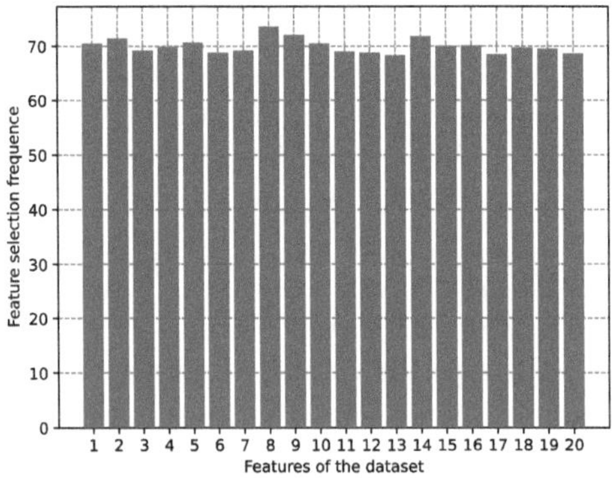

Fig. 4. The selection frequency of each feature among features of a dataset

Figure 5 compares the feature subset size in NSHI and the traditional random initialization. Clearly, Fig. 5a shows that NSHI has a fixed size, unlike the variable subset size in Fig. 5b. Moreover, nshi-n parameter is highly influential since it defines the starting point of the search, and it is designed as a fraction of the original feature set size. Assuming the number of the features is dim, several variations are proposed in this paper, such as Eq. 1, Eq. 2 and Eq. 3.

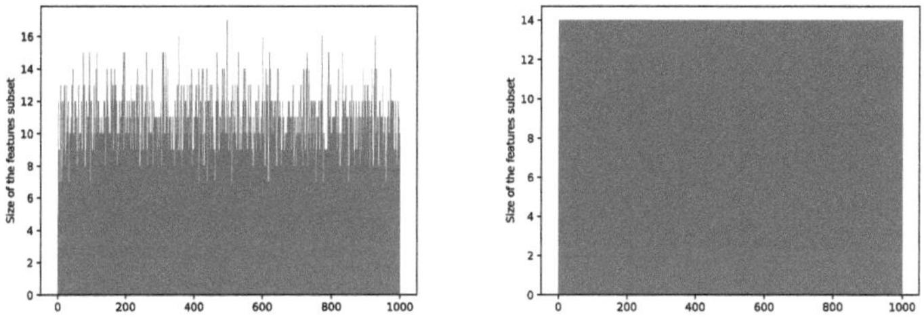

(a) Random initialization and its features subset size

(b) NSHI and its features subset size

Fig. 5. Comparison between the traditional random initialization and NSHI feature subset size assuming the experiment is repeated 1000

$$N = dim - round(dim/2) \quad (1)$$
$$N = dim - round(dim/3) \quad (2)$$
$$N = dim - (round(dim/3) - 1) \quad (3)$$

3.5 The Proposed Exploration Operator for Enhancing Exploration

The exploration operator is shown in Eq. 4, where each feature value is modified based on its value compared to a uniform random number($rand()$). When the value of $x_{i,j}$ is less than or equal to $rand()$, its value is changed to 1; otherwise, it is changed to 0. The randomness of the operators comes from comparing $x_{i,j}$ to a stochastic value, and based on that, the value of $x_{i,j}$ is changed to either zero or one. Reducing the variance of continuous data to a lower variance (zero and one values only) smooths out the values of the variables, eliminates outlier-related issues, and provides a better measurement for any upcoming operations, such as computing the arithmetic mean in GWO.

$$x_{i,j} = \begin{cases} 1 & \text{if } rand() \leq x_{i,j} \\ 0 & \text{otherwise} \end{cases} \quad (4)$$

This operator is simply and easily embedded in all metaheuristic algorithms to promote exploration and maintain diversity within the population. Moreover, the operator is not exclusive to binary domain problems, as its main purpose is not to transform to the binary domain. Instead, it simulates uninformed modification of the solution as a means to escape performance stagnation, unlike transfer functions that work mainly as a binarization technique [19].

3.6 The Proposed Models

We plan to benchmark the performance of nine proposed models representing all possible combinations of three machine learning models (SVM, KNN, XGBoost) and three proposed metaheuristic algorithms for feature selection (BGWO, BJaya, BGJ). The testing is planned to be performed using the NSL-KDD and the sampled version of UNSW-NB15 as well as three IoT-focused NIDS datasets BoT-IoT, ToN-IoT, and CICIoT 2023. The purpose of the planned experiments is to identify the superior machine learning and metaheuristic algorithm combination among the nine possible combinations across the five datasets.

4 Conclusion

IoT technologies elevate the quality of life. However, they expose security gaps and vulnerabilities as they continuously collect sensitive or confidential information from their surroundings. That is, security attacks are evolving and exploiting sophisticated attack patterns to pose a threat to assets. The primary objective of this work is to address the vulnerabilities in IoT systems by designing an anomaly-based NIDS that identifies novel IoT attacks based on normal behavior, rather than relying on known attack patterns and signatures.

This paper has proposed a GWO hybridized with the Jaya model using a novel implementation in the binary domain and feature selection problem, up to the best of our knowledge. The proposed model has been enhanced by swapping the traditional random population initialization with a novel heuristic population initialization (NSHI) based on selective population initialization. Moreover, the paper has addressed the GWO's weak exploration as well as Jaya's lack of diversity by introducing a stochastic exploration operator that is integrated into GWO, Jaya, and BGJ. The datasets used are NSL-KDD, UNSW-NB15, BoT-IoT, ToN-IoT, and CICIoT2023.

In the next phase of this work, as future work, the positive impact of both novel additions will be validated using the Friedman test to assess the classification performance on the following criteria: accuracy, precision, recall, F1 score, FPR, FNR, reduction of selected features, and execution time. Moreover, extensive experiments will be conducted to validate the metaheuristic algorithm, parameter values, and machine learning classifier.

References

1. Alghanam, O.A., Almobaideen, W., Saadeh, M., Adwan, O.: An improved PIO feature selection algorithm for IoT network intrusion detection system based on ensemble learning. Expert Syst. Appl. **213**, 118745 (2023)

2. AlJarrah, M.N., Yaseen, Q.M., Mustafa, A.M.: A context-aware android malware detection approach using machine learning. Information **13**(12) (2022). https://doi.org/10.3390/info13120563
3. Alsaedi, A., Moustafa, N., Tari, Z., Mahmood, A., Anwar, A.: TON_IoT telemetry dataset: a new generation dataset of IoT and IIoT for data-driven intrusion detection systems. IEEE Access **8**, 165130–165150 (2020)
4. Asadi, M., Jamali, M.A.J., Parsa, S., Majidnezhad, V.: Detecting botnet by using particle swarm optimization algorithm based on voting system. Futur. Gener. Comput. Syst. **107**, 95–111 (2020)
5. Dey, A.K., Gupta, G.P., Sahu, S.P.: Hybrid meta-heuristic based feature selection mechanism for cyber-attack detection in IoT-enabled networks. Procedia Comput. Sci. **218**, 318–327 (2023)
6. Eiben, A.E., Smith, J.E.: Introduction to Evolutionary Computing. Springer (2015)
7. Elmasry, W., Akbulut, A., Zaim, A.H.: Evolving deep learning architectures for network intrusion detection using a double PSO metaheuristic. Comput. Netw. **168**, 107042 (2020)
8. Escorcia-Gutierrez, J., et al.: Sea turtle foraging algorithm with hybrid deep learning-based intrusion detection for the internet of drones environment. Comput. Electr. Eng. **108**, 108704 (2023)
9. Fatani, A., Dahou, A., Al-Qaness, M.A., Lu, S., Elaziz, M.A.: Advanced feature extraction and selection approach using deep learning and aquila optimizer for IoT intrusion detection system. Sensors **22**(1), 140 (2021)
10. García, S., Luengo, J., Herrera, F.: Data Preprocessing in Data Mining, vol. 72. Springer (2015)
11. Halim, Z., et al.: An effective genetic algorithm-based feature selection method for intrusion detection systems. Comput. Secur. **110**, 102448 (2021)
12. Kareem, S.S., Mostafa, R.R., Hashim, F.A., El-Bakry, H.M.: An effective feature selection model using hybrid metaheuristic algorithms for IoT intrusion detection. Sensors **22**(4), 1396 (2022)
13. Koroniotis, N., Moustafa, N., Sitnikova, E., Turnbull, B.: Towards the development of realistic botnet dataset in the internet of things for network forensic analytics: BoT-IoT dataset. Futur. Gener. Comput. Syst. **100**, 779–796 (2019)
14. Mafarja, M., Heidari, A.A., Habib, M., Faris, H., Thaher, T., Aljarah, I.: Augmented whale feature selection for IoT attacks: structure, analysis and applications. Futur. Gener. Comput. Syst. **112**, 18–40 (2020)
15. Makhadmeh, S.N., et al.: Recent advances in grey wolf optimizer, its versions and applications. IEEE Access (2023)
16. Mishra, N., Pandya, S.: Internet of things applications, security challenges, attacks, intrusion detection, and future visions: a systematic review. IEEE Access **9**, 59353–59377 (2021)
17. Moustafa, N., Slay, J.: UNSW-NB15: a comprehensive data set for network intrusion detection systems (UNSW-NB15 network data set). In: 2015 Military Communications and Information Systems Conference (MilCIS), pp. 1–6. IEEE (2015)
18. Neto, E.C.P., Dadkhah, S., Ferreira, R., Zohourian, A., Lu, R., Ghorbani, A.A.: Ciciot2023: a real-time dataset and benchmark for large-scale attacks in IoT environment (2023)
19. Nssibi, M., Manita, G., Korbaa, O.: Advances in nature-inspired metaheuristic optimization for feature selection problem: a comprehensive survey. Comput. Sci. Rev. **49**, 100559 (2023)

20. Odat, E., Alazzam, B., Yaseen, Q.M.: Detecting malware families and subfamilies using machine learning algorithms: an empirical study. Int. J. Adv. Comput. Sci. Appl. **13**(2) (2022). https://doi.org/10.14569/IJACSA.2022.0130288
21. Samdekar, R., Ghosh, S., Srinivas, K.: Efficiency enhancement of intrusion detection in IoT based on machine learning through bioinspire. In: 2021 Third International Conference on Intelligent Communication Technologies and Virtual Mobile Networks (ICICV), pp. 383–387. IEEE (2021)
22. Shitharth, S., Kshirsagar, P.R., Balachandran, P.K., Alyoubi, K.H., Khadidos, A.O.: An innovative perceptual pigeon galvanized optimization (PPGO) based likelihood Naïve Bayes (LNB) classification approach for network intrusion detection system. IEEE Access **10**, 46424–46441 (2022)
23. Tatsat, H., Puri, S., Lookabaugh, B.: Machine Learning and Data Science Blueprints for Finance. O'Reilly Media (2020)
24. Tavallaee, M., Bagheri, E., Lu, W., Ghorbani, A.A.: A detailed analysis of the KDD cup 99 data set. In: 2009 IEEE Symposium on Computational Intelligence for Security and Defense Applications, pp. 1–6. IEEE (2009)
25. Wolpert, D.H., Macready, W.G.: No free lunch theorems for optimization. IEEE Trans. Evol. Comput. **1**(1), 67–82 (1997)
26. Zitar, R.A., Al-Betar, M.A., Awadallah, M.A., Doush, I.A., Assaleh, K.: An intensive and comprehensive overview of jaya algorithm, its versions and applications. Arch. Comput. Methods Eng. **29**(2), 763–792 (2022)

RansomSphereGuard: Early Detection of Zero-Day Ransomware Attacks Using Deep Multi-sphere Support Vector Data Description

Masoumeh Sharafi, Mahdi Abadi(✉), Amin Sahati, and Ali Esmaeili

Faculty of Electrical and Computer Engineering, Tarbiat Modares University, Jalal Al-e Ahmad Hwy, Tehran, Iran
{masoumeh.sharafi,abadi,a.sahati,a_esmaeili}@modares.ac.ir

Abstract. Zero-day ransomware attacks pose a severe cybersecurity threat, as they can evade traditional detection mechanisms and inflict irreversible damage, even if ransom payments are made. Current techniques often fail to detect such attacks early due to their reliance on known signatures or insufficient sensitivity to pre-encryption behavioral anomalies. To address these challenges, we propose RansomSphereGuard, a novel deep anomaly detection technique that detects zero-day ransomware during the pre-encryption stage by analyzing the initial I/O Request Packet (IRP) logs per process. RansomSphereGuard extracts high-level features—including operation types, IRP flags, and buffer length statistics—and trains a multi-sphere decision boundary exclusively on benign IRP traces to detect ransomware attacks by looking for abnormal IRP logs. Unlike conventional single-sphere models, RansomSphereGuard captures nuanced patterns in legitimate activity, enabling accurate detection of subtle deviations indicative of ransomware. The experimental results demonstrate that RansomSphereGuard can achieve a high detection rate with low false positives, highlighting its practical efficacy in real-world deployments. By focusing on initial behavioral traces and leveraging advanced anomaly detection, RansomSphereGuard provides a proactive defense against zero-day ransomware, significantly reducing potential damage while maintaining low operational overhead.

Keywords: Zero-day ransomware · Early detection · Deep anomaly detection · Deep learning · I/O request packet

1 Introduction

Malware, short for malicious software, is a generic term for any software that adversely affects computers [2]. In recent years, malware has become one of the major threats to Internet security.

Ransomware is a special type of malware that exploits security mechanisms such as cryptography to prevent or restrict users from accessing their systems.

H. R. Arabnia et al. (Eds.): CSCI 2025, CCIS 2969, pp. 253–269, 2027.
https://doi.org/10.1007/978-3-032-28460-0_18

There are two main types of ransomware: locker and crypto. Locker ransomware restricts user access to the infected machine until a ransom is paid to unlock it. Crypto-ransomware encrypts user data and demands payment in exchange for decryption. Due to the increasing threat of crypto-ransomware, we focus on this specific type of ransomware. For convenience, we will refer to crypto-ransomware simply as ransomware.

Unlike traditional malware, it is impossible to completely mitigate the effects of ransomware attacks without the help of its creator. The statistics show that ransomware attacks have become a significant cybersecurity threat globally. According to Statista [24], in 2023, organizations all around the world detected more than 317 million ransomware attempts. Another report from Cybersecurity Ventures [21] states that global ransomware damage costs are expected to reach around $275 billion annually by 2031, up from $325 million in 2015. Based on blockchain analysis of identifiable transactions, FinCEN (a bureau of the U.S. Department of the Treasury) claimed it had identified approximately $5.2 billion in outgoing Bitcoin transactions potentially tied to ransomware payments [14].

Zero-day malware is any malware that exploits unknown vulnerabilities or flaws for malicious purposes. The term "zero-day" refers to the fact that developers have just learned of the flaw, and thus they have only zero days to fix it [12].

Researchers have recently proposed various techniques [9] to detect ransomware attacks. Most of these techniques use known ransomware characteristics, which are not suitable for detecting zero-day ransomware attacks. Also, they often fail to detect ransomware attacks early, which is crucial for mitigating potential damage. Hence, detecting this type of malware still faces many challenges.

Early detection of ransomware attacks, also known as pre-encryption detection, is critical to minimizing the damage caused by these attacks and reducing the chances of successful ransomware infections. Detecting an attack in its early stages allows organizations to isolate affected systems, prevent the attack's spread to unaffected resources, and minimize the impact on critical data and systems.

Anomaly or outlier detection is a technique for finding unusual patterns in data that do not conform to the expected behavior [10]. Anomalies are usually caused by errors in data, but sometimes represent a previously unknown or fundamentally new trend. Often, anomaly detection is used to solve real-world problems in fault, fraud, and intrusion detection. In recent years, deep learning has shown promising performance in various domains and tasks. Deep learning for anomaly detection, also known as deep anomaly detection, aims to identify abnormal data using a deep neural network model trained on normal data [23].

I/O Request Packet (IRP) driven analysis is a new research paradigm for defense against ransomware attacks [6,19]. IRP logs are one of the footprints that ransomware leaves on the victim's machine. Anomaly detection techniques can analyze IRP logs to identify abnormal behavior, effectively detecting zero-day attacks.

Deep Support Vector Data Description (Deep SVDD or DSVDD) [25] is a powerful deep anomaly detection technique that creates a compact hypersphere boundary around normal data within a latent feature space. This technique employs a neural network architecture that simultaneously learns both feature representation and hypersphere boundary. A deep neural network model transforms input data into a meaningful latent space while minimizing the volume of a hypersphere that tightly encloses normal training samples in the latent space. Input samples that fall outside this learned hyperspherical decision boundary are classified as anomalies. DSVDD addresses the limitations of traditional kernel-based one-class classifiers [29] by replacing fixed kernel mappings with learnable deep neural networks. This allows the model to discover intricate patterns and relationships in data that would be difficult to capture with predefined kernel functions.

Traditional SVDD-based techniques rely on a single hypersphere to model latent normal data, limiting their effectiveness when the data exhibits multi-modal or clustered patterns—a common trait in complex systems. To address this challenge, we present RansomSphereGuard, a novel deep anomaly detection technique for the early detection of zero-day ransomware attacks, in which multiple adaptive hyperspheres are trained, each capturing a distinct latent cluster. By leveraging these localized hyperspheres, RansomSphereGuard establishes more accurate decision boundaries, improving anomaly detection in diverse and structured data. Specifically, RansomSphereGuard trains a deep multi-sphere SVDD model on benign (or normal) IRP traces and then uses this model to detect ransomware attacks by looking for abnormal IRP logs.

The main contributions of this paper are summarized as follows:

- We present RansomSphereGuard, a novel deep anomaly detection technique to detect zero-day ransomware attacks early enough to mitigate the damage as much as possible. For early-stage detection, RansomSphereGuard analyzes only the initial IRP logs per process, extracting high-level features—such as operation type information, IRP flag characteristics, and buffer length statistics—to minimize potential damage and improve response time.
- Unlike conventional deep single-sphere SVDD models, RansomSphereGuard employs multiple adaptive hyperspheres to better capture the latent structure of benign data. This improvement enables better discrimination between normal and anomalous patterns, particularly for data with inherent cluster-based distributions.
- We conducted extensive experiments to evaluate the performance of RansomSphereGuard for detecting zero-day ransomware in the early stages of its execution. The experimental results demonstrate the promising detection performance of RansomSphereGuard while having a low detection time, validating its practicality for real-world deployment.

The remainder of this paper is organized as follows: Sect. 2 provides the necessary background. Section 3 briefly discusses the related work on ransomware detection. Section 4 presents our technique for the early detection of zero-day ransomware attacks. Section 5 reports the experimental settings and results. Finally, Sect. 6 concludes the paper and summarizes our findings.

2 Background

In this section, we briefly explain the background of our work.

2.1 I/O Request Packet

I/O Request Packets (IRPs) are kernel-mode structures that store information needed to process I/O requests. When a program requests an I/O operation, the I/O manager constructs an IRP to represent the operation as it progresses through the I/O system. Then, the I/O manager passes the IRP to the file system driver, from where the requested I/O operation is completed. Finally, after receiving the I/O status from the IRP, the I/O manager frees the memory and returns a handle for the file object to the program if the I/O operation is successful. Otherwise, it returns the appropriate status to the program [4].

IRP is different from Fast I/O (FIO). In particular, IRP is the default mechanism for requesting I/O operations and can be used for synchronous or asynchronous I/O. By contrast, FIO is specifically designed for rapid synchronous I/O on cached files. In FIO operations, data is transferred directly between user buffers and the system cache [20].

There are four special types of IRP flags: `IRP_NOCACHE`, `IRP_PAGING_IO`, `IRP_SYNCHRONOUS_API`, and `IRP_SYNCHRONOUS_PAGING_IO`. Additionally, several other important flag attributes are available in IRP logs: the major operation type, which signifies the type of IRP operation, and the minor operation type, which is dependent on the major operation type. For example, when requesting an open operation, the major operation type is `IRP_MJ_CREATE`.

2.2 Minifilter

Filter drivers are one of the most powerful architectural features of the Windows I/O subsystem. A filter can add value to the functionality of an existing device by simply attaching itself to that device. One of the most common places to insert a filter in a Windows operating system is over a file system. File system filters can intercept and manipulate I/O operations before they reach the file system [22].

The filter manager is a kernel-mode driver that implements and exposes functionality commonly required in file system filter drivers. This functionality allows third-party developers to write minifilter drivers, which are easier to develop than legacy file system filter drivers. A minifilter driver can filter IRP-based I/O operations as well as FIO callback operations. For each of the I/O operations, a minifilter can register a preoperation callback routine, a postoperation callback routine, or both.

2.3 Deep SVDD

Support Vector Data Description (SVDD) [29] is a prominent machine learning technique for anomaly detection. It detects anomalies by constructing a minimal enclosing hypersphere around normal training data within a high-dimensional feature space. The optimization objective simultaneously minimizes the hypersphere's volume while encompassing the majority of normal samples [29]. This formulation makes SVDD particularly effective for high-dimensional spaces where traditional distance-based techniques underperform.

A key advantage of SVDD lies in its ability to model linear and non-linear data distributions through kernel functions, enabling the capture of complex data structures. Additionally, SVDD demonstrates robustness to noise and outliers by adapting to the overall data distribution rather than being overly sensitive to individual samples. These characteristics have led to its successful deployment in diverse domains [18,28].

Deep SVDD (DSVDD) [25] extends SVDD by using a deep neural network, which is jointly trained to map input training data to a minimal-volume hypersphere in a latent feature space. Formally, let $\mathcal{X} \subseteq \mathbb{R}^d$ be the input space, $\mathcal{F} \subseteq \mathbb{R}^p$ be the feature space, and $\varphi(\cdot; \mathcal{W}) : \mathcal{X} \to \mathcal{F}$ be a deep neural network with a set of weights $\mathcal{W}$. For an input sample $\mathbf{x} \in \mathcal{X}$, $\varphi(\mathbf{x}; \mathcal{W}) \in \mathcal{F}$ denotes its learned feature representation. The DSVDD objective jointly optimizes the network parameters while minimizing the volume of a hypersphere in $\mathcal{F}$ centered at $\mathbf{c} \in \mathcal{F}$. Given some training data $X \subseteq \mathcal{X}$, the optimization problem is defined as

$$\min_{\mathcal{W}} \frac{1}{N} \sum_{i=1}^{N} \|\varphi(\mathbf{x}_i; \mathcal{W}) - \mathbf{c}\|^2 + \frac{\lambda}{2} \sum_{l=1}^{L} \|\mathbf{W}^l\|_{\mathrm{F}}^2, \tag{1}$$

where N is the number of training samples, L is the number of network hidden layers, $\mathbf{x}_i \in X$ is a training sample, and $\mathbf{W}^l \in \mathcal{W}$ are the weights of layer l. The second term corresponds to a network weight decay regularizer with a hyperparameter $\lambda > 0$ and $\|\cdot\|_{\mathrm{F}}^2$ denotes the Frobenius norm.

3 Related Work

In this section, we provide a brief description of state-of-the-art ransomware detection techniques.

Kharraz et al. [16] were the first to show that by monitoring abnormal file system activity through IRP logs, it is possible to design a practical defense mechanism that could stop a large number of ransomware attacks. Also, their analysis suggested that by examining IRP requests and protecting the Master File Table (MFT) in the NTFS file system, it is possible to detect and prevent a significant number of zero-day ransomware attacks. Then, Scaife et al. [26] introduced CryptoDrop, an early-warning system for ransomware attacks that monitors user data for changes. Using a set of behavioral indicators, CryptoDrop can stop a process that appears to be tampering with a large amount of user data.

However, CryptoDrop cannot determine the intent of the changes it inspects. For example, it cannot distinguish whether the user or ransomware is encrypting a set of documents. Afterward, Continella et al. [11] proposed ShieldFS, an add-on driver that protects the Windows file system from ransomware attacks. For each running process, ShieldFS dynamically toggles a protection layer that acts as a copy-on-write mechanism according to the result of its detector. Internally, ShieldFS monitors low-level file system activity to continuously update adaptive models that profile the system activity over time. Whenever one or more processes violate these models, their operations are considered malicious, and the resulting side effects on the file system are rolled back transparently. The authors implemented the protection layer as a Windows minifilter driver that monitors file modifications by registering a callback for `IRP_MJ_CREATE` operations. Later, Mehnaz et al. [19] introduced RWGuard, a real-time ransomware detection mechanism. RWGuard consists of three monitoring modules: decoy monitoring, process monitoring, and file change monitoring. The decoy monitoring module deploys some decoy files within the system. Under normal circumstances, a decoy file should not be written, but whenever a process attempts to write to one of these decoy files, RWGuard detects it as ransomware. The process monitoring module uses some IRP metrics (e.g., number of `IRP_MJ_OPEN`, `IRP_MJ_CLOSE`, `IRP_MJ_CREATE`, `IRP_MJ_READ`, and `IRP_MJ_WRITE` requests) for building a baseline profile for each running process and utilizes these baseline profiles to detect anomalous process behavior. The file change monitoring module checks all changes made to the files (e.g., create, delete, and write operations) to determine anomalous file changes.

Agrawal et al. [1] identified small repeating patterns in long sequences of ransomware API calls, potentially corresponding to repeated encryption operations. They presented an enhanced neural cell, known as ARI (Attended Recent Inputs), to employ attention in learning from ransomware sequences. While processing the input sequence, the ARI cell also learns from recent history in the form of a subsequence. The authors presented an implementation of the ARI cell with LSTM networks, called ARI-LSTM, and used the resulting neural network for sequence learning with ransomware. The experimental results showed that ARI-LSTM improves the performance of LSTM in detecting Windows ransomware. Then, Al-rimy et al. [3] proposed RCGU, a redundancy coefficient gradual up-weighting technique to accurately estimate the significance of features when the amount of data is insufficient in the pre-encryption stage of ransomware attacks. They also integrated RCGU into the goal function of the Mutual Information Feature Selection (MIFS) algorithm and devised two improved feature selection techniques for the early detection of ransomware attacks. Afterward, Zahoora et al. [30] presented a zero-shot learning framework to detect zero-day ransomware attacks. They trained a customized deep contractive autoencoder to extract core features of known and unknown ransomware. The original and core features were then used to construct a heterogeneous ensemble of diverse supervised base classifiers. The probability distribution outputs from the base classifiers were combined in an inference stage using four combination rules to obtain the final predictions.

Alqahtani and Sheldon [5] proposed a temporal data correlation technique that associates cryptographic APIs with IRPs based on their timestamps to define the pre-encryption boundary accurately. The authors employed this technique to determine the pre-encryption boundary for each trace file in the corpus. The data before this boundary served as input for the feature extraction phase. The extracted features were then used to train five different classifiers, including Support Vector Machine (SVM), Logistic Regression (LR), Deep Belief Network (DBN), Convolutional Neural Network (CNN), and Multi-layer Perceptron (MLP). Then, Ayub and Siraj [7] investigated unique patterns in ransomware IRP logs that are common across all ransomware samples and distinctly distinguishable from benign IRP logs. To this end, they selected the IRP major operation type as a feature and derived five different sequences of IRP operations. These sequences were then used to compute some statistical measures from the ransomware IRP logs over a time window. The authors analyzed the statistical measures to examine whether they could create a common cluster that would enable them to discover new ransomware variants. Therefore, they used four one-class classification algorithms, including One-Class SVM (OCSVM), Isolation Forests, Local Outlier Factor (LOF), and Robust Covariance, to create the automated boundary of such statistical measures by training them with observations from 16 ransomware families. They investigated how well the derived clusters could fit the observations from the remaining five ransomware families. Afterward, Cen et al. [8] proposed RansoGuard, a ransomware detection technique that identifies malicious activity by analyzing sensitive API call patterns during the pre-encryption stage, enabling timely defense. RansoGuard dynamically monitors behavioral data before encryption and employs natural language processing to model and extract key features from API call sequences. These features are then used to train a Recurrent Neural Network (RNN) classifier that effectively distinguishes ransomware from benign programs.

However, most existing techniques either fail to detect ransomware attacks early or are limited to identifying previously known variants. Developing techniques capable of detecting zero-day ransomware attacks during their early stages remains a significant challenge that has received little attention.

4 RansomSphereGuard

In this section, we present RansomSphereGuard, a novel deep anomaly detection technique for the early detection of zero-day ransomware attacks. RansomSphereGuard trains a deep multi-sphere SVDD model on benign IRP traces and then leverages this model to detect ransomware attacks by looking for abnormal IRP logs. The core idea is to construct multiple adaptive hyperspheres, each representing a distinct cluster of benign data. This design captures the intrinsic structure of complex data distributions more effectively than conventional single-sphere approaches.

RansomSphereGuard comprises three main phases: preprocessing, model construction, and anomaly detection. The overall workflow is illustrated in Fig. 1, and each phase is described in detail in the following subsections.

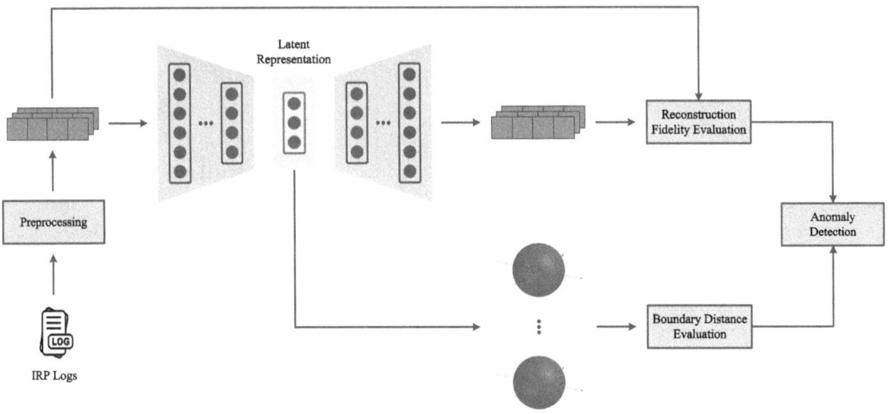

Fig. 1. Overall workflow of RansomSphereGuard

4.1 Preprocessing

The initial phase of RansomSphereGuard involves extracting meaningful features from low-level IRP logs to serve as a basis for downstream learning and anomaly detection. These logs capture metadata like target process, timestamp, operation type, IRP flags, buffer length, and status.

For early-stage detection, we partition the collected IRP logs into intervals. Since benign programs, especially system processes (e.g., services, update managers), frequently access large portions of files in system folders (e.g., `Windows` or `Program Files`), we exclude the IRP logs of these folders from each interval. The remaining logs in each interval are then processed to extract high-level features for each process. These features include operation type information (e.g., number of IRP operations and number of unique major/minor operation types), IRP flag characteristics (e.g., number of unique IRP flags), file system information (e.g., number of unique file objects accessed), buffer length statistics (e.g., min, max, mean, and standard deviation), and entropy characteristics of write operations (e.g., min, max, mean, and standard deviation). This interval-based approach strikes a balance between computational efficiency and the ability to detect anomalies in the early stages of a ransomware attack, where rapid intervention is crucial.

4.2 Model Construction

This phase focuses on learning discriminative representations of benign data while simultaneously modeling its underlying distribution through multiple adaptive hyperspheres. The process comprises three interconnected components: latent representation learning, latent space partitioning, and adaptive boundary learning.

Latent Representation Learning. To obtain robust and compact representations, we employ latent representation learning through a deep autoencoder architecture. The encoder, denoted as $f(\cdot;\phi)$ or simply $f_\phi(\cdot)$, maps an input feature vector $\mathbf{x}_i \in \mathbb{R}^d$ to a compact latent representation $\mathbf{z}_i \in \mathbb{R}^p$, while the decoder, denoted as $g(\cdot;\theta)$ or simply $g_\theta(\cdot)$, reconstructs the original input from this latent representation:

$$\mathbf{z}_i = f_\phi(\mathbf{x}_i), \quad \hat{\mathbf{x}}_i = g_\theta(\mathbf{z}_i), \tag{2}$$

where ϕ and θ denote the trainable parameters of the encoder and decoder, respectively. The autoencoder is first pre-trained in an unsupervised manner by minimizing the reconstruction loss over a subset of benign data, thereby suppressing noise and redundancy while preserving essential structural characteristics.

Latent Space Partitioning. After pre-training, we partition the latent space into K clusters using a clustering algorithm to capture the complex distribution of benign data. The centers of these clusters serve as the initial hypersphere centers in the latent space. Each cluster thus corresponds to a candidate benign data region to be modeled by a hypersphere.

Adaptive Boundary Learning. During fine-tuning, we map latent representations to their nearest hypersphere center. Formally, for each cluster $k \in \{1, 2, \ldots, K\}$, a dedicated hypersphere with center $\mathbf{c}_k$ is optimized by minimizing a boundary loss. This multi-sphere approach overcomes the limitations of conventional SVDD by learning cluster-specific decision boundaries that better model local data distributions, thereby capturing heterogeneous topological structures inherent in benign data.

The total training loss synergistically integrates both representation learning and boundary learning. By training the entire model end-to-end, we jointly optimize the autoencoder parameters and hypersphere centers, enabling accurate modeling of complex, multimodal benign IRP behaviors while mitigating representation collapse.

Let $X = \{\mathbf{x}_1, \mathbf{x}_2, \ldots, \mathbf{x}_N\}$ be the set of benign feature vectors. We partition X into two disjoint subsets: X_1 for fine-tuning the autoencoder parameters (ϕ, θ), and X_2 for updating the hypersphere centers $\mathcal{C} = \{\mathbf{c}_1, \mathbf{c}_2, \ldots, \mathbf{c}_K\}$. During each training epoch, the autoencoder is fine-tuned on X_1 with fixed hypersphere centers using an optimization algorithm such as Stochastic Gradient Descent (SGD) or Adam [17], minimizing both the reconstruction loss and the boundary loss:

$$\min_{\phi,\theta} \sum_{i=1}^{N_1} \left\|\mathbf{x}_i - g_\theta\left(f_\phi(\mathbf{x}_i)\right)\right\|^2 + \gamma_k \left\|f_\phi(\mathbf{x}_i) - \mathbf{c}_k\right\|^2, \tag{3}$$

where N_1 is the number of feature vectors in X_1, $\mathbf{x}_i \in X_1$ is a benign feature vector, $\mathbf{c}_k$ is the hypersphere center assigned to $f_\phi(\mathbf{x}_i)$, and γ_k is a hypersphere-specific weight balancing reconstruction fidelity against cluster cohesion. We employ the Nearest Prototype Rule (NPR) for hypersphere center assignment:

$$k = \arg \min_{j=1,\ldots,K} \left\| f_\phi(\mathbf{x}_i) - \mathbf{c}_j \right\|^2, \tag{4}$$

where k is the index of the nearest hypersphere center assigned to $f_\phi(\mathbf{x}_i)$.

Subsequently, hypersphere centers are updated on X_2 using adaptive optimization methods such as AdaGrad [13], while keeping the autoencoder parameters frozen:

$$\min_{\mathcal{C}} \sum_{i=1}^{N_2} \left\| f_\phi(\mathbf{x}_i) - \mathbf{c}_k \right\|^2, \tag{5}$$

where N_2 is the number of feature vectors in X_2, and $\mathbf{x}_i \in X_2$ is a benign feature vector.

At the end of each epoch, two refinement steps are applied: (1) hyperspheres containing fewer than ν percent of the total training samples are pruned, and (2) training samples are reassigned in the latent space to updated hypersphere centers via NPR. This iterative process ensures stable cluster structures and prevents degenerate hyperspheres, yielding robust characterization of multimodal benign IRP behaviors.

4.3 Anomaly Detection

Once trained, RansomSphereGuard evaluates input feature vectors using a dual-criteria anomaly score that combines reconstruction fidelity with hypersphere-based boundary distances. This integrated scoring mechanism leverages both the autoencoder's reconstruction capability and the multi-sphere architecture's adaptive boundary modeling. Formally, the anomaly score for an input feature vector $\mathbf{x}$ is defined as

$$\mathcal{S}(\mathbf{x}) = \left\| \mathbf{x} - g_\theta(f_\phi(\mathbf{x})) \right\|^2 + \gamma_k \left\| f_\phi(\mathbf{x}) - \mathbf{c}_k \right\|^2, \tag{6}$$

where f_ϕ and g_θ are the encoder and decoder networks, respectively, $\mathbf{c}_k$ is the hypersphere center assigned to $f_\phi(\mathbf{x})$ by NPR, and γ_k is the hypersphere-specific weight, maintaining consistency with the training objective in (3). The second term computes the squared Euclidean distance between the latent representation $f_\phi(\mathbf{x})$ and its nearest hypersphere center, weighted by γ_k. A feature vector $\mathbf{x}$ is flagged as ransomware if $\mathcal{S}(\mathbf{x}) > \tau_k$, where τ_k is a threshold calibrated on validation data.

This formulation provides several key advantages for heterogeneous benign data distributions:

- **Accommodating diverse patterns**: Each feature vector is evaluated against its most relevant hypersphere, naturally supporting diverse benign behaviors.

- **Enhanced boundary sensitivity**: Hypersphere-specific weights γ_k improve sensitivity to subtle anomalies near decision boundaries.
- **Reduced false alarms**: Restricting comparisons to the nearest hypersphere avoids mismatches with distant clusters, thereby reducing false alarms.

5 Experimental Evaluation

In this section, we experimentally evaluate the detection performance of RansomSphereGuard. We first describe the dataset, then define the evaluation metrics, and finally report the experimental results.

5.1 Experimental Setup

We conducted experiments on the IRP dataset originally introduced in [6,11]. The dataset consists of two categories of IRP logs: (1) benign IRP logs collected from 11 voluntary users running Microsoft Windows 7 to 10 on their machines, and (2) ransomware IRP logs obtained from the controlled execution of 383 ransomware samples on a Windows 7 machine. Due to unexpected formatting errors in the raw IRP logs (e.g., timestamp inconsistencies and string-to-integer conversion issues), we were able to utilize only 316 ransomware samples in our analysis.

To evaluate the performance of RansomSphereGuard in early-stage ransomware detection, we partitioned the collected IRP logs into fixed-size intervals. Within each interval, we first excluded IRP logs related to system folders and then grouped the remaining logs by process ID and process name to reconstruct a complete activity trace for each process. From these traces, we extracted the high-level features described in Sect. 4.1. This interval-based approach allows the model to capture early-stage I/O behavioral patterns while preserving the critical discriminative information necessary for timely ransomware detection.

All features were normalized to the $[0, 1]$ range using min–max scaling to ensure stable training and prevent bias toward features with larger magnitudes. The obtained benign feature vectors were then divided into training (70%), validation (15%), and test (15%) sets. Conversely, all ransomware feature vectors were reserved exclusively for testing to evaluate the model's ability to detect ransomware. Consequently, the final test set included feature vectors from both benign and ransomware samples, enabling a thorough evaluation of the model's generalization performance and detection accuracy.

For latent representation learning, we employed a fully connected autoencoder architecture. The encoder consisted of two hidden layers with 16 and 8 units, respectively, mapping the input feature vectors to a latent space of dimension 8. To mitigate the vanishing gradient problem and enhance training stability, Leaky ReLU activation functions were applied across all layers. The model was trained using mini-batches of size 128.

During training, each mini-batch was partitioned into two distinct subsets: 30% for fine-tuning the autoencoder and 70% for updating the hypersphere centers. The autoencoder parameters were optimized using Adam [17] with a learning rate of 0.001. For updating the hypersphere centers, AdaGrad [13] was employed with an initial learning rate of 1 and a decay factor of 0.1.

5.2 Evaluation Metrics

We evaluated the detection performance of RansomSphereGuard using three key metrics. The Detection Rate (DR) measures the proportion of ransomware samples correctly identified, with higher values indicating stronger detection capability. Conversely, the False Alarm Rate (FAR) quantifies the fraction of benign samples misclassified as ransomware; a low FAR is crucial to minimizing false alarms and ensuring model reliability. Finally, we computed an Anomaly Score (AS) for each sample by combining its reconstruction error with its distance from the corresponding hypersphere center in the latent space. This composite score enables more precise detection of subtle deviations characteristic of ransomware activity.

To ensure statistical reliability, we repeated each experiment 10 times with different random initializations and reported the average results across these independent runs.

5.3 Experimental Results

We conducted a series of experiments to evaluate the detection performance of RansomSphereGuard. To analyze the training dynamics, we plot the total loss alongside its constituent components—the reconstruction loss and the boundary loss—over training epochs, as shown in Fig. 2. The results confirm that the total loss converges during training. Initially, the total loss closely follows the boundary loss; however, as training progresses, it gradually aligns more closely with the reconstruction loss. This characteristic behavior stems from our training strategy, where the hypersphere centers are updated using an optimizer with a high initial learning rate and a large decay parameter. Consequently, the boundary loss drops sharply in the early epochs and then gradually decays with the learning rate.

We further analyzed the anomaly score distributions for both benign and ransomware samples, as shown in Fig. 3. Figure 3a reveals that benign samples exhibit anomaly scores tightly clustered near zero, indicating accurate reconstruction and effective hypersphere boundaries around benign activities. Conversely, Fig. 3b shows that ransomware samples produce significantly higher anomaly scores, clearly separated from benign ones. This distinct separation highlights the model's ability to identify ransomware patterns effectively while lowering false positives.

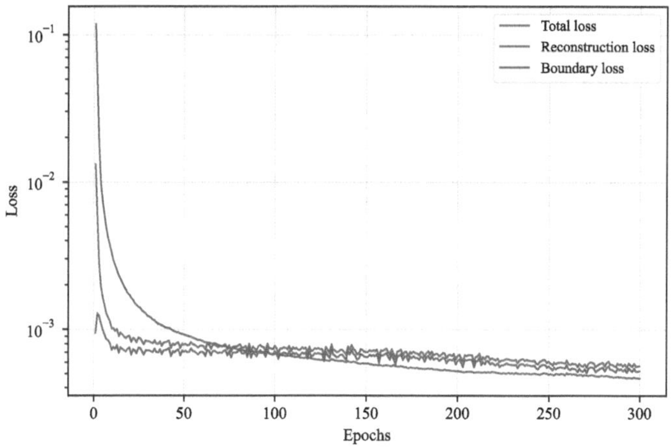

Fig. 2. Training dynamics of RansomSphereGuard illustrating the evolution of total loss, reconstruction loss, and boundary loss over the training epochs

To comprehensively evaluate the detection capabilities of RansomSphereGuard, we conducted a comparative analysis with two baseline techniques: OCSVM [27], a classical one-class classification technique, and DASVDD [15], a deep one-class classification technique based on a deep autoencoding single-sphere SVDD architecture. As reported in Table 1, RansomSphereGuard achieves a high DR of 93.972% with a low FAR of 2.921%. In contrast, DASVDD attains a lower DR of 76.123% and a slightly higher FAR of 3.011%, highlighting the limitations of its single-sphere architecture in modeling diverse benign behaviors. OCSVM exhibits even weaker detection performance, further demonstrating the advantages of deep, multi-sphere modeling for capturing complex benign patterns.

Table 1. Detection performance comparison of RansomSphereGuard against DASVDD and OCSVM

Model	DR (%)	FAR (%)
OCSVM [27]	71.551	2.973
DASVDD [15]	76.123	3.011
RansomSphereGuard	93.972	2.921

The performance advantage of RansomSphereGuard arises from its architecture, which integrates a pre-trained deep autoencoder with multiple hyperspheres. This design not only enhances latent representations through autoencoder pre-training but also enables more accurate modeling of complex benign patterns and supports finer-grained decision boundaries, leading to more effective anomaly detection.

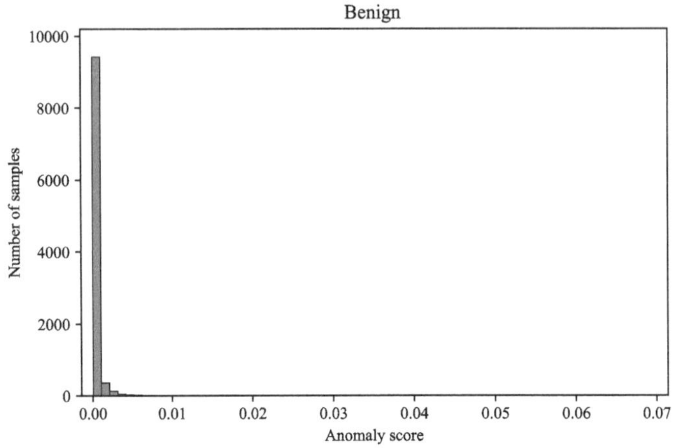

(a)

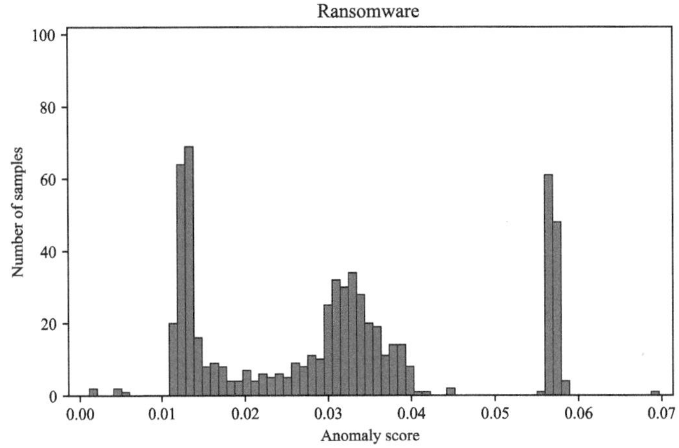

(b)

Fig. 3. Anomaly score distributions **(a)** benign and **(b)** ransomware samples. Benign samples are tightly clustered near zero, while ransomware samples show significantly higher scores with a clear separation

6 Conclusion, Discussion, and Future Work

In this work, we introduced RansomSphereGuard, a novel deep anomaly detection technique for early detection of zero-day ransomware during its pre-encryption stage by analyzing IRP logs at the process level. By leveraging high-level behavioral features—such as operation types, IRP flags, and buffer length statistics—and modeling benign activity through multiple adaptive hyperspheres in a pre-trained autoencoder latent space, RansomSphereGuard achieves highly sensitive and specific detection without relying on known attack signatures. The experimental results confirm its superiority over existing classical and deep one-class classification techniques.

The core strength of RansomSphereGuard lies in its hybrid architecture: the pre-trained autoencoder provides robust latent representations, while the multi-sphere strategy enables fine-grained modeling of diverse benign patterns. Furthermore, NPR-based hypersphere assignment and pruning of underpopulated hyperspheres further refine the decision boundaries, enhancing sensitivity to subtle anomalies and improving model stability.

As future work, we plan to extend RansomSphereGuard to incorporate complementary data modalities—such as API call sequences, file system metadata, and network traffic. Additionally, exploring lightweight model variants for resource-constrained environments will be crucial for enhancing scalability, trust, and practical deployment.

Disclosure of Interests. The authors have no competing interests to declare.

References

1. Agrawal, R., Stokes, J.W., Selvaraj, K., Marinescu, M.: Attention in recurrent neural networks for ransomware detection. In: Proceedings of the 2019 IEEE International Conference on Acoustics, Speech and Signal Processing, pp. 3222–3226. IEEE, Brighton, UK (2019). https://doi.org/10.1109/ICASSP.2019.8682899
2. Ahsani Somarin, A., Abadi, M., Sharafi, M.: MalDualCNN: a novel malware detection technique using binary file visualization and dual convolutional neural network. In: Bhateja, V., Patel, P., Tang, J. (eds.) Evolution in Computational Intelligence. Smart Innovation, Systems and Technologies, pp. 15–29. Springer, Singapore (2025). https://doi.org/10.1007/978-981-96-2124-8_2
3. Al-rimy, B.A.S., et al.: Redundancy coefficient gradual up-weighting-based mutual information feature selection technique for crypto-ransomware early detection. Future Gener. Comput. Syst. **115**, 641–658 (2021). https://doi.org/10.1016/j.future.2020.10.002
4. Allievi, A., Ionescu, A., Russinovich, M.E., Solomon, D.A.: Windows Internals, Part 2, 7th edn. Microsoft Press, Redmond (2021)
5. Alqahtani, A., Sheldon, F.T.: Temporal data correlation providing enhanced dynamic crypto-ransomware pre-encryption boundary delineation. Sensors **23**(9) (2023). https://doi.org/10.3390/s23094355
6. Ayub, M.A., Continella, A., Siraj, A.: An I/O request packet (IRP) driven effective ransomware detection scheme using artificial neural network. In: Proceedings of the 2020 IEEE 21st International Conference on Information Reuse and Integration for Data Science, pp. 319–324. IEEE, Las Vegas, NV, USA (2020). https://doi.org/10.1109/IRI49571.2020.00053
7. Ayub, M.A., Siraj, A.: Understanding the behavior of ransomware: an I/O request packet (IRP) driven study on ransomware detection against execution time. In: Proceedings of the 2023 IEEE 9th International Conference on Collaboration and Internet Computing, pp. 1–10. IEEE, Atlanta, GA, USA (2023). https://doi.org/10.1109/CIC58953.2023.00018
8. Cen, M., Jiang, F., Doss, R.: RansoGuard: a RNN-based framework leveraging pre-attack sensitive APIs for early ransomware detection. Comput. Secur. **150**, 104293 (2025). https://doi.org/10.1016/j.cose.2024.104293

9. Cen, M., Jiang, F., Qin, X., Jiang, Q., Doss, R.: Ransomware early detection: a survey. Comput. Netw. **239**, 110138 (2024). https://doi.org/10.1016/j.comnet.2023.110138
10. Chandola, V., Banerjee, A., Kumar, V.: Anomaly detection: A survey. ACM Comput. Surv. **41**(3) (2009). https://doi.org/10.1145/1541880.1541882
11. Continella, A., et al.: ShieldFS: a self-healing, ransomware-aware filesystem. In: Proceedings of the 32nd Annual Conference on Computer Security Applications, pp. 336–347. ACM, Los Angeles, CA, USA (2016). https://doi.org/10.1145/2991079.2991110
12. Deldar, F., Abadi, M.: Deep learning for zero-day malware detection and classification: a survey. ACM Comput. Surv. **56**(2) (2024). https://doi.org/10.1145/3605775
13. Duchi, J., Hazan, E., Singer, Y.: Adaptive subgradient methods for online learning and stochastic optimization. J. Mach. Learn. Res. **12**(61), 2121–2159 (2011)
14. FinCEN: Financial trend analysis: Ransomware trends in bank secrecy act data between January 2021 and June 2021 (2021). https://www.fincen.gov/news/news-releases/ransomware-trends-bank-secrecy-act-data-between-january-2021-and-june-2021
15. Hojjati, H., Armanfard, N.: DASVDD: deep autoencoding support vector data descriptor for anomaly detection. IEEE Trans. Knowl. Data Eng. **36**(8), 3739–3750 (2024). https://doi.org/10.1109/TKDE.2023.3328882
16. Kharraz, A., Robertson, W., Balzarotti, D., Bilge, L., Kirda, E.: Cutting the gordian knot: a look under the hood of ransomware attacks. In: Almgren, M., Gulisano, V., Maggi, F. (eds.) DIMVA 2015. LNCS, vol. 9148, pp. 3–24. Springer, Cham (2015). https://doi.org/10.1007/978-3-319-20550-2_1
17. Kingma, D.P., Ba, J.L.: Adam: a method for stochastic optimization. In: Proceedings of the 3rd International Conference for Learning Representations, pp. 1–15. San Diego, CA, USA (2015). https://doi.org/10.48550/arXiv.1412.6980
18. Lee, S.W., Park, J., Lee, S.W.: Low resolution face recognition based on support vector data description. Pattern Recognit. **39**(9), 1809–1812 (2006). https://doi.org/10.1016/j.patcog.2006.04.033
19. Mehnaz, S., Mudgerikar, A., Bertino, E.: RWGuard: a real-time detection system against cryptographic ransomware. In: Bailey, M., Holz, T., Stamatogiannakis, M., Ioannidis, S. (eds.) RAID 2018. LNCS, vol. 11050, pp. 114–136. Springer, Cham (2018). https://doi.org/10.1007/978-3-030-00470-5_6
20. Microsoft Learn: IRPs are different from fast I/O (2023). https://docs.microsoft.com/en-us/windows-hardware/drivers/ifs/irps-are-different-from-fast-i-o. Accessed 30 June 2025
21. Morgan, S.: Global ransomware damage costs predicted to exceed $275 billion by 2031 (2025). https://cybersecurityventures.com
22. OSR Open Systems Resources: An introduction to standard and isolation minifilters (2020). https://www.osr.com/nt-insider/2017-issue2/introduction-standard-isolation-minifilters/. Accessed 30 June 2025
23. Pang, G., Shen, C., Cao, L., Hengel, A.V.D.: Deep learning for anomaly detection: a review. ACM Comput. Surv. **54**(2) (2022). https://doi.org/10.1145/3439950
24. Petrosyan, A.: Number of ransomware attempts per year from 2017 to 2023 (2024). https://www.statista.com/statistics/494947/
25. Ruff, L., et al.: Deep one-class classification. In: Proceedings of the 35th International Conference on Machine Learning, vol. 80, pp. 4393–4402. PMLR, Stockholm, Sweden (2018)

26. Scaife, N., Carter, H., Traynor, P., Butler, K.R.B.: CryptoLock (and drop it): stopping ransomware attacks on user data. In: Proceedings of the 2016 IEEE 36th International Conference on Distributed Computing Systems, pp. 303–312. IEEE, Nara, Japan (2016). https://doi.org/10.1109/ICDCS.2016.46
27. Schölkopf, B., Platt, J.C., Shawe-Taylor, J., Smola, A.J., Williamson, R.C.: Estimating the support of a high-dimensional distribution. Neural Comput. **13**(7), 1443–1471 (2001). https://doi.org/10.1162/089976601750264965
28. Shin, J.H., Lee, B., Suk Park, K.: Detection of abnormal living patterns for elderly living alone using support vector data description. IEEE Trans. Inf Technol. Biomed. **15**(3), 438–448 (2011). https://doi.org/10.1109/TITB.2011.2113352
29. Tax, D.M., Duin, R.P.: Support vector data description. Mach. Learn. **54**(1), 45–66 (2004). https://doi.org/10.1023/B:MACH.0000008084.60811.49
30. Zahoora, U., Rajarajan, M., Pan, Z., Khan, A.: Zero-day ransomware attack detection using deep contractive autoencoder and voting based ensemble classifier. Appl. Intell. **52**(12), 13941–13960 (2022). https://doi.org/10.1007/s10489-022-03244-6

Investigating the Control Flow Graphs of IoT Binaries for Malware Detection

Haydar Teymourlouei[1], Arya Teymourlouei[2(✉)], and Vareva Harris[3]

[1] Department of Computer Science and Engineering Technology, University of Maryland Eastern Shore, Princess Anne, MD 21853, USA
hteymourlouei@umes.edu

[2] Department of Computer Science, University of Maryland College Park, College Park, MD 20742, USA
ateymo2@umd.edu

[3] Richardson Family SmartState Center for Economic Excellence in Tourism and Economic Development, College of Hospitality, Retail and Sport Management, University of South Carolina, Columbia, SC, USA
vareva@mailbox.sc.edu

Abstract. The rising popularity of embedded devices, particularly Internet of Things (IoT) devices, has led to concerns about malicious IoT executable files (binaries) which can cause unwanted or unintended behavior. Malware detection in the IoT network is performed in real-time using analysis of the network traffic. However, it is possible that additional insights into the behavior of malicious IoT binaries can be gained through analysis of the control flow graphs (CFGs) of the binaries. CFGs are directed graphs encoding the possible paths of execution in a binary executable. This study aims to analyze the CFGs of binaries arising from the Mirai botnet malware and compare to benign binaries to assess to what extent the CFGs are effective for malware detection. The results show that graph-theoretic features of CFGs yield excellent performance when used as inputs to machine learning algorithms, with over 90% average on F1 score for testing data and over 98% in the best case. Examination of the distribution of the number of nodes and density of the CFGs shows a normal-like distribution for benign binaries, but irregular patterns for Mirai binaries. The results of this study are consistent with and extend the findings of prior literature in malware detection.

Keywords: Malware detection · Internet of Things (IoT) · control flow graph · binary executable files · classification

1 Introduction

The Internet of Things (IoT) refers to a collection of devices and systems which communicate and exchange data over a network [1]. The IoT has allowed vast forms of data to be collected and used for many practical purposes, such as continuous streaming of weather information, health information based on wearable devices, and control smart home devices. However, there is an increased risk of IoT devices being exposed

H. R. Arabnia et al. (Eds.): CSCI 2025, CCIS 2969, pp. 270–281, 2027.
https://doi.org/10.1007/978-3-032-28460-0_19

to malicious software (malware), for several reasons. This includes insecure passwords (e.g. easy to guess), insecure network protocols and services (e.g., HTTP), poor use of cryptographic mechanisms for authorization, lack of software and hardware updates, and several others [2].

For these reasons, preventing malware on IoT devices is a challenging problem, and it is inevitable that malware will infect some parts of IoT networks in different applications. Therefore, accurate and robust malware detection is a crucial task which must be achieved to improve the resilience of the IoT network. In prior work, several methods for this purpose were identified: Real-time monitoring of the network protocols behavior, machine-learning based classification of network traffic, and custom firewalls for IoT networks [3]. These methods were shown to be effective in *online* malware detection for IoT networks. Their main purpose was to identify malicious behavior through the characteristics observed in network traffic. However, an *offline* approach that analyzes the malicious binary executable files (called "binaries") may also be helpful in the malware detection process.

Offline analyses of malicious executables sometimes get less attention than online analyses since they are less used in real-time. However, they are still extremely useful as they allow cybersecurity experts to get a more detailed understanding of the underlying behavior and mechanisms of malicious executables. This, in turn, could be used to develop more effective methods for real-time malware detection and prevention. To this extent, a recent dataset was published [4] which curated over 65,000 binaries, most originating from well-known IoT malware like Mirai and Gafygt. The authors of [4] thoroughly examined the binary files through an analysis of function signatures and features (symbol tables and strings).

The common methods for offline, static analysis of malware includes control flow graphs (CFGs), operation codes (opcodes), and string and header analysis [5]. Opcode analysis looks at the machine-level CPU instruction sequences used by the malicious program, as these will naturally have distinct characteristics compared to a traditional program. A recent paper analyzed the opcode sequences in the binaries in the previously mentioned dataset [4] and showed that such features could perform malware detection with classification accuracy over 97% [6]. This research will focus instead on CFGs, as they provide a graph-theoretical perspective to the analysis of binaries which are different in nature from opcode and similar methods. A recent review showed that CFGs have been used for malware detection in prior work, including for IoT malware [7]. There are also methods for dynamic analysis and symbolic execution [8], where the malicious software runs in a contained environment, but these will be omitted from the current work.

In this work, we will analyze the properties of CFGs of malicious and benign (non-malicious) binary executable files from a publicly available dataset [4]. This dataset has primarily been analyzed in terms of function signatures and opcodes, which differ in nature from CFGs. The local and global structure of the graphs will be analyzed. Local structure refers to the individual components of the CFG which correspond to the functions which make up the entire executable. Global structure refers to the properties of the entire CFG. This research has three main objectives: (1) To determine to what extent CFG-based features can be used for IoT malware detection with the new dataset; (2) To determine the local and global structure of CFGs arising from malicious executables, and (3) To determine whether it is more effective to analyze local structure or global structure for malware detection based on CFGs. We hypothesize, based on prior work [7], the proposed approach will be able to accurately detect malware, and both local and global features will have equal importance in the detection process.

2 Background

2.1 Internet of Things (IoT)

The IoT is a connected network of embedded systems and technologies that enable rapid communication both within-network and to the cloud [9]. IoT is a rapidly growing phenomenon, but so are the associated risks and dangers. IoT networks which handle sensitive data, such as those in healthcare, commerce, and the military, can be particularly vulnerable and must set up defenses against the many risks. However, they are also popular for home networks which consist of simple embedded devices [3]. Cyber threats can come from malicious software such as malware, ransomware, adware, and spyware [10]. Also, risks to sensitive data can arise from hackers intercepting IoT devices holding sensitive data.

2.2 Malware Detection

Software security has become more complex across the development of technology due to the growing scale and complexity of software used across a range of applications. Prior work shows that every 1000 lines of code in a software can have a range of 5–20 bugs [11]. Considering that even one bug can cause severe damage, such as Heartbleed [12], malware detection is a crucial aspect of the analysis of binary files. There are three main types of malware detection: Static analysis, dynamic analysis (including symbolic execution, an extension of dynamic analysis), and hybrid approaches. Static analyses do not execute the malicious program, while dynamic analyses do. A hybrid approach utilizes both static and dynamic analyses. Figure 1 illustrates an example of a hybrid model for real-time malware detection on the IoT network. Infected devices on the network are identified by analyzing collected network traffic for malicious behavior (step 1). Then, files on the infected device are quickly scanned using a static analysis approach (step 2). To identify the malware, suspicious files are more thoroughly examined using symbolic execution or a similar dynamic approach (step 3). Our prior work [3] can be seen as step 1 of the hybrid model, and the current work can be seen as step 2 of this model.

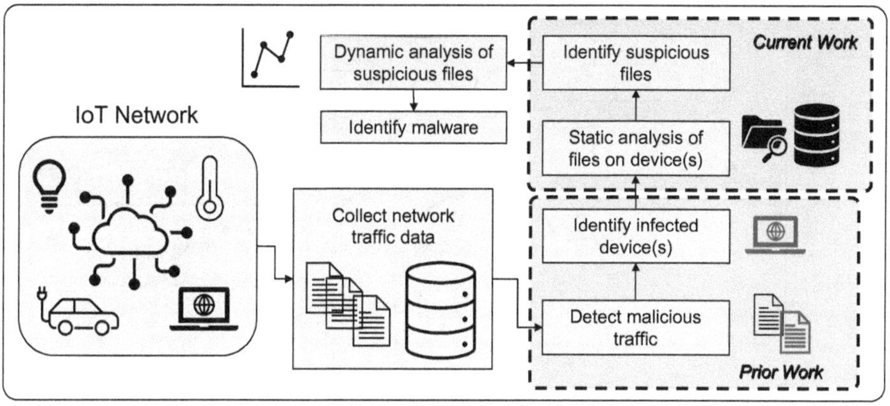

Fig. 1. Example of a hybrid model for malware detection on the IoT network. The infected IoT devices are identified by analyzing the network traffic data for malicious activity (for example, see our prior work [3]). The files on these devices are then quickly scanned using a static analysis (the current work). Suspicious files are dynamically analyzed to find the malware.

2.3 Control Flow Graphs

CFGs represent the different possible paths of execution of a program. One run of the program would represent a traversal through the graph. Formally, a CFG is a directed graph where nodes are basic blocks in the program [13]. A basic block is defined as any sequence of code which is linear and has only one entry and exit. For example, two sequential assignment statements could form a basic block. An edge forms between two nodes if there exists a control flow path in the program between the two basic blocks. For example, an edge could form between the declaration of a counter variable (one block) and the subsequent for-loop (one block). A toy example of a CFG for a simple program is presented in Fig. 2, generated with the py2cfg library [14]. The example in Fig. 2 demonstrates how different types of code, which lead to different forms of control flow, change the structure of the graph. The for-loops in Fig. 2 create cycles, and the if-statement creates a branching point. Therefore, as the complexity of the control flow in a program increases, the number of nodes and edges will increase.

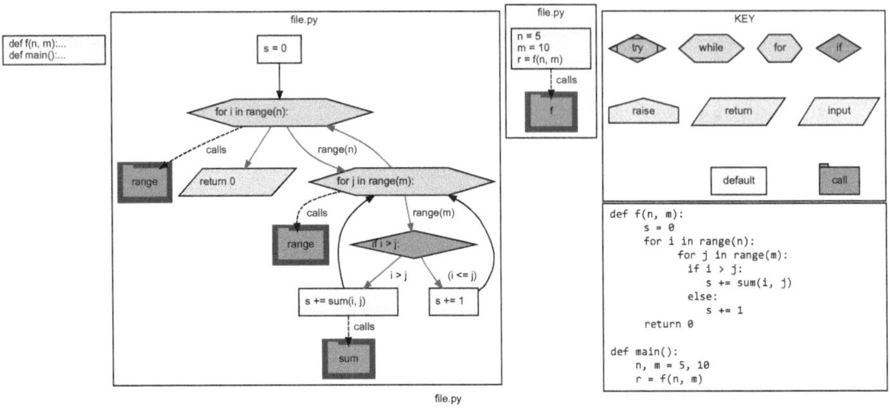

Fig. 2. Toy example of a control flow graph generated by py2cfg [14]. Graph legend is in the upper-right box. The Python source code is in the bottom-right text box.

3 Methodology

3.1 IoT Binaries Dataset

A publicly available dataset of 65,956 binary files used in IoT malware is used [4]. The binaries were curated mostly from VirusShare, MalwareBazaar, VxUnderground, and Contagio. The dataset contains a roughly equivalent amount of stripped and non-stripped binaries. Most of the files (89.44% of the entire dataset) are statically linked, while the remaining are either dynamically linked (10.20%) or unspecified (0.35%). This study will consider files regardless of these characteristics.

The authors of the dataset have constructed threat labels for the binaries. The labels include Mirai, Gafgyt, Tsunami, Rootkit, and other types of malware. To reduce the complexity of the experiment, we will only consider the binaries which are labeled benign (not malicious) or Mirai. Mirai is a malware which generates a botnet attack, turning normal IoT devices (such as those in the home network) into bots [15]. The bots can then be used in an overwhelming DDoS attack. Mirai binaries make up more than half (54.51%) of the dataset. The dataset collected binaries from several different IoT architectures, including ARM, i386, MIPS, Motorola, Sparc, and AMD64 (x86–64). To simplify this study, only ARM, i386, and AMD64 are considered.

3.2 Control Flow Graph Generation

To analyze the characteristics of the binaries from the perspectives of graph theory and software engineering, the control flow graph (CFGs) of each binary file is created using the Python library angr [16]. CFG recovery is done using a static recovery of the control flow, known as the CFGFast mode in angr. This is done as opposed to dynamic analysis via symbolic execution (CFGEmulated mode). Indirect jump resolution is also turned off. These choices were made to reduce the computational cost of constructing the CFGs.

3.3 Analysis of Control Flow Graphs

To analyze the CFGs, two methods are considered. The first is a global approach, where the structure of the entire CFG is analyzed. For this approach, the first measure computed is the number of nodes N in the CFG. The second is the density D of the CFG, which is defined as

$$D = \frac{|E|}{(N(N-1)/2)} \quad (1)$$

where $|E|$ is the number of edges in the CFG. The second method is a local approach, where the functions in the binaries are studied individually. The first measure starts by computing the number of nodes N_f for the function. This is done for each function in the binary and then the average value of N_f is taken. The second measure computes the graph density D_f for each function, taking the average value across all functions.

Other possibilities for graph metrics were considered, including the average shortest path length, clustering coefficient, and metrics related to dominance. However, due to the significant cost (time and memory) of computing these metrics, they were excluded from the analysis. This was particularly a concern for the larger CFGs, which had hundreds of thousands of nodes. All network analysis was conducted with the NetworkX library [17].

3.4 Prediction of Malicious Binaries

Aside from the CFG analysis, we study the problem of detecting malicious (Mirai) binaries from benign ones. This is a binary classification problem. Three different machine learning (ML) algorithms are used for this problem. They are random forest (RF), support vector machine (SVM), and k-nearest neighbors (KNN). These algorithms have demonstrated excellent classification performance in other types of malware detection [3]. RF is an ensemble learning method, as the final prediction is the aggregation of predictions made by 50 decision trees. KNN is a non-parametric model where the prediction for a non-labeled data point is done by utilizing the k-nearest neighbors of the data point ($k = 5$). SVM is a linear classifier (unlike RF and KNN) which creates a hyperplane as the decision boundary between the two classes of binaries.

The classification followed a five-fold cross-validation procedure (80%-20% training-testing ratio). 10 repetitions of the cross-validation were done for each split. Model scoring was done with F1, loss, and training (fit) time. F1 is the harmonic mean of precision and recall. F1 is used as opposed to accuracy (the fraction of correct predictions) since the dataset is unbalanced. Loss is the distance of the incorrect prediction to the correct result using logarithmic function. Fit time is the amount of time needed to train the algorithm prior to testing. The mean and standard deviation for each metric, for each algorithm, is reported in the results. Classification was implemented via Scikit-Learn [18].

4 Results

Figures were generated with Matplotlib [19] and Seaborn [20].

4.1 IoT Malware Detection Performance

In this section the performance of the classification algorithms for detecting malware is shown. Figure 3 shows the average F1 scores for testing data from each of the three IoT architectures. For all situations, on average, the RF classifier (RFC) had the best performance. Generally, SVM had the worst performance. However, for the ARM and i386 architectures, the difference between the classifiers was not more than 10% on average. The results also show that for the ARM and i386 architectures, the average F1 was over 90% for all algorithms. This was true for AMD64 except for the SVM classifier. When comparing the different feature subsets used to train the algorithms, using all the features (full) had the best average performance. For ARM and i386, the difference between global vs local features was not large, but for AMD64 there was a 5–10% drop in average F1 score when using local features (as opposed to global). The results in Fig. 3 show AMD64 had a larger standard deviation than ARM and i386, this was likely since AMD64 had a smaller set of data.

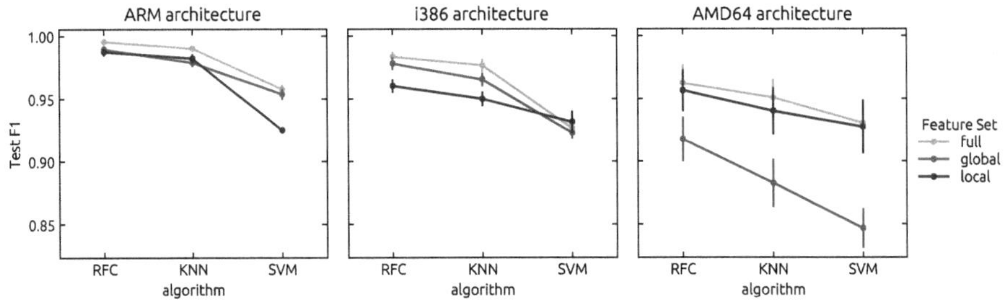

Fig. 3. Average F1 scores for each of the three classifiers, for each architecture. The legend indicates the subset of features used: Full (all features), global only, and local only. Error bars denote the standard deviation from the mean.

Figure 4 shows the average loss scores for testing data for each of the three IoT architectures. The graph shows that on average, the loss is the highest for SVM, which is expected as it had the lowest average F1 scores. Figure 4 also shows that for two out of the nine cases (KNN for ARM architecture and SVM for i386 architecture), using local features resulted in a lower loss than global features. In the remaining seven cases, using global features provided better performance. The other trends observed in Fig. 3 are reflected with consistency in Fig. 4.

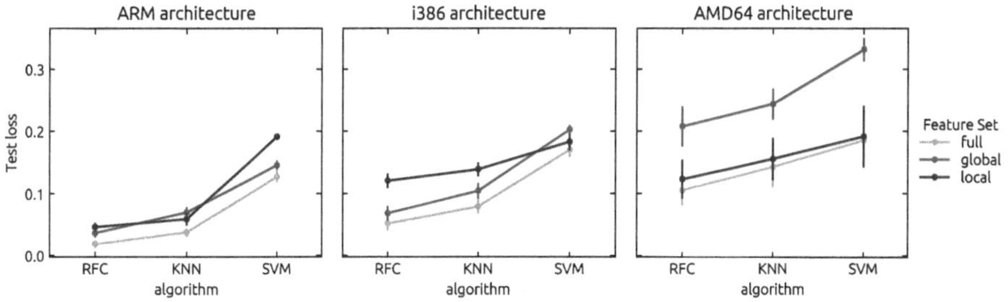

Fig. 4. Average loss scores for each of the three classifiers, for each architecture. The legend indicates the subset of features used: Full (all features), global only, and local only. Error bars denote the standard deviation from the mean.

Finally, Fig. 5 shows the average time taken to train each of the ML algorithms. RFC has, on average, the largest training time. KNN and SVM took less time to train, but the difference between them is relatively smaller. Using the full feature set resulted in a larger training time than using either global or local features, which is expected.

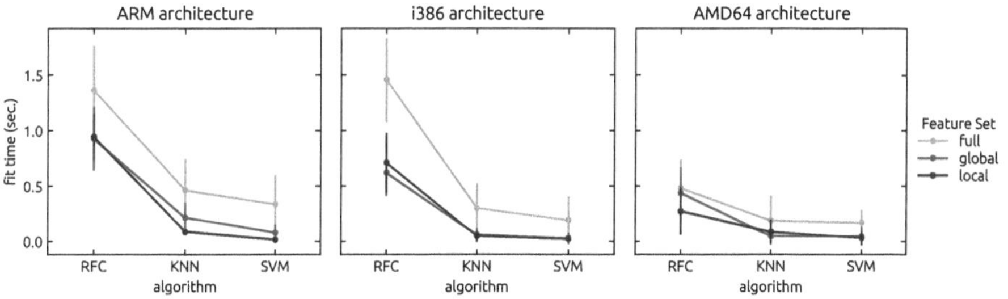

Fig. 5. Average training time scores for each of the three classifiers, for each architecture, in seconds. The legend indicates the subset of features used: Full (all features), global only, and local only. Error bars denote the standard deviation from the mean.

Figure 5 also shows that the ARM and i386 architectures, which had larger number of binaries, took longer to train on average in comparison with the AMD64 data.

4.2 Analysis of the Control Flow Graphs

In this section, an analysis of the CFGs of malicious and benign binary executable files is presented. Figure 6 shows the global properties, in particular, the distribution of the number of nodes N and the density D of the CFGs. It is shown that for the benign binaries, the distribution of N and D follows an approximately lognormal distribution (note the log x-axes on Fig. 6) with the ARM and i386 architectures having more than one peak. For the malicious (Mirai) binaries the distribution is much more multimodal, with data points densely concentrated around two or three peaks.

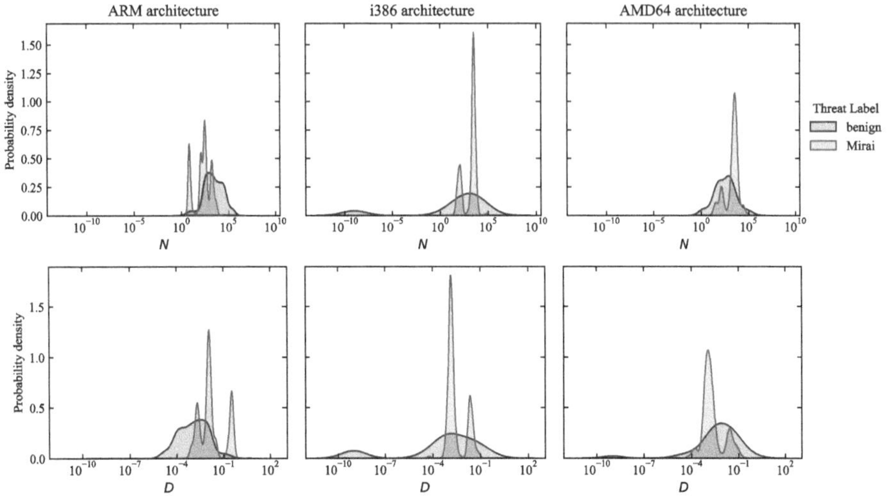

Fig. 6. Analysis of global structure of the CFGs. Top: Distribution of the total number of nodes in the CFG. Bottom: Distribution of the density of the entire CFG. Note that this density and the probability density are completely different. The legend denotes the classification of the binary file.

However, for some distributions in Fig. 6, there is a small density of extremely small (e-10) N or D. Constructing accurate CFGs for very small binaries in the dataset was challenging, and often the algorithm constructed these with 0 nodes and 0 edges. This inaccuracy was a limitation of the CFG generation process. When plotting the distribution of such points, e-10 was used instead of 0 to allow for a logarithmic axis.

Figure 7 shows the local properties of CFGs. The top row shows the distribution of the average number of nodes in each function N_f and the bottom row shows the average density of each function D_f. The benign binaries show, the distribution of D_f is approximately lognormal, but the variance of D_f is larger for the i386 and AMD64 architecture when compared to the ARM architecture. The distribution of N_f for the benign binaries varies significantly across the different architectures. For the Mirai binaries, the distributions are generally multimodal with tight peaks, similar to the patterns seen in Fig. 6. However, in general, the difference in variance and concentration of data points (as indicated by the probability density) is larger in Fig. 7 than it is in Fig. 6 when comparing across architectures. Also, it is important to note the same issue regarding very small binary files was present in the local analysis.

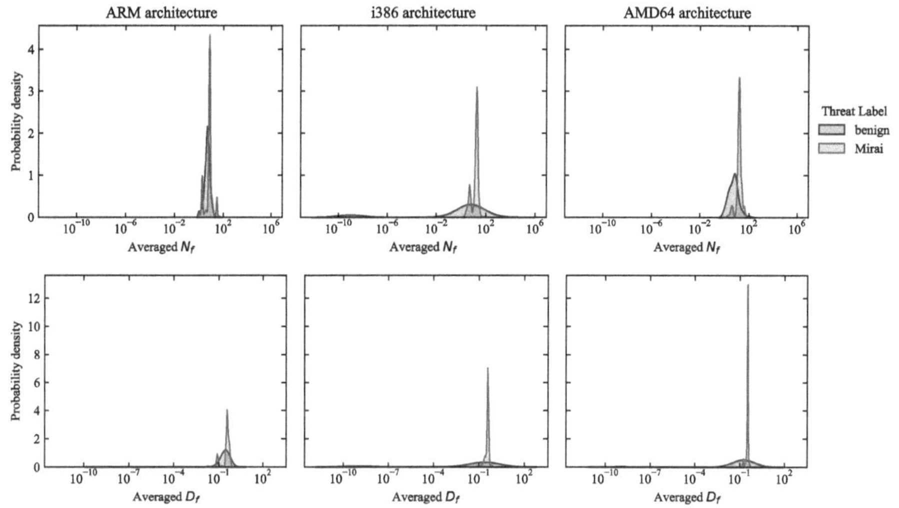

Fig. 7. Analysis of local patterns in CFGs. Top: Distribution of the average number of nodes in each function in the binaries. Bottom: Distribution of the average CFG density of each function in the binaries. Note that this density and the probability density are completely different. The legend denotes the classification of the binary.

5 Discussion and Conclusions

In this study, the effectiveness of using CFG-based features for malware detection was assessed conducting an analysis of the properties of the CFGs. The results showed that CFGs were generally effective for malware detection, as the average F1 score was always over 90% in the best case. In particular, for binaries from the ARM architecture, the average F1 score exceeded 98%. These findings are consistent with our hypothesis and compare well to the other study which looked at malware detection with this dataset; although, there were methodological differences [6]. RFC generally had the best classification performance; however, it also generally had the most computational cost, which is consistent with prior work [3]. When comparing local vs global features for classification it was shown that generally the performance was similar, but in some cases global features provided a slight advantage. Overall, however, the results suggest that analyzing both the local and global structure of the CFGs provides useful information for accurate malware detection, and this is also consistent with our hypothesis.

The analysis of the CFGs showed that the structural properties (local and global) of normal binary executables generally followed a lognormal distribution, with some outliers due to issues in the CFG generation process. This finding could possibly be explained by the fact that the dataset studied in this work generally had many binaries which were very small in size (only a few kb), and some which were very large in size, but relatively few that would fall into the middle of this range. This observation was not necessarily true across the entire dataset, but it was seen for some architecture and threat labels. When studying the structure of the malicious executable CFGs, the distribution of the data took a different form than the benign files. The data was tightly concentrated around two or three peaks and had little variance. We hypothesize that this may be due to the nature of malware, which is rapidly modified to avoid signature detection (e.g., as with polymorphic and metamorphic viruses). It is possible that each peak observed in the distribution corresponds to one predominant variant of the Mirai malware, with many similar copies generated for each variant. Further study of these distributions and their relationship to the evolution of the Mirai malware over time could offer additional insight.

One of the primary limitations of this study was the computational cost of generating CFGs. First, 18,600 binaries were analyzed from the 65,956 available in [4], and a CFG analysis had to be performed for each binary. This was particularly time-consuming for larger binaries which had hundreds of thousands of nodes and edges. At the other end, analyzing very small binaries was also challenging. The CFGs of such binaries consisted of very few nodes and edges (e.g., less than 10), and due to the settings of the algorithm used to generate CFGs, this often resulted in the algorithm saying that the CFG had zero nodes and edges. This was reflected in the statistical distributions presented in the results, particularly for the benign binaries. Another limitation of this research was inability to conduct a more rigorous analysis of the CFGs. The measures used were simplistic, and it is possible that other graph algorithms could provide additional insight.

Future work should consider applying more efficient procedures for graph algorithms to enable full characterization of the CFGs, such as with spectral graph theory. We also recommend that future work utilize emerging paradigms and advancements such as automatic differentiation, just-in-time compilation via computational graphs, and GPU-based program acceleration to enable faster analyses of CFGs.

References

1. IBM: What is the Internet of Things (IoT)? IBM (2023). https://www.ibm.com/think/topics/internet-of-things
2. Ferrara, P., Mandal, A., Cortesi, A., Spoto, F.: Static analysis for discovering IoT vulnerabilities. Int. J. Softw. Tools Technol. Transfer **23**(1), 71–88 (2021)
3. Teymourlouei, H., Stone, D.: Malware Detection in the IoT Home Network. In: 25th International Conference on Internet Computing & IoT, July 22–25. Springer Nature Switzerland, Las Vegas, USA (2024)
4. Olsen, S., OConnor, T.: Toward a Labeled Dataset of IoT Malware Features. In: 2023 IEEE 47th Annual Computers, Software, and Applications Conference. IEEE (2023)
5. Ngo, Q., Nguyen, H., Le, V., Nguyen, D.: A survey of IoT malware and detection methods based on static features. ICT Express **6**(4), 280–286 (2020)
6. Ramamoorthy, J., Gupta, K., Shashidhar, N., Varol, C.: Linux IoT malware variant classification using binary lifting and opcode entropy. Electronics **13**(12), 2381 (2024)
7. Mitra, S., Torri, S., Mittal, S.: Survey of malware analysis through control flow graph using machine learning. In: 2023 IEEE 22nd International Conference on Trust, Security and Privacy in Computing and Communications. IEEE (2023)
8. Namani, N., Khan, A.: Symbolic execution based feature extraction for detection of malware. In: 2020 5th International Conference on Computing, Communication and Security. IEEE (2020)
9. What is IoT? (n.d.). Oracle. https://www.oracle.com/internet-of-things/what-is-iot/
10. Wang, A., et al.: An inside look at IoT malware. In: Industrial IoT Technologies and Applications: Second EAI International Conference, Industrial IoT 2017, Wuhu, China. Springer International Publishing (2017)

11. Libicki, M., Ablon, L., Webb, T.: The defender's dilemma: Charting a course toward cybersecurity. RAND Corporation (2015)
12. Durumeric, Z., et al.: The matter of heartbleed. In: Proceedings of the 2014 ACM Internet Measurement Conference (2014)
13. Allen, F.: Control flow analysis. ACM Sigplan Notices **5**(7), 1–19 (1970)
14. Studer, J., Taylor, Lai, K.: Py2cfg: Python3 control flow graph generator (2020). https://py2cfg.readthedocs.io/en/latest/#
15. Antonakakis, M., et al.: Understanding the Mirai botnet. In: 26th USENIX Security Symposium (2017)
16. Shoshitaishvili, Y., et al.: Sok:(state of) the art of war: Offensive techniques in binary analysis. In: 2016 IEEE Symposium on Security and Privacy. IEEE (2016)
17. Hagberg, A., Swart, P., Schult, D.: Exploring network structure, dynamics, and function using NetworkX (No. LA-UR-08-05495; LA-UR-08-5495). Los Alamos National Laboratory (LANL), Los Alamos, NM, USA (2008)
18. Pedregosa, F., et al.: Scikit-learn: Machine learning in Python. The J. Machine Learn. Res. **12**, 2825–2830 (2011)
19. Hunter, J.: Matplotlib: A 2D graphics environment. Comp. Sci. Eng. **9**(03), 90–95 (2007)
20. Waskom, M.: Seaborn: Statistical data visualization. J. Open Sour. Softw. **6**(60), 3021 (2021)

CryNet: An Empirical Study Toward a Versatile IRC Botnet Framework for Realistic Cybersecurity Assessments

Sean Naughton, Bingyu Liu(✉), Ashar Neyaz, and Leonidas Deligiannidis

Wentworth Institute of Technology, Boston, MA 02115, USA
{naughtons,liub,neyaza,deligiannidisl}@wit.edu

Abstract. As digital infrastructure faces increasing cyber-threats, penetration testing and offensive security tactics to test security measures are invaluable. Botnets pose a significant challenge to network security, particularly regarding the integrity of the digital infrastructure. In this work, we present a novel approach to analyzing botnet behaviors through the CryNet botnet framework. A custom botnet environment leveraging an IRC C&C structure is presented to simulate realistic attack scenarios. Designed and implemented as an open-source tool, CryNet is an advanced IRC botnet intended for penetration testing, offering a versatile and realistic testing environment. It symbolizes the inexorable growth of sophistication and accessibility of open-source cybersecurity toolsets. In our experiments, we conduct various simulated attack scenarios to assess the capabilities and operational performance of CryNet. The evaluation demonstrates that CryNet is capable of effectively launching various types of attacks while highlighting key system vulnerabilities that need to be addressed in modern networks. These findings underscore the importance of proactive security assessments and offer actionable insights for enhancing network defenses.

Keywords: IRC Botnet Detection · Penetration Testing · Network Security · Open-source Tool

1 Introduction

Botnets [5,23,25] have long been recognized as a major security threat on the Internet. By exploiting a command-and-control (C&C) channel, attackers can take control of an army of compromised hosts (Bots) to carry out illegal activities such as click fraud, email spam, and DDoS attacks. Security through obscurity is a security principle that relies on the secrecy of its design. Moreover, it is the practice of withholding information about the machinations of a concept's design in order to reduce the chance of an attacker finding vulnerabilities to exploit. While this idea may sound effective to an extent, one may notice some glaring flaws. Specifically, security by obscurity naturally discourages peer review and public scrutiny. This is because if the security mechanism is disclosed to the

H. R. Arabnia et al. (Eds.): CSCI 2025, CCIS 2969, pp. 282–297, 2027.
https://doi.org/10.1007/978-3-032-28460-0_20

public, the integrity of said security is virtually nullified. Therefore, this strategy is difficult to maintain and contain within a field which inevitably brings impeachment of security by determined attackers. Unfortunately, many industries persist with this approach towards security. Considering that some physical lock manufacturers design their locks with a predefined sequence of inputs, the solution can potentially be reverse-engineered.

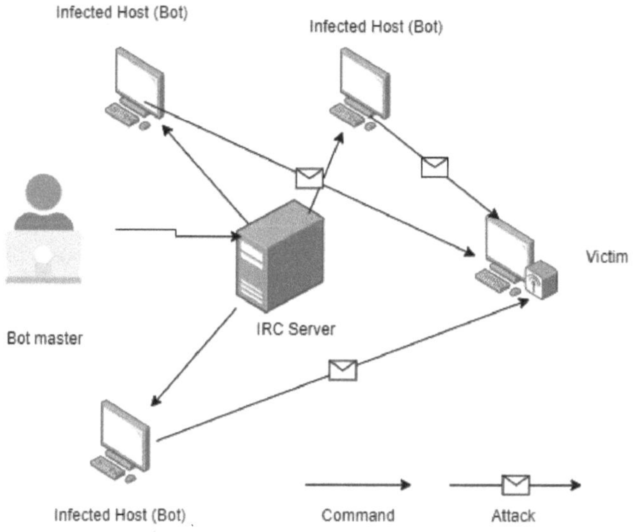

Fig. 1. General Structure of an IRC Botnet

While this approach does prevent the average person from cracking the lock, an experienced locksmith, or any determined individual with internet access, could find the solution via reverse engineering or simply discovering leaked information. On the contrary, the information technology, or more specifically cybersecurity, industry has taken a more comprehensive and aggressive approach. Overall, the tech industry attempts to make all information about security mechanisms known to the public. CryNet is a versatile botnet that not only performs expected botnet functions, but also has the capability to perform other malicious actions on a secure device. This work significantly emphasizes the need for proactive defenses and facilitates future research into machine learning-based detection systems [4,7,14,17,18] and advanced countermeasures to combat evolving botnet threats.

Despite many developments in detection and mitigation strategies [6,13,24], botnets remain one of the more challenging threats to the digital security landscape. Botnets are a collection of compromised devices that perform actions based on the commands from a singular master or controller. While they can come in different models such as distributed or peer-to-peer, the most common model is a Command and Control model (C&C). The C&C model is a traditional model where a single centralized leader exerts complete control over the actions of all connected nodes. There are many methods of communication used to achieve a functional C&C model, but CryNet uses Internet Relay Chat (IRC) [10,15]. IRC is a text-based instant messaging protocol that facilitates group communication in the form of channel based forums. IRC has other capabilities, such as one-on-one messaging, but CryNet primarily utilizes one-to-many communication. Once a Bot is running on an infected device, it simply connects to a specified IRC Server, joins a channel, and listens for messages sent to the channel. The general structure of an IRC botnet is illustrated in Fig. 1. It depicts an instance where the Master and "n" Bots have already connected and authenticated to the IRC Server. The Master sends IRC messages to a given IRC channel and any connected Bots will receive the message. Upon receiving the message, each Bot will parse it as a command, determine if the command was directed to them, and if so, attempt to execute the request. In this instance, the request was to send some sort of attack to a victim device. While DDoS attacks [21,22] are the primary use of botnets, CryNet is designed to add another layer of complexity to the concept. CryNet Bots can be compiled with varying capabilities. These capabilities range in the penetration testing concepts of remote reconnaissance, file encryption, resource throttling, packet creation/sending, and more. In our work, CryNet Bots compiled with file system encryption have the ability to remotely encrypt an entire infected device's file system. Currently, CryNet includes a built-in local file decryption mechanism. Note that the Bots in this work are used strictly for security testing purposes. The results of our analysis emphasize the importance of adopting robust cybersecurity measures, such as encryption, firewalls, and intrusion detection systems, to mitigate the effects of botnets. CryNet's versatility in deploying attacks showcases the potential for extensive damage, underlining the need for proactive defenses. More importantly, this work paves the way for future research on defense mechanisms based on Machine Learning, Large Language Model (LLM) detection systems, and advanced countermeasures to combat evolving botnet threats [7,9,11,16,18]. The insights gained into the dynamic landscape of network security help guide the development of more effective strategies to safeguard critical infrastructure.

2 Design

2.1 Architectural Overview

CryNet is composed of three main components: the IRC Bot, the IRC Server, and the Master program. The IRC Server functions as a middle man between the Master program and the IRC Bot(s). Furthermore, all communication is facilitated through the IRC Server's channels. Bots are compiled with an assigned IRC channel to join when executed. From there, the Master program can choose what channels they would like to join and send commands. Once the IRC Bot(s) receive a message on the channel they've joined, the message is parsed and determined if it is actionable. If the message is deemed actionable, then the action

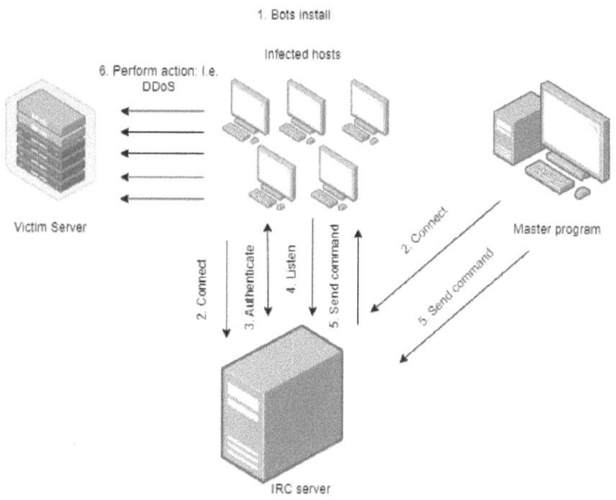

Fig. 2. Workflow of CRYNET

is carried out. Figure 2 showcases a high-level flow from Server initialization to actions being executed by the Bots.

As illustrated in Fig. 2, the Bots and Master connect and authenticate and then begin to communicate. At the micro level, careful design of each component was required to achieve a seamless macro-level workflow. The workflow can be demonstrated in the protocol that articulates how the System runs on a high level, as shown in Algorithm 1. It succinctly describes the expected process that takes into account all three CRYNET components and dictates that $n * k$ actions will be performed. Meaning, that for all n Bots connected to the IRC Server, a given action command will result in k actions (See Sect. 4 for more details). As one can see, the greater amount of Bots, the greater the amount of carried actions across all infected devices occur.

2.2 IRC Server

The IRC Server is the most integral component of CRYNET. It permits for a C&C model and allows for a singular entity to manage communication, minimizing difficulty. The server is highly object-oriented and designed to allow for several hundred concurrent connections. The Server functions like a typical server, awaiting connections, parsing connection input, and performing internal and external actions as needed. CRYNET's server is open on IRC standard port (port 6667). IRC channels are virtual spaces on an IRC Server where users can communicate with each other. These channels, identified by a "#" prefix, serve as chat rooms where multiple participants can join, leave, and send messages visible to all members of the channel. Channels facilitate organized communication by allowing users to gather in specific rooms for different topics or purposes.

Algorithm 1 Procedure for CryNet

```
INIT IRCSERVER
n_bots CONNECTS to IrcServer
n_bots JOINS #BOTNET_CHANNEL
Master CONNECTS to IrcServer
while Master is connected do
    Master JOINS #BOTNET_CHANNEL
    Master SENDS ActionCommand
    for all Bot in n_bots do
        Bot PARSES ActionCommand
        if ActionCommand is applicable then
            PERFORM k actions
        end if
    end for
end while
```

Table 1. Partial listing of the Standard IRC and Custom Server Commands

CryNet IRC Server Command List
JOIN: Used to join a channel or create one if it does not currently exist
KICK: Removes a user from a channel
LIST: Lists all requested channels or all if specified
NAMES: Lists the nicknames of users currently in specified channels
NICK: Sets a nickname for a connection
PART: Allows a user to remove itself from a specified channel
PRIVMSG: Allows a user to send a direct message to specified users and or channels
QUIT: Allows the connection to disconnect from the IRC Server
TOPIC: Sets the topic for a specified channel
VALIDATE: Allows for only the Master program and CryNet Bots to validate themselves, no other connections are allowed
SHUTDOWN: Allows the Master program to shut down the server

Note: Underlined commands are CryNet*-specific extensions to the IRC protocol.*

In the context of CryNet , they function as a meeting point for Bots and the Master program.

CryNet's IRC Server is closely aligned with the RFC of IRC, but it has some additional features and omits some IRC functions in order to more effectively fit our work's needs. For instance, a core concept of IRC Servers are their commands. IRC Server commands are denoted by a forward slash, "/", followed by a command type and all of the necessary parameters. Much like other IRC Servers, commands have the following syntax:

```
/COMMAND < channels, >< additional_param > : < command_message >
```

The above syntax shows how a typical IRC command flows, where "COMMAND" is the name of the command, "<channels,>" is a comma separated list of channels to operate the command on, "<additional_param>" provides any required arguments, and "<command_message>" contains an optional message payload. Although not all IRC commands adhere to this exact syntax, this hypothetical command is a good example of what is generally seen on IRC Servers. Table 2 shows an example taken from CRYNET's ServerCommands.md documentation file, the KICK command.

Table 2. KICK Command Syntax and Description

/KICK <CHANNEL> <NICKNAME> :<COMMENT>
CHANNEL: The name of the channel the user will be kicked from
NICKNAME: The nickname of the user to be kicked from the channel
COMMENT: Optional comment that will be sent to the kicked user

In addition, CRYNET has added some other IRC Server commands for project-specific needs, as shown in Table 1. The VALIDATE command is integral because it ensures, through cryptographic means, that only valid Bots, generated via the CRYNET repository, and the Master program can connect to the server. The VALIDATE command takes in a passphrase from the connection, and if it matches the hash digest of the server-stored password hash, the connection is validated as the Master or a Bot. Yet, these non- standard IRC commands are a good example of how CRYNET uses the IRC protocol to its advantage, but not blindly adhering to it.

Remark. When the Server receives input from a connection, it parses the data to determine whether it is a general message or a server command. A general message is broadcast to all channels the connection has joined, while a command triggers a specific server action. The parser validates the input format before execution, ensuring that only well-formed messages and commands are processed.

2.3 IRC Bot

CRYNET's IRC Bot is the component which provides the actual force of our work. It is designed to connect to the Server and execute commands as directed from the Master. The Bot has capabilities from packet crafting and sending to data exfiltration. It is important to note that the victim device is unable to use

Table 3. CryNet Bot File Sizes with Different Actions

Bot Actions	File Size (Bytes)
None	51520
P, E, F, GE, R	276808
P	112960
E	207176
F, GE, R	67904

Note: Action abbreviations - (P)ING, (E)NCRYPT, (F)ORK_BOMB, (GE)T_PASSWD, (R)EPORT. Bots compiled using g++, GNU Compiler Collection. Executables were stripped.

the program or provide it any inputs. All actions taken by the Bot are strictly from the remote Master program with IRC messages.

Once executed on an infected device, the Bot is daemonized (unless a specific command-line parameter is used) and runs in the background, repeatedly attempting to connect to the IRC Server until successful. After establishing a connection, it authenticates, sets its preconfigured nickname and username, and joins its designated IRC channel. Then, the Bot waits and listens for messages. Upon receiving a message, the Bot parses it to determine whether it contains a valid instruction addressed to itself. If so, the Bot attempts to execute the request. In CryNet, these instructions are defined as "Actions"—one or more operations performed on either the infected host or a target device. The syntax for action requests is described in Table 4. Unlike IRC Server commands, Bot action commands are not denoted by a leading symbol because they are sent through the IRC Server as a general message. Furthermore, Bot commands are merely messages sent to the IRC Server as text, but with a specific syntax, allowing the Bot to interpret the message accordingly. Once the action is completed, the Bot continues to listen for more commands on the IRC channel.

Table 4. Bot Command Syntax and Types

`ACTION <PARAM_1> <PARAM_2> ... <PARAM_N>`
PING: Craft a packet(s) with a specific protocol & send to a target IP address and port
ENCRYPT: Encrypts the entire infected devices file system using a specified symmetric algo.
FORK_BOMB: Runs a fork bomb on the infected device.
GET_PASSWD: Returns the contents of the "passwd" or "shadow" file on the infected device
REPORT: Returns vital system information about the infected device

Remark. Another important design feature of CRYNET's IRC Bot, is its ability to be configured with varying levels of complexity. At compilation time, C++ macros are used to determine whether a given Bot is allowed to execute certain actions. C++ macros are labels defined in C++ source code replaced by a value prior to compilation. Thus, one can compile a CRYNET Bot to have a variable amount of actions, such as the ability to PING and ENCRYPT, REPORT and FORK_BOMB, or, the base case, no actions. Consequently, file size for CRYNET Bots can vary drastically. Although showcasing all Bot compilation action combinations would be too lengthy and frankly trivial, Table 3 showcases some differing Bot file sizes in bytes. As shown in Table 3, there is a significant change in file size depending on what actions are possible on a given CryNet Bot. Note that actions such as PING and ENCRYPT have a much greater impact on file size than other actions such as REPORT. Since PING and ENCRYPT require more logic in the source code and require external libraries such as Crypto++ [1] and libtins [2]. Note that different levels of Bot complexity allow for some Bots to be less detectable than others. Bots that have the capability to PING a device may cause noise and require greater resources on the victim. In addition, the larger file size required for a PING action may allow systems to detect a Bot easier. On the contrary, Bots compiled just for reconnaissance and or data exfiltration, actions such as REPORT and or GET_PASSWD, will not require many system resources to perform said actions and will have a reduced file size. A CRYNET user has the option to test their security environments by deploying Bots that differ in detectability and power.

2.4 Master Program

The Master program serves as the core of the CRYNET ecosystem, functioning as the central control hub within its botnet framework. It orchestrates the activities of distributed Bots through a built-in IRC client. Additionally, the Master program allows CRYNET users to compile Bots with varying levels of complexity and assign them specific botnet identifiers.

When the program is launched, the user is given the option to connect to the IRC Server, enter the IRC console, or compile a Bot. If the user wishes to access the IRC console, they must first connect and authenticate to the Server, which is all handled programmatically. Then, the IRC console is unlocked which is where the Master can enter both IRC Server commands or Bot action commands. The user is able to issue a "QUIT_NOW" command to the IRC console and return to the main menu. If the user were to choose the Bot compilation option, then they are guided through the process of setting the Bot macro information such as Bot nickname, assigned IRC channel, etc.

3 Implementation

CRYNET consists of three major components: Server, Bot, and Master. Each was developed independently using an object-oriented programming model, allowing any component to compile and execute in isolation. However, meaningful operation requires all three to function together. For example, a Bot may run without errors on its own, but cannot receive or execute instructions without an active Server (see Sect. 2 for details). All three components follow a similar initialization sequence: start up, read configuration, initialize subcomponents, and await input. The nature of this input differs by role: the Master waits for user commands, the Server listens for incoming connections and their messages, and the Bot monitors for instructions on its assigned channel. Once initialization completes, inter-component communication can begin.

3.1 Server

The IRC Server is designed using object-oriented principles, with each component implemented as a separate class. Key classes include:

IrcServer. Functions as a hub for all server processes to run through including accepting connections, parsing input, and executing server commands

Connection. Interface to support connections which helps facilitate waiting for connection input, sending output to the connection, and handling IRC user information. ConnectionHandler is Subtype.

Channel. Class used for handling an IRC channel

Message: Supertype for all received input from connections. Raw input from a connection is placed within a `Message` object and then it is parsed to determine the necessary "Message" subtype, i.e., ExternalMessage, InternalErrorMessage. ExternalMessage is a message that will produce some sort of effect. Subtypes contains CommandMessage (for executing server commands), GeneralMessage (general messages sent to IRC channels). Parsers includes CommandBuilder and MessageParser. CommandBuilder determines the type of a received server command and creates the necessary command. MessageParser determines what type of message has been received from connection input.

3.2 Bot

The IRC Bot is a C++ based program that also takes an object-oriented approach. There is an associated header file that dictates the structure of the class. For the more sophisticated CryNet Bot actions, such as PING or ENCRYPT, the Master is able to specify what sort of algorithm or protocol should be used when carrying out the action. In the case of PING actions, a given target URI and port are sent the specified packet type a specific amount of times or indefinitely. For ENCRYPT actions, all files on the infected device are encrypted using Advanced Encryption Standard (AES) with the specified key length (128, 192, or 256 bits). Note that, currently Bots are only being compiled for Unix systems using x86_64 CPU architecture. More details on the commands and their syntax can be viewed in the CryNet repository [3].

BotMain. Starting function for the Bot, which daemonizes the process and attempts to connect to the IRC Server

IrcClient. Class that functions as a hub for all Bot operations. It handles connecting to the IRC Server, listening for messages, and creating actions

MessageParser. Static class that parses each message and informs the program if any received messages are actionable

Action. Supertype for all action commands. This allows for a command to be executed as an Action despite being a specific allocation of a subtype action, i.e., PingAction, EncryptAction, ForkBombAction, ReportAction, GetPasswdAction.

PacketLauncher. Used within a PingAction to craft packets and send them

CryptoManager. Used for file encryption and decryption

3.3 Master Program

The Master is a straightforward command-line experience that provides a means to send commands to CryNet Bots. We have different key classes, i.e., MasterProgram, MasterIrcClient, TerminalBuffer and BotCompiler.

MasterProgram. Main program for the Master. It provides a user menu with different options such as connecting to the IRC Server, entering the IRC console, or compiling a Bot.

MasterIrcClient. Connects to the IRC Server and handles messages to &from

TerminalBuffer. Custom terminal buffer that handles input and output operations. This is used to allow for seamless messaging in the Master's IRC client

BotCompiler. Allows the user to compile a customized Bot with a specific set of CryNet actions

3.4 Configuration and Deployment

This section outlines the methodologies used to simulate various scenarios, the parameters set for each test, and the criteria for assessing the outcomes. Although resource constraints prevented testing at full scale (several hundred Bots in diverse environments), the experiments presented here effectively showcase CryNet's current functionality and operational potential. CryNet requires proper configuration of all three components before deployment. Specifically, the Server's IP address must be configured identically in the Server, Master, and Bot configuration files to enable proper initialization and connection establishment. Here are the configuration methods for each component:

- Server: Reads in configuration from a JSON file and stores it inside a Config class. This JSON configuration file can either be manually mutated or set through the setup Python script in the root of the CryNet repository.

- Bot: C++ macros are used, which are predefined variables or sets of instructions that are expanded and replaced by their corresponding values or code during the preprocessing phase of compilation. For instance, macros define the IRC Server address, port number, etc.
- Master: Much like the Server, configuration is read in from a JSON file. It can also be manually set or set through the setup Python script.

Automation and Scripting. CryNet utilizes scripting languages such as Bash and Python in order to perform various tasks such as building and running components, facilitating configuration setup, and setup cleanup. The use of Bash and Python scripting in CryNet provides a flexible, powerful, and efficient way to manage the lifecycle of the system's component.

Listeners and Trackers. CryNet also provides a "tools" folder that holds different listeners and trackers. These are scripts that function as endpoints that can track incoming packets, packet size, system resource usage, and more.

4 Experiments

4.1 Experimental Setup

The IRC Bots were running on local virtual machines (VMs) imitating remote infected devices. It is important to note that the victim machine only was used for simulated DDoS attacks.

Table 5. Five attack scenarios on the IRC Bot actions

	Attack Scenarios
Scenario 1:	DDoS attack on a victim device
	Using UDP packets & TCP-XMAS packets
Scenario 2:	Remote "fork bomb" execution
Scenario 3:	Remote file system encryption using AES-256-CBC
Scenario 4:	Data exfiltration of critical local password hash files
Scenario 5:	Reporting of vital victim machine host information

- IRC Server on Ubuntu 24.04 (64-bit)
  - Hardware: AWS EC2 instance, Intel Xeon CPU E5-2676 @ 2.40GHz, 1 GB RAM
- IRC Bots (3x) on Ubuntu 22.04.4 (64-bit)
  - Hardware: Intel i7-9700k @ 3.60GHz, VMs with 4 GB RAM each
- Master Program on Windows 10
  - Hardware: Intel i7-9700k @ 4.8GHz, 16 GB RAM
- Victim Machine on Windows 10
  - Hardware: Virtual Machine Intel i7-9700k @ 3.6GHz, 4 GB RAM

4.2 Attack Scenarios

There were five attack scenarios included in Table 5. Each scenario is accompanied by one of the Bot Actions on a victim machine. All scenarios follow the CRYNET procedure, Algorithm 1, where $n * k$ actions are completed. Moreover, when 3 Bots are used for an experiment, there are $3 * k$ actions per scenario. The k is dependent on the action command. For instance, if an ENCRYPT action is sent to a single Bot, k will be however many files were encrypted via that action command.

4.3 Evaluation

We conducted experiments for multiple attack scenarios. In each experiment, the process began with the IRC Server being initialized and active, followed by the connection of the three Bots to the Server. The Master then connected and issued the respective commands for the attacks. The effectiveness of each attack was assessed using different metrics.

DDoS Attack Scenario. This scenario is the only one in which damage is inflicted to a device that is not the host machine of the Bot (except for the necessary system resources required to send mass amounts of packets). In addition, two different packet types were invoked: UDP and TCP-XMAS. To track the success of this attack, a Python script was created that opens a port on the victim machine and awaits packets. Once a packet is received, the script saves the current change in time from execution and increments a received packets counter. The script was initially set up to receive UDP packets on port 9999 and kept running for 60 s. The attack was carried out with one Bot receiving the command to PING, two Bots, and then three. The following command was issued:

PING UDP < IP_OF_VICTIM > 9999 LOOP

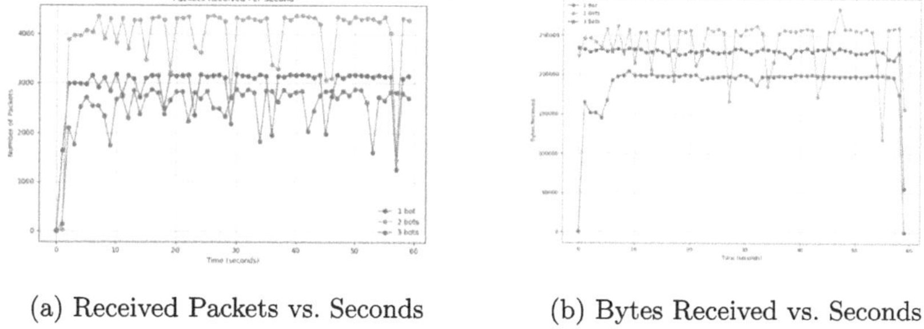

(a) Received Packets vs. Seconds (b) Bytes Received vs. Seconds

Fig. 3. DDoS Attack Scenario

The IPv4 address of the victim node was supplied and the final parameter, "LOOP", informs the Bot to send packets indefinitely. Therefore, Bots were manually halted once the 60 s of gathering data had elapsed. Then, another script was used to calculate how many packets were received per second, producing 60 data points. The script did this for the data collected for when one, two, and three Bots were pinging and then plotted next to each other. Figure 3a is the plot of the UDP flooding attack. The plot reveals a rather analogous pattern of received packets over time for each number of Bots, but one may notice some interesting aspects about the plot. For example, when one Bot is pinging the victim a steady 3000 packets per second were recorded. When two Bots were introduced, a significant jump of about 1000 more packets per second occurred. However, when three Bots pinged the victim, the received packet rate was significantly lower than when only one or two Bots were active. This could occur for multiple reasons, but it is most likely caused by the intended consequences of a DDoS attack. Furthermore, DDoS attacks intend to congest a network by flooding a given endpoint with packets, rendering it unable to send or receive other network requests due to the backup caused by incoming packets. CRYNET appears to be achieving this as a sizeable amount of packets did not get picked up by the victim device's open endpoint. DDoS attacks can come in many different forms and use a plethora of protocols. The exact scenario was completed again, but this time with TCP-XMAS packets. A TCP XMAS attack is when TCP packets are crafted to have all TCP flags set in the packet. This causes the packet to be greater in size and difficult to handle for ports accepting TCP packets. Additionally, the modified script computes the amount of bytes per second the endpoint was receiving. The attack was carried out on the open TCP port 80, as shown in Fig. 3b. The same trends experienced in the UDP attack are seen in the bytes versus seconds TCP-XMAS attack. Thus, CRYNET's botnet appears to be functioning as expected with even a limited amount of active Bots.

Remote File System Encryption. This experiment tracked how long it took on average to encrypt entire file systems using the CRYNET Bots. It involved the ENCRYPT Bot action, which encrypts the entire infected device's file system. The AES-256 was picked to perform this experiment. Tracking was done by having a script scan the Master program's terminal waiting for text that appeared stating the number of files encrypted. Once the script detected the text, a timer was stopped and the amount of time passed since execution was placed in a JSON file. This is possible because the ENCRYPT command has the Bot send a PRIVMSG to the Master program when all files are encrypted. This message includes the number of encrypted files. Before the Bots were sent the encryption command, the number of files on each of the infected devices were changed. The process was repeated multiple times to ensure sufficient data points. The following command was issued to the three Bots:

```
ENCRYPT AES-256 ROOT ENC
```

The "ROOT" parameter indicates that all files, starting from the root directory, should be encrypted. Also, the files were remotely decrypted via the same command, but instead of "ENC", "DEC" is specified to decrypt any found encrypted files. The number of files on each run varied from approximately 500 to 8000. Figure 4 displays the collected data. The results show a nearly linear relationship between seconds and the number of files. This linearity suggests that the encryption algorithm scales predictably with the number of files, handling increased workloads without significant drops in performance. CryNet achieves this efficiency through optimized cryptographic operations and resource management, ensuring that each file is processed quickly and systematically. The consistent performance across varying numbers of files demonstrates that CryNet can be relied upon for timely encryption in diverse scenarios, from small-scale deployments to large-scale operations involving thousands of files.

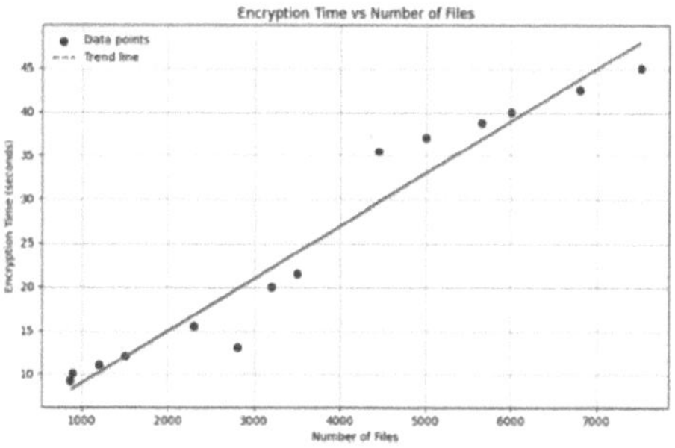

Fig. 4. Encryption Time vs. Number of Files

5 Conclusion and Future Work

This paper proposes the CryNet framework as a novel means of studying and analyzing botnet behavior. Our work leverages an IRC command-and-control structure to simulate realistic attack scenarios. We evaluate CryNet's capabilities and performance in controlled environments using various pseudo-malicious actions. The results demonstrate that CryNet can effectively launch multiple types of attacks, underscoring key system vulnerabilities that must be addressed in modern networks and supporting the advancement of cybersecurity research.

In future work, CryNet will undergo further development, along with the integration of new enhancements. Meanwhile, future research will incorporate AI-based botnet detection using Large Language Model [8,12,16,20] approaches to identify botnet TLS certificates. Using TLS certificates [11,19] as input, LLMs can leverage embeddings to generate vector representations of their textual content. New certificates can then be classified as malicious or benign by comparing their similarity to a known set of botnet-related and legitimate certificates.

Ethical Concerns and Responsible Disclosure. This research was conducted exclusively for scientific and educational purposes. All experiments were carried out in controlled environments, and no real-world systems or users were affected. The methods, models, and results presented are intended to support the advancement of cybersecurity research and are not designed for or applicable to malicious activities. Responsible disclosure and ethical standards were followed throughout the study.

References

1. Crypto++ (2023). https://github.com/weidai11/cryptopp.git
2. Libtins (2023). https://github.com/mfontanini/libtins.git
3. Crynet. GitHub (2024). https://github.com/s1-dev/CryNet
4. Abraham, B., Mandya, A., Bapat, R., Alali, F., Brown, D.E., Veeraraghavan, M.: A comparison of machine learning approaches to detect botnet traffic. In: IJCNN (2018)
5. Aljabri, M.S., Zagrouba, R., Shaahid, A., Alnasser, F., Saleh, A., Alomari, D.M.: Machine learning-based social media bot detection: a comprehensive literature review. Soc. Netw. Anal. Min. **13**(1), 20 (2023)
6. Feily, M., Shahrestani, A., Ramadass, S.: A survey of botnet and botnet detection. In: 2009 Third International Conference on Emerging Security Information, Systems and Technologies, pp. 268–273. IEEE (2009)
7. Gadelrab, M.S., ElSheikh, M., Ghoneim, M.A., Rashwan, M.A.: Botcap: machine learning approach for botnet detection based on statistical features. Int. J. Commun. Networks Inf. Secur. **10**(3) (2018)
8. Gambini, M., Fagni, T., Falchi, F., Tesconi, M.: On pushing deepfake tweet detection capabilities to the limits. In: ACM WebSci, pp. 154–163 (2022)
9. Garre, J.T.M., Pérez, M.G., Ruiz-Martínez, A.: A novel machine learning-based approach for the detection of SSH botnet infection. Future Gener. Comput. Syst. **115**, 387–396 (2021)
10. Kalt, C.: RFC2810: internet relay chat: architecture (2000)
11. Kumar, S., Hahn, F., Millar, S., Ou, X.: Using LLM embeddings with similarity search for botnet TLS certificate detection. In: Pintor, M., Chen, X., Jagielski, M. (eds.) 2024 Proceedings AISec, pp. 173–183 (2024)
12. Li, S., Qiao, B., Li, K., Lu, Q., Lin, M., Zhou, W.: Multi-modal social bot detection: learning homophilic and heterophilic connections adaptively. In: Proceedings of 31st ACM International Conference Multimedia, pp. 3908–3916 (2023)
13. Liu, J., Xiao, Y., Ghaboosi, K., Deng, H., Zhang, J.: Botnet: classification, attacks, detection, tracing, and preventive measures. EURASIP J. Wirel. Commun. Network. 1–11 (2009)
14. McDermott, C.D., Majdani, F., Petrovski, A.: Botnet detection in the internet of things using deep learning approaches. In: 2018 International Joint Conference on Neural Networks, IJCNN, pp. 1–8. IEEE (2018)
15. Oikarinen, J., Reed, D.: Internet relay chat protocol. Technical report RFC 1459, Internet Engineering Task Force (1993)
16. Qiao, B., Li, K., Zhou, W., Li, S., Lu, Q., Hu, S.: Botsim: LLM-powered malicious social botnet simulation. In: Walsh, T., Shah, J., Kolter, Z. (eds.) AAAI (2025)
17. Samani, E.B.B., Jazi, H.H., Stakhanova, N., Ghorbani, A.A.: Towards effective feature selection in machine learning-based botnet detection approaches. In: Proceedings of IEEE Conference on Communications and Network Security CNS, pp. 247–255 (2014)

18. Shinan, K., Alsubhi, K., Alzahrani, A., Ashraf, M.U.: Machine learning-based botnet detection in software-defined network: a systematic review. Symmetry (2021)
19. Theofanous, A., Papadogiannaki, E., Shevtsov, A., Ioannidis, S.: Fingerprinting the shadows: unmasking malicious servers with machine learning-powered TLS analysis. In: Proceedings of the ACM on WWW 2024, pp. 1933–1944 (2024)
20. Tourille, J., Sow, B., Popescu, A.: Automatic detection of bot-generated tweets. In: Proceedings of the 1st International Workshop on Multimedia AI Against Disinformation (2022)
21. Tuan, T.A., Long, H.V., Son, L.H., Kumar, R., Priyadarshini, I., Son, N.T.K.: Performance evaluation of botnet DDoS attack detection using machine learning. Evol. Intell. **13**(2), 283–294 (2020)
22. Wang, P., Sparks, S., Zou, C.C.: An advanced hybrid peer-to-peer botnet. IEEE Trans. Dependable Secur. Comput. **7**(2), 113–127 (2010)
23. Xing, Y., Shu, H., Zhao, H., Li, D., Guo, L.: Survey on botnet detection techniques: classification, methods, and evaluation. Math. Probl. Eng. **2021**(1), 6640499 (2021)
24. Zeidanloo, H.R., Shooshtari, M.J.Z., Amoli, P.V., Safari, M., Zamani, M.: A taxonomy of botnet detection techniques. In: 2010 3rd International Conference on Computer Science and Information Technology, vol. 2, pp. 158–162. IEEE (2010)
25. Zhao, Z., Li, Z., Zhang, F., Li, T., Yin, J.: Poster: combine topology and traffic to calibrate P2P botnet identification in large-scale network. In: Proceedings of the ACM SIGCOMM 2024 Conference: Posters and Demos, Sydney, pp. 16–18 (2024)

Quantum-PCA for Malware Detection in Supply Chains, Healthcare, and Smart Manufacturing Communications

Sthefanie J. G. Passo(✉), Vishal H. Kothavade, and John J. Prevost

Computer Engineering Department, University of Texas at San Antonio, San Antonio, USA
{Sthefanie.Passo,Vishal.Kothavade,Jeff.Prevost}@utsa.edu

Abstract. Quantum computing is reshaping cybersecurity in supply chain, healthcare, and smart manufacturing systems. This study presents a simulator-based comparison between classical Principal Component Analysis (PCA) and Quantum PCA (QPCA) for malware and anomaly detection across these domains. Within 10-dimensional feature spaces, hybrid QPCA models achieved performance comparable to classical PCA, with precision peaks of 100% (PCA + cKernel), 96.7% (PCA + qKernel), and 90.78% (PCA + cKernel + qDataMatrix). While classical PCA attained the highest precision and specificity, QPCA variants demonstrated greater balance across metrics—accuracy up to 91.27%, sensitivity 92.63%, and F1-score 91.38%. These findings indicate that hybrid quantumclassical approaches can match classical state-of-the-art detection while operating under simulator-level constraints. Future work should advance hardware-efficient encodings, reduced-qubit circuits, and on-device testing to assess scalability and robustness under realistic quantum noise.

Keywords: Quantum Principal Component Analysis · Malware Detection · Supply Chains · Healthcare · Smart Manufacturing · Quantum Machine Learning · Cybersecurity

1 Introduction

The digitalization of supply chains (SC), healthcare, and manufacturing has delivered operational gains but also introduced new vulnerabilities—particularly in cybersecurity. Malware targeting communication channels in SC and healthcare can disrupt operations, compromise sensitive data, and impact patient safety or production continuity [1,2]. Classical Machine Learning (ML) methods for dimensionality reduction (DR) like Principal Component Analysis (PCA) are widely used for anomaly and malware detection as they reveal suspicious patterns reducing the data complexity. However, as data volumes and complexity increase, classical PCA faces computational bottlenecks, limiting its real-time effectiveness [3–5].

H. R. Arabnia et al. (Eds.): CSCI 2025, CCIS 2969, pp. 298–318, 2027.
https://doi.org/10.1007/978-3-032-28460-0_21

Quantum computing, leveraging phenomena like superposition and entanglement, offers a new paradigm. Quantum Principal Component Analysis (QPCA) promises to process high-dimensional data exponentially faster than classical PCA, enabling more rapid and accurate malware detection, especially for complex or large-scale datasets. Recent research has shown that Quantum Machine Learning (QML) models, including QPCA, quantum neural networks, and Quantum Support Vector Machines (QSVMs), can outperform classical models in certain cybersecurity tasks, offering improved accuracy and resilience against evolving threats [3,6].

In healthcare, quantum computing frameworks are indeed being piloted for analyzing large-scale medical datasets, with applications spanning early disease detection, biomarker discovery, and personalized treatment planning. For instance, QML methods have been explored for cancer subtype classification and genomics [7,8]. Recent work also demonstrates quantum-enhanced clustering, kernel methods for precision medicine and early diagnosis in oncology [9]. This strengthens the motivation for our study as the SC and healthcare domains share similar challenges in handling large, high-dimensional datasets where quantum methods may provide advantages in scalability and accuracy [1].

Despite these advances, practical deployment of quantum solutions faces challenges: current hardware limitations, algorithmic complexity, and the need for integration with existing systems. The move toward post-quantum cryptography, as recommended by organizations like NIST, underscores the urgency of preparing digital infrastructure for quantum-era threats [10].

This article provides a concise comparative study of the state-of-the-art in QPCA for malware detection in SC, healthcare, and smart manufacturing, comparing it with classical PCA and outlining key directions for future research and deployment.

2 Related Work

2.1 Quantum PCA and Dimensionality Reduction in Cybersecurity

PCA and its quantum counterpart, QPCA, have emerged as key tools for feature extraction and dimensionality reduction in ML pipelines. Classical PCA, while effective, faces computational bottlenecks as datasets become increasingly high-dimensional in domains such as network security, healthcare, and industrial communications. Recent advancements in QML propose QPCA as a scalable alternative, leveraging the exponential state-space of qubits for faster eigendecomposition and feature transformation [11,12].

Bellante et al. [3] evaluated QPCA for anomaly and malware detection using benchmark datasets such as KDDCUP99 and CICIDS2017. Their findings show that QPCA outperforms classical PCA when data dimensionality exceeds classical processing thresholds, achieving improved robustness and lower reconstruction error. Moreover, studies such as [13,14] demonstrate that hybrid quantum-classical PCA approaches can enhance classification accuracy and reduce feature redundancy when combined with Logistic Regression (LR) and Support Vector

Machines (SVMs). These results collectively establish QPCA as a strong candidate for high-dimensional threat detection.

2.2 Hybrid Quantum Algorithms in Real-World Applications

Evaluating hybrid quantum algorithms, such as Quantum Kernel PCA and Quantum Circuit PCA, is crucial for near-term applicability due to hardware limitations. Research by Havlíček et al. [15] introduced the use of quantum feature maps (e.g., ZZFeatureMap) to embed classical data into quantum Hilbert space, showing significant improvements in non-linear data separability. This concept underpins QPCA's kernel-based variant, where quantum feature encodings enhance sensitivity to complex correlations in data.

In cybersecurity, hybrid quantum approaches have been applied to malware classification, network intrusion detection, and anomaly identification in IoT and SCADA systems [16]. In healthcare, quantum frameworks are being piloted for analyzing large-scale biomedical datasets, improving early diagnosis and personalized treatment [17]. Within supply chain management (SCM) and manufacturing, QML models have been proposed for real-time risk analysis, supplier anomaly detection, and fault prediction in smart manufacturing environments [18].

2.3 Identified Gaps and Research Motivations

Despite promising advances, current literature highlights several critical gaps. Most quantum PCA experiments are conducted on simulators or small-scale hardware, limiting their generalizability to large operational datasets. Additionally, comparative studies directly benchmarking QPCA against classical PCA using standardized metrics remain limited. The scalability of classical and quantum kernels, circuit depth optimization, and noise resilience continue to be open challenges [19].

The need for evaluating hybrid quantum algorithms—such as QPCA—stems from both computational and security imperatives. In high-dimensional anomaly detection tasks, QPCA can capture subtle nonlinear interactions that classical PCA overlooks. Furthermore, supply chain, healthcare, and manufacturing communications demand resilient, low-latency detection systems to protect critical infrastructure from evolving cyber threats. This study contributes to this growing research area by systematically evaluating hybrid QPCA models—integrating classical and/or quantum kernels—against classical baselines, and by quantifying their robustness, detection accuracy, and scalability in real-world scenarios.

3 Methodology

This study investigates PCA malware detection for SC, healthcare and manufacturing communications using both classical and quantum approaches. We compare three models: PCA with a classical Kernel (cKernel), PCA with a Quantum Kernel (qKernel), and PCA with a cKernel but Quantum-Data derived similarity Matrix (qDataMatrix). Our aim is to determine if QPCA can better identify complex threats in large, noisy datasets, building on our previous research [10,20]. So, ultimately, the research question guiding this work is: *Can hybrid quantum dimensionality reduction methods, implemented through currently available quantum simulators, demonstrate measurable benefits in detection robustness and feature representation compared to classical PCA?*

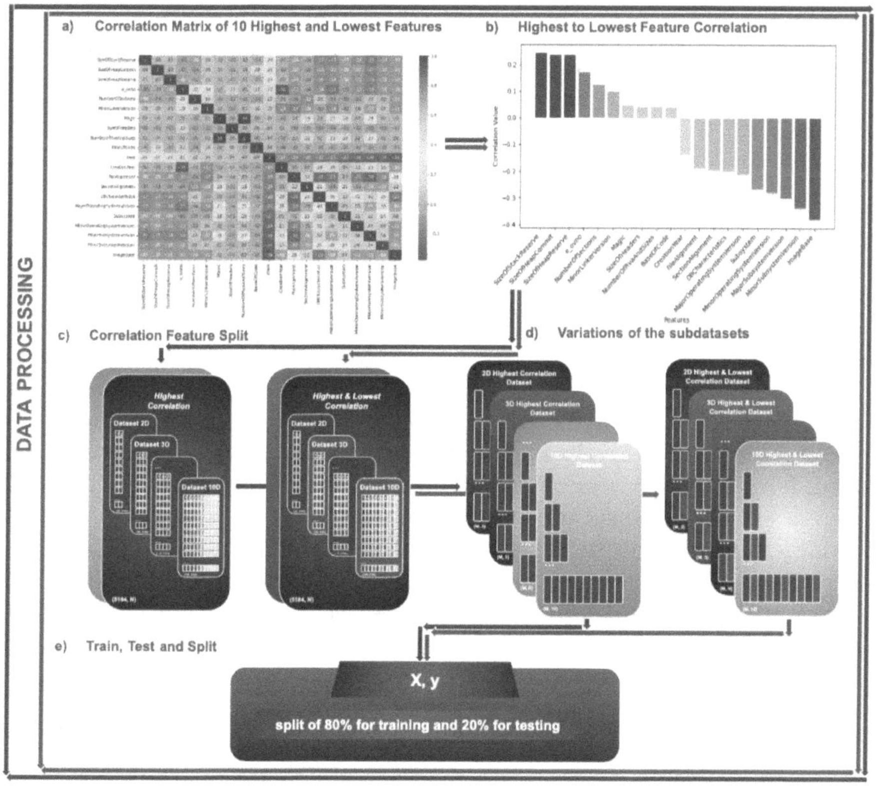

Fig. 1. Data Processing Method: a) Correlation Matrix of 10 Highest and Lowest Attribute Correlations from target = "class". b) The data is correlated to the target "class," where the 1st attribute (SizeOfStackReserve) has the HC and the 16th attribute (ImageBase) has the lowest. c) Subdatasets are created using attributes with the HC, ranging from 2 dimensions (1st and 2nd attributes) to 10 dimensions on the left. Similarly, subdatasets combining the HLC range from 2 dimensions (1st and 16th attributes) to 10 dimensions on the right [20]. d) Each subdataset is split to contain 10%, 20%, up to 100% of the dataset. e) Each variation of the subdataset is divided into training and test data.

3.1 Dataset and Feature Selection

We use the ClaMP dataset, which contains 5,814 samples and 108 features relevant to malware detection in SC, healthcare, and manufacturing communications contexts [10,20]. To ensure robust model evaluation and assess dimensionality reduction performance across varying feature correlation structures, we apply three complementary feature selection strategies guided by correlation analysis (Fig. 1a) and feature ranking (Fig. 1b):

- **Highest Correlation (HC):** Selects the top features most strongly correlated with malware labels, providing a focused feature set with maximum discriminative power (Fig. 1c, left panel). This strategy establishes a baseline for evaluating how classical and quantum PCA leverage high-correlation features for variance capture and cluster separation.
- **Highest and Lowest Correlation (HLC):** Alternates between the most and least correlated features to test model stability and generalization across heterogeneous feature spaces (Fig. 1c, right panel). This approach evaluates PCA robustness when mixing highly predictive and low-signal features, simulating real-world data noise and imbalance conditions.
- **Hierarchical Clustering Correlation (HCC):** Constructs a hierarchical feature correlation matrix where $1 - \rho$ (with ρ as the Pearson correlation coefficient) serves as the distance metric. Features are clustered hierarchically into correlation-based groups using Ward's linkage method, identifying nested relationships, redundancy, and multicollinearity patterns. Recent studies demonstrate that correlation-matrix-based hierarchical clustering effectively uncovers latent structures in high-dimensional data, outperforming independent component analysis for sensitivity to weak, spatially distributed dependencies [10]. HCC ensures selected features maximize informational diversity while minimizing redundancy, improving the interpretability of dimensionality reduction outcomes and enabling assessment of PCA performance on features with complex interdependencies.

For each strategy, feature subsets ranging from 2 to 10 dimensions are extracted to maintain compatibility with quantum circuit constraints on Noise Intermediate Scale Quantum (NISQ) devices, where circuit depth and qubit count directly impact decoherence and computational fidelity [1]. As described by Acharya et al. [21], decoherence occurs when quantum information degrades through environmental interactions, affecting result reliability. Attempts to encode more than 10–12 features lead to kernel computation failures—consistent with findings in [4]—attributed to increased circuit depth, higher gate noise, and reduced fidelity.

The ClaMP dataset consists of 2,501 benign and 3,313 malware-labeled samples. Although PCA is an unsupervised dimensionality reduction technique, labels are retained for external validation of cluster quality in the reduced feature space. A stratified 80–20 train-test split is applied to ensure balanced class representation across all experimental configurations. This dual-strategy approach, combined with progressive dimensionality evaluation (2D to 10D), enables systematic assessment of how feature correlation structure influences classical versus quantum PCA performance in malware detection applications.

3.2 Data Processing and Scaling

For each experiment, we extract between 2 and 10 features (Fig. 1c), keeping within the limits of current quantum hardware in a Noisy Intermediate Scale Quantum (NISQ) [22]. Data is normalized and incrementally scaled, starting with 10% of the dataset and increasing to 100% (Fig. 1d). While PCA is unsupervised, we retain labels for external validation. An 80/20 train-validation split ensures reliable assessment of PCA quality (Fig. 1e). This staged approach also helps us understand how model performance evolves as data volume grows.

3.3 Quantum Kernel and Hardware Implementation

A key idea in our work is the development of a hybrid QPCA with cKernel, qDataMatrix and qKernel executed on a quantum simulator. We use a ZZFeatureMap [15] as the quantum feature map, which encodes classical data into quantum states via layers of Hadamard gates and ZZ rotations (Fig. 2a). QKernels are constructed using fidelity measurements between quantum states (Fig. 2b and c), capturing nuanced similarities that cKernels may miss (Fig. 2d).

We selected the ZZFeatureMap for our QPCA research because it provides an expressive yet hardware-efficient way to encode classical data into quantum states. Its layered structure of parameterized single-qubit rotations combined with entangling ZZ interactions effectively captures pairwise correlations among features, which is essential in detecting subtle and nonlinear patterns typical in SC and healthcare communication logs. Unlike simpler feature maps that treat features independently, the ZZFeatureMap ensures higher-order interactions are represented in the Hilbert space, allowing QPCA to extract principal components that better separate malware-induced anomalies from normal operations. Additionally, it is natively supported in Qiskit and optimized for NISQ devices, making it a practical choice for both simulation and near-term hardware execution.

Our analysis employs a variable number of qubits ranging from 2 to 10, corresponding to the dimensionality of the input feature space (2D to 10D). Each dimension d requires d qubits to encode the feature vector, with circuits executed independently for each dimensionality. For instance, the 2D feature space is analyzed using 2-qubit circuits, while the 10D space requires 10-qubit circuits. This scalable approach allows us to systematically evaluate how quantum advantage manifests across different feature dimensions and to identify the optimal trade-off between quantum resource requirements and classification performance.

Furthermore, for each subdataset from 2D to 10D, we trained and tested both classical and quantum models using varying proportions of the dataset, ranging from 10% to 100%. This procedure allowed us to compare the precision

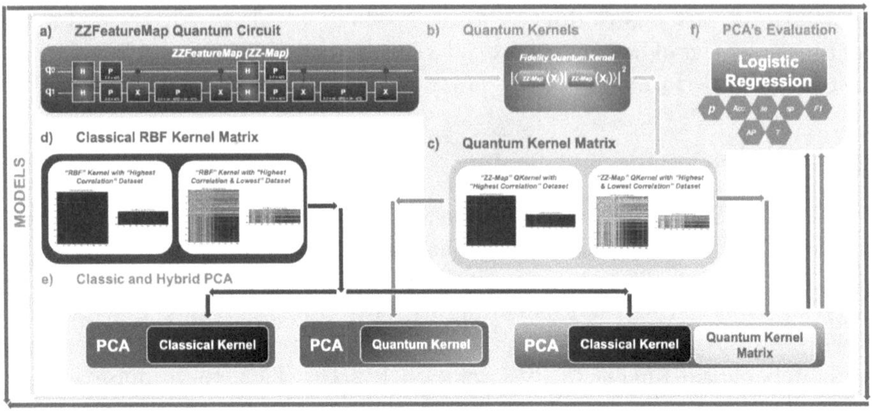

Fig. 2. (a) A quantum ZZFeatureMap with two repetitions was instantiated. The sub-dataset used for the plots had a two-dimensional shape of (5184, 2) for HC, HLC and HCC. Subsequently, the ZZFeatureMap was re-implemented for higher-dimensional datasets, ranging from 3D to 10D. b) The feature map was sent to a Fidelity qKernel, which structured the circuit for the quantum simulator. The feature map represented σ in the mathematical equation. c) The Fidelity evaluated the data, creating two qKernel for HC, HLC and HCC. The first dataset appears to exhibit multicollinearity [23], where features are highly correlated, leading to overfitting, redundancy, poor interpretability, and unstable coefficients. The qKernel matrices show increasingly scattered and irregular patterns, indicating their ability to capture complex correlations in the quantum feature space. Darker regions signify higher similarity between data points, reflecting redundancy within the kernel space. d) The cKernel matrix, built using an RBF kernel, presents a more uniform and regular structure with distinct dark stripes, highlighting dominant correlations but also higher redundancy. The differences, especially in off-diagonal regions, demonstrate the ZZ-Map's ability to capture nuanced, non-linear, group-structured relationships while reducing redundancy. e) The cKernel is sent to a classical PCA (1st DR) and hybrid PCA (3rd DR), while the qKernel is sent to the two hybrid and PCA. The difference is that in one of the hybrid models (2nd DR) we import the qKernel as a callable function and the second hybrid model (3rd DR) we send the qKernel as a transition matrix. f) The data was sent to the three DR approaches and submitted to a classical Logistic Regression with 7 metrics of evaluation - Sect. 3.7 - to evaluate the performance of each of the DR methodologies and their capability to translate data complexity optimally.

of classical versus quantum classifiers and to determine which model achieved superior predictive performance under different data availability conditions.

To maximize simulator efficiency and result fidelity, circuits are transpiled with advanced optimization, mapped to the hardware's topology, and executed with error mitigation. The largest kernel matrix (10 features/qubits) requires significant compute time, reflecting the current constraints of quantum devices.

3.4 Model Overview and Evaluation

Our methodology includes:

- **PCA + cKernel:** Standard classical PCA (Fig. 2e to the left).
- **PCA + qKernel:** PCA using quantum-derived kernel matrices (Fig. 2e in the middle).
- **PCA + cKernel + qDataMatrix:** hybrid QPCA computed directly from quantum metrics (Fig. 2 to the right).

Model performance is evaluated using five metrics and hardware performance is evaluated using elapse time to assess practical deployment feasibility.

3.5 Models Implementation

This research explores three PCA techniques for malware detection in SC, healthcare and manufacturing communications: PCA + cKernel, PCA + qKernel, and PCA + cKernel + qDataMatrix. Each method operates on the ClaMP dataset [18], leveraging different computational paradigms.

We selected PCA combined with a cKernel as our baseline because kernel-enhanced PCA is widely recognized as one of the strongest classical extensions of standard PCA for nonlinear anomaly detection tasks. Unlike linear PCA, which is limited in capturing nonlinear feature interactions, kernel PCA can map data into higher-dimensional feature spaces, making it more competitive for anomaly detection in high-dimensional datasets such as SC and healthcare telemetry. Prior studies [24] report that kernel PCA significantly outperforms linear PCA in malware detection tasks, showing accuracy improvements up to 1015% in some benchmark datasets. Additionally, recent SC security work using ClaMP datasets has demonstrated kernel PCA's advantage as a robust baseline for classification and anomaly detection.

PCA with cKernel. We implement PCA + cKernel using `scikit-learn`, with a Radial Basis Function (RBF) kernel and `n_components=2` to capture nonlinear feature interactions (Fig. 2d). The RBF affinity matrix highlights complex data distributions and is defined as:

$$K_{\mathrm{RBF}}(\mathbf{x}_i, \mathbf{x}_j) = \exp\left(-\gamma \|\mathbf{x}_i - \mathbf{x}_j\|^2\right) \tag{1}$$

where γ is a scaling factor and $\|\mathbf{x}_i - \mathbf{x}_j\|$ is the Euclidean distance between data points. PCA is applied on the eigenvectors of the Laplacian of this matrix to reduce dimensionality. The reduced feature set is then passed to a LR classifier for evaluation.

PCA with qKernel. The second model, PCA + qKernel, uses a precomputed qKernel based on the Fidelity Quantum Kernel (FQK) method (Fig. 2c). This kernel captures richer data relationships through quantum state overlaps, given by:

$$K_{\text{Quantum}}(\mathbf{x}, \mathbf{x}') = |\langle \psi(\mathbf{x}) | \psi(\mathbf{x}') \rangle|^2 \tag{2}$$

where $\psi(\mathbf{x})$ is a quantum state encoded using a $ZZFeatureMap$ with two repetitions. The resulting kernel matrix is used as the affinity matrix for PCA, after which dimensionality-reduced data is sent to a LR classifier.

PCA with cKernel and qDataMatrix. This hybrid method, PCA + cKernel + qDataMatrix, integrates both classical and quantum kernels. It combines the RBF kernel function with the FQK, resulting in an enriched affinity matrix. PCA is applied to this combined matrix, and the reduced data is classified using LR. This fusion aims to leverage the strengths of both quantum and classical representations.

3.6 Logistic Regression

LR is a widely used statistical method for binary classification that models the probability of an outcome based on one or more predictor variables. In this study, we applied LR to the outputs of three DR techniques to evaluate how effectively each method captures meaningful patterns for malware detection. By feeding the reduced feature representations into the LR model, we could directly compare the performance impact of each DR approach, highlighting the improvements in detection robustness and anomaly separation offered by quantum-enhanced methods over classical baselines.

All model experiments were implemented using Python (v3.9.1), Qiskit (v1.2.4), Qiskit-IBM-runtime (v0.31.0), and Qiskit-Machine-Learning (v0.7.2). Quantum circuits were designed and tested on a local MacBook Pro with Apple M1 Pro chip, 16 GB RAM, and macOS Sonoma 14.6.1.

3.7 Metrics of Evaluation

The performance of the PCA algorithms applied to the ClaMP dataset is evaluated using five classical evaluation metrics commonly adopted in classification tasks: precision, accuracy, sensitivity, specificity, and F1-score [25] as shown on Fig. 2f. Although PCA is inherently unsupervised, ground truth labels available in ClaMP allow for validation through these metrics.

Precision, shown in Eq. (3), evaluates the proportion of correctly identified positive samples out of all predicted positives.

Accuracy, presented in Eq. (4), reflects the overall effectiveness of the model by measuring the ratio of correct predictions over the total number of instances.

Sensitivity (also known as recall), in Eq. (5), quantifies the ability to detect true positive cases among all actual positives.

Specificity, described in Eq. (6), assesses the model's effectiveness in correctly identifying true negatives among actual negatives.

F1-score, detailed in Eq. (7), provides a balanced measure that combines precision and sensitivity, particularly useful in cases of class imbalance.

$$p = \frac{TP}{TP + FP} \tag{3}$$

$$Acc = \frac{TP + TN}{TP + TN + FP + FN} \tag{4}$$

$$se = \frac{TP}{TP + FN} \tag{5}$$

$$sp = \frac{TN}{TN + FP} \tag{6}$$

$$F1 = 2 \times \frac{p \times se}{p + se} \tag{7}$$

In these equations, true positives (TP) and true negatives (TN) refer to correctly classified samples, while false positives (FP) and false negatives (FN) denote misclassifications. These values are derived from the confusion matrix.

For runtime analysis, we measure the total training time T of classical models [4,10]. This is calculated by adding the duration of all training epochs t_i, as expressed in Eq. (8):

$$T = \sum_{i=1}^{N} t_i \tag{8}$$

4 Experiments and Results

This section presents a comparative analysis of three PCA methods applied to malware detection in SC, healthcare, and manufacturing communications contexts. All experiments were conducted on classical computers using a statevector simulator implemented via the Qiskit Aer backend due to the current limitation that Qiskit does not support PCA training on quantum hardware. This choice allows for exact simulation of quantum circuits in a noise-free setting, ensuring that the proof-of-concept evaluation focuses on algorithmic feasibility rather than hardware noise effects. The ClaMP dataset with correlation feature split (Fig. 1), containing labeled malware and benign samples, was used for both training and testing in classical and hybrid QPCA models.

Figure 3a shows performance variations of up to 20% across the evaluated models. The highest precision values achieved were 100% for PCA + cKernel, 96.70% for PCA + qKernel, and 90.78% for PCA + cKernel + qDataMatrix. When models are trained using the HC configuration, their behavior is notably unstable, frequently yielding either the best or the worst precision values. In contrast, the HLC configuration consistently produces the highest precision across models, while the HCC configuration results in intermediate performance, avoiding both extreme best- and worst-case outcomes.

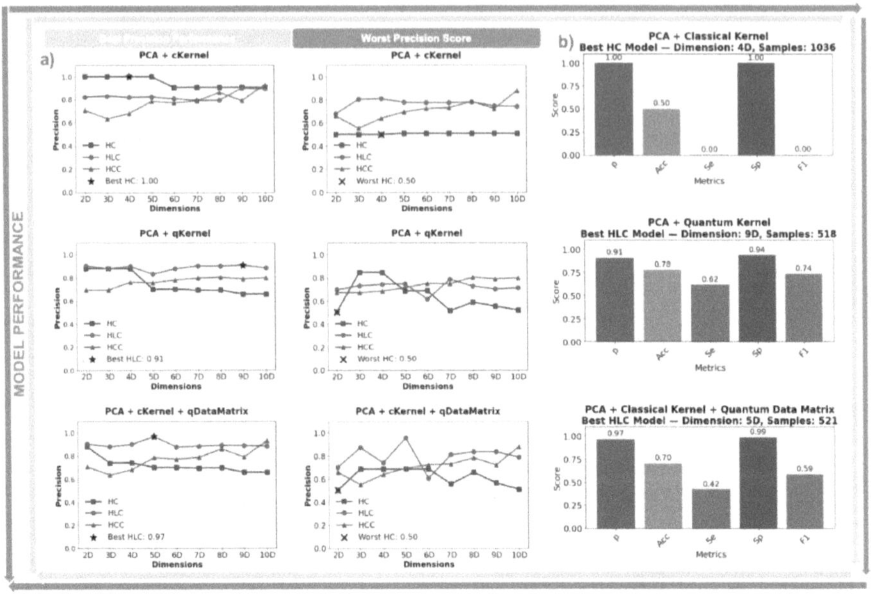

Fig. 3. Model Performance: a) For each model type—PCA + cKernel, PCA + qKernel and PCA + cKernel + qDataMatrix - best (first column) and worst (second column) performances are presented across dataset ranging from 2D to 10D. b) For the best-performing models, all evaluation metrics are plotted to assess model robustness.

The lowest precision of 50% is observed for all three approaches in the 2D and 4D settings, indicating that overly compressed feature spaces lack sufficient discriminative power. Overall, these results suggest that hybrid QPCA-based methods can closely approach classical state-of-the-art performance, even under simulator constraints and injected noise designed to emulate realistic quantum hardware conditions.

Figure 3b further investigates the robustness of the best-performing configurations by jointly analyzing accuracy, sensitivity, specificity, and F1-score. While the classical PCA + cKernel model attains perfect precision in some cases, this result is frequently accompanied by extremely low sensitivity and moderate accuracy, indicating that the model predominantly predicts the majority class. In contrast, the hybrid quantum approaches demonstrate a more balanced trade-off among metrics. PCA + qKernel improves sensitivity and overall accuracy relative to the classical baseline, while PCA + cKernel + qDataMatrix yields the most consistent performance across all metrics, achieving higher accuracy and F1-score despite slightly reduced precision. These results suggest that quantum-enhanced feature representations better capture nonlinear correlations inherent in malware communication patterns.

4.1 Detailed Performance Analysis Across Dimensionality

Tables 1, 2, and 3 summarize the average performance metrics (mean ± standard deviation) for PCA + cKernel, PCA + qKernel, and PCA + cKernel + qDataMatrix, respectively, across dimensions ranging from 2D to 10D and multiple data splits. Reporting averaged results rather than best-case outcomes provides a more reliable assessment of model stability and generalization capability under varying feature selection strategies and data partitioning schemes.

Table 1. Classical PCA + cKernel Performance Metrics

Dimension	Data Type	Accuracy	Precision	Sensitivity	Specificity	F1 Score	Time (s)
2	HLC	**0.8062 ± 0.0304**	**0.7877 ± 0.0511**	0.8497 ± 0.0327	**0.7626 ± 0.0900**	**0.8154 ± 0.0180**	1.37 ± 0.57
2	HC	0.5027 ± 0.0027	0.5517 ± 0.1495	**0.9002 ± 0.2995**	0.1035 ± 0.2988	0.6018 ± 0.1996	1.86 ± 0.78
2	HCC	0.6748 ± 0.0176	0.6787 ± 0.0167	0.6748 ± 0.0176	0.7105 ± 0.0213	0.6748 ± 0.0176	**0.81 ± 0.26**
3	HLC	**0.8214 ± 0.0089**	**0.8169 ± 0.0100**	**0.8292 ± 0.0100**	**0.8135 ± 0.0103**	**0.8229 ± 0.0091**	1.34 ± 0.56
3	HC	0.5052 ± 0.0036	0.6827 ± 0.1868	0.5087 ± 0.4910	0.4998 ± 0.4966	0.3516 ± 0.3173	1.84 ± 0.76
3	HCC	0.5947 ± 0.0247	0.5961 ± 0.0267	0.5947 ± 0.0247	0.4962 ± 0.0341	0.5911 ± 0.0250	**0.76 ± 0.24**
4	HLC	**0.8174 ± 0.0034**	**0.8130 ± 0.0038**	**0.8252 ± 0.0043**	**0.8096 ± 0.0042**	**0.8190 ± 0.0034**	1.23 ± 0.44
4	HC	0.5064 ± 0.0032	0.7408 ± 0.1656	0.3129 ± 0.4495	0.6981 ± 0.4549	0.2264 ± 0.2904	1.81 ± 0.80
4	HCC	0.6507 ± 0.0143	0.6587 ± 0.0122	0.6507 ± 0.0143	0.4967 ± 0.0295	0.6424 ± 0.0159	**0.81 ± 0.25**
5	HLC	**0.8025 ± 0.0242**	**0.8061 ± 0.0144**	**0.7964 ± 0.0425**	0.8084 ± 0.0072	**0.8010 ± 0.0287**	1.22 ± 0.43
5	HC	0.5068 ± 0.0028	0.7823 ± 0.1214	0.1151 ± 0.2951	**0.8965 ± 0.2988**	0.0969 ± 0.1930	1.67 ± 0.62
5	HCC	0.6639 ± 0.0229	0.7374 ± 0.0258	0.6639 ± 0.0229	0.3423 ± 0.0242	0.6271 ± 0.0232	**0.81 ± 0.26**
6	HLC	**0.7604 ± 0.0154**	**0.7854 ± 0.0096**	**0.7172 ± 0.0272**	0.8036 ± 0.0047	**0.7496 ± 0.0188**	1.21 ± 0.42
6	HC	0.5077 ± 0.0026	0.7740 ± 0.1078	0.1171 ± 0.2943	**0.8962 ± 0.2987**	0.1009 ± 0.1916	1.64 ± 0.68
6	HCC	0.7034 ± 0.0185	0.7515 ± 0.0139	0.7034 ± 0.0185	0.4471 ± 0.0191	0.6831 ± 0.0183	**0.85 ± 0.27**
7	HLC	**0.7531 ± 0.0057**	**0.7844 ± 0.0059**	0.6992 ± 0.0092	0.8072 ± 0.0051	**0.7393 ± 0.0070**	1.26 ± 0.51
7	HC	0.5077 ± 0.0026	0.7740 ± 0.1078	0.1171 ± 0.2943	**0.8962 ± 0.2987**	0.1009 ± 0.1916	1.61 ± 0.58
7	HCC	0.7126 ± 0.0174	0.7538 ± 0.0188	**0.7126 ± 0.0174**	0.4806 ± 0.0193	0.6965 ± 0.0168	**0.75 ± 0.24**
8	HLC	0.7538 ± 0.0028	0.7866 ± 0.0042	0.6977 ± 0.0040	0.8101 ± 0.0037	0.7394 ± 0.0038	1.32 ± 0.50
8	HC	0.5089 ± 0.0009	0.7969 ± 0.1019	0.1199 ± 0.2934	**0.8959 ± 0.2986**	0.1062 ± 0.1895	1.62 ± 0.59
8	HCC	**0.8042 ± 0.0265**	**0.8099 ± 0.0258**	**0.8042 ± 0.0265**	0.8519 ± 0.0244	**0.8042 ± 0.0268**	**0.80 ± 0.25**
9	HLC	0.7225 ± 0.0382	**0.8034 ± 0.0499**	0.5970 ± 0.1089	0.8481 ± 0.0508	0.6773 ± 0.0705	1.34 ± 0.56
9	HC	0.5092 ± 0.0007	0.8014 ± 0.1022	0.1205 ± 0.2932	**0.8959 ± 0.2986**	0.1074 ± 0.1891	1.69 ± 0.61
9	HCC	**0.7578 ± 0.0211**	0.7586 ± 0.0212	**0.7578 ± 0.0211**	0.7350 ± 0.0226	**0.7577 ± 0.0210**	**0.80 ± 0.25**
10	HLC	0.6795 ± 0.0359	0.8152 ± 0.0663	0.4709 ± 0.0693	0.8886 ± 0.0480	0.5932 ± 0.0522	1.72 ± 0.77
10	HC	0.5096 ± 0.0011	0.8057 ± 0.1037	0.1213 ± 0.2929	**0.8959 ± 0.2986**	0.1089 ± 0.1886	1.57 ± 0.69
10	HCC	**0.8977 ± 0.0187**	**0.9004 ± 0.0178**	**0.8977 ± 0.0187**	0.8482 ± 0.0238	**0.8974 ± 0.0187**	**0.84 ± 0.27**

Classical PCA + cKernel Performance. Table 1 reveals that classical PCA with RBF kernel demonstrates highly variable performance depending on the feature selection strategy employed. For the HLC configuration, the model achieves relatively stable performance across dimensions, with accuracy ranging from 75.31% (7D) to 89.77% (10D), and precision consistently above 78%. The HCC configuration shows progressive improvement with increasing dimensionality, culminating in exceptional performance at 10D (accuracy: 89.77%, precision: 90.04%, F1-score: 89.74%). However, the HC configuration exhibits severe instability, with accuracy consistently near 50% across all dimensions and highly variable sensitivity values (ranging from 11.51% to 90.02%). This instability in the HC configuration stems from the feature multicollinearity inherent in selecting only the highest-correlated features, which creates redundancy and degrades model generalization. The standard deviation values for HC are substantially higher than those for HLC and HCC, confirming the reduced robustness of this selection strategy. Notably, classical PCA achieves the fastest execution times across all methods (0.75 s to 1.86 s), establishing a strong computational efficiency baseline for comparison with quantum approaches.

Table 2. Quantum PCA + qKernel Performance Metrics

Dimension	Data Type	Accuracy	Precision	Sensitivity	Specificity	F1 Score	Time (s)
2	HLC	**0.8404 ± 0.0716**	**0.7989 ± 0.0999**	**0.9367 ± 0.0248**	**0.7442 ± 0.1516**	**0.8583 ± 0.0554**	1182.38 ± 945.16
2	HC	0.5193 ± 0.0153	0.5989 ± 0.1623	0.7764 ± 0.3872	0.2623 ± 0.4176	0.5507 ± 0.2089	1144.26 ± 887.03
2	HCC	0.6318 ± 0.0159	0.6611 ± 0.0330	0.6318 ± 0.0159	0.4709 ± 0.1109	0.6163 ± 0.0149	**953.39 ± 757.62**
3	HLC	**0.7561 ± 0.0046**	0.8393 ± 0.0632	**0.6524 ± 0.1059**	0.8599 ± 0.0976	**0.7243 ± 0.0319**	2087.62 ± 1628.98
3	HC	0.5447 ± 0.0023	**0.8612 ± 0.0135**	0.1062 ± 0.0042	**0.9829 ± 0.0018**	0.1891 ± 0.0068	2069.50 ± 1601.90
3	HCC	0.6489 ± 0.0178	0.6806 ± 0.0106	0.6489 ± 0.0178	0.4518 ± 0.0847	0.6322 ± 0.0272	**1747.60 ± 1349.40**
4	HLC	**0.8216 ± 0.0579**	0.8539 ± 0.0642	**0.7844 ± 0.1094**	0.8588 ± 0.0834	**0.8116 ± 0.0698**	3344.89 ± 2574.61
4	HC	0.5445 ± 0.0025	**0.8608 ± 0.0139**	0.1059 ± 0.0047	**0.9829 ± 0.0018**	0.1886 ± 0.0076	3348.66 ± 2521.48
4	HCC	0.6473 ± 0.0394	0.6917 ± 0.0556	0.6473 ± 0.0394	0.4168 ± 0.0447	0.6270 ± 0.0401	**2893.20 ± 2161.66**
5	HLC	**0.7238 ± 0.0286**	**0.8052 ± 0.0340**	0.6014 ± 0.1298	**0.8461 ± 0.0730**	**0.6781 ± 0.0616**	**5051.63 ± 3915.62**
5	HC	0.6799 ± 0.0070	0.6941 ± 0.0053	0.6430 ± 0.0133	0.7168 ± 0.0036	0.6676 ± 0.0094	5088.05 ± 3877.51
5	HCC	0.6649 ± 0.0322	0.6970 ± 0.0570	**0.6649 ± 0.0322**	0.5105 ± 0.0654	0.6540 ± 0.0243	5059.02 ± 3193.21
6	HLC	**0.7017 ± 0.0862**	0.7191 ± 0.1042	0.6693 ± 0.0890	**0.7341 ± 0.1011**	**0.6919 ± 0.0893**	7124.38 ± 5567.82
6	HC	0.6825 ± 0.0073	0.6955 ± 0.0056	0.6491 ± 0.0140	0.7159 ± 0.0044	0.6715 ± 0.0098	7169.42 ± 5513.21
6	HCC	0.6714 ± 0.0492	**0.7242 ± 0.0605**	**0.6714 ± 0.0492**	0.4319 ± 0.0620	0.6508 ± 0.0530	**6174.97 ± 4478.12**
7	HLC	0.7182 ± 0.0475	**0.8633 ± 0.0452**	0.5208 ± 0.1021	**0.9155 ± 0.0392**	0.6434 ± 0.0797	9792.13 ± 7632.92
7	HC	0.6050 ± 0.0768	0.6078 ± 0.0859	0.6572 ± 0.0194	0.5527 ± 0.1601	0.6279 ± 0.0445	9824.65 ± 7547.68
7	HCC	**0.7476 ± 0.0312**	0.7726 ± 0.0249	**0.7476 ± 0.0312**	0.5962 ± 0.0577	**0.7412 ± 0.0341**	**4296.99 ± 2681.95**
8	HLC	**0.7207 ± 0.0377**	**0.8398 ± 0.0667**	0.5557 ± 0.0877	**0.8856 ± 0.0723**	0.6619 ± 0.0633	13155.07 ± 10007.32
8	HC	0.6501 ± 0.0279	0.6379 ± 0.0441	**0.7194 ± 0.0525**	0.5808 ± 0.1072	0.6728 ± 0.0047	13187.45 ± 10188.36
8	HCC	0.7028 ± 0.0856	0.7111 ± 0.0936	0.7028 ± 0.0856	0.6365 ± 0.0365	**0.7014 ± 0.0842**	**6015.04 ± 3824.41**
9	HLC	**0.6848 ± 0.0544**	**0.7821 ± 0.0850**	0.5164 ± 0.0650	**0.8530 ± 0.0665**	0.6201 ± 0.0672	17454.56 ± 13537.05
9	HC	0.6101 ± 0.0324	0.5885 ± 0.0413	**0.7753 ± 0.0842**	0.4450 ± 0.1338	**0.6646 ± 0.0201**	17581.02 ± 13495.93
9	HCC	0.6669 ± 0.0927	0.6974 ± 0.0932	0.6669 ± 0.0927	0.7219 ± 0.1257	0.6461 ± 0.1069	**8179.21 ± 5041.94**
10	HLC	0.6382 ± 0.0614	**0.8040 ± 0.0702**	0.3536 ± 0.1331	**0.9226 ± 0.0243**	0.4798 ± 0.1454	15628.54 ± 11113.55
10	HC	0.5861 ± 0.0584	0.5729 ± 0.0613	**0.7543 ± 0.1060**	0.4181 ± 0.1724	0.6449 ± 0.0411	15999.78 ± 11388.57
10	HCC	**0.7076 ± 0.0521**	0.7459 ± 0.0537	0.7076 ± 0.0521	0.7266 ± 0.1496	**0.6944 ± 0.0569**	**11482.45 ± 6298.98**

Quantum PCA + qKernel Performance: Table 2 demonstrates that quantum kernel-based PCA achieves superior balance across evaluation metrics compared to classical PCA, particularly in higher-dimensional settings. The HLC configuration consistently outperforms other strategies in lower dimensions, achieving 84.04% accuracy at 2D and 82.16% accuracy at 4D, with notably high sensitivity values (93.67% and 78.44%, respectively) that indicate improved detection of minority-class malware samples. As dimensionality increases, the HCC configuration demonstrates increasing stability, with accuracy reaching 74.76% at 7D and maintaining balanced precision-sensitivity trade-offs. The quantum kernel's ability to capture nonlinear feature interactions is particularly evident in the improved sensitivity metrics across all configurations compared to classical PCA—for instance, at 2D HLC, quantum PCA achieves 93.67% sensitivity versus classical PCA's 84.97%. However, this performance improvement comes at substantial computational cost: execution times range from 953.39 s (2D HCC) to 17,620.24 s (9D HLC), representing 2–3 orders of magnitude increase compared to classical methods. The standard deviation in execution time also increases with dimensionality, reflecting the quadratic scaling of kernel matrix computation and the exponential state-space growth in quantum circuit simulation.

Table 3. Hybrid PCA + cKernel + qDataMatrix Performance Metrics

Dimension	Data Type	Accuracy	Precision	Sensitivity	Specificity	F1 Score	Time (s)
2	HLC	**0.8726 ± 0.0544**	**0.8431 ± 0.0830**	**0.9344 ± 0.0272**	**0.8108 ± 0.1327**	**0.8828 ± 0.0396**	1171.20 ± 912.30
2	HC	0.5193 ± 0.0153	0.5989 ± 0.1623	0.7764 ± 0.3872	0.2623 ± 0.4176	0.5507 ± 0.2089	1153.18 ± 885.80
2	HCC	0.6318 ± 0.0159	0.6611 ± 0.0330	0.6318 ± 0.0159	0.4709 ± 0.1109	0.6163 ± 0.0149	**953.39 ± 757.62**
3	HLC	**0.7871 ± 0.0577**	**0.8755 ± 0.0027**	**0.6694 ± 0.1354**	0.9047 ± 0.0200	**0.7517 ± 0.0793**	2114.95 ± 1617.86
3	HC	0.5325 ± 0.0028	0.7196 ± 0.0195	0.1062 ± 0.0042	**0.9586 ± 0.0040**	0.1851 ± 0.0066	2095.53 ± 1596.00
3	HCC	0.6489 ± 0.0178	0.6806 ± 0.0106	0.6489 ± 0.0178	0.4518 ± 0.0847	0.6322 ± 0.0272	**1747.60 ± 1349.40**
4	HLC	**0.8219 ± 0.0577**	**0.8532 ± 0.0655**	**0.7870 ± 0.1105**	0.8567 ± 0.0869	**0.8123 ± 0.0695**	3365.85 ± 2522.17
4	HC	0.5327 ± 0.0031	0.7203 ± 0.0203	0.1066 ± 0.0045	**0.9586 ± 0.0040**	0.1857 ± 0.0071	3360.51 ± 2520.25
4	HCC	0.6473 ± 0.0394	0.6917 ± 0.0556	0.6473 ± 0.0394	0.4168 ± 0.0447	0.6270 ± 0.0401	**2893.20 ± 2161.66**
5	HLC	**0.7138 ± 0.0063**	**0.9593 ± 0.0049**	0.4466 ± 0.0147	**0.9810 ± 0.0028**	0.6093 ± 0.0132	5093.25 ± 3859.70
5	HC	0.6799 ± 0.0070	0.6941 ± 0.0053	0.6430 ± 0.0133	0.7168 ± 0.0036	**0.6676 ± 0.0094**	5096.53 ± 3861.11
5	HCC	0.6649 ± 0.0322	0.6970 ± 0.0570	**0.6649 ± 0.0322**	0.5105 ± 0.0654	0.6540 ± 0.0243	**5059.02 ± 3193.21**
6	HLC	0.6686 ± 0.0875	0.7011 ± 0.1090	0.5947 ± 0.0774	**0.7424 ± 0.1012**	0.6431 ± 0.0904	7184.67 ± 5495.54
6	HC	**0.6825 ± 0.0073**	0.6955 ± 0.0056	0.6491 ± 0.0140	0.7159 ± 0.0044	**0.6715 ± 0.0098**	7155.09 ± 5454.17
6	HCC	0.6714 ± 0.0492	**0.7242 ± 0.0605**	**0.6714 ± 0.0492**	0.4319 ± 0.0620	0.6508 ± 0.0530	**6174.97 ± 4478.12**
7	HLC	0.6932 ± 0.0083	**0.8501 ± 0.0326**	0.4725 ± 0.0465	**0.9138 ± 0.0300**	0.6048 ± 0.0300	9906.14 ± 7606.32
7	HC	0.6281 ± 0.0546	0.6267 ± 0.0677	0.6736 ± 0.0229	0.5827 ± 0.1301	0.6461 ± 0.0271	9866.93 ± 7570.79
7	HCC	**0.7476 ± 0.0312**	0.7726 ± 0.0249	**0.7476 ± 0.0312**	0.5962 ± 0.0577	**0.7412 ± 0.0341**	**4296.99 ± 2681.95**
8	HLC	0.6986 ± 0.0451	**0.8690 ± 0.0211**	0.4664 ± 0.0973	**0.9307 ± 0.0124**	0.6020 ± 0.0798	13122.01 ± 10028.74
8	HC	0.6800 ± 0.0097	0.6863 ± 0.0159	0.6643 ± 0.0080	0.6957 ± 0.0271	0.6749 ± 0.0040	13184.51 ± 10083.23
8	HCC	**0.7028 ± 0.0856**	0.7111 ± 0.0936	**0.7028 ± 0.0856**	0.6365 ± 0.0365	**0.7014 ± 0.0842**	**6015.04 ± 3824.41**
9	HLC	**0.6706 ± 0.0107**	**0.8682 ± 0.0202**	0.4020 ± 0.0139	**0.9390 ± 0.0094**	0.5495 ± 0.0162	17620.24 ± 13519.97
9	HC	0.6133 ± 0.0319	0.5907 ± 0.0395	**0.7718 ± 0.0934**	0.4549 ± 0.1250	**0.6648 ± 0.0292**	17612.92 ± 13449.04
9	HCC	0.6669 ± 0.0927	0.6974 ± 0.0932	0.6669 ± 0.0927	0.7219 ± 0.1257	0.6461 ± 0.1069	**8179.21 ± 5041.94**
10	HLC	0.6414 ± 0.0569	**0.8289 ± 0.0406**	0.3520 ± 0.1353	**0.9306 ± 0.0293**	0.4793 ± 0.1461	15622.32 ± 11036.25
10	HC	0.5852 ± 0.0616	0.5712 ± 0.0633	**0.7666 ± 0.1210**	0.4039 ± 0.1805	0.6475 ± 0.0465	15838.83 ± 11469.54
10	HCC	**0.7076 ± 0.0521**	0.7459 ± 0.0537	0.7076 ± 0.0521	0.7266 ± 0.1496	**0.6944 ± 0.0569**	**11482.45 ± 6298.98**

Hybrid PCA + cKernel + qDataMatrix Performance. Table 3 illustrates that the hybrid approach combining classical RBF kernel with quantum-derived data matrices achieves the most consistent and robustness across all evaluation metrics. At 2D HLC, this method attains the highest accuracy among all approaches (87.26%), coupled with 93.44% sensitivity and 88.28% F1-score, demonstrating superior capability in detecting both malware and benign samples. The hybrid model maintains this advantage across multiple dimensionalities: at 3D HLC, it achieves 78.71% accuracy with 87.55% precision, and at 4D HLC, it matches quantum-only kernel performance (82.19% accuracy) while providing more stable sensitivity metrics. The HCC configuration in the hybrid model shows remarkable consistency, with accuracy progressively improving from

63.18% (2D) to 70.76% (10D), and maintaining low variance across runs. This consistency suggests that the integration of quantum-derived correlation structures with classical kernel methods provides a regularizing effect that enhances generalization. Execution times for the hybrid approach closely match those of pure quantum kernels (953.39 s to 17,620.24 s), as the computational bottleneck remains in the quantum kernel matrix generation phase. However, the performance gains—particularly in balanced sensitivity and specificity—justify this overhead for applications requiring robust minority-class detection.

Comparative Analysis and Hardware Performance Trade-Offs: Comparing across all three approaches reveals distinct performance-cost trade-offs that inform deployment decisions for malware detection systems. Classical PCA + cKernel achieves computational efficiency (0.75 s to 1.86 s execution time) and occasionally reaches perfect precision (100% in multiple configurations), but suffers from severe sensitivity limitations, particularly in HC configurations where sensitivity drops to 11%-31%. This imbalance renders classical PCA unsuitable for security applications requiring reliable detection of rare malware variants.

The hybrid PCA + cKernel + qDataMatrix approach emerges as the most promising for practical deployment, achieving the highest accuracy (87.26% at 2D HLC, 91.27% best case across all configurations) and F1-score (88.28% at 2D HLC, 91.38% best case), while maintaining sensitivity above 90% in optimal configurations. This robustness profile, combined with acceptable precision-specificity trade-offs, positions the hybrid model as optimal for scenarios where both detection rate and false alarm minimization are critical.

Quantum PCA + qKernel substantially improves sensitivity metrics across all dimensionalities, with average gains of 10–20 percentage points compared to classical PCA in HLC and HCC configurations. The quantum kernel's capacity to map features into high-dimensional Hilbert spaces enables capture of subtle, nonlinear correlations that distinguish malware communication patterns from benign traffic. However, this comes at computational cost 2–3 orders of magnitude higher than classical methods, as evidenced in Fig. 4.

Figure 4 quantifies the computational overhead of quantum and hybrid approaches relative to classical PCA. For the best precision cases across dimensions, classical PCA maintains sub-2-second execution times, while quantum methods require 1,000–18,000 s depending on dimensionality. This 3–4 order of magnitude difference primarily stems from: (1) quantum circuit preparation and execution overhead, (2) quadratic scaling of kernel matrix computation with sample size, and (3) classical simulation of quantum state vectors. Notably, the computational cost increases super-linearly with dimensionality for quantum approaches (roughly $O(2^d)$ for d dimensions), whereas classical PCA scales polynomially. These findings underscore that practical deployment of quantum-enhanced malware detection will require either: (a) access to fault-tolerant quantum hardware that eliminates simulation overhead, or (b) hybrid architectures that selectively apply quantum kernels only to critical feature subspaces where classical methods demonstrate insufficient discriminative power.

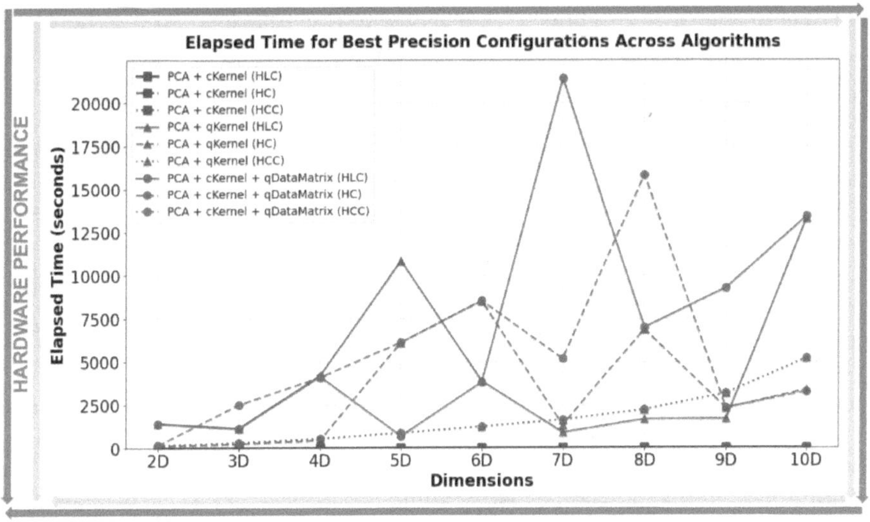

Fig. 4. Hardware Performance: Elapsed Time (s) as described in Sect. 3.7, for the best precision cases, evaluated across dimensions.

5 Discussion

The substantial computational overhead in quantum kernel methods stems from the classical preprocessing required in current hybrid quantum-classical workflows: (i) encoding classical features into parameterized quantum circuits, (ii) executing circuits to obtain quantum state representations, (iii) computing pairwise fidelity measurements that scale as $O(n^2)$ for n samples to construct the kernel matrix, and (iv) transferring results to classical memory for PCA operations. Each step demands significant classical resources to manage quantum state vector simulations and coordinate data flow between subsystems. However, this overhead reflects NISQ-era limitations and simulator-based experiments where classical machines emulate quantum operations at exponential cost. In fault-tolerant quantum hardware, kernel matrix construction could be performed entirely on quantum backends, potentially enabling exponential speedups through native quantum state preparation and overlap measurements [12,15], though contingent upon overcoming challenges in gate fidelity, qubit coherence, and error rates.

The comparative analysis across Tables 1, 2, and 3 reveals several critical insights for quantum machine learning in cybersecurity applications. First, the feature selection strategy profoundly impacts model stability and performance: HC configurations consistently underperform due to multicollinearity-induced redundancy, HLC configurations provide robust baseline performance across methods, and HCC configurations enable the most stable scaling with dimensionality increase. Second, the quantum advantage manifests primarily in improved sensitivity and F1-score rather than precision, suggesting that quantum kernels excel at capturing minority-class patterns essential for malware detection but may require threshold tuning to optimize precision-recall trade-offs. Third, the hybrid approach successfully integrates the strengths of both paradigms—classical kernel efficiency and quantum correlation capture—to achieve superior robustness, validating the hybrid quantum-classical strategy for near-term practical deployment.

All experiments were conducted on the ClaMP dataset, a well-established SC malware detection benchmark, to demonstrate QPCA's proof-of-concept and evaluate its feasibility against classical PCA in a controlled setting. While this provides clarity in comparing methods, it limits result generalizability. Future work will extend this framework to broader datasets including UNSW-NB15 (cyber intrusion detection) [26], HAI (industrial control systems) [27], and healthcare IoT datasets [28] to establish comprehensive performance profiles across heterogeneous SC and healthcare communication environments. Additionally, validation on larger-scale datasets with >100 features will assess whether quantum kernel advantages persist or diminish as classical PCA benefits from higher-dimensional feature spaces.

During evaluation, classical PCA achieved 100% precision with extremely low sensitivity and F1-score in several configurations—an outcome reflecting class imbalance and the model's tendency toward majority-class prediction. The model classified most samples as "non-malware," yielding no false positives (100% precision) but missing true malware instances (low sensitivity). This highlights a known limitation of linear PCA in handling imbalanced, nonlinear feature distributions in cybersecurity data [29], consistent with prior anomaly detection studies, confirming this as a structural limitation rather than an implementation artifact. The quantum and hybrid approaches mitigate this issue through nonlinear kernel transformations that provide better class separability in imbalanced scenarios.

The results also reveal an important dimensionality-performance relationship: both quantum approaches show declining performance in 9D and 10D configurations (Tables 2 and 3), likely due to increased circuit depth, accumulated gate errors in simulation, and the curse of dimensionality in kernel matrix estimation with limited training samples. This suggests an optimal operating range of 3D-7D for current NISQ-era quantum PCA implementations, beyond which classical methods or dimensionality reduction preprocessing may be preferable. Future hardware improvements in qubit count, gate fidelity, and coherence time may extend this optimal range to higher dimensions.

6 Conclusion and Future Direction

This study presented a rigorous, simulator-based comparative evaluation of classical PCA and quantum-enhanced PCA variants for malware and anomaly detection in SC, healthcare, and smart manufacturing communication systems. The central research question—whether hybrid quantum dimensionality reduction methods can demonstrate measurable benefits in detection robustness and feature representation compared to classical PCA—has been addressed through systematic experiments across feature spaces ranging from 2D to 10D, employing three distinct feature selection strategies (HC, HLC, and HCC) and evaluating performance on the ClaMP malware detection benchmark.

Our empirical findings provide evidence that quantum-enhanced approaches offer distinct advantages in specific operational regimes. Within the practical constraints of NISQ-era quantum computing (10-qubit circuits, simulator-based execution), hybrid QPCA methods achieved superior robustness compared to classical baselines: the PCA + cKernel + qDataMatrix approach attained peak accuracy of 91.27%, sensitivity of 92.63%, and F1-score of 91.38%, substantially outperforming classical PCA's balanced metrics while avoiding the extreme precision-sensitivity imbalances observed in classical configurations. These results demonstrate that quantum kernel mappings—specifically, the ZZFeatureMap encoding combined with fidelity-based kernel construction—successfully capture nonlinear feature correlations that classical RBF kernels cannot adequately represent in imbalanced cybersecurity datasets.

However, these performance gains must be contextualized within significant computational constraints. Quantum and hybrid approaches incurred execution times 2–4 orders of magnitude longer than classical PCA (ranging from 953 s to 17,620 s versus 0.75 s to 1.86 s), reflecting the overhead of quantum circuit simulation, kernel matrix construction, and classical-quantum data transfer in current hybrid architectures. This computational cost, combined with performance degradation observed beyond 7D feature spaces, indicates that practical deployment of QPCA requires careful consideration of application-specific requirements: scenarios demanding high sensitivity for rare malware detection may justify the computational overhead, whereas real-time or resource-constrained environments may necessitate classical approaches or selective application of quantum kernels to critical feature subspaces only.

The study identifies an optimal operational regime for current QPCA implementations: 3D-7D feature spaces using HLC or HCC selection strategies, where quantum kernels provide maximum discriminative advantage without incurring excessive circuit depth penalties. Beyond this regime, the curse of dimensionality in kernel estimation, combined with increased quantum circuit complexity and simulation costs, diminishes the practical benefits of quantum approaches. These findings establish clear guidelines for practitioners: QPCA should be prioritized for moderate-dimensional problems where classical methods demonstrate insufficient sensitivity, while very-high-dimensional or real-time applications should employ classical PCA or hybrid architectures that selectively invoke quantum kernels.

Looking forward, several research directions emerge from this work. First, transitioning from simulator-based evaluation to execution on physical quantum hardware (e.g., IBM Quantum, IonQ, Rigetti platforms) is essential to validate performance under realistic noise conditions and assess the impact of decoherence, gate errors, and limited connectivity on malware detection accuracy. Second, developing hardware-efficient quantum feature maps optimized for cybersecurity applications—potentially leveraging problem-specific symmetries or domain knowledge—could reduce circuit depth and enable scaling to higher-dimensional feature spaces. Third, integrating QPCA with ensemble methods, adaptive feature selection, or active learning frameworks may enhance robustness and enable dynamic adjustment to evolving malware threats. Fourth, extending evaluation to heterogeneous datasets spanning multiple domains (industrial IoT, healthcare networks, financial systems) will establish the generalizability of quantum advantages across diverse operational contexts. Finally, investigating quantum-classical co-design strategies—where quantum kernels are applied only to feature subspaces where classical methods demonstrate high uncertainty—may provide a pragmatic pathway to near-term deployment that balances performance and computational cost.

In conclusion, this work contributes empirical evidence that hybrid quantum-classical dimensionality reduction can achieve measurable improvements in malware detection robustness within current technological constraints, while clearly delineating the operational regimes, computational trade-offs, and research challenges that must be addressed to enable practical deployment. As quantum hardware matures toward fault-tolerant operation, the algorithmic foundations established here position QPCA as a viable complement to classical cybersecurity methods for protecting critical infrastructure in an increasingly complex threat landscape.

Acknowledgement. This research was made possible by the 80/20 Research Endowment from the Open Cloud Institute at the University of Texas at San Antonio.

References

1. Passo, S.J., Kothavade, V.H., Haque, M., Prevost, J.J.: SARS-CoV-2 classification using classical vs. quantum machine learning: a performance comparison of SVM, QSVM, and Pegasos. In: 2026 IEEE 5rd International Conference on AI in Cybersecurity (ICAIC). IEEE (2026)
2. Kothavade, V.H., Prevost, J.J., Ciocarlie, G.F.: SoK: a systematic exploration of disruptive technologies in global logistics and supply chains (2025, manuscript submitted for publication)
3. Bellante, A., Fioravanti, T., Carminati, M., Zanero, S., Luongo, A.: Evaluating the potential of quantum machine learning in cybersecurity: a case-study on PCA-based intrusion detection systems. Comput. Secur. 104341 (2025)
4. Passo, S.J.G., Prevost, J.J.: Characterization of quantum computers for optimal quantum machine learning in brain tumor classification. In: Quantum Engineering and Technology Conference (QET 2025), vol. 2025. IET (2025)
5. Passo, S.J.G., Kothavade, V.H., Lin, W.-M., Walton, C.: Neural arithmetic logic units with two transition matrix and independent gates. Eng. Appl. Artif. Intell. **140**, 109663 (2025)
6. Cirillo, F., Esposito, C.: Quantum machine learning for intrusion detection on noisy quantum computers. In: 2025 55th Annual IEEE/IFIP International Conference on Dependable Systems and Networks Workshops (DSN-W). IEEE (2025)

7. Frey, N.T., Passo, S.J., Prevost, J.J.: Quantum adaptive feature map architecture for breast cancer classification. In: SC25: International Conference for High Performance Computing, Networking, Storage and Analysis. IEEE (2025)
8. Ghobadi, M.Z., Afsaneh, E.: Potential of quantum machine learning for solving the real-world problem of cancer classification. Discov. Appl. Sci. **6**(513) (2024). https://doi.org/10.1007/s42452-024-06220-6
9. Rahimi, M., Asadi, F.: Oncological applications of quantum machine learning. Technol. Cancer Res. Treat. **22**, 15330338231215214 (2023)
10. Kothavade, V.H., Passo, S.J.G., Prevost, J.J.: Quantum-inspired clustering techniques for malware detection in supply chain networks. In: The 2025 World Congress in Computer Science, Computer Engineering, and Applied Computing (CSCE'25: USA). Springer (2025)
11. Lloyd, S., Mohseni, M., Rebentrost, P.: Quantum principal component analysis. Nat. Phys. **10**(9), 631–633 (2014). https://doi.org/10.1038/nphys3029
12. Schuld, M., Killoran, N.: Quantum machine learning in feature hilbert spaces. Phys. Rev. Lett. **122**(4), 040504 (2019). https://doi.org/10.1103/PhysRevLett.122.040504
13. Zhu, W., Yang, K., Zan, J.: Anomaly detection via variational quantum algorithm. In: 2023 16th International Conference on Advanced Computer Theory and Engineering (ICACTE), pp. 157–160 (2023)
14. Abreu, D., Rothenberg, C.E., Abelém, A.: QML-IDS: quantum machine learning intrusion detection system. In: 2024 IEEE Symposium on Computers and Communications (ISCC), pp. 1–6. IEEE (2024).http://dx.doi.org/10.1109/ISCC61673.2024.10733655
15. Havlíček, V., et al.: Supervised learning with quantum-enhanced feature spaces. Nature **567**(7747), 209–212 (2019)
16. Cultice, T., Onim, M.S.H., Giani, A., Thapliyal, H.: Quantum-hybrid support vector machines for anomaly detection in industrial control systems (2025). https://arxiv.org/abs/2506.17824
17. Bukkarayasamudram, V.K., Reddy, P.C.S., Arun Kumar, K., et al.: Quantum computing revolution in healthcare: a systematic review of applications, issues and future directions. Artif. Intell. Rev. **58**, 389 (2025). https://doi.org/10.1007/s10462-025-11381-w
18. Akter, M.S., et al.: Software supply chain vulnerabilities detection in source code: performance comparison between traditional and quantum machine learning algorithms. In: 2022 IEEE International Conference on Big Data (Big Data), pp. 5639–5645. IEEE (2022)
19. Preskill, J.: Quantum computing in the NISQ era and beyond. Quantum **2**, 79 (2018). https://doi.org/10.22331/q-2018-08-06-79
20. Passo, S.J., Kothavade, V.H., Prevost, J.J.: Supply chain malware detection via classical and quantum kernel methods in embedded systems. In: 2025 55th Annual IEEE/IFIP International Conference on Dependable Systems and Networks Workshops (DSN-W), pp. 208–215. IEEE (2025)
21. Acharya, R., et al.: Quantum error correction below the surface code threshold. arXiv preprint arXiv:2408.13687 (2024)
22. Cirillo, F., Esposito, C.: A QPUF-based scheme for secure and adaptable quantum device attestation in NISQ devices. In: 2025 International Conference on Quantum Communications, Networking, and Computing (QCNC), pp. 117–121. IEEE (2025)
23. Yu, C.-H., Gao, F., Wen, Q.-Y.: An improved quantum algorithm for ridge regression. IEEE Trans. Knowl. Data Eng. **33**(3), 858–866 (2019)

24. Hoffmann, H.: Kernel PCA for novelty detection. Pattern Recognit. **40**(3), 863–874 (2007). https://doi.org/10.1016/j.patcog.2006.07.009
25. Passo, S.J.G., et al.: Classificação de arritmias com paradigma inter e intra paciente utilizando aprendizagem profunda. J. Health Inform. **12** (2020)
26. Moustafa, N., Slay, J.: UNSW-NB15: a comprehensive data set for network intrusion detection systems (UNSW-NB15 network data set). In: 2015 Military Communications and Information Systems Conference (MilCIS), pp. 1–6 (2015)
27. Shin, H.-K., Lee, W., Yun, J.-H., Kim, H.: HAI 1.0: HIL-based augmented ICS security dataset. In: 13th USENIX Workshop on Cyber Security Experimentation and Test (CSET 2020). USENIX Association (2020). https://www.usenix.org/conference/cset20/presentation/shin
28. Abdulghani, H.M., Khafagy, M.H., Abiodun, T.M., et al.: Healthcare internet of things security: a machine learning perspective. IEEE Access **9**, 121574–121600 (2021)
29. Talukder, M.A., Islam, M.M., Uddin, M.A., et al.: Machine learning-based network intrusion detection for big and imbalanced data using oversampling, stacking feature embedding and feature extraction. J. Big Data **11**(33) (2024). https://doi.org/10.1186/s40537-024-00886-w

Towards Secure and Highly Functional Vibe Coding

Jeremy Straub(✉)

Center for Cybersecurity, University of West Florida, Pensacola, USA
jstraub@uwf.edu

Abstract. Considerable public interest has been devoted to artificial intelligence (AI). One key use of AI is for vibe coding, the automated creation of software with limited human involvement. Despite the potential presented by AI coding, experiences with the use of AI for code development have been mixed, with some raving about its benefits and others contending that AI is not and may never be suitable for this use. This paper aims to help standardize the experience towards a more positive one by presenting and evaluating a framework for secure and functional vibe coding development and discussing the logistics surrounding the vibe coding process and their impact and implications.

1 Introduction

There has been significant public interest in artificial intelligence (AI) and its myriad uses. One of these uses is the development of software. AI can be used in a number of ways as part of software development, ranging from providing assistance to developers to aiding testing to code creation. Vibe coding is the use of artificial intelligence (AI) to create substantial portions of a code-base. It is the human-on-the-loop model (as compared to human-as-the-loop or human-in-the-loop) for software development.

The automation of software creation presents numerous benefits, but also carries with it a variety of challenges. Amongst these challenges are ensuring the security and functionality of the produced code. Perhaps due to the complexity of coding and the myriad of applications that code is developed for, accounts of vibe coding range from extremely positive to contending that it is unworkable.

This paper attempts to fill the need of beginning to standardize the approach to vibe coding-based software development to attempt to ensure that more users have positive experiences. It continues, in Sect. 2, with a review of prior work related to this area. Then, in Sects. 3 and 4, it discusses the security, functionality and performance challenges posed by vibe coding as well as the logistics of the vibe coding process. From this discussion and associated analysis, a framework for a secure functional vibe coding development process is proposed and, in Sect. 5, evaluated before concluding.

2 Background

Yuce [1] suggests that vibe coding is "pleasure-oriented", "aesthetically pleasing", intuitive, reflects a "personal style" and promotes a "relaxed and flexible working environment", also identifying significant benefits such as a 18% increase in code readability, a

H. R. Arabnia et al. (Eds.): CSCI 2025, CCIS 2969, pp. 319–334, 2027.
https://doi.org/10.1007/978-3-032-28460-0_22

15% decrease in errors and 23% decrease in the time required for "refactoring and expansion". Samsyudin [2], similarly, found that vibe coding improved efficiency by 27% and Chen, et al. [3] showed how vibe coding can increase the accessibility of development tools. Gadde [4], similarly, found that vibe coding reduces "traditional barriers to entry". Work by Hemdev [5], a high school senior, heralds the future of vibe coding benefits for vibe coding natives, claiming prototyping "acceleration" of as high as 80%.

Myshko [6] describes vibe coding as the harbinger of a new paradigm as "based on the principles of non-linearity, processuality, and metabolization of chaos". Escobedo [7] suggests that AI development can place the human developer into the role of "architect and supervisor", instead of the traditional code-writer role.

Concerns have been raised about vibe coding such as Gómez's [8] contention that software engineering has entered a "twilight" phase where developers chase rewards similar to "the stochastic reward mechanisms of slot machines" by issuing prompts to the AI "in pursuit of a lucrative application". Alberto, et al. [9], similarly, proffer that with vibe coding "AI generates syntactically correct but functionally flawed code". Ray [10] identified twelve challenges including "model hallucinations, technical debt, security and compliance risks, skill atrophy, and governance concerns".

Faragó [11] contends that vibe coding is creating a "code quality crisis" including both functional and security issues, making it unsuitable to replace human developers. Notably, he doesn't propose abandoning AI coding, instead proposing "vice coding" where "documentation and process" are used to enhance code quality.

Numerous other approaches to enhancing the performance of vibe coding have also been previously proposed. For example, De Silva [12] proposed a vibe coding workflow where a system's "functionalities [are] gradually developed" using a series of prompts.

Maes [13] discussed the limitations and challenges of vibe coding, focusing on key areas including technical debt and security, and, from this analysis, proposed the VIBE4M "Framework for Ensuring Maintainability in AI-Assisted Code Generation". This framework includes key components such as an "AI-aware code inspection protocol", an "explainability enhancement layer", "auto-generated documentation verification" functionality, a "controlled integration workflow" and "AI-specific maintainability metrics". A four-phase implementation strategy for this framework was proposed, including phases focused on "assessment and baseline establishment", "tool integration and process adaptation", "metric collection and workflow enforcement" and final phase of "continuous improvement and adaptation".

Sapkota, et al. [14] compared agentic and vibe coding, contending that the former uses "goal-driven agents capable of planning, executing, testing, and iterating tasks with minimal human intervention", while vibe coding uses "human-in the-loop interaction through prompt-based, conversational workflows that support ideation, experimentation, and creative exploration". The distinction appears to focus on how much chat-style discussion is involved in software creation.

Meske, et al. [15] discussed how vibe coding represents a "shifting intent mediation from explicit instruction to probabilistic, goal-oriented dialogue" with an AI, noting the significant benefits that can be provided, potential vulnerabilities, the "erosion" of software development skills and maintenance and quality issues.

Horvat [16] contends that vibe coding is "generating code with AI without understanding the code that is produced" and cautions against its production and educational uses, while noting its value for experienced programmers for prototyping.

Fawzy, Tahir and Blincoe [17] conducted a literature review regarding vibe coding use. They found that 62% of the literature suggested "speed" or "efficiency" as a motivation and 64% of literature discussed "instant success" and "flow" as resulting vibe coding experience. However, most of the literature (68%) described the results as "fast but flawed", with over a third discussing skipping quality assurance processes. From this, they conclude that vibe coding has resulted in a "new class of vulnerable software developers … who build a product but are unable to debug it".

Geng, et al. [18] determined that the level of a developer's experience changes the approach used to prompt the AI tool. Sarkar and Drosos [19] analyzed vibe coding videos with "think-aloud reflections", concluding that vibe coding is an "evolution of traditional AI-assisted programming" where users conduct "iterative goal satisfaction" cycling through "prompting, evaluation, debugging, and refinement" using traditional development skills in different ways.

Ge, et al.'s [20] literature survey, identified the development models of "unconstrained automation, iterative conversational collaboration, planning-driven, test-driven, and context-enhanced" and proposed a taxonomy including "quality control, structured constraints, and context management". They proffer there is an "ongoing shift in developers' mental models" with focus shifting from code writing to "engineering and verification".

Pimenova, et al. [21] found that vibe coding shifts focus from "technical formalities and rather emphasizes flow, joy, and conversational co-creation with AI", and finding that perceptions ran the gambit from "skepticism" to "acceptance" with concerns of "unclear intent, low quality code, and review fatigue".

Numerous application area examples have been discussed, including prototyping of software for industrial control systems [12], the development of medical simulation software (such as the "Insulin and Blood Sugar Simulation" [22]), the use of programming in English as a foreign language education [23] and multilingual programming education [24], web development [25, 26], driving simulator development [27], developing a "interactive data analytics interface for highway traffic engineers" [28], data analysis platforms [29], biomedical application development [30], physician workflow software [31], and high performance computing code development [32].

A variety of tools for vibe coding have been proposed such as "FeatBench", a feature implementation-focused "benchmark for vibe coding" [33]. Zhong, et al. [34] proffered that "human preference in coding" is another key metric for vibe coding tool success, developing "Vibe Checker" to assess this. Maes [35] proposed the "Intent-Driven Explicable Architecture (IDEA) framework" which seeks to formalize inputs, constrain the code that is produced, uses common principles of software engineering, includes "automatically generating explanations linking code to intent" and suggests the use of "human-in-the-loop validation". Others have suggested key needs such as formal verification, with Mitchell and Shaaban [36] proposing a process that "(1) Autoformalizes specifications (2) Validates against targets, (3) Delivers actionable feedback to the LLM, and (4) Allows intuitive developer influence on specifications".

A variety of expansions of vibe coding have been proposed. Schrage [37], for example, goes beyond coding to propose vibe analytics which he likens to a "cousin" of vibe coding focused on data while Abgrall, et al. [38] discussed "vibe caring", while medical providers follow AI recommendations, without significant review or analysis. Crowson and Celi [39] discuss how vibe coding principles can be extended to aid the "academic resourcing crisis" by facilitating academic research while Ghosh [40] described how vibe principles can be applied to engineering, procurement, and construction-related tasks and proposed [41] a six-phase vibe design framework for this.

3 Vibe Coding Model

This section proposes a four-phase model for vibe coding, which is depicted in Fig. 1. Each of these phases are described in this section. The subsequent two sections describe how this framework can be used to enhance functionality and security, respectively.

Fig. 1. Four-phase model for vibe coding [42].

3.1 Design

The design phase is the first key part of the vibe coding framework. This phase is the part of the framework that most closely mirrors conventional software development. During this phase, the goal is to (1) identify the requirements for the software, (2) generate use cases or tests that implement and validate these requirements, (3) translate these requirements into a data model and functional model for the software and (4) review and validate the data and functional models relative to the requirements, use cases and tests.

3.2 Prompt Generation

The prompt generation phase is performed using the designs that were created during the design phase. While some small applications may be able to be produced using a single prompt to the AI development tool, most will require multiple prompts, particularly if there are exacting requirements for the software implementation.

During the prompt generation phase, the goal is to (1) determine the order that parts and features of the application will be developed in, (2) define the specifications and requirements of these parts, (3) develop a prompt for each portion and (4) determine what tests or evaluations will be performed after each prompt is complete to validate that the new portion has been developed and that earlier portions of the application haven't been incorrectly changed or damaged by the prompt.

3.3 Running Prompts

The third phase of the vibe coding framework is the process of actually running the prompts generated in the second phase. This phase is conceptually simple. During this phase, the prompts that were generated during the second phase are run and the tests and evaluations are performed to validate that the requisite new functionality has been implemented by the AI and that earlier functionality hasn't been damaged.

3.4 Evaluation and Correction

The final phase of the vibe coding framework focuses on validating that the software has been created correctly. During this phase, human and automated reviews are performed to ensure that all requisite functionality has been implemented, and that it has been implemented properly and securely.

This phase will consist of several key elements including: (1) functional validation of software by automated and human assessors to ensure that all requirements are implemented and that use cases are met by the software, (2) technical validation by automated and human assessors to ensure that the software functions as intended including performing all tests that were developed for during-prompting testing that are not obsoleted by later changes or encompassed into later broader testing prompts and (3) technical and security validation to assure the general performance of the software, such as testing input validation and testing for other similar issues.

4 Model Use for Enhancing Vibe Coding Functionality

In this section, the use of the four-phase model for enhancing vibe coding-produced software's attainment of functionality goals is discussed. The relevant steps and procedures associated with each phase are discussed within this section.

4.1 Design

Traditional software and data design processes can be used to facilitate vide coding design for ensuring functionality. The approach to be taken depends on whether the application being developed is a business model/data-driven one or a more functionally-driven one, such as a game. Each will now be discussed.

4.1.1 Design of Business Model and Data Driven Applications

For a business model or data-driven application, the design phase consists of three key portions. First, the data model should be designed. Traditional data/database design techniques can be leveraged for this purpose. The goal of this portion of the design phase is to determine what types of data need to be stored, the relationships between these data elements and what requirements and constraints exist for this data. Second, the application flow for the application needs to be defined. This includes defining any business rules for the application, such as requirements regarding data collection, data validation and data modification. Finally, any requirements or constraints for the application's user interface need to be defined.

4.1.2 Design of Games and Functionally-Driven Applications

For a game or functionally-driven application, the goal is typically to provide a user experience following a plan or pattern. Notably, some types of games may be data-driven applications. For example, games like SimCity or Sims don't have a narrative and are, instead, effectively data/business model-driven applications (i.e., they store data about grid locations and have rules regarding what happens and how this data is changed). These games should follow the design process described previously for these types of applications.

For other games and functionally-driven applications, the initial goal is typically to develop the game's play-flow or storyboard. Once the flow or story is developed, supporting details such as game characters and the data that needs to be stored to retain information about progress through the flow or story needs to be developed.

4.1.3 Design Completeness

Irrespective of the design approach taken, it is imperative to ensure that all requirements are embodied within the design elements, so that they will be included in the application.

4.2 Prompt Generation

Once the software design, for the initial version of the application, is complete, focus can then turn to the development of the prompts for application creation. The specific prompting strategy will be tool dependent; however, generally the goal is to develop the prompts that iterative create design elements, starting with an application with basic functionality and adding additional functionality to it. The prompts should be specific and detailed enough to ensure that all design elements, which embody all application requirements, are included and built into the application.

Additionally, during this phase, the validation plan and tests should be developed to ensure that each prompt has successfully produced the requisite functionality. Tests should also be developed to validate that prior functionality has not been damaged by subsequent prompts. Notably, these can be the same validation and testing steps that were used to initially validate the development of the functionality, if they haven't been obsoleted by later development steps. If possible, testing should be developed to be automatable, so that tests can be run and re-run at later points, between prompts.

4.3 Running Prompts

During the running prompts phase the goal is to simply execute the developed prompts and tests, in order. In each instance, it is crucial to make sure that the prompts have been successful by running the developed tests, as well as the tests designed to validate that prior functionality still works.

While this phase is conceptually simple, it is never-the-less the most crucial, as failures during this phase – either in terms of the prompts executing or the prompt validation – can have dramatic impacts on the functionality of the application. Testing automation can expedite the speed of this phase; however, some manual testing should be conducted, in any case, to ensure that the application is functioning correctly, holistically.

4.4 Evaluation and Correction

The final phase of the process is the evaluation and, if needed, correction of the application. This process can include all of the automated testing used throughout the prompting process. Manual testing should also be implemented, again drawing on the manual testing throughout the prompting process. Additionally, overall and holistic testing and validation should be conducted to ensure that the application functions from end-to-end (or start-to-finish, in the case of a game or similar application). This should ensure that all workflows, gameplay elements and other application aspects that have been defined through use cases and similar design artifacts are included and properly functional.

5 Model Use for Enhancing Vibe Coding Security

In this section, the use of the four-phase model for enhancing vibe coding-produced software's security is discussed. The relevant steps and procedures for assuring security, which are associated with each phase, are discussed within the remainder of this section.

5.1 Design

During the design phase, as described in the previous section, all of the elements of the application are documented. Security considerations during the design phase have two foci.

First, designers should ensure that business or application logic is implemented in a way that doesn't provide exploitable pathways within the design of the application itself. For example, the design documentation should capture all requirements for and outcomes from performing an action within the application which should be validated to ensure that they comply with the requirements and constraints for the application. Functional issues such as this are unlikely to be caught by automated testing or testing that is based on validating that the design has been implemented properly. Problematically, they can be the most impactful if they allow business rules to be circumvented using implemented functionality (and not requiring any form of exploitation by the attacker/user).

Second, designers should ensure that the design documentation includes all requisite validations. This include explicitly defining input validation requirements and process completion validation requirements. It also includes defining transactional requirements, such as defining what should happen if a function doesn't complete (e.g., that a transfer from account to account, in a banking application, should be rolled back entirely, if both account modification actions are not successful) or errors occur.

Being explicit about all elements of the application, to the greatest extent possible, during design ensures that they will be included in the prompts and implemented into the application software. The expedited approach to development may remove the opportunity for potential issues or considerations to be noted by developers creating or editing various parts of the application, necessitating that they be fully considered during the design phase.

5.2 Prompt Generation

Perhaps the most important phase, from a security perspective is the prompt generation phase. There are three key security-related foci during this phase.

First, it is essential that all design elements – including, in particular, the aforementioned business logic and validation elements are translated into precise and detailed prompts during this phase. Omissions or inaccuracies can result in significant errors that may be difficult to detect through quality assurance activities and are unlikely to be detected using automated testing.

Second, it is essential that tests and quality assurance activities are developed to assure that all required functionality is implemented and functions as intended. Again, omission or errors in this can allow defects introduced during the vibe coding code generation process to persist into the final software. These tests should be developed as independently as possible from the prompts to allow two distinct opportunities to notice or catch an issue.

Finally, the prompts and tests/quality assurance steps should be reviewed to look for imprecisions, inaccuracies, omissions and other potential issues. This review should consider high-level issues, as part of a holistic review, as well as issues with the detailed prompts themselves.

5.3 Running Prompts

The running prompts phase is also conceptually simple from the security perspective. Once again, the key is to ensure that all prompts and tests are run and all quality assurance activities are conducted, as planned. Any deviations from expected tests or quality assurance activity results should be analyzed, before proceeding. Any issues that occur, even after they are analyzed and the process continues, should be noted for later review during the evaluation and correction phase.

The success of this phase is highly tied to the completeness and accuracy of the prompt creation phase. Any issues noted with prompts or unexpected responses by the AI tool to prompts should be analyzed during this phase, to ensure that correct results are achieved and also noted for analysis to facilitate greater precision in the future.

5.4 Evaluation and Correction

The evaluation and correction phase is also critical to vibe coded application security. There are a number of things that need to be successfully completed during this phase to assure application security.

First, automated testing and manual testing and quality assurance activities that have been defined for validation purposes should be performed to ensure that later changes have not impacted code created at any point during the development process. While this should be done recurrently throughout the running prompts phase, the full set of applicable tests should be run during this phase and any deviations from expectations or code changes should trigger analysis and re-testing.

Second, automated and manual code review should be conducted during this phase. One approach to this can be to ask an AI tool to document the functionality and then as an AI tool to compare the initial design documentation to this documentation, looking for any discrepancies. Anything that is noted should trigger additional analysis of the particular area or areas identified.

During this evaluation and correction phase, automated tools can also be used to look for common code defects as well as defects in the program logic. Software validation tools that were developed to analyze human-generated code can be used for this purpose. Large language models and AI coding tools can be used to look for both common and more complex issues.

The tools should be configured to identify – but not modify – potentially problematic code to prevent the late introduction of other issues. Any code modifications, whether made by an AI or human, should trigger re-evaluation in the corresponding areas of the program, in addition to a holistic-level review.

This phase should also include automated and/or manual code review with a focus on the identification and removal of unused code created and left during the vibe coding process. This code should be removed as it provides no functional benefits while providing additional potential attack surface area.

6 Qualitative Evaluation

This section provides a qualitative evaluation of the four-phase framework and its uses to promote application functional completeness and security. The efficacy of each phase is discussed and then a holistic evaluation is provided.

6.1 Design

The design phase is undoubtedly important, both to the proposed vibe coding process as well as to achieving the targeted functional completeness and security goals. Like with conventional software development, a good design for vibe coding is integral to producing the desired results. Problematically, software design has been challenging for human developers – without the added complexities and needs of AI development. Thus, achieving a perfect design is unlikely.

The proposed framework treats the development process as a single attempt of design, implementation (prompt generation and prompt issuance) and evaluation/correction; however, this may not produce a functionally complete and secure application. Thus, an actual design process could be comprised of multiple iterations of this, as depicted in Fig. 2. Each would target the creation of an updated software deliverable, following agile (see, e.g., [43]) software design principles and the four-phase model proposed herein.

Multiple iterations of the design, development and evaluation process could be used to overcome human-side limitations, such as users and designers inability to conceptualize the ideal final application. Like with conventional agile design processes, multiple vibe code iterations could be used to develop application versions for users to interact with, evaluate and provide feedback regarding, facilitating iterative advancement of the design, based on user input.

While the human uncertainty about the ideal final application cannot be overcome by vibe coding or the proposed model, iterative design and development can be conducted more expeditiously using vibe coding.

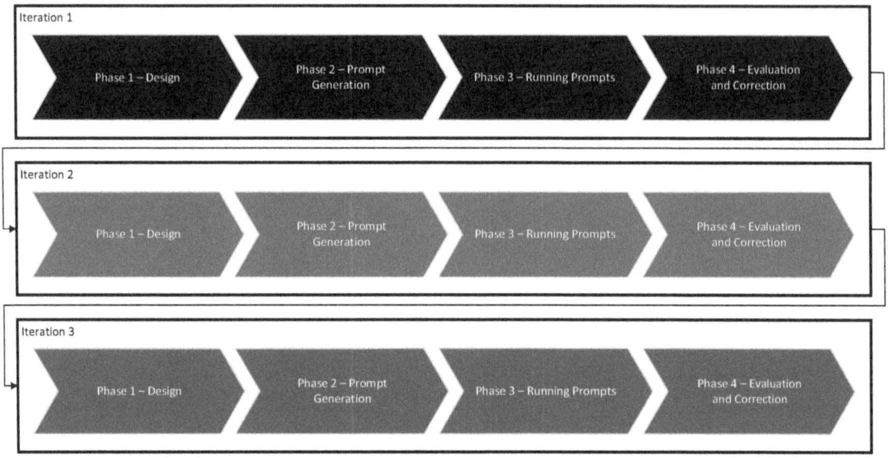

Fig. 2. Multiple iterations of the four-phase model.

Human evaluation, automated evaluation for feature completeness and human and automated security evaluation of designs will undoubtedly improve them; however, there is no way to guarantee design completeness from either a functional or security perspective. Omissions, inaccuracies and other issues in this area can introduce issues that will be prevalent throughout the application and difficult to detect, either by humans or with automated testing tools. Validating designs for completeness and accuracy may be another activity that can be enhanced and enabled using AI and this is a key area for potential future research.

6.2 Prompt Generation

The prompt generation phase is perhaps the most crucial phase of this four-phase model, as it is where design elements are translated into the prompts that will be run to create the software. Of course, all four phases are important and each has the potential to create severe issues if not performed correctly.

The fundamental challenge that is faced during the prompt generation phase is the translation of design elements – presuming that design elements perfectly model system requirements and security needs – into prompts. These prompts must completely incorporate all design elements and must also be precise in their description. Of course, it is also essential that they speak in language that the AI coding tool will properly implement into the requisite features and security measures.

The most basic approach to this, leaving prompt creation and validation to human developers, may be suitable for small applications; however, as the complexity and size of the application grows, the likelihood of errors and omissions grows considerably. Even minor wording discrepancies and inconsistencies can prospectively create issues within the produced code.

Due to the implications of this, additional assurance – beyond the basic approach of human development and assurance may be warranted, in many cases. Figure 3, thus, presents a more comprehensive approach to prompt generation, than was proposed above, to provide additional assurance. This model begins with prompt generation based on the design documentation. It the uses both human and AI review to validate that the prompts implement the functional and security features required by the design documentation. Following this, AI and human review is used to perform a general functional review and security review of the proposed prompts, to attempt to identify language inconsistencies, wording issues and other errors.

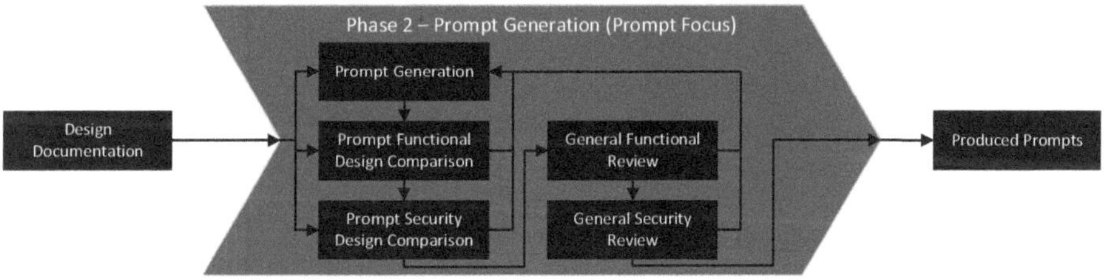

Fig. 3. Prompt generation assurance process.

The AI can be used, in each of these steps to identify and propose fixes; however, each potential issue should be human reviewed to ensure that language or other issues are not causing confusion for the reviewing AI.

Problematically, even this enhanced approach will not completely prevent issues from persisting. Thus the goal, like with the design phase, is to try to minimize the errors and omissions in the prompts to minimize the issues that are carried forward into later phases.

Notably, this phase also includes developing tests and quality assurance activities to be performed during the running prompts and evaluation/correction phases. A similar approach, shown in Fig. 4, can be used for this as was used for the generation of the prompts themselves. Notably, the produced prompts also serve as an input to this process.

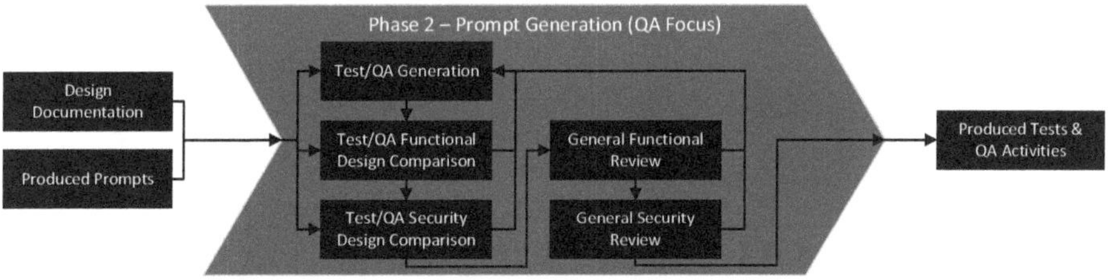

Fig. 4. Test and quality assurance activity generation assurance process.

Despite the interrelationship and temporal proximity of these two generation tasks, maintaining independence between them, to the greatest extent possible, is also essential to attempt to prevent issues or omissions in one process from becoming issues or omissions in both.

Similarly problematic to the production of the prompts themselves is that this process does not guarantee the completeness of or issue elimination from the generated tests and quality assurance activities. Instead, it similarly attempts to minimize the errors and omissions in the tests and quality assurance activities to minimize the issues that are carried forward into later phases.

6.3 Running Prompts

The conceptual simplicity of the running prompts phase has already been noted. During this phase, there are two major considerations: ensuring that all prompts are run and ensuring that all tests and quality assurance activities are carried out and documented. A simple checklist (or similar) can be effective at ensuring that nothing is missed, so this is perhaps the phase that is least likely to introduce issues, if a reasonable level of diligence is employed.

The one area that may require special consideration is noting and responding to any errors, test failures and unexpected occurrences. These should all be documented for review in the evaluation phase (in addition to taking whatever responsive actions are required). Responding to these unexpected occurrences is the area of the framework that is least constrained and most dependent on the skills and capabilities of the human developers. An AI tool could, prospectively, be implemented to aid in this response process and provide consistency in terms of the response activities. This is a potential topic for evaluation in future work.

6.4 Evaluation and Correction

The evaluation and correction phase presents the last opportunity to detect and correct issues before the application (or application iteration) is completed. By this point, the developers have a trove of tests and quality assurance activities to perform, as any test or activity created for use in prompt verification (that has not been obsoleted by later prompts' results) can be performed again during this phase to validate the specific performance of each area of the application. This is the more controlled part of this phase.

The less controlled portion of the phase and the area that is more dependent on the human developers and, potentially, human testing team is the holistic evaluation of the software. This holistic evaluation should attempt to identify any issue present in the application (functional or security). This process is similar to the process that would be used for a human-coded application and, while it can be potentially aided by AI and conventional tools, it is highly dependent on the skills and abilities of the testing team, creating another source for potential errors and omissions in the process.

This phase can (and for crucial applications, should) include a human code review as well. This process can be as detailed as allowed for by the schedule and budget of the development process. Again, AI and conventional tools can be used to aid in the review, by indicating areas of particular focus for human reviewers; however, this process is similarly highly dependent on the skills and abilities of the testing team, and thus, similarly, creates another source for potential errors and omissions in the process.

6.5 Holistic Evaluation

Despite incorporating numerous checks and balances, the proposed framework is not able to deliver the precision or perfection that would be desired from a completely automated process. While human involvement introduces imperfections, omissions and other issues, this is not the only source of potential defects. AI models' performance can be erratic, particularly as numerous prompts are generated and run. Small errors and omissions in early phases can persist and grow in impact in later phases and design phase errors and omissions can readily produce issues that are difficult to detect and resolve in later phases.

While the proposed framework cannot guarantee code completeness for features or absolute security, neither can human development processes. The proposed framework does bring together human and AI capabilities in an attempt to assure functionality and security to the greatest extent possible and having both human and AI 'eyes' on the process may catch issues that would be missed by one source of review on its own. A key topic for future work is, thus, to quantify the benefits and/or detriments produced by AI code generation, as compared to human-alone performance of this process and the benefits and/or detriments of AI code review, as compared to human-alone performance of this process. Presumably, the combination of the two – particularly in the review aspect – will outperform either alone (as the benefits of both are combined); however, the overarching performance of AI generation and various review approaches needs to be quantified and will drive approach selection for future vibe coding projects.

7 Conclusions and Future Work

Vibe coding is poised to change the way that software is developed. It facilitates greater abstraction and allows focus to be placed on user-desired results instead of implementation. While it offers great future potential, it is a new and not fully matured technology and paradigm. Significant study is needed to advance it to allow it to deliver its full potential.

To use vibe coding effectively, developers must understand both its strengths and weaknesses. Currently, successful vibe coding benefits greatly from prior software design and development knowledge and skills, though the specific benefit and general level of benefit that these provide may change somewhat in the future.

This paper has attempted to advance the use of vibe coding through the discussion of the benefits to be gained, and the challenges posed by vibe coding. It has reviewed the existing literature related to vibe coding and presented a four-phase model for vibe coding, explained how this model can be used to enhance functionality and security. Then, this model was qualitatively assessed.

The model presented herein may be simple; however, it brings together key elements required for successful vibe coding and provides a reminder of the things that should be continuously considered during the vibe coding development process. Future work will focus on understanding the current uses and limitations of vibe coding across industries as well as the technical limitations that may preclude its use in some areas. Understanding where vibe coding fails to deliver the targeted results allows technical focus to be placed on overcoming these limitations and facilitating even greater use of vibe coding.

Acknowledgements. The author currently also holds an appointment at North Dakota State University and an ownership interest in an AI educational technology company.

References

1. Yüce, A.: Vibe Coding: Analysis of a Post-Modernist Programming Paradigm - Yapay Zeka, Bilişim, Eğitim, Yazılım (2025)
2. Samsyudin, I.: Vibe Coding and AI-Led Conversational Programming: Emerging Trends in Software Development. SSRN (2025). https://doi.org/10.2139/SSRN.5469367
3. Chen, N., Qiu, L.K., Wang, A.Z., Wang, Z., Yang, Y.: Screen Reader Users in the Vibe Coding Era: Adaptation, Empowerment, and New Accessibility Landscape. arXiv:2506.13270, 1 (2025). XXXXXXX.XXXXXXX
4. Gadde, A.: Democratizing Software Engineering through Generative AI and Vibe Coding: The Evolution of No-Code Development. J. Comput. Sci. Technol. Stud. **7**, 556–572 (2025). https://doi.org/10.32996/JCSTS.2025.7.4.66
5. Hemdev, N.: VIBECODING: A Mixed-MethodsCase Study on Conversational AIProgramming and ApplicationDevelopment. IJSAT - Int. J. Sci. Technol. 16 (2025). https://doi.org/10.71097/IJSAT.V16.I3.7765
6. Myshko, A.: Vibe-coding as Geodesic Transition: Ontology of Digital Culture in the Paradigm of Metamonism. PhilPapers (2025)
7. Escobedo, S.D.: Single Conversation Methodology: A Human-Centered Protocol for AI-Assisted Software Development. arXiv:2507.12665 (2025)
8. Gómez, O.S.: The Twilight of Software Engineering in the Age of Vibe Coding. Res. Gate
9. Rose, I., et al.: The Rise of "Vibe Science": How AI Threatens Scientific Rigor Available online: https://doi.org/10.36227/techrxiv.175606229.90717623. Accessed on 23 Oct 2025
10. Ray, P.P.: A Review on Vibe Coding: Fundamentals, State-of-the-art, Challenges and Future Directions Available online: https://doi.org/10.36227/techrxiv.174681482.27435614. Accessed on 23 Oct 2025
11. Faragó, D.: From Vibe to Vise Coding: Addressing the AI-Generated Code Quality Crisis
12. De Silva, D., Mills, N., Issadeen, Z., Moraliyage, H., Jennings, A., Manic, M.: Generative AI Vibe Coding for Prototyping Industrial Systems. IEEE Int. Symp. Ind. Electron. (2025). https://doi.org/10.1109/ISIE62713.2025.11124737
13. Maes, S.H.: The Gotchas of AI Coding and Vibe Coding. It's All About Support And Maintenance. Res. Gate (2025)
14. Sapkota, R., Roumeliotis, K.I., Karkee, M.: Vibe Coding vs. Agentic Coding: Fundamentals and Practical Implications of Agentic AI. arXiv:2505.19443 (2025)
15. Meske, C., Hermanns, T., von der Weiden, E., Loser, K.-U., Berger, T.: Vibe Coding as a Reconfiguration of Intent Mediation in Software Development: Definition, Implications, and Research Agenda. arXiv:2507.21928 (2025)
16. Horvat, M.: What is Vibe coding and when should you use it (or not)? TechRxiv 175459780.03758839/v1 (2025)
17. Fawzy, A., Tahir, A., Blincoe, K.: Vibe Coding in Practice: Motivations, Challenges, and a Future Outlook -- a Grey Literature Review. 1 (2025). https://doi.org/10.1145/nnnnnnn.nnnnnnn
18. Geng, F., et al.: Exploring Student-AI Interactions in Vibe Coding. arXiv:2507.22614, 1 (2025). https://doi.org/10.1145/nnnnnnn.nnnnnnn
19. Sarkar, A., Drosos, I.: Vibe coding: programming through conversation with artificial intelligence. arXiv:2506.23253 (2025)

20. Ge, Y., et al.: A Survey of Vibe Coding with Large Language Models. arXiv:2510.12399, 4 (2025). https://doi.org/10.34133/icomputing.0124
21. Pimenova, V., Fakhoury, S., Bird, C., Storey, M.-A., Endres, M.: Good Vibrations? A Qualitative Study of Co-Creation, Communication, Flow, and Trust in Vibe Coding. arXiv:2509.12491, 1 (2025)
22. Chow, M., Ng, O.: From technology adopters to creators: Leveraging AI-assisted vibe coding to transform clinical teaching and learning. Med. Teach. (2025). https://doi.org/10.1080/0142159X.2025.2488353
23. Woo, D.J., Guo, K., Yu, Y.: A vibe coding learning design to enhance EFL students' talking to, through, and about AI (2025). arXiv:2509.08854
24. Chilambarasan, N.R., Kumar, K.N.: Vibe Coding in Vernacular Contexts: A Comprehensive Study on Tamil and Global Implications for Multilingual Programming Education. Res. Sq. (2025). https://doi.org/10.21203/RS.3.RS-7593886/V1
25. Heinonen, J.: Transforming Web Development with NLP and Vibe Coding Techniques (2025)
26. Horvat, M., Kralj, B., Gledec, G.: A Comparative Study of Vibe Coding with ChatGPT and Gemini in Front-end Web Development. Proc. Cent. Eur. Conf. Inf. Intell. Syst. (2025)
27. Fortes-Ferreira, M., Alam, M.S., Bazilinskyy, P.: Vibe Coding in Practice: Building a Driving Simulator Without Expert Programming Skills. Adjun. Proc. 17th Int. Conf. Automot. User Interfaces Interact. Veh. Appl. 60–66 (2025). https://doi.org/10.1145/3744335.3758482
28. Li, T., University, P., Tanay Maheshwari, U.: User-Centered Design with AI in the Loop: A Case Study of Rapid User Interface Prototyping with "Vibe Coding." Proc. ACM Collect. Intell. (CI '25), 1 (2025)
29. Meyer, J.G.: Rapid Development of Omics Data Analysis Applications through Vibe Coding. arXiv:2510.09804 (2025)
30. Moore, J.H., Tatonetti, N.: Vibe coding: a new paradigm for biomedical software development. BioData Min. **18**, 46 (2025). https://doi.org/10.1186/S13040-025-00462-9
31. Cohen, N., Tzezana, R.: Can Physicians Code? The Use of Vibe Coding AI in IVF Clinical Practice. Fertil. Steril. (2025)
32. Hayashi, S., Morita, K., Mukunoki, D., Hoshino, T., Katagiri, T.: VibeCodeHPC: An Agent-Based Iterative Prompting Auto-Tuner for HPC Code Generation Using LLMs. arXiv:2510.00031 (2025)
33. Chen, H., Li, C., Li, J.: FeatBench: Evaluating Coding Agents on Feature Implementation for Vibe Coding. arXiv:2509.22237 (2025)
34. Zhong, M., et al. : Vibe Checker: Aligning Code Evaluation with Human Preference. arXiv:2510.07315 (2025)
35. Maes, S.H.: Ensuring the Maintainability and Supportability of "Vibe-Coded" Software Systems: A Framework for Bridging Intuition and Engineering Rigor. Res. Gate (2025)
36. Mitchell, J., Shaaban, Y.: Position: Vibe Coding Needs Vibe Reasoning: Improving Vibe Coding with Formal Verification. Proc. 1st ACM SIGPLAN Int. Work. Lang. Model. Program. Lang., 84–90 (2025). https://doi.org/10.1145/3759425.3763390
37. Schrage, M.: Vibe Analytics: Vibe Coding's New Cousin Unlocks Insights Available online: https://sloanreview.mit.edu/article/vibe-analytics-vibe-codings-new-cousin-unlocks-insights/. Accessed on 23 Oct 2025
38. Abgrall, G., Chelly Dagdia, Z., Monnet, X., Zeitouni, K., Arora, A.: From vibe coding to vibe caring: what clinicians can learn from programmers. Lancet **406**, 1561–1562 (2025). https://doi.org/10.1016/S0140-6736(25)01807-0
39. Crowson, M.G., Mbi, M., Celi, L., Celi, A., Mpa, M.D.: Academic Vibe Coding: Opportunities for Accelerating Research in an Era of Resource Constraint. arXiv:2508.00952 (2025)
40. Ghosh, D.P.: Vibe Engineering Automation (VEA) and Orchestration (VEO): An AI-Driven Framework for Design Integration in EPC Projects (2025)

41. Ghosh, D.P.: Vibe Design: Democratizing AI in EPC Engineering Through Natural Language-Driven Collaboration. Res. Gate (2025)
42. Straub, J.: AI, Vibe Coding and Cybersecurity: How AI Coding Can Actually Improve Cybersecurity. NSF Cybersecurity Summit (2025)
43. Matharu, G.S., Mishra, A., Singh, H., Upadhyay, P.: Empirical study of agile software development methodologies. ACM SIGSOFT Softw. Eng. Notes **40**, 1–6 (2015). https://doi.org/10.1145/2693208.2693233

Predicting Maritime Piracy: Timing, Trends, and Risk Assessment

John Michael Mizerany(✉) and Jing Chen

Athens State University, Athens, AL 35611, USA
jmizeran@my.athens.edu, jing.chen@athens.edu

Abstract. Where there is demand for goods and services, there is always some form of trade involved before civilians are able to buy those goods or services. One of the major forms of transportation to transport goods and services is by water; however, with this transportation of goods and services, there are also acts to try and steal these goods or disrupt transportation. This is known as maritime piracy, which poses threats to the economy and can slow down trading. In this research paper, we will dive into maritime piracy and how maritime piracy has taken an economic toll on regions where attacks are most prominent. We aim to highlight how piracy affects the economy with different machine learning techniques, which include regression and ARIMA models.

Keywords: Piracy · Southeast Asia ship hijackings · Contemporary Maritime Piracy

1 Introduction

Maritime piracy is the act of when a ship in international waters is boarded and hijacked. In addition, crew members on said ships are taken captive and, in some cases, are murdered because of the hijacking. In addition, maritime piracy also targets goods instead of people stealing resources. Some of these resources include food, oil, and even currency. The dataset we used in this research is from The National Geospatial Intelligence Agency (NGA). The dataset has 8,828 unique records, with yearly dates spanning from 1978 to 2023. This dataset includes information such as location, date, and details of hostility. This paper uses data analysis and data mining techniques with publicly available piracy data to answer the following questions:

- What were the most common types of attacks per geographical region?
- Based on the provided data, can we predict future attacks based on geographical locations?
- If we can predict future attacks, can we also predict victims by geographical area?
- Can we predict the number of attacks for future years?

H. R. Arabnia et al. (Eds.): CSCI 2025, CCIS 2969, pp. 335–346, 2027.
https://doi.org/10.1007/978-3-032-28460-0_23

The paper is structured as follow: Sect. 2 describes the literatures used as references, Sect. 3 is a deep dive into the research methodologies used, Sect. 4 are the research results, Sect. 5 is the results from Sect. 4, and Sect. 6 is the conclusion.

2 Literature Review

In this section, we will give an overview of the selected literature associated with the data set, explanations as to why the topics in the literature are relevant, and a restatement of our hypothesis.

The main piece of literature that we will reference throughout this research paper is "Contemporary Maritime Piracy in Southeast Asia" by Peter Chalk. The overview goal of the chosen literature is to show the incidents, types of attacks, geographical location, and facilitatory factors that are associated with maritime piracy in Southeast Asia [1]. He also, in his paper, aims to give remediation against these attacks, and how we can prevent piracy from growing.

The other piece of literature that we will also reference is "Piracy studies coming of age: a window on the making of maritime intervention actors" by Katja Lindskov Jacobsen and Jessica Larsen. The reason for selecting this literature is to bring in a different perspective and focus more on the legal issues associated with piracy studies, as well as international politics and global security governance. Although this is a more quantitative research paper, it is good to bring in non-quantitative data to back up and show the reasoning behind the results. This literature also introduces efforts to battle maritime piracy.

We also used different articles outside the two listed above that we found helpful to reference in this research paper. More details about these articles can be found in the Resources section.

3 Methodology

In this research paper, we used the following algorithms to answer the research questions in Sect. 1.

1. **Random Forest Regression**, which we can use for classification and uses averages to improve predictive accuracy and control over-fitting [3]. The research questions that would benefit the most from this methodology would be how we can predict future attacks, and if we can then predict victims. The benefit of using this methodology is that it is highly accurate, can handle missing data points and can handle noise from our data source that may lead to inaccurate results. On the other hand, some of the cons that come with using this methodology also deal with the potential of overfitting, where the nature of Random Forest Regression is trained to fit the data closely, which can result in inaccurate predictions for new data.

2. **Supervised learning** is due to the nature of the dataset provided. More specifically, we will focus on regression algorithms. The benefits from this type of analysis are that we can predict continuous values and use it for a wide range of applications; however, one of the main cons is that it is sensitive to outlying data, which can lead to inaccurate results. Similarly, we can also use k-means clustering algorithms to help us solve the question of what the most common attack from the dataset is per geographical region.
3. **Prophet for forecasting data**. Prophet was developed by Facebook (Meta) and is an open-source library in Python. There are other models, such as Autoregressive Integrated Moving Average (ARIMA), that can also forecast future trends in time series data, which we also tried and compared with our dataset. The reason we chose Prophet over using an ARIMA model is the overall ease of use. We also found it to be faster with predictions compared to ARIMA, and it also includes nice functionality like holiday effects and seasonality detection. On the other hand, some of the cons associated with using Prophet include overfitting, which leads to inaccurate predictions, and it is also not suited for complex patterns.

4 Results

In this section, we will go over the results from our experiments relating to our research questions.

1. What were the most common types of attacks per geographical region? The location information of the dataset is latitude and longitude data. Figure 1 shows the map of where these attacks have happened. The majority of piracy is near Southeast Asia, Nigeria, and the Southern Red Sea Region.

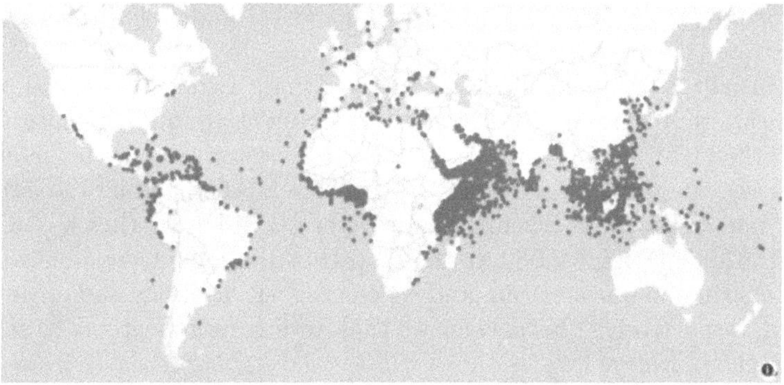

Fig. 1. Map of historical maritime piracy attacks

To determine the precise number of clusters, we utilized the Elbow Method, which is a graphical technique used to determine the optimal number of clusters. From Fig. 2, we need to use four clusters.

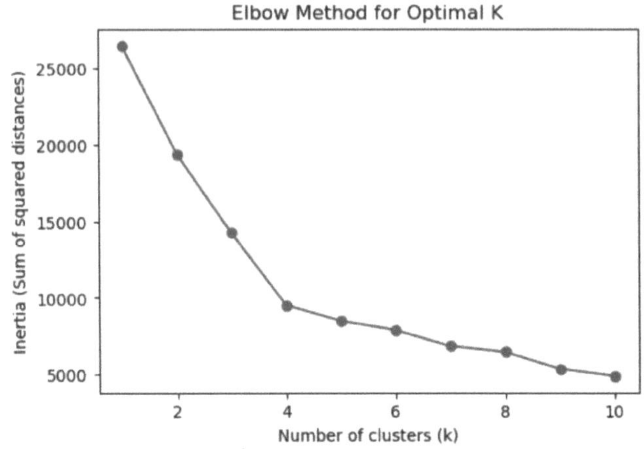

Fig. 2. Elbow Method for Optimal K

Given that "Pirate Assault" is the most common hostility type, we focused on the first data block in Fig. 2. The first cluster is located at (9.491938, 51.845983), the second at (6.410880, 105.492909), the third at (3.136968, −68.437589), and the fourth at (5.636800, 2.900518).

Shifting to "Naval Engagement," we narrowed it down to coordinates (6.410880, 105.492909) in the South China Sea. Both hostility types fall within the same regions shown in Fig. 1.

2. Based on the provided data, can we predict future attacks based on geographical locations? We employed a Random Forest Regressor to predict the latitude and longitude of future attacks and a Random Forest Classifier to predict attack types in our dataset. The subregion was included as a key feature for predicting latitude and longitude, with 100 decision trees set for the Regressor.

The initial results yielded a Root Mean Square Error (RMSE) of 6.45 and a Classification Accuracy of 78.88 next experiment by adding more features, including the Date. This required splitting the timestamp into year, month, day, and weekday categories to enhance our feature set. After adding the date to the training model, the Root Mean Square Error (RMSE) for this was 6.18 with a Classification Accuracy of 83.81 the map that the model was able to predict more points than in our first run, and we can see our accuracy and error went up and down, respectively. This also shows that adding more features to the model was a direct improvement.

Rather than just showing the map's geographical data, we also wanted to map those points to attack types. To accomplish this, we wrote the results from our predictions into a new spreadsheet with the following columns and their descriptions: latitude predictions, longitude predictions, and hostility type predictions.

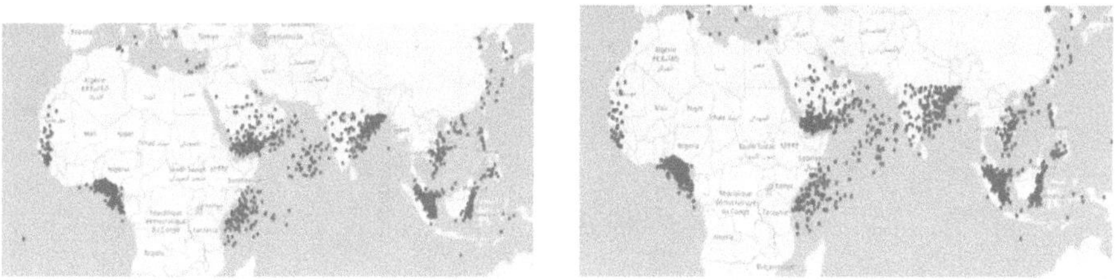

Fig. 3. Map of Future Predicted Attacks

From the new spreadsheet, the top three attack type categories were "Pirate Assault" at 1513, "Unknown" at 77, "Naval Engagement", and "Attempted Boarding", both at 65. Originally, we set the model to have 100 decision trees; we wanted to see if lowering this number could improve our results. By dropping the decision trees to 50, we saw that our Classification accuracy remained the same at 83.81 6.07. Running our experiment again, we dropped the decision trees count to 10, which resulted in the RMSE dropping to 5.84 and our Classification Accuracy rising to 84.14

Interestingly, there were no significant changes in the attack types, as "Pirate Assault", "Unknown", "Naval Engagement", and "Attempted Boarding" were all in the top three attacks. One may ask why dropping the decision tree count resulted in a slightly higher accurate model? Referring back to our Research and Methodologies section of this paper, we talked about the pros and cons of each methodology, which includes using a Random Forest Regressor. One of the cons of using the model is overfitting, in which the model learns too close to our dataset and results in inaccurate results. Although this is not necessarily the case for every single situation, we can see in our case that having our decision tree count be 100 was causing some overfitting in our model. Thus, dropping the count, we were able to achieve some slight improvement and reduce overfitting.

In conclusion, our model choice proved to be accurate in predicting attack types and geographical locations of the attacks.

3. If we can predict future attacks, can we also predict victims by geographical area? In the previous section, we were able to successfully determine that it is possible to predict future attacks based on geographical data, latitude, and longitude, from our dataset. Now, we are going to apply the same methodologies to predict victims in geographical regions. We will use a Random Forest Regressor in addition to a Random Forest Classifier to predict victims by geographical region; however, this time we will be using victims from our dataset to accomplish this experiment. We again chose subregion as a feature for our model, to run against latitude and longitude for our first test. Since we determined that we could optimize this model by adding dates to our features, we will fast forward and start with subregion, year, month, day, and weekday as our features for the model. The results of this first run are displayed in Fig. 4.

We decided to first start out at 10 decision trees for this model run as we initially thought this was more accurate from our previous experiments; however,

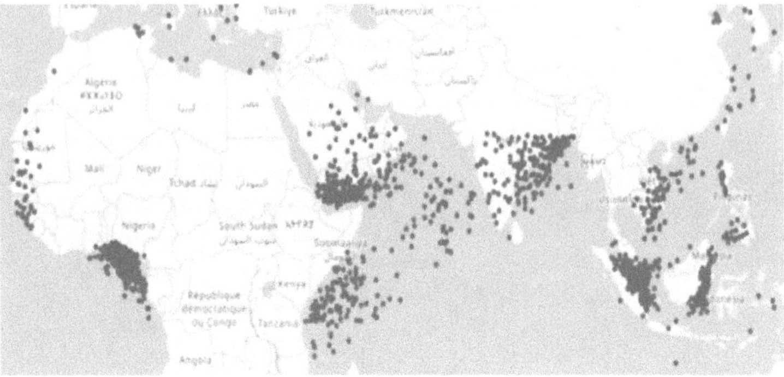

Fig. 4. Map of Predicted Victims: 10 Decision Trees

when we raised our decision trees to 100, we noticed a slight increase in accuracy. The results of these changes can be found in Fig. 5. The Root Mean Square Error (RMSE) for this was 6.18, and the model now has a Classification Accuracy of 55.15 more accurate model since the number of nodes increased, allowing for a more granular analysis of the data.

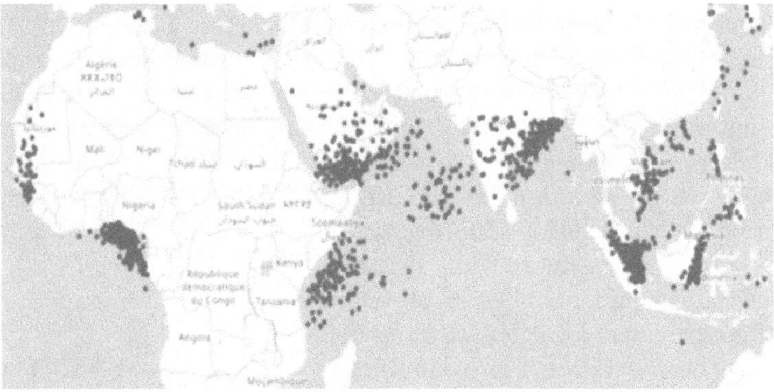

Fig. 5. Map of Predicted Victims: 100 Decision Trees

Just for clarity, we decided to run the experiment again this time running with 150 decision trees for our model. The results can be found in Fig. 6. The Root Mean Square Error (RMSE) was 6.21, with a Classification Accuracy of 55.21 increase, it was not significant. Raising the decision tree count to 200 resulted in an RMSE of 6.20 and a Classification Accuracy of 54.76 decrease in accuracy, highlighting that more is not always better.

In our research, we noted that the Random Forest Regressor can suffer from overfitting, which occurs when the model closely learns the training data, including noise and null values, leading to poor performance. This was evident in our

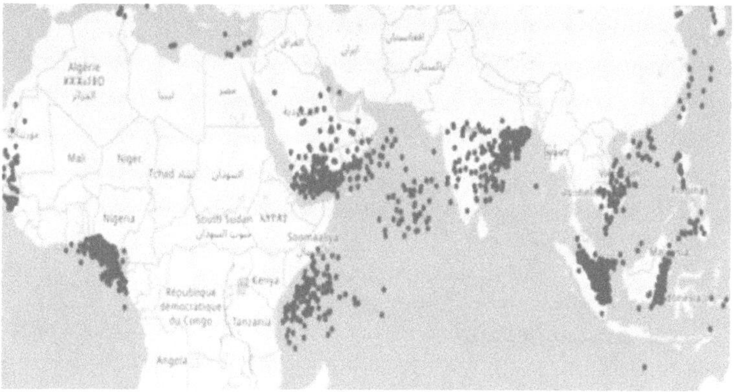

Fig. 6. Map of Predicted Victims: 150 Decision Trees

experiment as an increase in decision trees introduced overfitting and reduced model accuracy. We concluded that optimizing the model is best achieved with 100 to 150 decision trees.

We documented our model predictions in a spreadsheet, revealing that the top three predicted victims were "Merchant Vessel" (669), "Cargo Ship" (456), and "Tanker" (355). The prediction of "Merchant Vessel" aligns with our hypothesis about maritime piracy impacting economies in heavily affected regions. Despite the model's accuracy being just above 50 can still effectively predict victim-related attacks by geographical area.

4. Can we predict the number of attacks for future years? In previous sections, we focused on predicting future attacks and victims without directly analyzing the dates associated with these events. In this section, we aim to explore trends related to dates and attacks.

We will present a high-level overview of trends extending into early 2025, based on our dataset, which recorded data up to 2023. Using Facebook's Prophet for time series predictions, we isolated key columns: latitude, longitude, Date, and hostility type. We ran the analysis with default seasonality and holiday settings. The results of this initial run can be found in Fig. 7.

For clarity, our dataset's latest date is 2023–05-11, leading to predictions beginning in 2023. From 2023 to 2025, we observe an initial trend of "Attempted Boarding," which transitions to "Mothership Activity" by 2024. This trend shift aligns with our dataset, which reveals over seven thousand instances of "Pirate Assault".

To address concerns about seasonality and overfitting in our forecasting method, we specified quarterly seasonality. Figure 8 illustrates the results of this approach.

In 2023, the trend began with "Naval Engagement" and continued to trend towards "Pirate Assault" through 2025. While this seems aligned with our original dataset, it's important to note that these predictions depend on the "Date". To determine the best approach, we graphed both historical and predicted val-

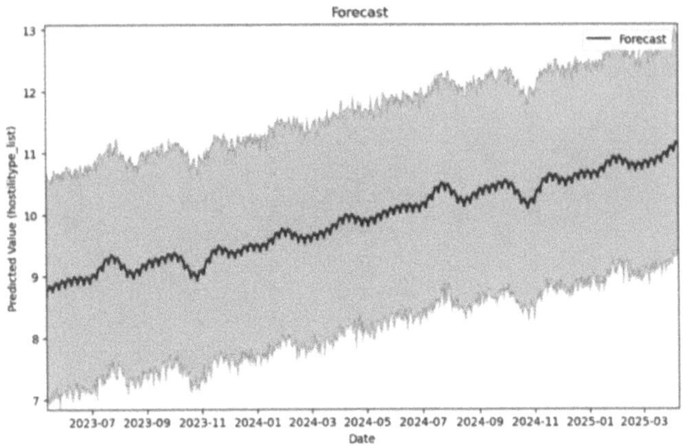

Fig. 7. Forecast of Attack Types (No Seasonality)

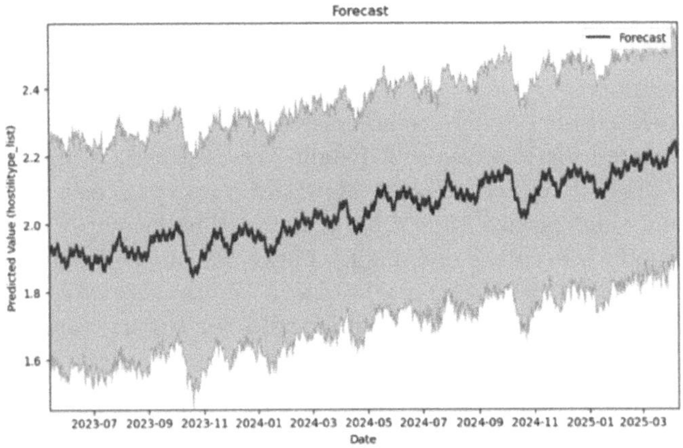

Fig. 8. Forecast of Attack Types (Seasonality)

ues, focusing solely on future predictions. For clarity, we did not account for seasonality and used the default settings in Prophet. The combined results are displayed in Fig. 9.

The graph illustrates that red represents historical data and blue denotes predicted data. From 1980 to 2010, attack types remained consistent, but in 2020, there was a noticeable shift towards "Attempted Boarding" and "Mothership Activity." We will later explore the reasons for this shift, which has led to higher predicted values. Out of 700 predicted events over the next 700 days, we estimate 238 for "Attempted Boarding," 304 for "Mothership Activity," and 158 for "Robbery." This raises questions about the causes of the shift in attack types. We will address these questions later in the paper.

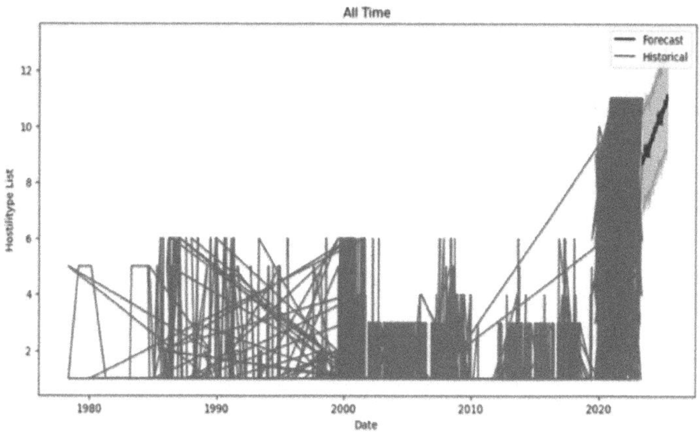

Fig. 9. Historical and Forecasted Attack Types

Using the same methods to predict future victims, we configured our procedure, Prophet, to project 700 days from our last dataset date, 2023–05-11. We maintained default seasonality and holidays, focusing on latitude, longitude, Date, and hostility type. The results are shown in Fig. 10.

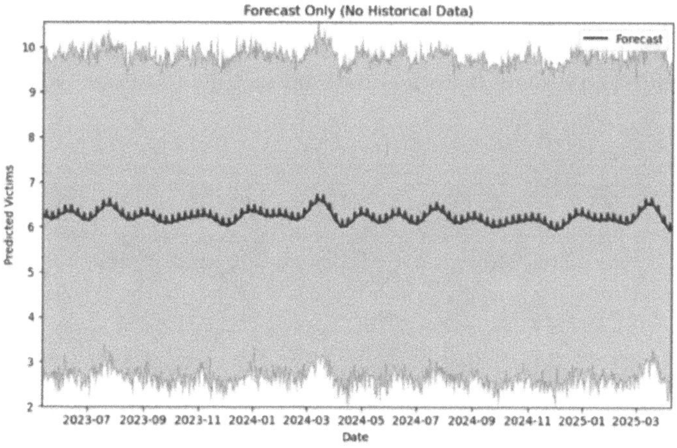

Fig. 10. Forecast of Victims (No Seasonality)

From 2023 to 2025, there is a consistent trend of victims in the categories "Offshore Vessel" and "Passenger Ship." While our initial results seem accurate, we recall that "Merchant Vessel" was the most frequent victim type, coded with a value of 5, in our dataset. To further investigate, we applied the Prophet procedure, accounting for quarterly seasonality, as noted in our Research Methodologies section. Figure 11 presents the results of this analysis.

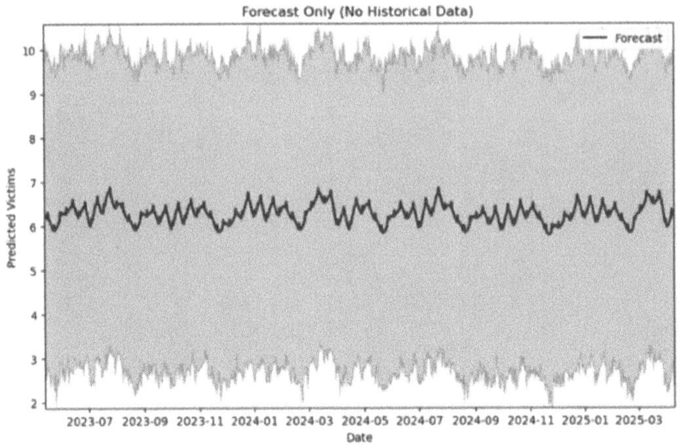

Fig. 11. Forecast of Victims (Seasonality)

We found that adding our own seasonality did not significantly change the prediction compared to the first run, but it did increase the variation between the predictions of 6 or 7 from the victim. Next, we aim to combine our historical dataset with the new predicted values in one graph, focusing only on events after 2020–01-01. Our recent analysis shows that historical data from 2020 onward aligns with predicted victims. We selected 2020 because it marked a shift in common attack types, which we believe also influenced the typical victims. Figure 12 presents historical data on victims from our dataset.

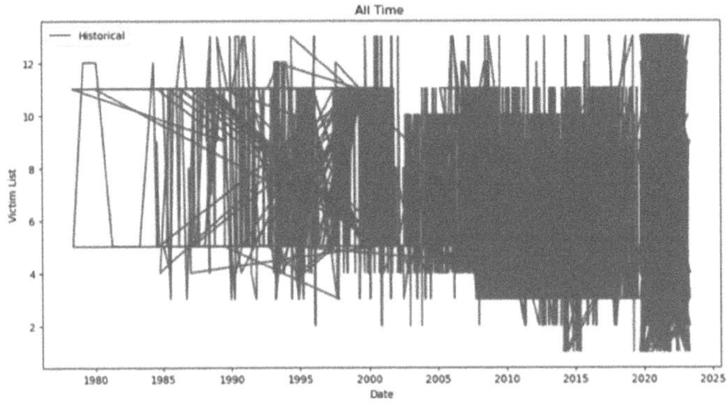

Fig. 12. Historical and Forecasted Victims

Plotting this, we can see our assumption held true as something did happen in 2020 where other victim types were more common.

5 Discussion

In the previous section, we analyzed our research on the economic impact of maritime piracy in regions with high incidence of attack. Our findings show that piracy is most prevalent in the Southeast Asia Sea, Red Sea, and areas near Africa. For instance, Chalk [1] notes that Southeast Asia accounted for significant global piracy incidents in the early 1990s. Attacks typically occur close to shore, allowing quick escapes for attackers.

Chalk [1] also highlights that economic changes in the late 1900s pushed some civilians into piracy for survival. Our research indicates that piracy negatively affects the economy, with "Merchant Vessels" being the primary victims, threatening essential trade routes. Frequent attacks on these vessels can lead to delivery failures and heightened fear among operators.

Additionally, "Tankers," which transport critical resources like petroleum, are also targeted. Since oil constitutes around 3 increases due to reduced supply amid rising demand [4]. According to Gong, Jiang, and Yang, maritime piracy results in losses of approximately 25 billion dollars annually from theft, ransoms, and insurance costs, further straining economies, especially since over 90 relies on maritime transport [5].

Overall, the evidence supports our hypothesis of a detrimental economic impact from maritime piracy.

6 Conclusion

In this research paper, we aimed to show how maritime piracy has a direct effect on the economy. We used a quantitative analysis approach to show predictions on future attacks and victims, common relations between said attacks and victims, and to show time series analysis on when attacks could happen. We based each of our methodologies around research-inspired questions, which we then used to tie back a quantitative number to our original hypothesis. Although some of these research methodologies introduced new questions from our findings, we found that they were useful in providing more in-depth discussions to help explain what happened. Interestingly, we found that there are two sides to the story of maritime piracy. Although there is proof that maritime piracy does affect the economy, we found there is also the side on how an increase in the cost of the economy drives up maritime piracy.

References

1. Chalk, P.: Contemporary maritime piracy in Southeast Asia. Stud. Conflict Terrorism **21**(1), 87–112 (1998). https://doi.org/10.1080/10576109808436055
2. Jacobsen, K.L., Larsen, J.: Piracy studies coming of age: a window on the making of maritime intervention actors. Int. Aff. **95**(5), 1037–1054 (2019). https://doi.org/10.1093/ia/iiz099
3. Scikit-learn: RandomForestRegressor. https://scikit-learn.org/stable/modules/generated/sklearn.ensemble.RandomForestRegressor.html

4. World Economic Forum: Why do oil prices matter to the global economy? An expert explains (2022). https://www.weforum.org/stories/2022/02/why-oil-prices-matter-to-global-economy-expert-explains/
5. Gong, X., Jiang, H., Yang, D.: Maritime piracy risk assessment and policy implications: a two-step approach. Mar. Policy **150**, 105547 (2023). https://doi.org/10.1016/j.marpol.2023.105547

Detection of Identity Theft in Critical Sectors: A Comprehensive Literature Review

Shikhar Pandey(✉), Ali Al-Sinayyid, Luice Khamboo, Samuel Adewumi, Vishnu Vardhan Reddy Sheri, and Rohith Reddy Battula

Department of Computer Science and Mathematics, West Virginia State University, Institute, WV, USA
{spandey,ali.alsinayyid,lkhamboo,sadewumi,vsheri, rbattula}@wvstateu.edu

Abstract. Identity theft is a major threat to both individuals and critical sectors such as financial services, government sector and so on where stolen personal data can cause serious financial and operational damage. This study reviews past research on identity theft detection, focusing on methods, performance, and application areas. Advanced techniques like Graph Neural Networks (GNN) are highlighted for their high effectiveness in detecting attacks such as phishing, hacking, and social engineering. The review also shows that detection performance varies with the type of attack, the targeted sector and potential damage, and that response time is crucial for timely mitigation. These findings emphasize the need for adaptive, context-aware detection strategies. By synthesizing existing studies, this review provides a foundation for developing robust systems that enhance the detection of identity theft and strengthen protection of sensitive data in critical sectors.

Keywords: Identity theft · cybercrime · critical sectors · financial fraud · phishing · Graph Neural Networks (GNN) · detection systems · data protection · cybersecurity · response time

1 Introduction

In recent years, Identity theft has become one of the most widespread and damaging cybercrimes [1], affecting millions of people around the world. This crime involves stealing personal information such as Social Security numbers, bank account details, and credit card information, and using it for fraudulent purposes. As digital technology continues to grow, cybercriminals are developing more advanced techniques, making it harder to protect personal and organizational data. Identity theft threatens not only individuals but also critical sectors such as banking, healthcare, and government services [2]. Criminals can exploit stolen data to disrupt operations, steal resources, or commit large-scale fraud. Reports show that financial losses from fraud have increased significantly in recent years, and demand for identity protection services is also rising sharply. Most cases of identity theft are committed for financial gain. Criminals may

H. R. Arabnia et al. (Eds.): CSCI 2025, CCIS 2969, pp. 347–360, 2027.
https://doi.org/10.1007/978-3-032-28460-0_24

open fake credit accounts, drain bank balances, file false tax returns, or misuse medical insurance information. In some cases, stolen identities are used to gain employment or hide criminal activities. Victims often face not just financial loss but also emotional and psychological stress. The process of recovering your identity and repairing credit can be long and frustrating. Businesses also suffer consequences such as reputational damage, legal action, and loss of public trust when data breaches occur [3]. This research paper aims to examine the scope and impact of identity theft, explore its various forms, and emphasize the urgent need for effective detection techniques and preventive measures to protect both individuals and national infrastructure.

Identity theft brings serious consequences for both individuals and organizations. One of the most common problems is financial loss. Thieves may use stolen identities to make purchases, open unauthorized bank accounts, or take out loans, causing longterm financial strain for the victim. Victims often spend months or even years trying to fix the damage. In addition to financial issues, many victims suffer from emotional stress. They may feel anxious, overwhelmed, or unsafe. Damage to credit scores can make it harder to rent a home, apply for jobs, or qualify for a loan. Some criminals also target medical records, using them to receive treatment or medications under another person's name [4]. This can lead to false entries in a victim's health records, which may cause serious problems later. In other cases, stolen identities are used in criminal activities, putting innocent people at risk of legal trouble. With the rise of digital platforms, hackers use various techniques like phishing emails, fake websites, and data breaches to access large amounts of personal data. Many people are unaware of how to protect their information online, and some companies and government systems still lack strong security practices [5]. As the use of the internet, artificial intelligence, and smart devices continues to grow, it becomes easier for criminals to collect and misuse personal data. That's why it's important for individuals, companies, and governments to work together, promote cybersecurity awareness, and implement strong protection strategies.

Identity theft does not only affect individuals. It can also cause serious problems for critical infrastructure, which includes systems like power grids, water supplies, hospitals, airports, and communication networks [6]. If someone gains access to these systems using a stolen identity, the consequences can be extremely dangerous. For example, hackers could disable power lines, interfere with hospital equipment, or manipulate transportation systems. Since many of these systems are now connected to the internet, they are more exposed to cyberattacks. It is vital to protect the personal data of people who work in these sectors. Organizations should use secure passwords, limit access, verify users, and ensure that only trusted individuals can interact with sensitive systems [7]. Protecting identities in these areas is essential for maintaining public safety and national stability.

As identity theft continues to grow in scale and sophistication, traditional prevention methods alone are no longer sufficient. To effectively combat this evolving threat [8] the use of intelligent detection algorithms has become essential. These algorithms are designed to analyze user behavior, transaction patterns, access logs, and other digital indicators to identify suspicious or unusual activities that may signal identity theft [9]. Machine learning and artificial intelligence techniques are increasingly being applied to improve detection performance. For example, models can learn from historical data to recognize fraud patterns, detect anomalies in real time, and automatically flag highrisk activities for further investigation. However, detection algorithms must strike a balance between sensitivity and precision. High false positive rates can cause frustration and unnecessary alerts, whereas false negatives allow fraud to go undetected. Furthermore, cybercriminals constantly adapt their tactics, which means algorithms must be regularly re-trained and updated to remain effective [10]. This study aims to critically evaluate the effectiveness of various identity theft detection algorithms, highlight their challenges, and explore potential improvements. By focusing on the algorithmic side of identity theft prevention, this research contributes to building smarter, faster, and more reliable solutions to secure personal and institutional data.

2 Literature Review

The paper Preventing Identity Theft: Perspectives on Technological Solutions from Industry Insiders by Piquero et al. [11] reports an estimated 26 million American citizens per year have been victims of an identity-based crime. It emphasizes the need for a shift toward recovery. It highlights financial sector were more effected. Technologies such as biometric scanning, dark web monitoring, and AI-based tools show promise, highlighting the essential balance between security and ease of transactions. However, the study reveals gaps in prior research, particularly the limited understanding of how protective technologies are applied in practice.

The study by Zou et al. [12] highlight Users frequently struggle to adhere to expertrecommended security and privacy practices, indicating a need for better recommendations and alignment with user needs. It shows 11.3% had been victims of identity theft among all cyberattack Also study show security practices were adopted more widely than privacy or identity theft practices. It may be limited by social desirability bias and a focus on only 30 predefined practices, while its findings are primarily relevant to online safety, cybersecurity, and identity theft protection for general consumers only.

The paper proposed by Wang et al. [13] focuses on addressing the difficulty of detecting online identity theft in scenarios where available user data is scarce or of low quality, specifically within Online Social Networks (OSNs).The authors found an"insightful result" demonstrating a complementary effect among these different dimensions of records (online content and offline location) for modeling a user's behavioral patterns and the comprehensive performance (efficacy, response latency, and robustness) of the model is validated by extensive evaluations. The relative anomalous score (Sr) approach outperformed the logarithmic anomalous score (Sl). The method only concentrates on the dependency between a user's check-in location and tip-posting content, disregarding the impact of specific occurring time due to low resolution/missing time intervals in real-life OSN datasets.

The paper Modeling Access Environment and Behavior Sequence for Financial Identity Theft Detection in E-Commerce Services paper by Ye et al. [14] proposed hybrid approach named EnvIT which is introduced for detecting financial identity theft in Online-to-Offline (O2O) e-commerce platforms. This method integrates heterogeneous graph-based relationships (capturing access environment factors) with historical behavioral sequence analysis to enhance detection accuracy. The paper round around whether a verified account in an O2O e-commerce service is being used by a fraudster by modeling the access environment and behavior sequence, framed as a binary classification problem or not. From the approach EnvIT outperformed all baselines (LSTM, GRU, GAT, HAN, etc.) got overall good performance The models heavily relies on its components parts and the future research were not explicitly detailed beyond the overall effectiveness demonstrated.

Study done by Wyre et al. [15] examines the identity theft response system from the victim's perspective by applying the sociotechnical systems methodology to analyze the core social actors, tasks, and information flows involved in recovery. Paper is around to establish empirical details regarding its social, task, and information requirements to address the needs of victims. It employs the Event Analysis of Systemic Teamwork (EAST) methodology, which examines three interconnected networks that is Social (actors), Task (activities), and Information (data communicated) and this applies content analysis alongside social network analysis metrics (e.g., sociometric status). The study found no coordinated response system, leaving victims to link disconnected actors. That must detect fraud, investigate, enhance security, and report incidents often proving identity with the same compromised credentials. A key limitation is that two-thirds of clients declined participation, reducing representativeness.

Another gap lies in the low emphasis on proactive measures (e.g., credit bans), reflecting limited public awareness of effective protections. Additionally, the task network is only a simplified representation.

A comparative study of identity theft protection frameworks enhanced by machine learning algorithms by Sriram et al. [16] comparative analysis of identity theft protection frameworks, focusing on their enhancement using machine learning (ML) algorithms (supervised, unsupervised, and hybrid methods) to improve detection and minimize fraud. It shows traditional security methods often fail, making ML algorithms necessary to detect and mitigate evolving threats. Paper proposed three identity protection frameworks (A, B, C) and multiple Machine Learning approaches (Supervised, Unsupervised, Reinforcement Learning) and comparative metric (Cj) was used, integrating Accuracy (Aj), Real-time Responsiveness (Rj), and False Positive Rate (FPRj) the findings show that ML enables advanced authentication, but long-term resilience demands integrated, proactive strategies. The key challenge lies in designing generalizable models that balance recognition accuracy with performance trade-offs.

Bilge et al. [17] research highlights social networking sites are rapidly gaining popularity and house millions of registered users who share personal information. This growth attracts criminals, making security solutions critical. It developed a prototype attack system called iCloner, which includes a Crawler, Profile Creator, Message Sender, and a CAPTCHA Analyzer. To identify identical users across sites, the system used Google searches to compare the top three hits for matching names/institutions the experiments demonstrated that both Profile Cloning and Cross-Site Profile Cloning attacks are effective and feasible in practice. The automated system was successful at breaking CAPTCHAs used by sites like StudiVZ, MeinVZ, and Facebook. But the attacks rely on the high degree of trust users place in social networks which only offers suggestions on how social networking sites can improve security.

The paper examines the growing threat of cyberattacks on critical infrastructure across pivotal sectors more focusing on vulnerability assessment and the value of compromised data it also highlights importance of maintaining trust in financial institutions. The author shaji et al. [18] show Cyberattacks are rapidly increasing in scale and sophistication (quadrupling from 2010–2021) with projected losses reaching $10.5 trillion annually by 2025. Critical infrastructure sectors (representing over 50% of U.S. GDP) are highly vulnerable paper analysis comparative Framework based on five criteria: 1) Financial Loss Exposure, 2) Sensitivity of Compromised Data, 3) Reliability and Safety Impacts, 4) Cyber Readiness, and 5) Regulatory Requirements. The framework highlights acute vulnerabilities in the finance sector, where compromised customer data drives identity and new account fraud, with mitigation through measures like NIST standards, MFA, segmentation, and encryption. The gaps such as legacy systems, talent shortages, and the inability of a single framework to fully address cross-sector cyber risks remain unresolved.

The Research done by Alarfaj et al. [19] utilizes Graph Neural Networks (GNNs) and Autoencoders to enhance business practices and reduce fraudulent activities in large organizations, demonstrating that these deep learning methods effectively detect fraud with a balance of precision and recall and improve the efficiency of banking systems and it proposes a methodology that combines architectural frameworks and advanced neural network models to handle both massive historical data and continuous transaction streams more focused in critical sector specially financial.The GNN model performed optimally with two layers. The model achieved high classification under that model, However it fell short on addressing the critical need for real-world accuracy demanded by financial institutions, as both the GNN and Autoencoder models demonstrated insufficient accuracy, explicitly noted as less in terms of accuracy which cannot be used in the real world.

Enhanced credit card fraud detection based on attention mechanism and LSTM deep model research done by Benchaji et al. [20] focuses on developing a novel and robust system for credit card fraud detection (CCFD) that effectively addresses the major challenges in the payment industry, particularly the sequential and imbalanced nature of transactional data it uses a hybrid deep learning and preprocessing methodology to develop a novel credit card fraud detection system this paper proposed LSTM-Attention deep model is a highly effective and superior alternative to existing classification methods for credit card fraud detection (CCFD), specifically due to its ability to model sequential data and selectively focus on critical transactions. Based on paper result its heavy reliance on a multi-stage preprocessing pipeline (feature selection, SMOTE, UMAP) for feature robustness, which complicates real-time operational deployment, and the implied architectural sub-optimality of the LSTM foundation, which the authors acknowledge by planning future work solely based on non-recurrent attention/transformer architecture.

The study by Saikrishna et al. [21] paper highlights the limitation of current research, which often focuses on only a single facet of user behavior (e.g., keystroke, click stream, UGC), making detection susceptible to insufficient data quality. It emphasizes a necessary paradigm change in favor of composite behavioral modeling to effectively detect oblique patterns suggestive of identity theft. The study employs a Composite Behavior Model (CBM), a Bayesian network–based generative model that integrates online

(UGC) and offline (check-in) activities to detect identity theft in OSNs. It delivers strong results with high AUC scores (0.956 on Foursquare, 0.947 on Yelp) and recall (up to 72.2%), while keeping the false positive rate under 1%. However, the evaluation is limited to established OSN datasets (Yelp, Foursquare) it lacking on validation proprietary or dynamic real-world data. Moreover, the model has not been thoroughly assessed for computational complexity or real-time latency, raising concerns about its feasibility in production-grade response systems.

The Paper proposed by Abdelhalim et al. [22] examine the threat of "black hat Google hacking" that is mining publicly available online data for committing application fraud (e.g., fraudulent applications for credit cards or passports). They emphasize that identity theft is rapidly growing, fueled by the Internet's vast repository of sensitive PII, while prior research has largely focused on transactional fraud, leaving application fraud underexplored. The authors adopt a two-pronged approach: an exploratory study using "white hat" Google hacking and the design of an Application Fraud Detector tool. Findings highlight the alarming accessibility and volume of personal data online, confirming the threat's severity; however, the proposed solution remains a blueprint rather than a fully developed system.

A supervised machine learning algorithm for detecting and predicting fraud in credit card transactions done by Afriyie et al. [23], focuses on studying and comparing the performance of three supervised machine learning models used to classify, predict, and detect fraudulent credit card transactions. The analysis utilized a dataset of simulated credit card transactions from the western side of the United States of America spanning January 1, 2020, to December 31, 2020. Based on the finding on methodology, Random Forest was determined to be the most appropriate machine learning algorithm for predicting and detecting fraud in credit card transactions however, caking in the current findings are a comparison against a broader range of machine learning algorithms and the use of national or inter-regional real-world data.

The Study done by Mitchell et al. [24] Develops an identity theft prediction model using supervised machine learning and historical U.S. identity theft victim data, employing a design science research approach to address the gap in proactive prediction. Paper main concern is, Can supervised machine learning be used to successfully create an identity theft prediction model to detect and prevent future identity theft by using historical victim data or not for that they employ a Design Science Research (DSR) methodology supported by inferential statistics. The key artifact is a newly constructed dataset used to train a supervised ML model. To analyze relationships, multiple linear regression was applied between the total number of victims and the three most common fraud types, with random sampling conducted on the derived dataset of 1,300 records. The study found a statistically significant relationship, with the training dataset showing strong potential for identifying identity theft victims. However, it did not fully capture the complexity of variable interactions and lacked gender-based analysis, pointing to the need for validation on test data and practical application development.

The research done by Sholademi et al. [25] analyzes the risks posed by AI-generated deep fakes (hyper-realistic fake video, image, and audio content) in facilitating financial crimes and identity theft, while also evaluating AI-driven countermeasures. It highlights the problem of AI's rapid evolution, which has introduced deep fake technology as a potent tool for fraud, leading to significant financial losses, reputational damage, and erosion of trust in digital financial systems.AI-driven systems, such as anomaly detection algorithms and behavioral analytics, are shown to enhance real-time detection and mitigation of deep fakes. The paper also notes limitations of existing technologies, ethical dilemmas arising from the dual-use nature of AI, and the necessity for robust regulatory frameworks.

The study by Chakraborty et al. [26] compares traditional classifiers (SVM, Neural Networks, Logistic Regression, Decision Tree, Bayesian Network, CBR) with a proposed Multilayer Mining Algorithm, comprising Territory Detection and Suspicion Score Detection to detect synthetic identity fraud in credit applications. The approach emphasizes resilience, adaptivity, and quality data, aiming to complement existing detection systems although the algorithm is specific to credit application fraud, and traditional classifiers are unsuitable for real-time detection due to reliance on known fraud labels.

The Paper by Al-Smadi et al. [27] proposes a secure online payment system for credit cards, combining a one-time credit card number generation method with a Machine Learning based fraud detection algorithm to combat database breaches and identity theft. The system was tested on a real European credit card dataset and six artificially generated datasets, integrating ML algorithms (Decision Tree, SVM, Random Forest, Logistic Regression, Na¨ıve Bayes) for robust detection. Logistic Regression achieved the best performance, while Random Forest had the worst time efficiency. Previous systems often addressed only one security factor. SVM and Na¨ıve Bayes showed poor performance and efficiency, and future work should tackle emerging cyber threats like Man-in-the-Middle attacks and database breaches.

The evolution of fraud: ethical implications in the age of large-scale data breaches and widespread artificial intelligence solutions deployment [28] paper examines the evolution of fraud in the context of large-scale data breaches and advanced deep learning techniques, predicting the emergence of "Identity Theft 2.0." Through conceptual analysis of data brokers, the mosaic effect, and AI, it highlights risks from combining disparate datasets for targeted fraud and emphasizes the need for global collaboration, public awareness, and stricter privacy regulations although its hard to creating a global regulatory body with enforcement powers is challenging, and the opacity of deep learning decisions complicates accountability.

Medapati et al. [29] study the application of machine learning (ML) and deep learning (DL) algorithms for identity theft detection, particularly in credit card fraud, emphasizing pattern recognition and outlier detection. It discusses a wide range of algorithms, including SVM, Random Forest, Logistic Regression, Decision Tree, ANN, Naive Bayes, XGBoost, CNN, and specialized models like LDA, highlighting that ML classifiers often outperform traditional approaches. The study omits implementation challenges, lacks country-specific datasets, and does not provide a comprehensive comparison of alternative algorithms beyond autoencoders and CNNs.

The paper by Lorimer et al. [30] proposes a decentralized, participatory framework for identity theft detection on mobile social platforms using mobile edge computing. Less-complex computational tasks are delegated to a user's connections/followers, and results are aggregated by a trusted node, employing a Decision Tree ensemble to achieve efficient and responsive detection. Simulations show high detection rates (¿90% ITDR) and early anomaly detection within the second time window. The framework assumes a trusted aggregator node, and the effects of varying feature sets, participant numbers, weights, and time windows remain underexplored.

3 Methodology

This study uses a structured literature review to consolidate existing research on identity theft detection and prevention. Rather than introducing new datasets or algorithms, it synthesizes prior findings with a focus on methodologies, performance outcomes, and application domains. A systematic search was conducted in IEEE Xplore, ACM Digital Library, ScienceDirect, SpringerLink, and Google Scholar using keywords such as "identity theft," "fraud detection," "machine learning," and "cybersecurity. "Most of the selected works are recent peer-reviewed journal articles and conference papers, while purely theoretical studies were excluded.

The table provides a comparative overview of various studies on identity theft and cybersecurity attacks, highlighting the algorithms used, targeted sectors, attack types, attack damage, response time, and performance metrics. Algorithms include KNN (KNearest Neighbors), GNN (Graph Neural Network), NN (Neural Network), SVM (Support Vector Machine), RF (Random Forest), and DT (Decision Tree). Targeted sectors span SOCIAL (social networks), E-COMMERCE, and FINANCIAL institutions, with attack types such as PHISHING, HACKING, and SOCIAL ENGINEERING. Attack damage ranges from low to high, and response time reflects the system's mitigation speed. Performance metrics, namely accuracy, precision, and sensitivity, are reported in percentages. Overall, GNN and SVM exhibit consistently high accuracy, precision, and sensitivity, while NN and RF perform well in financial sector attacks. KNN and DT are applied mainly in social sectors with moderate effectiveness.

3.1 Comparison of Algorithm and Accuracy

Based on the provided table, a comparison of algorithms with respect to accuracy shows the following insights. The table lists multiple algorithms including K-Nearest Neighbors (KNN), Graph Neural Networks (GNN), Neural Networks (NN), Support Vector Machines (SVM), Random Forest (RF), and Decision Trees (DT). The accuracy for each algorithm varies across different attack types and targeted sectors. Focusing on GNN and SVM, both algorithms achieve 99% accuracy, indicating that in terms of raw accuracy, GNN and SVM perform similarly, making it difficult to definitively state which algorithm is superior solely based on this metric. Other algorithms like KNN, NN, RF, and DT also perform well, with accuracies ranging from 92 to 97%, showing that while they are effective, they are slightly lower than GNN and SVM in this context. Overall, the analysis suggests that while accuracy is a useful measure, it alone is insufficient to determine the best algorithm, especially when multiple algorithms achieve comparable values, as in the case of GNN and SVM (Fig. 1).

Table 1. Comparison Table based on various matrixes

Cite	Algorithm	Accuracy (%)	Precision (%)	Sensitivity (%)	Response Time	Attack Type	Targeted Sector	Attack Damage
13	KNN	97	79	72	Low	Phishing	Social	Identity Theft
14	GNN	94	90	93	Low	Phishing	Business	Financial loss
16	NN	92	89	91	Low	Hacking	Financial	Identity Theft
17	GNN	99	-	-	High	Social Engineering	Financial	Identity Theft
19	GNN	99	92	99	High	Phishing	Financial	Financial loss
20	SVM	97	98	91	High	Hacking	Financial	Financial loss
21	GNN	97	79	72	Low	Phishing	Social	Identity Theft
23	RF	96	94	97	Low	Phishing	Financial	Financial loss
24	LR	97	-	-	-	Hacking	Financial	Identity Theft
25	GNN	-	92	93	High	Phishing	Financial	Identity Theft
26	GNN	98	89	91	High	Hacking	Financial	Financial loss
27	LR	98	-	97	High	Hacking	Financial	Financial loss
28	GNN	-	99	-	-	Data Breach	Social	Identity Theft
29	SVM	99	92	91	Low	Phishing	Financial	Identity Theft
30	DT	95	99	90	High	Social Engineering	Social	Identity Theft

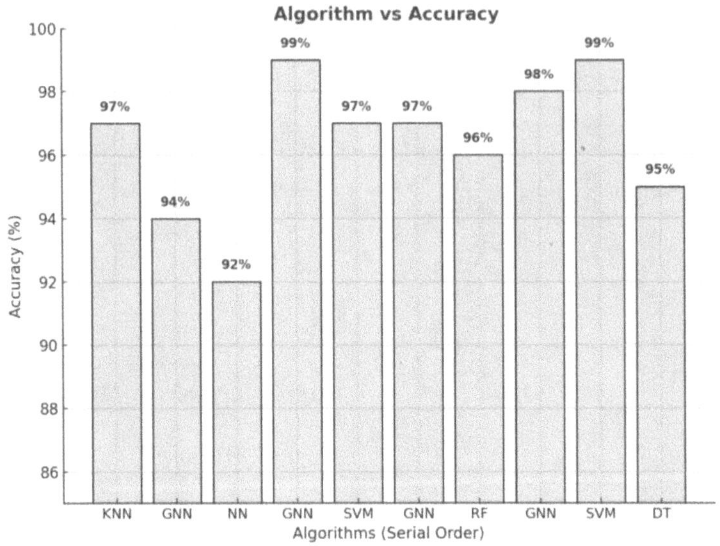

Fig. 1. Algorithm vs Accuracy

3.2 Comparison of Algorithm and Precision

As our analysis of algorithm performance shows, precision varies significantly across different detection methods. Algorithms such as SVM, DT, and RF achieve higher precision, reflecting their effectiveness in accurately identifying identity theft, phishing, and other attacks with minimal false positives. In contrast, KNN, NN, and GNN exhibit slightly lower precision but often compensate with faster response times. This demonstrates that choosing the right algorithm requires balancing precision with detection speed, depending on the severity of the attack and the criticality of the targeted sector (Fig. 2).

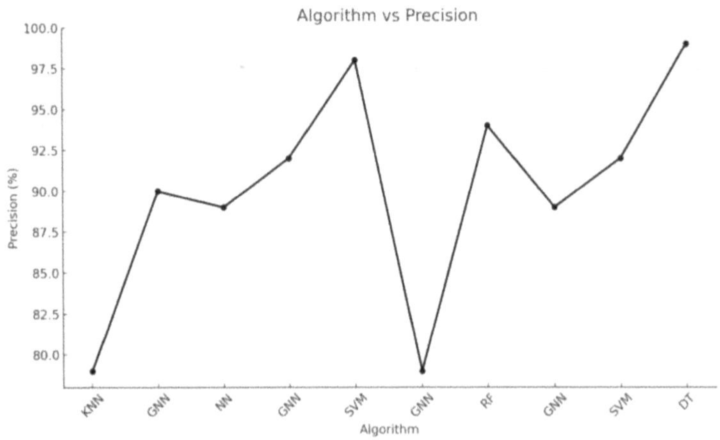

Fig. 2. Algorithm vs Precision

3.3 Algorithm vs Sensitivity

When comparing algorithms based on sensitivity, the studies show that Graph Neural Networks (GNN) perform exceptionally well, achieving 99% sensitivity in some cases. This indicates that GNN is highly effective at correctly identifying true positive cases, making it the best-performing algorithm in terms of sensitivity according to several research studies (Fig. 3).

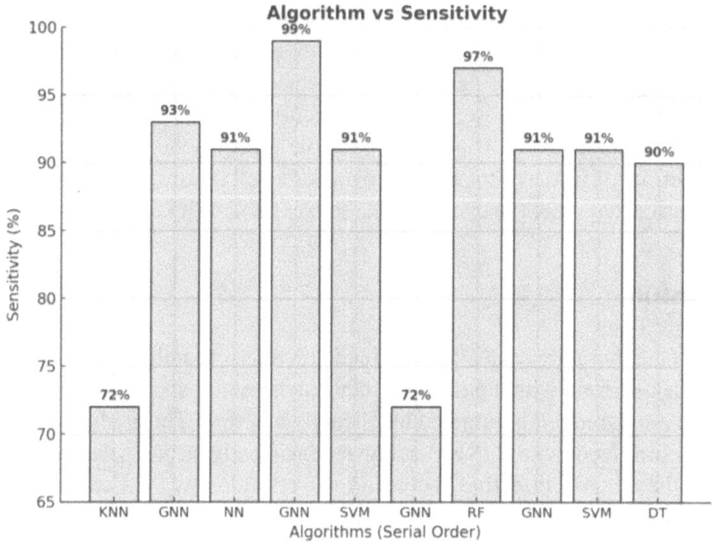

Fig. 3. Algorithm vs Sensitivity

However, it is important to note that performance can vary depending on the type of attack or targeted sector. While GNN excels in certain contexts, other algorithms may perform better for specific scenarios such as hacking detection or social engineering attacks. Therefore, although GNN demonstrates overall superior sensitivity, the optimal choice of algorithm may depend on the specific use case and attack type being addressed.

3.4 Targeted Sectors

From the data Table 1, it is clear that the financial sector and e-commerce sector are among the most affected. Many algorithms, including GNN, SVM, and NN, are applied to detect attacks in these sectors. The primary attacks involve phishing, hacking, and identity theft, which can lead to financial loss or compromise of sensitive personal information. The recorded attack damage is generally low to moderate, but even lowlevel attacks in financial systems can have significant consequences due to the volume of transactions and the sensitivity of data. In contrast, sectors like social networks are also affected, particularly by phishing and identity theft, but the damage tends to be less direct in monetary terms, often impacting personal information or reputation. Overall, while all sectors experience threats, the financial and e-commerce sectors are more critical due to the potential for high-impact losses, making accurate and responsive algorithms essential for detection and mitigation.

3.5 Future Work

To better protect critical infrastructure, future work should prioritize the development of scalable, real-time identity theft detection systems that can effectively process complex, imbalanced data from diverse industrial environments. Enhancing model transparency and explain ability will be vital for adoption by security teams and compliance officers. Industry-driven solutions combining behavioral analytics, graphbased techniques, and deep learning hold promise against sophisticated threats such as synthetic identity fraud and deepfakes. Furthermore, aligning detection systems with regulatory requirements and privacy standards will ensure responsible deployment. Collaboration between industry practitioners, researchers, and policymakers will be key to creating resilient and adaptive defenses that meet the evolving challenges of identity theft.

4 Conclusion

This review of the literature critically examined previous research on identity theft detection in high-risk sectors, with a particular focus on financial and e-commerce domains, which are the most frequently targeted and heavily affected. The analysis demonstrates that Graph Neural Networks (GNNs) achieve consistently strong sensitivity and accuracy, making them highly effective in detecting diverse identity theft attempts, including phishing and financial fraud. However, algorithms such as Support Vector Machines (SVM), Decision Trees (DT), and Random Forests (RF) often deliver higher precision, highlighting an important trade-off between accuracy/sensitivity and precision across different methods. Simpler attacks such as phishing and low-damage identity theft are generally identified rapidly, while more complex attacks, including hacking and social engineering, require extended processing to ensure reliable detection. These findings indicate that no single method offers universal effectiveness; instead, the efficiency of detection strategies depends on the attack type, the targeted sector, and the potential level of damage. Overall, this synthesis underscores the importance of adopting context-aware and adaptive detection approaches that balance sensitivity, accuracy, and precision, while also considering response time. Such balanced strategies provide a solid foundation for developing robust identity theft detection systems, ultimately strengthening the protection of critical sectors and safeguarding sensitive personal and financial information.

References

1. Koops, B.J., Leenes, R.: Identity theft, identity fraud and/or identity-related crime. DuD **30**, 553–556 (2006). https://doi.org/10.1007/s11623-006-0141-2
2. Anifalaje, K.: A legal approach to the protection of customers of banks and other financial institutions from identity theft in nigeria. Northern Ireland Legal Q. 187–214 (2024). https://doi.org/10.53386/nilq.v75i2.1163

3. Borketey, B.: Identity theft – the root cause of widespread fraudulent activities. Int. J. Auto. Artif. Intel. Machine Learn. **5**, 26–28 (2025). https://doi.org/10.61797/ijaaiml.v5i1.406
4. Anderson, K.B., Durbin, E., Salinger, M.A.: Identity theft. J. Econ. Perspect. **22**(2) (2008)
5. Whitley: A review of the complex interplay between information systems, identification and identity. European J. Info. Sys. **23**(1) (2014)
6. Al Smadi, B., Min, M.: A critical review of credit card fraud detection techniques. In: 2020 11th IEEE Annual Ubiquitous Computing, Electronics & Mobile Communication Conference (UEMCON), pp. 0732–0736. IEEE (2020)
7. Guedes, I., Martins, M., Cardoso, C.S.: Exploring the determinants of victimization and fear of online identity theft: An empirical study. Security J. 1 (2022)
8. Maher, C.A., Hayes, B.E.: Nonfinancial consequences of identity theft revisited: Examining the association of out-of-pocket losses with physical or emotional distress and behavioral health. Crim. Justice Behav. **51**(3), 459–481 (2024)
9. Salman, H.M.: Identity theft in the banking system. In: Online Identity-An Essential Guide. IntechOpen, ??? (2024)
10. Sriram, H.K.: A comparative study of identity theft protection frameworks enhanced by machine learning algorithms. SSRN 5236625 (2024)
11. Piquero, N.L., et al.: Preventing identity theft: Perspectives on technological solutions from industry insiders
12. Zou, Y., et al.: Examining the adoption and abandonment of security, privacy, and identity theft protection practices. In: CHI '20: Proceedings of the 2020 CHI Conference on Human Factors in Computing Systems (CHI). Honolulu, HI, USA (2020). https://doi.org/10.1145/3313831.3376570
13. Saikrishna, D., Manisha, D.: Composite behavioral modeling for identity theft detection in online social networks 14 (2024)
14. Ye, Q., et al.: Modeling access environment and behavior sequence for financial identity theft detection in e-commerce services
15. Wyre, M., Lacey, D., Allan, K.: The identity theft response system. Trends and Issues in Crime and Criminal Justice (592) (2020)
16. Sriram, H.K.: A comparative study of identity theft protection frameworks enhanced by machine learning algorithms. MSW Manage. Multidiscipl. Sci. Work and Manage. J. **34**, 1080–1101 (2024)
17. Bilge, L., Strufe, T., Balzarotti, D., Kirda, E.: All your contacts are belong to us: Automated identity theft attacks on social networks. In: Proc. 18th Int. Conf. World Wide Web (WWW) (2009)
18. George, D.A.S., Baskar, D.T., Srikaanth, D.P.B.: Cyber threats to critical infrastructure: Assessing vulnerabilities across key sectors. Partners Univ. Int. Innov. J. (PUIIJ) 02 (2024). https://doi.org/10.5281/zenodo.10639463
19. Alarfaj, F.K., Shahzadi, S.: Enhancing fraud detection in banking with deep learning: Graph neural networks and autoencoders for real-time credit card fraud prevention. IEEE Access (2024). https://doi.org/10.1109/ACCESS.2024.3466288
20. Benchaji, I., Douzi, S., El Ouahidi, B., Jaafari, J.: A novel system for credit card fraud detection based on sequential modeling of data, using attention mechanism and lstm deep recurrent neural networks. J. Big Data **8**, 151 (2021). https://doi.org/10.1186/s40537-021-00541-8
21. Saikrishna, D., Manisha, D.: Behavioral analytics for identity theft detection in social networks: A composite modeling approach 14 (2024)
22. Abdelhalim, A., Traore, I.: The impact of google hacking on identity and application fraud
23. Afriyie, J.K., et al.: A supervised machine learning algorithm for detecting and predicting fraud in credit card transactions. Deci. Analy. J. **6**, 100163 (2023). https://doi.org/10.1016/j.dajour.2023.100163

24. Mitchell, J., Charles, D., Sambasivam, S.: Predictive modeling for identity theft detection: A design science approach using machine learning and historical data. Issu. Info. Sci. Info. Technol. **22**(Article 3) (2025). https://doi.org/10.28945/5506
25. Sholademi, D.B.: Financial crime in the age of ai: Deep fakes and identity theft risks. Int. Res. J. Modernizat. Eng. Technol. Sci. (IRJMETS) 06 (2024). https://doi.org/10.56726/IRJMETS65507
26. Bhattacharya Chakraborty, S., Shaikh, M.Z.: A comprehensive and relative study of detecting deformed identity crime with different classifier algorithms and multilayer mining algorithm. Int. J. Adv. Res. Comput. Comm. Eng. 3 (2014)
27. Al-Smadi, B.: Credit card security system and fraud detection algorithm, Dissertation. Louisiana Tech University (2021)
28. The evolution of fraud: Ethical implications in the age of large-scale data breaches and widespread artificial intelligence solutions deployment. ITU Journal: ICT Discoveries (2018)
29. Medapati, S.N.: Identity theft detection using machine learning. Int. J. Res. Pub. Rev. **4**, 5033–5037 (2023)
30. Lorimer, P.A.K., Diec, V.M.-F., Kantarci, B.: Participatory detection of identity theft on mobile social platforms

Auditable Feedback Filtering for Autonomous Agents: A Privacy-Preserving Defense Against Feedback Poisoning Using zk-Proofs, OPA, and Post-quantum Cryptography

Raghu Katikeri[1,2](✉), Rachita Barla[1,2], and Amit Vaid[1,3]

[1] Artificial Intelligence, Emerging Tech CoE, State Street Corporation, Hyderabad, India
{rkatikeri,rbarla,avaid}@statestreet.com
[2] Hyderabad, India
[3] Bengaluru, India

Abstract. Active learning and continual adaptation expose autonomous agents to a critical vulnerability: malicious or low-quality feedback that can poison model updates. Unlike static dataset noise, adversarial feedback introduced during operation directly reshapes model behavior, leading to risks such as semantic drift, financial misinterpretation, or regulatory non-compliance. We propose an auditable, privacy-preserving architecture for feedback filtering and staging that ensures only high-quality, verifiable feedback contributes to training. The architecture combines deterministic policy enforcement, cross-validation checks, and influence-based staging. At ingest, Open Policy Agent (OPA) enforces rules such as schema validity, data access control, and structural consistency. Each validation is paired with zero-knowledge proofs (zk-proofs) to guarantee that checks were performed without exposing sensitive content. Beyond rule-based screening, feedback is corroborated through multi-perspective validation, including execution against Merkleized database snapshots, semantic equivalence checks between inputs and outputs, and verification of factual claims against authoritative sources. To mitigate subtle poisoning, the system applies influence estimation techniques—gradient alignment, distributional divergence, and sandbox retraining experiments—to quantify the downstream impact of feedback. Items with disproportionately high influence are quarantined in a candidate store, where they undergo deeper automated analysis and, if needed, human intervention. Finally, the architecture is secured with post-quantum cryptography (PQC). Dilithium signatures provide authenticity for certificates, policies, and attestations, while Kyber enables quantum-resistant key encapsulation for secure key distribution. Hybrid classical + PQC modes ensure interoperability during migration. This layered pipeline—combining OPA-based policy checks, zk-proof verifiability, Merkle-backed execution validation, influence-aware staging, consortium-based human oversight, and PQC-hardened security—provides a robust

H. R. Arabnia et al. (Eds.): CSCI 2025, CCIS 2969, pp. 361–371, 2027.
https://doi.org/10.1007/978-3-032-28460-0_25

defense against feedback poisoning. It enables agents to safely learn from user interactions while preserving trust, auditability, and resilience in sensitive domains.

Keywords: Feedback poisoning · Zero-knowledge proof · Merkle snapshot · OPA · Post-quantum cryptography · Influence estimation

1 Introduction

1.1 Motivation

The increasing adoption of autonomous agents in data-driven environments—has elevated the importance of continual learning. In such systems, feedback from users or other agents is continuously ingested to refine model behavior. While this adaptive mechanism enables rapid improvement and contextual awareness, it also introduces a new vulnerability: feedback poisoning. Unlike traditional data corruption in static datasets, poisoned or low-quality feedback in active learning systems exerts a direct and immediate influence on model weights, potentially distorting decision boundaries and propagating systemic errors. Feedback poisoning can manifest in multiple ways. It may arise from unintentional human corrections that misrepresent domain logic or from adversarial manipulations designed to bias system outputs. For agents operating in sensitive domains—such as financial systems, enterprise databases, or legal and regulatory decision-support—such corrupted feedback can result in semantic drift, financial inaccuracies, compliance violations, or reputational damage. This necessitates a systematic approach to filter, validate, and stage feedback before it impacts learning pipelines.

1.2 Problem Statement

Current feedback-handling mechanisms rely primarily on heuristic or manual vetting, which scales poorly and lacks cryptographic assurances. Moreover, many existing systems treat feedback validation as a binary acceptance or rejection process, ignoring the varying influence each feedback item may exert on the learning model. As a result, even minor but influential miscorrections can propagate widely through the model's internal representations. To address these limitations, there is a need for an architecture that:

1. Deterministically validates feedback against well-defined policies and schemas.
2. Preserves privacy while proving validation integrity.
3. Quantifies influence of each feedback instance on downstream learning.
4. Stages high-risk feedback through automated quarantine and human oversight.
5. Maintains security and auditability of the entire process against tampering and future cryptographic threats.

1.3 Proposed Approach

This paper proposes a privacy-preserving and auditable architecture for feedback filtering in active learning agents. The architecture unifies multiple layers of validation and assurance: **Deterministic Policy Enforcement:** Feedback is first processed through OPA, which enforces domain-specific rules, schema integrity, and access controls. **Zero-Knowledge Verifiability:** Each policy decision is paired with a zero-knowledge proof (zk-proof) that confirms successful validation without disclosing sensitive content. This enables trust between agents or modules operating across security boundaries. **Cross-Validation and Data Integrity:** Structural checks are complemented by execution validation against Merkleized snapshots, semantic equivalence assessments between original and corrected outputs, and corroboration against authoritative data sources. [15] **Influence-Aware Staging:** The system employs influence estimation techniques—such as gradient alignment, distributional divergence, and sandbox retraining—to estimate the potential behavioral shift caused by a feedback instance. High-influence feedback is quarantined for further review or human evaluation before integration. **Human Oversight via Consortium Model:** Domain experts intervene only when defined thresholds are crossed. For instance, in a Text-to-SQL agent, Database Administrators (DBAs) validate schema-impacting feedback, while in a Domain QA agent, Subject Matter Experts (SMEs) and stakeholders verify factual corrections related to legal or financial compliance.

1.4 Security and Cryptographic Assurance

To ensure long-term authenticity and tamper-resistance, the proposed system integrates PQC primitives. Dilithium signatures [14] provide integrity for feedback attestations, policies, and model updates, while Kyber [9] facilitates key encapsulation for secure communication between system components. The system supports hybrid cryptographic modes, combining classical and PQC algorithms to ensure backward compatibility and smooth transition.

1.5 Key Components

The key components of this work are as follows: A unified feedback filtering architecture combining deterministic OPA policies, zk-proof validation, and influence estimation for secure active learning. A staging and quarantine mechanism that isolates high-risk feedback for analysis, minimizing model poisoning risk. Integration of cryptographic integrity and privacy guarantees, employing zk-proofs for verification and PQC primitives for long-term security. Application and validation through two reference agents—a Text-to-SQL translator and a Domain QA system—to demonstrate adaptability across distinct domains.

enforcement, ensuring that identical feedback yields identical outcomes across deployments. [7]

2. **Zero-Knowledge Proof Integration** After OPA validation, a zero-knowledge proof (zk-SNARK or zk-STARK) is generated summarizing the policy execution trace. The proof confirms that policy predicates evaluated to true without revealing the actual feedback or data content. This mechanism allows downstream verifiers—such as other agents or auditing modules—to trust the validation outcome without gaining access to sensitive information. Proof verification requires only lightweight computation, typically milliseconds per check, enabling near-real-time operation.

Each OPA decision is compiled into a zk-circuit where policy predicates act as public inputs and feedback metadata as private inputs, enabling verifiable policy compliance proofs. Together, OPA and zk-proofs form a verifiable deterministic baseline that ensures every accepted feedback instance is compliant by construction.

3.3 Probabilistic and Influence-Based Validation

Beyond deterministic policy enforcement, the architecture incorporates a probabilistic layer that estimates how much each feedback instance may alter model behavior.

1. **Gradient Alignment** The cosine similarity between gradients induced by candidate feedback and those from historical trusted data is computed. A similarity score below a predefined threshold (e.g., 0.5) signals potential misalignment or adversarial intent.
2. **Distributional Divergence** Embedding-space statistics are monitored to detect out-of-distribution feedback. If the embedding distance exceeds two standard deviations (2σ) from the training mean, the feedback is flagged as anomalous.
3. **Sandbox Retraining** Selected feedback is temporarily applied in a sandboxed environment where the model is fine-tuned on a validation subset. Performance changes on benchmark tasks are measured; a degradation beyond 2 "%" or instability in prediction entropy indicates high influence.
4. **Thresholds and Escalation** Thresholds are domain-specific and adjustable through policy configuration. For example, in a Text-to-SQL agent, feedback that modifies monetary or personally identifiable fields automatically triggers escalation. In a Domain QA agent, updates altering regulatory thresholds or legal constraints require human verification.

These probabilistic checks provide a quantitative signal of influence, complementing deterministic validation with behavioral risk estimation.

infrastructure having a larger footprint than on-premise infrastructure, poor planning can result in exploitation and the creation of more security holes [6]. Insider threats could leverage their permissions for financial gain or revenge, and users who are unaware of their power could accidentally leak sensitive information. While many companies are switching to a zero-trust model to counteract these threats, many companies that leverage the cloud are falling behind due to the complex nature of the shared responsibility model. The best measure to counteract the dynamic and complicated threats of cloud environments is a well-tuned SIEM.

While the threats to any organization become a vastly different situation once you switch from on-premise to cloud, threats become even more prevalent when it comes to cyberwarfare. SIEMs can serve as the critical point for detecting the tactics, techniques, and procedures (TTPs) that nation-state attackers use in cyberwarfare. Furthermore, many cyberwarfare tactics rely on compromised credentials or insiders to operate undetected. Including UEBA within a SIEM will create a baseline for all of the users, and when unusual action occurs, like with a cyberwarfare attack, it can be detected much more easily, even with the complexity included with cloud environments.

3 Central Cybersecurity Command Center (4C)

3.1 Single Pane of Glass: How to Make Cloud Security Easier

A SIEM is the heart of the Security Operations Center (SOC); it is the point where all security information and IPS/IDS logs are collected [7]. The SIEM takes in a high volume of real-time data, then normalizes it into a single form of data that makes it easier to process in a SOC. This centralized approach to security is often called a "single pane of glass" that makes it much easier to assess the security of an organization and eliminate blind spots. Splunk, the largest enterprise SIEM provider, describes a SIEM tool as "essentially an analytics-driven security command center—it's often the centerpiece of a highly functional SOC" [8]. Interfaces provided by CSPs are often confusing and complicated enough that CSPs often recommend a certification just to understand their platform. Creating this "single pane of glass" can make the security of the cloud more secure, eliminating many of the challenges that are created by cloud infrastructure.

3.2 Unifying Multi-Cloud Environments

One of the most valuable aspects of a modern cloud-based SIEM is its ability to aggregate logs from complex and multi-cloud environments. Major cloud providers like AWS, Azure, and Google often include cloud native monitoring that can be further piped into an organization's well-tuned SIEM. These next-generation SIEMs are built to integrate naively with cloud platforms through API-based integrations. This allows the SIEM to ingest and analyze data from cloud services and APIs, ensuring full coverage with multi-cloud environments [9]. The best practices for multi-cloud security include: continuous monitoring and logging, AI threat detection, regular Vulnerability assessments, and threat intelligence sharing [10]. Combining all of these best practices into the "single pane of glass" creates not just a dashboard but a Central Cybersecurity Command Center

(4C). A properly done 4C could advance the security of multi-cloud environments to make them just as secure, if not more secure than traditional infrastructure. This leaves behind only one security vulnerability because you do not own the infrastructure, you are more at risk for supply chain attacks.

The 4C Workflow. The Core of the 4C is how the components work together. The SIEM is the foundation as the data aggregator and correlator, taking in all of the logs and creating one concise format. The User and Entity Behavior Analytics (UEBA) is the behavioral intelligence of the 4C, identifying when events happen outside the baseline. Then there is SOAR to automate the responses using "playbooks" to respond to simple threats without human intervention. A practical example of a 4C in action would start with the SIEM for detection. The SIEM ingests log data and flags several alerts for high volumes of failed login attempts on a critical server. That data from the SIEM moves onto the UEBA engine and gets analyzed to determine if the failed logins are from a known corporate IP address. The UEBA engine finds that the IP is unfamiliar and occurs at an unusual hour outside the normal work hours, and the alert becomes elevated to a higher threat level. SOAR then takes control of the event and runs a predefined playbook that is used for "suspicious logins". The playbook executes by banning the IP address, sending a notification out to the security team, resetting the user's password, and forcing a logout of all active sessions. This approach taken by the 4C shows how a security incident can be solved within seconds compared to the traditional approach that could take minutes to hours if the security team were not in the office. This approach significantly reduces the mean time to detect (MTTD) and mean time to respond (MTTR) to an incident.

4 Core 4C Features for Cloud Security

4.1 User and Entity Behavior Analytics (UEBA)

User and Entity Behavior Analytics (UEBA) is a newer feature of modern SIEMS. There are two types of UEBA systems: one that is standalone and the other that you can integrate into a SIEM, furthering the "single pane of glass" [11]. Every major SIEM provider now includes UEBA because of how useful this technology is. UEBA uses machine learning and behavioral analytics to detect anomalies that are not in traditional signature-based detection. UEBA collects user and entity data from all sources to build a baseline of normal behavior for every user and device. When activity deviates from the baseline, it is flagged so that further investigation can occur. While one flag is usually not an Indicator of Compromise (IoC), multiple anomalies occurring across the kill chain signal a high-risk event. UEBA is a core feature that should be included in every 4C to help better protect cloud infrastructure in normal organizations and in cyberwarfare.

4.2 Threat Intelligence Integration

Threat intelligence integration is another core feature of a 4C that enables SIEMs to have better for threat detection. With threat intelligence integration, organizations can integrate real-time data on known attacks to be as up to date as possible when it comes to known threats. Cloud infrastructure is still in its early phase, so it's more likely that a new threat will become prevalent, and having early detection capability provided from threat intelligence integration can help protect an organization's system. For example, a SIEM can correlate an internal event along with a malicious IP address from threat intelligence and create an alert for a rapid response. You can also use Threat Intelligence Platforms (TIP) as a way to automatically ingest threat data from many sources into your SIEM, then correlate that data to export to other parts of an organization's infrastructure to remediate threats [12]. TIP can rationalize threat data to automatically find Indicators of Compromise (IOCs) and compare them against other known threats to enable faster threat response from other 4C components.

Security Orchestration, Automation, and Response (SOAR), Security Orchestration, Automation, and Response (SOAR) is a crucial extension of SIEMs, designed for automation and orchestration of formerly manual tasks in SIEMs. When a SIEM detects a threat and generates an alert for it, SOAR can automatically respond based on a pre-made workflow known as a "playbook". This automation reduces the manual workload and becomes necessary for organizations on a large scale. In the case of cyberwarfare, not having SOAR included within a SIEM creates a huge vulnerability if an organization's team is not large enough or already overloaded. If a sophisticated attack hits, there will be too many alerts being set off to handle, leading to the attacker having more time to infiltrate the system before being stopped. Once you add SOAR to a 4C, all of the smaller repetitive tasks will automatically be solved, letting the cybersecurity team focus on the major alerts.

Regulatory Compliance. With a SIEM at the core of a 4C, it is much easier to ensure that compliance is met. The 4C can ensure the integrity of Protected Health Information (PHI), and it can also maintain detailed historical records of system and user activity that are required for HIPAA compliance. The 4C can also comply with PCI DSS with its tracking and monitoring of all access to network resources, which is required for compliance. For global organizations, a 4C can maintain compliance with GDPR. The 4C's ability to ensure that data handling, access, and storage are in adherence to GDPR principles. The automated logging can also provide clear audit trails of all the data processing for better compliance with standards.

5 Practical Implementation and Management Challenges

5.1 Economic Considerations

While creating a Central Cybersecurity Command Center does come with its many benefits, it also comes with a high cost. The cost of SIEM deployment has been a barrier for many companies, especially smaller companies and those whose main focus is not security. However, experts agree that the value of a SIEM is not in generating revenue but in "cost avoidance," which is the prevention of costly security disasters [13]. Cloud-native SIEMs often come as a subscription-based model, but that also creates a financial challenge with the cost normally associated with the data ingestion volume, unlike on-premise, where it's mainly a large upfront cost. When it comes to the scale of cyberwarfare, the price to ingest terabytes to petabytes of data comes at an egregious cost. The combination of the cost and the difficulty of setting up a system like this makes it not possible for all companies. Without full cooperation between all companies implementing this advanced security, it leads back to the problem of supply chain attacks, where one company can get attacked, and it can spread throughout many more companies. If a nation-state were to want to attack a large company or government, they could start from the bottom of the chain at a smaller company and work through the other systems to compromise their main target.

5.2 Human Factor

While a good SIEM can be very powerful, there is also the human factor to consider. The most significant challenge is alert fatigue, where analysts become so desensitized to wasted time looking at alerts that lead to nothing and genuine threats being overlooked. While including SOAR and alert tuning in a SIEM can help filter a number of the fatigue-causing alerts, the system needs continuous tuning and maintenance, and without it, the system's effectiveness will degrade over time. The issues that SIEMs have with the cost and the reliance on the human factor are fundamental problems that need to be solved in the next generation of SIEM technology.

Beyond just alert fatigue, there are many other significant challenges caused by the human factor. The global cybersecurity workforce gap has reached 4.8 million unfilled roles, which is a 19% year-over-year increase. This has direct quantifiable impacts on an organization's financial and operational resilience. Organizations with security staff shortages face data breach costs that are, on average, 1.76 million USD higher than fully staffed counterparts [14]. When a company is understaffed, between 35% and 44% of SOC analysts report symptoms of burnout, leading to the average tenure lasting only one to three years [15]. This constant need to recruit and train new hires creates a massive financial drain and leads to no long-term cybersecurity professionals who understand the company and its needs.

A 4C offers a solution to these vulnerabilities rooted in humans. The integration of its many components, including SOAR, AI, and ML, directly combats alert fatigue and workforce shortages. By automating repetitive tasks such as the initial alert triage and simple alerts, the 4C filters out much of the noise in SIEMs and reduces the amount of time that cybersecurity professionals waste on repetitive tasks. This reduces the amount of burnout, leading to longer tenures and more professionals understanding an organization's system at a professional level to provide the best levels of defense. In turn, this reduces the rate at which an organization experiences a costly attack, saving them money in the long term. The following table shows the key financial human costs that the 4C is designed to mitigate (Table 2).

Table 2. The Economic Need For 4C

Economic Costs Mitigated by 4C	Key Data Point
Global Average Data Breach Cost	4.88 million USD in 2024, a 10% increase from 2023 [16]

(continued)

Table 2. (*continued*)

Economic Costs Mitigated by 4C	Key Data Point
U.S. Average Data Breach Cost	10.22 million USD per incident [17]
Cost Increase from Staff Shortages	Data breach costs are 1.76 million USD higher for understaffed organizations [14]
Cloud Misconfiguration Prevalence	82% of cloud misconfigurations are due to human error [18]
Impact on Startups	89% of businesses impacted by cloud misconfigurations were startups [18]
Cybersecurity Workforce Gap	A global gap of 4.8 million unfilled roles, a 19% increase from the previous year [14]

6 The Future of SIEM

6.1 AI and Machine Learning

Artificial Intelligence (AI) and Machine Learning (ML) are the most revolutionary components of next-generation SIEMs. Using the AI tools creates more intelligent threat detection by directly processing the logs and analyzing what could be alerts. AI can significantly reduce the time it takes to detect and respond to threats in SIEMS [19]. That can lead to bolstering an organization's cybersecurity posture and greatly reduce the chance of an attack. Machine learning has major benefits when it comes to combating alert fatigue. ML can learn from alerts being resolved and find if they were a real threat or not, and use that information to determine if an alert should be sent in the future, reducing the overall number of alerts. AI and ML can work together to reduce false positives and automate repetitive tasks, resulting in a significant reduction in mean time to remediate (MTTR). This integration, paired with SOAR, is a critical part of a good 4C to create a professional defense.

6.2 Implementing Security as Code (SaC)

Security as Code (SaC) is a methodology that embeds security policies and controls directly into the software development lifecycle. This is known as "shifting left" and enables an organization to proactively identify and fix vulnerabilities early in the development phase, where it is much cheaper than later on in the cycle [20]. Adopting SaC has been rapidly becoming more popular for cloud users due to the cost savings and growing concern over security risks with the cloud transition. In a report done in 2022, 62% of respondents stated that adopting security as code was a priority for their organization [21]. SaC also ensures consistent security implementation by automating security checks, which reduces human error and ensures that compliance is always met. While SaC is different from SIEM, as it is more of a methodology compared to a direct security tool, it helps ensure that the end-to-end of the security lifecycle is correct. SaC handles the proactive security of the development pipeline by building security into the code, which prevents vulnerabilities from ever reaching production. SaC becomes much more powerful when combined with AI and ML. While SaC embeds security directly into the development pipeline, a 4C can leverage AI and ML to detect and respond to threats in real time, providing security from development to production.

7 Conclusion

The transition from on-premise to cloud infrastructure has reshaped the cybersecurity landscape. Switching from a static and clearly defined network to a complex borderless environment that is more vulnerable to cyberwarfare and supply chain attacks. This new style of computing needs a new approach to security from the traditional SIEM-centric SOC to a holistic Central Cybersecurity Command Center (4C).

The 4C represents the future of cloud defense, a unified and intelligent platform that aggregates and analyzes different events from across multi-cloud and hybrid environments. 4C is the core of a future security operation with advanced features like UEBA, threat intelligence, and SOAR to provide a comprehensive "single pane of glass" that eliminates blind spots and connects seemingly unrelated events into a cohesive attack chain. As the financial costs continue to rise and the human error levels grow worse, the future of the 4C lies in the further integration of AI, ML, and Security as Code. These technologies will be used to solve the fundamental problems of cost and alert fatigue, advancing the proactive and automated defense against cyber threats from small attacks to nation-state and APTs.

4C represents the future of cloud defense. It is an intelligent, unified platform that aggregates and analyzes events from across multi-cloud and hybrid environments to enable a "single pane of glass" security model. The C4 can connect seemingly unrelated events into a cohesive attack chain, discovering incidents that previous security systems were unable to find. By adopting 4C into an organization, levels of security can be achieved that surpass traditional security, taking us into a more cyber-secure future.

References

1. Byström, O.: A comparison between on-premise and cloud environments in terms of security: With an emphasis on Software-as-a-Service & Platform-as-a-Service. Master's thesis. Blekinge Institute of Technology, Karlskrona (2022). URN: urn:nbn:se:bth-22990
2. Samuel, A., Obajemu, A., Ogundoyin, P., Onabolu, O., Oluwadare, O.: Enhancing Cloud Security—Proactive Threat Monitoring and Detection Using a SIEM-Based Approach. Applied Sciences **13**(22), 12359 (2023). https://www.researchgate.net/profile/Adebis-Samuel/publication/386076652_Strengthening_Cloud_Security_Proactive_Threat_Monitoring_and_Detection_with_a_SIEM-Based_Approach/links/6742077527661f7ae6663325/Strengthening-Cloud-Security-Proactive-Threat-Monitoring-and-Detection-with-a-SIEM-Based-Approach.pdf

3. Mendoza, C., Reyes, C.: (PDF) Research Gate. In: EXPLORING THE IMPACT OF SHARED RESPONSIBILITY MODELS ON CLOUD SECURITY POSTURE AND VULNERABILITY MANAGEMENT (2023). https://www.researchgate.net/publication/386220026_EXPLORING_THE_IMPACT_OF_SHARED_RESPONSIBILITY_MODELS_ON_CLOUD_SECURITY_POSTURE_AND_VULNERABILITY_MANAGEMENT
4. Metibemu, O., Adesokan-Imran, T., Juliana Tiwo, A.J.A., et al.: (PDF) Research Gate, In: Developing Proactive Threat Mitigation Strategies for Cloud Misconfiguration Risks in Financial SaaS Applications (2025). https://www.researchgate.net/publication/389875755_Developing_Proactive_Threat_Mitigation_Strategies_for_Cloud_Misconfiguration_Risks_in_Financial_SaaS_Applications
5. Alquwayzani, A., Aldossri, R., Frikha, M.: (PDF) Research Gate: Prominent Security Vulnerabilities in Cloud Computing (2024). https://www.researchgate.net/publication/378721981_Prominent_Security_Vulnerabilities_in_Cloud_Computing
6. Puzas, D.: 12 cloud security issues: Risks, Threats & Challenges. In: CrowdStrike (2024). https://www.crowdstrike.com/en-us/cybersecurity-101/cloud-security/cloud-security-risks/#:~:text=Poor%20planning%20can%20manifest%20itself,Identity%20and%20access%20management
7. Perera, A., Rathnayaka, R.S.: This person is not on ResearchGate, or hasn't claimed this research yet. In: Che, S., Madushanka, C.W.W. (eds.) (PDF) Natural Language Processing for Cyber Threat Intelligence Analysis. In: Research Gate (2020). https://www.researchgate.net/publication/390113970_Natural_Language_Processing_for_Cyber_Threat_Intelligence_Analysis
8. Kidd, C.: Siem: Security Information & Event Management explained. In: Splunk (2025). https://www.splunk.com/en_us/blog/learn/siem-security-information-event-management.html
9. What is cloud siem? In: Palo Alto Networks. https://www.paloaltonetworks.com/cyberpedia/what-is-cloud-siem. Accessed 14 Sep 2025
10. Patel, S.: (PDF) Systematic Literature Review on Cloud Computing Security: Threats and mitigation strategies, Research Gate (2024). https://www.researchgate.net/publication/375185647_Systematic_Literature_Review_on_Cloud_Computing_Security_Threats_and_Mitigation_Strategies
11. Seppänen, M.: Theseus, jamk.fi (2021). https://www.theseus.fi/bitstream/handle/10024/503253/Seppanen_Mikko.pdf;jsessionid=A266C42EAE7E092FFC6C3C852413C452?sequence=2
12. Siem + threat intelligence: Quickly identify the threats that matter to you. In: ThreatConnect (2019). https://threatconnect.com/wp-content/uploads/ThreatConnect-SIEM-Threat-Intelligence-Whitepaper.pdf
13. Anderson, E.: Siem and return on investment: Four pillars for success | lumifi cybersecurity. In: lumifi (2023). https://www.lumificyber.com/blog/siem-and-return-on-investment-four-pillars-for-success/
14. Khalil, M.: Cybersecurity skills gap statistics for 2025: Record 4.8m roles unfilled. In: DeepStrike (2025). https://deepstrike.io/blog/cybersecurity-skills-gap. Accessed 14 Sep 2025
15. Soc Analyst burnout: How ai is changing the game. In: Abnormal AI (2025). https://abnormal.ai/blog/soc-analyst-burnout-ai
16. Bonnie, E.: 110+ of the latest Data Breach Statistics [updated 2025]. In: Secureframe (2025). https://secureframe.com/blog/data-breach-statistics. Accessed 14 Sep 2025
17. Koskey, A., White, M., Donelson, B.: Ten key insights from IBM's cost of a data breach report 2025. In: JD Supra (2025). https://www.jdsupra.com/legalnews/ten-key-insights-from-ibm-s-cost-of-a-9975485/
18. 50+ cloud security statistics in 2025. In: SentinelOne (2025). https://www.sentinelone.com/cybersecurity-101/cloud-security/cloud-security-statistics/. Accessed 14 Sep 2025

19. Pulyala, S.: From Detection to Prediction: AI-powered SIEM for Proactive Threat Hunting and Risk Mitigation. In: Semantic Scholar (2024). https://pdfs.semanticscholar.org/f443/bd7729170954df7b36ef6067180a09dda1f8.pdf
20. Hensley, M.: IAC Security Best Practices: Safeguarding Your Cloud Resources. In: Linux Security (2025). https://linuxsecurity.com/news/cloud-security/iac-security-best-practices
21. Katz, E.: What is security as code and how to get started implementing it. In: Spectral (2023). https://spectralops.io/blog/what-is-security-as-code/

Security Algorithms for Identifying Malicious E-mail

Woo Young Park, Sang Hyun Kim, Duy-Son Vu, Chang Han Song, Hee Soo Jung, and Hyeon Jo(✉)

RealSecu, IS Tower #809, 60 Centum Buk-Daero, Haeundae-Gu, Busan 48059, Republic of Korea
{wypark,shkim,vdson,chsong,sky}@realsecu.net, sineoriz@gmail.com

Abstract. The majority of cyberattacks are carried out via email. Existing security solutions block malicious emails by forming blacklists based on signatures; however, they are vulnerable to zero-day attacks and suffer from a cold-start problem. This study proposes a new security algorithm that leverages email communication information and Simple Mail Transfer Protocol (SMTP) data. The algorithm aims to identify and process malicious emails by analyzing the HELO domain, session domain, pointer (PTR) record, Sender Policy Framework (SPF) record, sending domain, receiving domain, and SMTP reply codes. Based on these features, the algorithm classifies malicious emails into private mail, spear-phishing, and impersonation mail. The performance of the proposed algorithm was validated through its implementation in three organizations. The results demonstrate that the algorithm can detect new types of malicious emails that were not identified by existing solutions.

Keywords: Security algorithm · malicious mail · impersonation mail · private mail · spear-phishing mail · simple mail transfer protocol

1 Introduction

Spam email has caused numerous problems with the widespread use of the Internet and advances in technology (Sumathi and Pugalendhi, 2021). It originated from advertising emails that are unilaterally delivered to unspecified users (Mahmoud, 2015). Nowadays, hackers exploit malicious emails to harm users. Such emails may be sent from private mail servers or by impersonating legitimate senders (Lee and Chen, 2021). They also facilitate phishing attacks targeting specific individuals or institutions. Existing email security systems have developed techniques to detect keywords and malicious code based on blacklists (Fang et al., 2019).

Machine learning approaches have also been introduced. For instance, a spam email detection system based on machine learning was developed using 800 Turkish email datasets, comparing seven different algorithms, where sequential minimal optimization showed the best performance (Eryilmaz et al., 2020). Another study proposed new technical features for malicious email detection and found that classification performance

H. R. Arabnia et al. (Eds.): CSCI 2025, CCIS 2969, pp. 383–391, 2027.
https://doi.org/10.1007/978-3-032-28460-0_27

improved when a random forest algorithm was applied (Cohen et al., 2018). In addition, static machine learning techniques have been applied to MS Office documents and compressed files to classify malicious emails, demonstrating superior performance over existing methods (Rudd et al., 2018).

However, blacklist-based email security technologies remain limited in detecting novel malicious patterns. Therefore, an advanced security algorithm is required to effectively identify emerging threats.

This study proposes a novel security algorithm that incorporates email communication information and Simple Mail Transfer Protocol (SMTP) response codes. The email communication features include the HELO domain, session domain, pointer (PTR) record, Sender Policy Framework (SPF) record, sending domain, and receiving domain. SMTP, a core protocol in email communication environments, has been utilized in various contexts, such as IoT device identification, large-scale mailing processing, and honeypot systems (Kumari et al., 2018; Wu et al., 2021; Zobal et al., 2020). Nevertheless, its application to the identification and classification of malicious emails remains limited. By jointly analyzing email communication information and SMTP response codes, this study aims to determine whether an email is malicious.

This study defines key terms as follows. Malicious email refers to abnormal emails that may cause economic loss to users. These include emails sent from private mail servers (hereafter, private mail), emails that impersonate legitimate senders (hereafter, impersonation mail), and spear-phishing emails. This study proposes a security algorithm designed to identify and process such malicious emails. The terms "security algorithm" and "security system" are used interchangeably to refer to the proposed approach.

2 Algorithm Architecture

The architecture of the proposed algorithm is illustrated in Fig. 1. It consists of two main components: a mail data collection engine and a classification engine. The algorithm records and manages email communication information as metadata, referred to as "mail data." The collection engine receives incoming emails, gathers relevant information to determine whether an email is malicious, and transmits the processed mail data to the classification engine. Specifically, the mail data collection engine extracts and processes information related to the HELO domain, session domain, pointer (PTR) record, Sender Policy Framework (SPF), sending domain, and receiving domain.

The classification engine categorizes malicious emails into domain mail, private mail, impersonation mail, and spear-phishing mail. It consists of two subcomponents: Classifier 1 and Classifier 2. Classifier 1 processes emails when the sending domain and receiving domain are identical, whereas Classifier 2 processes emails when the sending domain and receiving domain differ.

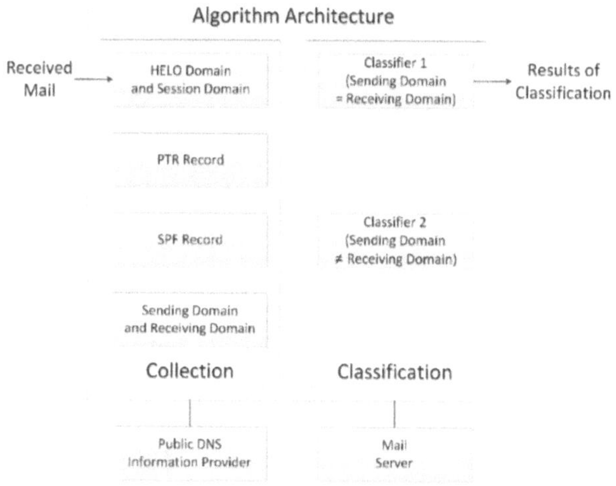

Fig. 1. Architecture of the Proposed Email Security Algorithm

3 Data Collection

3.1 HELO Domain and Session Domain

The proposed system collects HELO and session domain data. HELO is an email communication command that initiates the SMTP conversation. The system detects A records in both the HELO domain and the session domain from incoming mail. It then queries public DNS information using *nslookup* to verify the presence of A records in both domains. *Nslookup* is a network management command-line tool used to query DNS (Mandal and Jadhav, 2016). The system generates mail data to record whether an A record exists.

The presence of HELO domain records in public DNS indicates that A records are registered, suggesting that the SMTP conversation was legitimately initiated by a client. In contrast, the absence of HELO domain records implies that A records are not registered in public DNS or that the domain may have been deliberately forged or altered.

The session domain corresponds to the part of the SMTP command that specifies the sender's email address and signals the initiation of a mail transaction. The presence of a session domain record in public DNS indicates that the corresponding A record is registered. Conversely, the absence of such a record suggests that the domain may not be registered or may have been intentionally manipulated.

3.2 PTR Records

The proposed algorithm collects PTR records after processing the HELO and session domains. PTR records enable reverse DNS lookup by querying the domain associated with a sending IP address rather than a domain name. The presence of a PTR record indicates that a domain is properly associated with the sending IP in public DNS. In contrast, the absence of a PTR record implies that no such association exists. The algorithm queries public DNS using *nslookup* to determine whether an A record exists for the PTR of incoming mail. It then records the presence or absence of this information in the mail data.

3.3 SPF Records

The system collects Sender Policy Framework (SPF) records from incoming mail after processing PTR records. Since SPF processing is relatively complex, the procedure is described step by step:

1. The algorithm extracts the session domain A record and the header sending domain A record.
2. It checks whether an A record exists for the session domain.
   - If no A record exists, the system records the absence of an SPF TXT record for the session domain in the mail data.
   - If an A record exists, the algorithm extracts the SPF TXT record associated with the session domain.

     It queries the SPF TXT record of the session domain using *nslookup*. SPF TXT records are DNS entries that help prevent spoofing and phishing by verifying the authorized sending domain of an email.
     The system then records whether the SPF TXT record exists in public DNS.
3. The algorithm checks whether an a record exists for the header sending domain.
   - If no SPF TXT record exists, this absence is recorded in the mail data.
   - If an A record exists, the algorithm retrieves the SPF TXT record for the header sending domain and queries public DNS using *nslookup*. The results are then recorded in the mail data.

3.4 Sending and Receiving Domains

The proposed algorithm collects information on sending and receiving domains as follows. First, it extracts and compares the sending domain and receiving domain of the incoming email. The system records whether these domains match. It then compares the sending account and receiving account and records whether they correspond.

The relationships between domains are interpreted as follows. A match between the session sending domain and the receiving domain indicates consistency between sender and recipient domains during address specification. A mismatch suggests inconsistency between the sender and recipient domains during the SMTP session. Similarly, a match between the header sending domain and the receiving domain indicates alignment between sender and receiver domains during message transmission, whereas a mismatch indicates inconsistency during the SMTP session.

4 Classification

The proposed algorithm consists of two detailed classification processes. Classifier 1 is applied when the sending domain and the receiving domain are identical, whereas Classifier 2 is applied when they differ. Figure 2 illustrates the process of Classifier 1.

1. Classifier 1 first checks whether the sending and receiving domains match.
   - If the sending and receiving domains match, it verifies whether session SPF (SSPF) and header SPF (HSPF) are included in the mail data.

     If neither SSPF nor HSPF is included, the classification process is terminated, and a secondary process is initiated.
     If both SSPF and HSPF are included, the session domain is further examined within Classifier 1.

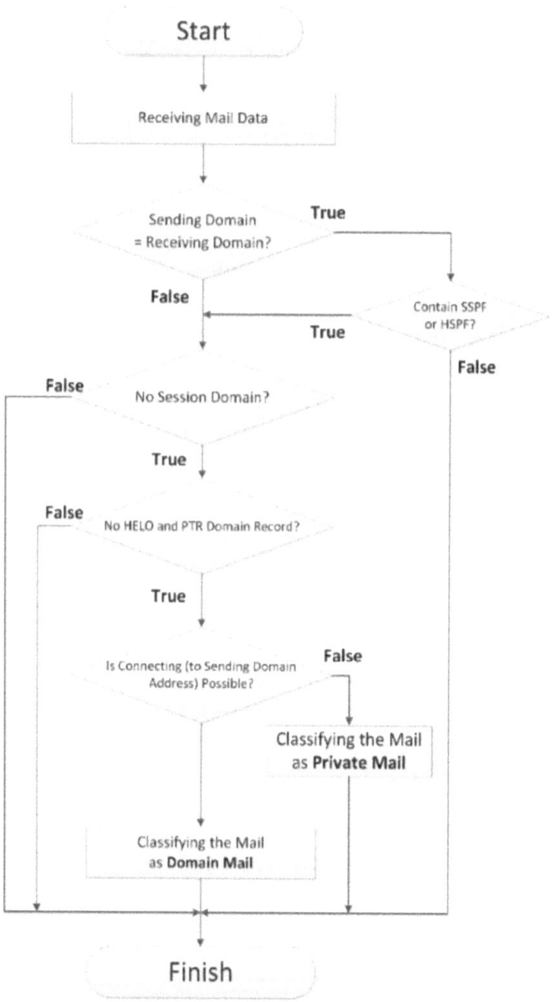

Fig. 2. Decision Flow of Classifier 1 for Same-Domain Email Analysis

2. If the sending and receiving domains do not match, or if the mail data includes SSPF or HSPF, the algorithm checks whether a session domain exists. It then verifies whether HELO and PTR domain records are included in the mail data. If no session domain exists and neither HELO nor PTR domain records are present, the system attempts to access the sending domain address. If the address is accessible, the email is classified as domain mail; otherwise, it is classified as private mail. If the mail data does not include a session domain, Classifier 1 is terminated. Likewise, if the mail data does not contain session domain, HELO, or PTR domain records, the classification process is terminated.

Figure 3 presents the process of Classifier 2. Classifier 2 is applied when the sending and receiving domains differ. When mail data is received, the algorithm first checks whether the sending and receiving domains match. If they do not match, the classification process is terminated. If they match, the algorithm verifies whether SSPF and HSPF are included in the mail data. If either SSPF or HSPF is present, the classification is terminated.

If neither SSPF nor HSPF is included, the algorithm checks whether the local user policy is allowed. The local user policy refers to a case in which the recipient account is not registered on the receiving mail server. If the local user policy is not allowed, the email is classified as impersonation mail. If the policy is allowed, the algorithm attempts to access the server IP of the incoming mail.

If the SMTP response code is 250 or 250′, indicating normal operation of the mail server, the email is classified as impersonation mail. If the SMTP response code is 550, indicating that the mail is unavailable, the email is classified as spear-phishing mail. If the SMTP response code differs from 250, 250′, or 550, the algorithm checks whether the sending and receiving accounts match. If they do not match, the email is classified as spear-phishing mail; otherwise, it is classified as impersonation mail. If the incoming mail server IP cannot be reached, the email is classified as impersonation mail.

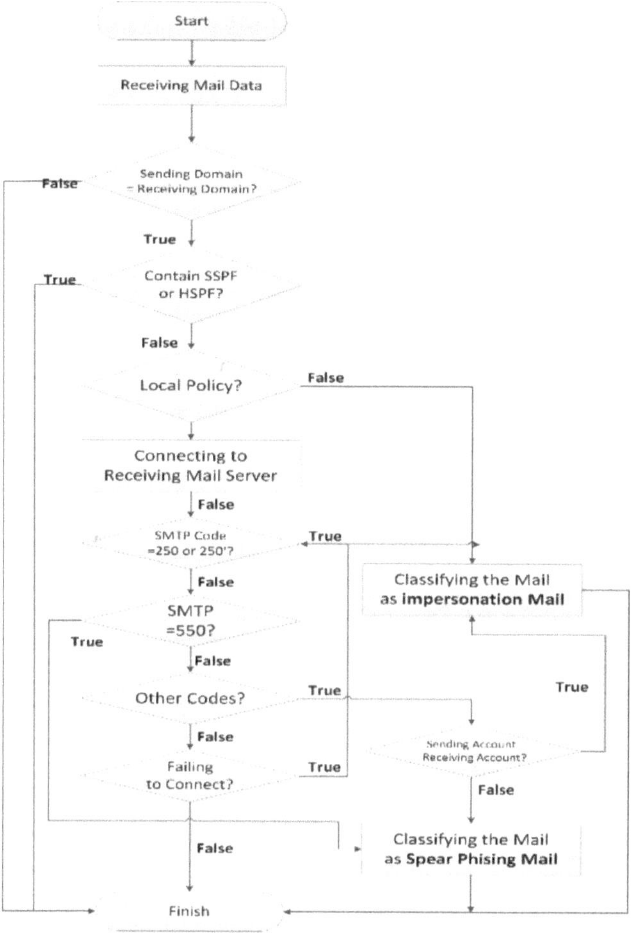

Fig. 3. Decision Flow of Classifier 2 for Cross-Domain Email Analysis

5 Evaluation

To evaluate the performance of the proposed algorithm, it was applied to organizations that use email systems. The tests were conducted in three institutions, each of which was already using an existing anti-spam solution. The proposed algorithm was introduced to analyze emails that had already been filtered by these existing solutions. Therefore, any abnormal emails detected by the proposed algorithm indicate its additional effectiveness over conventional systems.

The results show that the proposed algorithm identified a substantial number of malicious emails that were not detected by existing solutions, demonstrating its superior performance. The evaluation results are presented in Table 1.

Table 1. Results of Evaluation

Organization	Type	Period	Blocked Mail	Incoming Mail	Percent
A	Company	2021.9.9–2021.10.1	3,418	6,906	49.5%
B	University	2021.6.8 ~ 2021.7.16	38,448	563,111	6.8%
C	Company	2021.10.29 ~ 2021.11.19	6,746	55,487	12.2%

6 Conclusion

This study collected mail data to determine whether emails are malicious. Based on this approach, a security algorithm was proposed and empirically validated by applying it to real-world email environments.

The academic and practical implications of this research are as follows. First, this study contributes to the existing literature by detecting malicious emails using structured mail data and SMTP response codes. This approach opens new possibilities for future research to explore diverse security designs by extending the application of SMTP across broader contexts. Second, this study provides a systematic framework for analyzing malicious emails, offering practical insights that can help information system managers more effectively respond to email-based attacks.

Despite these contributions, several limitations should be acknowledged. First, this study does not analyze email content, such as message bodies or embedded links. Future research should consider integrating signature-based methods with SMTP-based identification to enhance detection accuracy. Second, the evaluation was conducted on a limited number of institutions in South Korea. Future studies should validate the proposed approach across more diverse organizational settings to improve the generalizability of the findings.

References

Cohen, A., Nissim, N., Elovici, Y.: Novel set of general descriptive features for enhanced detection of malicious emails using machine learning methods. Expert Syst. Appl. **110**, 143–169 (2018)

Eryilmaz, E.E., Şahin, D.O., Kılıç, E.: Machine Learning Based Spam E-mail Detection System for Turkish. 2020 5th International Conference on Computer Science and Engineering (UBMK), pp. 7–12 (2020)

Fang, Y., Zhang, C., Huang, C., Liu, L., Yang, Y.: Phishing email detection using improved RCNN model with multilevel vectors and attention mechanism. IEEE Access **7**, 56329–56340 (2019)

Kumari, A., Agrawal, N., Lilhore, U.: Clustering malicious spam in email systems using mass mailing. 2018 2nd International Conference on Inventive Systems and Control (ICISC), pp. 870–875 (2018)

Lee, C.-D., Chen, T.-H.: New Secure and Practical E-Mail Protocol with Perfect Forward Secrecy. Symmetry **13**(7), 1144 (2021)

Mahmoud, A.B.: E-mail advertising in Syria: Assessing beliefs, attitudes, and behaviors. J. Promot. Manag. **21**(6), 649–665 (2015)

Mandal, N., Jadhav, S.: A survey on network security tools for open source. IEEE International Conference on Current Trends in Advanced Computing (ICCTAC) **2016**, 1–6 (2016)

Rudd, E.M., Harang, R., Saxe, J.: Meade: Towards a malicious email attachment detection engine. IEEE International Symposium on Technologies for Homeland Security (HST) **2018**, 1–7 (2018)

Sumathi, S., Pugalendhi, G.K.: Cognition based spam mail text analysis using combined approach of deep neural network classifier and random forest. J. Ambient. Intell. Humaniz. Comput. **12**(6), 5721–5731 (2021)

Wu, J.-S., Hsu, F.-H., Lee, C.-H.: Identifying IoT Devices with SMTP. 2021 IEEE International Conference on Consumer Electronics-Taiwan (ICCE-TW), pp. 1–2 (2021)

Zobal, L., Kolár, D., Kroustek, J.: Exploring Current E-mail Cyber Threats using Authenticated SMTP Honeypot. 17th International Joint Conference on e-Business and Telecommunications (ICETE), pp. 253–262 (2020)

The Use of Artificial Intelligence and Data Science in Cyber Security

NNHammerGuard: Leveraging DNNs' Fault Tolerance for Efficient Row Hammer Defense

Lening Wang[1,2(✉)], Saide Zhu[3], Biresh Kumar Joardar[2], and Xin Fu[2]

[1] Department of Computer Science, Prairie View A&M University, Prairie View, USA
[2] Department of Electrical and Computer Engineering, University of Houston, Houston, USA
lenwang@pvamu.edu
[3] Department of Information Science Technology, Penn State Berks, Reading, USA

Abstract. Row Hammer is a hardware-based fault injection attack that flips bits in Dynamic Random-Access Memory (DRAM) by repeatedly accessing (hammering) adjacent memory rows. Previous studies have shown that carefully induced bit flips can degrade the prediction accuracy of DNN models to the level of random guessing. Traditional defense schemes often incur substantial overhead, requiring extra logic to count or monitor potential victim rows and increase their refresh rates. In this work, we leverage the intrinsic fault-tolerant properties of DNNs to propose NNHammerGuard, a lightweight method for protecting DNNs from Row Hammer attacks. We first characterize DNN fault tolerance capability and identify the most vulnerable memory cells in DRAM. Building on these insights, we introduce a significance-aware column-based weight mapping scheme that selectively loads DNN weights onto DRAM. Our approach ensures that significant weight bits are strictly protected from corruption, while less significant bit whose corruption has minimal impact on accuracy are safeguarded with lighter mechanisms using row-based offset randomization. Additionally, we incorporate a weight checksum mechanism to enable early detection of Row Hammer attacks. Leveraging the bit-slice nature of our approach, we further optimize the computation flow of NNHammerGuard to maximize efficiency. Our experimental results show that NNHammerGuard's selective protection strategy provides strong resilience against Row Hammer attacks with minimal overhead.

Keywords: DNN Security · Bit Flips · Row Hammer

1 Introduction

Row Hammer [9] is a hardware-based fault injection attack that induces bit flips in DRAM by repeatedly accessing (hammering) adjacent memory rows. This phenomenon is caused due to electrical interference between cells on adjacent

H. R. Arabnia et al. (Eds.): CSCI 2025, CCIS 2969, pp. 395–411, 2027.
https://doi.org/10.1007/978-3-032-28460-0_28

rows, enabling a malicious attacker to corrupt memory contents without requiring direct access. Row Hammer has more recently become a significant security concern for deep neural networks (DNNs). Carefully induced bit flips can dramatically degrade the accuracy and performance of DNN models [26]. As DNNs are increasingly deployed in safety-critical and high-assurance domains, their vulnerability to hardware-level attacks raises significant safety concerns.

Traditional defenses against Row Hammer aim to refresh victim rows before bit flips occur by implementing access monitoring along with targeted or probabilistic refresh strategies [10,16,18,20]. However, these approaches often introduce substantial overhead in latency, energy consumption, and hardware complexity [16,20]. More critically, although probabilistic defense mechanisms can reduce implementation overhead, they do not provide deterministic guarantees [10,18].

In this work, we leverage the intrinsic fault tolerance of DNNs [25] to propose NNHammerGuard, a lightweight and efficient method for protecting neural networks from Row Hammer attacks. We begin by studying the fault-tolerant capabilities of modern DNN architectures and characterizing the vulnerable memory cell properties related to Row Hammer attacks. It is well known that redundancy exists in neural network models [22] and that only a subset of bits (referred to as significant bits) are sensitive to corruption. We also observe that, in practice, the number and location of possible bit flipped on DRAM is limited, and only a small fraction of cells (referred to as vulnerable cells) are likely to be flipped [9,15]. Building on these insights, we introduce a significance-aware column-based weight mapping scheme that selectively loads DNN weights onto DRAM. Our approach ensures that the most significant weight bits are placed on the same DRAM rows, which are strictly protected from corruption by monitoring the access patterns. Less significant bits are safeguarded using a lighter-weight protection scheme, as bit flips in these positions typically have minimal impact on model predictions.

To further enhance resilience, we propose a Row-based Offset-Aware Randomized Mapping scheme to distribute remaining less significant bits across DRAM rows with weaker protection. This mapping intentionally misaligns the positions of remaining relative significant bits within the less significant regions and the locations of vulnerable DRAM cells, thereby reducing the likelihood that bit flips concentrate on remaining critical positions. In addition, we embed a weight checksum mechanism to enable early detection of Row Hammer attacks, allowing the system to trigger targeted refresh operations before significant damage occurs.

To maximize performance, we optimize the computational flow of NNHammerGuard. The bit-slicing nature of our approach naturally aligns with existing bit-sliced neural network accelerators [1,2], enabling efficient integration and high throughput. We evaluate NNHammerGuard on CIFAR-10, ImageNet with multiple DNN models. Experimental results demonstrate that NNHammerGuard's selective protection strategy, which combines strong protection for significant bits through significance-aware column-based mapping with lighter protection for less significant bits using Row-based Offset-Aware Randomized Mapping and checksum-based Row Hammer detection, achieves strong resilience against Row Hammer attacks while incurring negligible overhead.

2 Background

2.1 Row Hammer Attack and Defense Solutions

Row Hammer is a hardware-based fault injection attack that exploits electrical interference in modern DRAM memory. By repeatedly accessing ("hammering") certain memory rows (aggressor rows) at high frequency, an attacker can induce bit flips in adjacent rows (victim rows), changing stored bits from 1 to 0 or vice versa. First demonstrated around 2014 [10], Row Hammer exposed a fundamental vulnerability in commodity DRAM chips [15]. As chip density has increased, the number of accesses required to trigger bit flips (referred to as the Row Hammer threshold) has decreased significantly. For instance, approximately 140,000 accesses were needed to cause bit flips in DDR3 memory in 2014, while by 2020, as few as 5,000 accesses were sufficient for LPDDR4 [14]. Row Hammer represents a critical threat to system security. Traditional defenses against Row Hammer aim to refresh victim rows before bit flips occur. One approach is to increase the DRAM refresh rate, but this incurs significant energy overhead. Currently, most DRAM manufacturers employ Target Row Refresh (TRR) techniques [6] to proactively refresh potential victim rows. Another approach is to monitor the access pattern and to refresh the target row(s) only [16,20]. However, this requires substantial additional hardware to track access counts across high-density DRAM chips, and it is impractical to monitor every row. Techniques such as PARA [10] and RRS [18] introduce probabilistic refresh or randomized row swapping to reduce relate overhead, but they do not offer deterministic guarantees for safety-critical inference.

2.2 Bit Flips Attacks in DNN and Defense Solutions

Aiming to degrade the performance of deep neural network (DNN) models, Bit-Flip Attacks (BFA) exploit the Row Hammer phenomenon in commodity DRAM. Prior work [4,17] has developed techniques to identify the most vulnerable bits which are stored in victim cells with the largest gradients that most impact model accuracy. To mitigate the weight changes caused by bit flips, researchers have proposed modifying model parameters and structures through approaches such as weight reconstruction [13] and weight clipping [8]. However, deploying these defenses is challenging because they typically require retraining the model. Training high-performance DNNs demands massive computational resources [23,28], which presents a barrier for most practitioners. Even worse, altering the structural and bit-level representation of DNN models can lead to compatibility issues with existing systems and various accelerators.

3 NNHammerGuard

3.1 Threat Model

Without loss of generality, we adopt the threat model from prior Bit-Flip Attack (BFA) work [26]. In this model, an attacker exploits the Row Hammer vulnerability to flip bits in critical weight locations of a victim DNN model, assuming full knowledge of the model. Through reverse engineering and public sources, the attacker is also assumed to know the DRAM layout and virtual-to-physical address mapping. This allows targeting significant weight bits stored in physically vulnerable cells where "significant" bits are those whose corruption severely degrades accuracy, and "vulnerable" cells are prone to Row Hammer-induced flips due to manufacturing variation. Following standard practice, we assume the model uses 8-bit fixed-point quantization for weights to reflect common deployment scenarios and enable efficient attack simulation.

3.2 Key Observations

We introduce NNHammerGuard based on three key observations, with several preliminary experiments obtained from an 8-bit quantized ResNet-16 model with 91.2% prediction accuracy on the CIFAR-10 dataset.

Observation 1: The Number of Significant Bits Is Limited, and Most Significant Bits Tend to Be Concentrated in the Higher-Order Portions of the Weights. Figure 1 shows the accuracy degradation of DNN model under three different attack settings: a) a targeted bit-flip attack where the most significant bits are flipped, b) a targeted bit-flip attack where the most 100 significant bits are protected, and c) a random bit-flip attack. Each attack type was repeated 10 times using different random seeds, and the results were averaged. The significant bits were ranked and selected using the Gradient-Based Bit Ranking (GBR) method described in [26] which identifies the top p bits whose corruption most significantly increases the loss function. While the targeted attack in setting (a) reduces accuracy to nearly random guessing with as few as 10 bits flipped, both setting (b) and setting (c) have negligible impact on prediction accuracy, even when a larger number of bits are flipped.

We further recorded the position distribution of the 200 most significant bits in Fig. 2. Our analysis reveals that these bits are predominantly concentrated in the higher-order portions of the weights. In the context of fixed-point 8-bit quantization, the higher-order portion corresponds to the most significant bits (MSBs), specifically bit indices 7 and 6 in our representation.

Insights: It is well known that DNNs are inherently tolerant to certain types of random noise [25] and small changes. However, bit flips in the higher-order tend to induce larger deviations in computed outputs, leading to more pronounced degradation. Therefore, for an attack to be effective, the flipped weight bits must be carefully selected. Motivated by this observation, we propose to

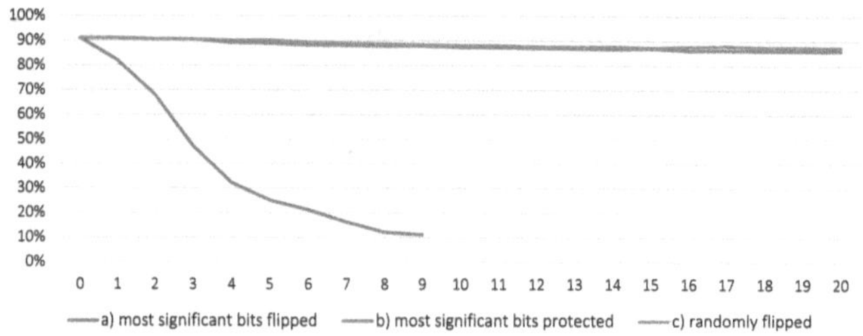

Fig. 1. Model prediction accuracy vs. the number of flipped bits under three attack settings.

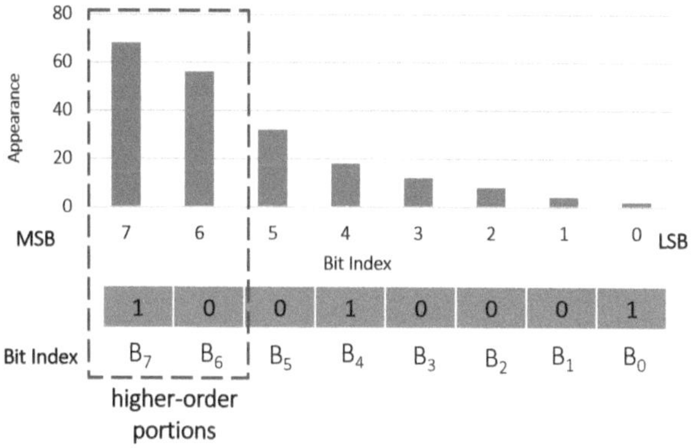

Fig. 2. The Distribution of Significant Bits

strictly protect the most significant bits located in the higher-order portions of the weights, while safeguarding less significant bits with a lower-cost, weaker protection scheme to achieve a balanced trade-off between resilience and efficiency.

Observation 2: DNN Weights Are Sensitive to the Direction of Bit Flips. We further analyzed the flip directions of the most significant bits across various DNN models and observed that the impact of bit flips depends strongly on whether the transition is from 0 to 1 or from 1 to 0. While the observed flip directions were approximately equally distributed overall, we found that 0-to-1 flips tend to have a greater impact on model accuracy when they occur in positive weight values, whereas 1-to-0 flips have a more pronounced effect in negative weight values. These results are summarized in Table 1.

Insights: In DNNs, intermediate layers extract and encode features that contribute to final classification decisions. During inference, these extracted features are represented by the activations of intermediate neurons. The activation values in layer l are computed through matrix multiplication and summation of the activations from the preceding layer and the corresponding weights. A nonlinear ReLU function is then applied to ensure that the resulting activations are non-negative. A positive activation indicates the presence of a feature, while its magnitude reflects the model's confidence in that feature.

Returning to our observation: 0-to-1 bit flips in negative weights and 1-to-0 bit flips in positive weights tend to reduce the value of weights. As a result, the corresponding activations become smaller, weakening or even eliminating the associated features. However, DNNs often exhibit resilience to such degradations due to redundant feature representations [5,22]. Similarly, decreases in neuron values may even have little to no impact due to the zeroing effect of the ReLU activation function. In contrast, bit flips that increase the magnitude of a weight can have a more harmful effect. Such flips may lead to the unintended activation of irrelevant features. For example, if a neuron representing the feature"leg" is inadvertently strongly activated in an input image of a car, the network may produce a high-confidence incorrect prediction. This phenomenon is consistent with findings from [25], which show that abnormally large neuron activations can trigger severe misclassification.

Table 1. The Distribution of Flip Directions

Flip Direction	1 to 0	0 to 1	Increase Weight(s)	Decrease Weight(s)
Distribution	43.8	56.2%	72.8%	27.2%

Observation 3: The Number of Flippable Cells, Along with Their Locations and Flip Directions Is Limited. We performed Row Hammer attacks on four real-world DRAM chips from two different manufacturers to evaluate the fundamental limitations imposed by Row Hammer, specifically focusing on (1) the number of cells that can be flipped and (2) the location of cells that can be flipped.

Although bit flips were observed, flippable cells were extremely rare, approximately 10^{-6} of all DRAM cells, consistent with prior work [10], confirming that only a small subset is vulnerable to Row Hammer. We analyzed vulnerability at the row and column levels since cell level behavior is not directly observable.

At the row level, We tested 100 random rows and repeated each experiment 20 times for statistical averaging. Figure 3 shows the minimum number of aggressor activations needed to trigger flips versus the percentage of victim rows affected. A small portion of highly vulnerable rows, about 1 %, flipped with few activations, while most required much higher counts. Table 2 shows the relationship between row level vulnerability and activation thresholds. As the number of activations increases, bit flips appear in more victim rows. To assess repeatability, we define Reproduce Capability as the percentage of rows that flipped in at least 15 of 20 trials. This metric captures the consistency of vulnerability under repeated Row Hammer and confirms that vulnerable rows are persistent and location fixed.

At the column level, 68% columns showed no flips, indicating that vulnerable cells are highly localized. Combining both analyses, we infer that flippable cells appear at intersections of vulnerable rows and columns. Lastly, we observed a clear bias in flip direction.

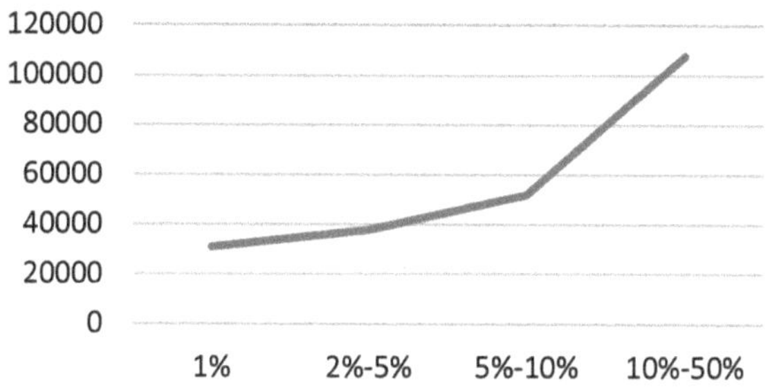

Fig. 3. Activation Count Required to Trigger Bit Flips (Horizontal Axis: Percentage of Victim Rows; Vertical Axis: Activation Count)

Table 2. Row-level Cell Vulnerability

Activation Count	40 K	100 K
Number of Rows	12	48
Reproduce Capability	75%	70%

Insights: Due to manufacturing variations, some DRAM cells exhibit weaker insulation or reduced charge margins, making them more susceptible to bit flips under Row Hammer attacks [26,27]. These vulnerable cells are more likely to experience charge leakage or disturbance. Modern DRAM devices with Row Hammer mitigation techniques also limits the extent of hammering, making successful bit flips more difficult. In practical scenarios, attackers can typically induce only 10 to 100 bit flips per session, and only under carefully tuned access patterns and system conditions.

3.3 Implementation of NNHammerGuard

Overview. Figure 4 shows the overall architecture of NNHammerGuard. From the application perspective, it operates in four phases:

1. Model Deployment Phase. The original DNN model is converted to unsigned 8-bit quantization to enable hardware-level bit protection. Quantization simplifies weight representation and allows bit-aware mapping to DRAM, which is essential for later protection and optimization.

2. Weight Loading Phase. To exploit the counter-based refresh defense in modern DRAM, Row Hammer mitigation mechanisms track row access frequency and refresh adjacent rows when a threshold is reached. NNHammerGuard places all significant weight bits on the same DRAM row so they can be monitored and refreshed together. Less significant bits are protected by a mismatch strategy: they are deliberately misaligned with known vulnerable cells using a row offset mechanism that randomizes their placement.

3. Model Acceleration Phase. To handle the fragmented data layout, NNHammerGuard adopts a bit-slice-oriented computation flow. It processes high- and low-order bits separately, enabling efficient accumulation and neuron reconstruction. This approach aligns well with bit-serial or bit-slice DNN accelerators [1,2] and maintains high execution efficiency.

4. Model Monitoring Phase. Since lower-order bits on less protected rows may still flip, NNHammerGuard integrates a lightweight monitoring mechanism. If flips are detected, the system reloads weights from trusted storage to restore integrity.

Overall, these four phases form a layered, efficient, and DNN-aware defense. NNHammerGuard selectively protects critical bits, preserves performance through bit-slice-friendly computation, and ensures reliability with fast recovery.

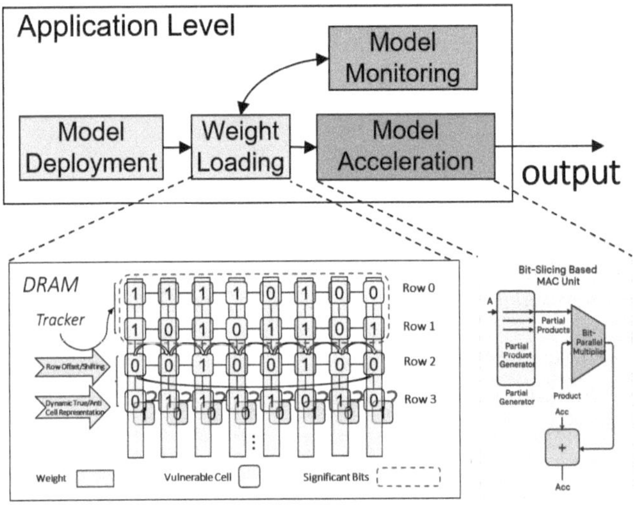

Fig. 4. Overview of NNHammerGuard

Model Deployment: Converting to Unsigned Quantization. In fixed-point weight representations, both signed and unsigned formats are common.

Unsigned weights are always non-negative, while signed weights can represent positive and negative values. Based on Observation 2, bit flips that increase a weight's value are more harmful to model accuracy, such as 0→1 flips in positive weights or 1→0 flips in negative weights. These flips can amplify existing features or spuriously activate unrelated ones, often causing misclassifications. To unify handling, we apply a shift transformation to signed weights, converting them to non-negative values before storage.

For 8-bit quantization, let q denote the quantized integer and s the scaling factor. In signed quantization, $q \in [0, 2^8]$, while in unsigned quantization $q \in [-2^7, 2^7 - 1]$. The scaling factor s is computed using Eq. 1.

$$s = \frac{\max\left(|w_{\min}|, |w_{\max}|\right)}{127} \tag{1}$$

let w denote the original weight. Then we can get q using Eq. 2 for 8-bit unsigned quantization.

$$q = \text{round}\left(\frac{w}{s}\right) \tag{2}$$

The quantized weight can be restored using Eq. 3.

$$\hat{w} = q \cdot s \tag{3}$$

To convert to an unsigned representation, the signed quantization value q is shifted by 128 to form a new unsigned integer q', where $q' \in [0, 255]$. In this case, Eq. 3 becomes Eq. 4:

$$\hat{w} = (q + 128 - 128) \cdot s = q' \cdot s - 128 \cdot s \tag{4}$$

In practice, when computing neuron outputs using the unassigned format, the shift must be compensated for. The complete neuron output computation becomes Eq. 5, where x demotes the input and y denotes the output.

$$y = \hat{w} \cdot x = (q' \cdot s - 128 \cdot s) \cdot x = (q' \cdot x - 128 \cdot x) \cdot s \tag{5}$$

This transformation ensures a uniform flip direction interpretation which bit flips from 0 to 1 now always increase the numerical value of effective weight, regardless of the original sign. This consistency simplifies the design of NNHammerGuard defense mechanisms.

Weight Loading: Column-Based Weight Mapping. The significant bits in DNN weights are crucial for maintaining model accuracy, so protecting them from Row Hammer-induced corruption is essential. Because DRAM refresh operates at the row level, grouping significant bits on the same row enables targeted refresh before they are hammered. Observation 1 shows that most significant bits lie in the higher-order positions of weights. To protect them, we monitor their access counts using a counter-based Row Hammer defense scheme, refreshing vulnerable rows before flips occur.

Figure 5 (a) shows a conventional row-based weight mapping which stores all bits of each weight on a single row. Accessing a weight activates its entire row, making it impossible to selectively protect only the significant bits. This requires extra hardware to track many rows and wastes energy refreshing cells that hold insignificant bits with little effect on accuracy.

To overcome this, we propose a column-based weight mapping scheme as shown in Fig. 5 (b). Instead of storing whole weights on one row, our method splits each weight across rows and groups the most significant bits (two leftmost bits, based on our experience) from many weights onto the same row. These rows are monitored by access counters [16], which trigger targeted refresh when activation counts exceed a threshold, preventing bit flips on critical parameters.

By concentrating protection on rows holding significant bits and spreading less critical bits elsewhere, this scheme provides selective protection with minimal hardware cost, reducing energy use and improving defense efficiency against Row Hammer attacks.

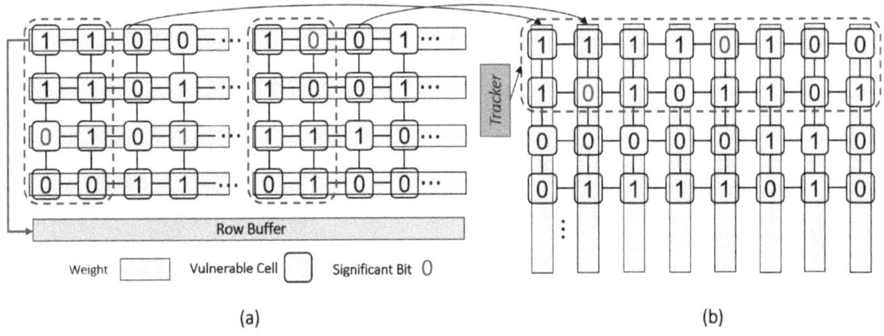

Fig. 5. Column-based Weight Mapping

Weight Loading: Row-Based Offset-Aware Randomized Weight Mapping. The lower-order portions of DNN weights are more tolerant to bit flips and do not require expensive counter-based Row Hammer defenses. Based on Observation 3, only a small fraction of DRAM cells are vulnerable to Row Hammer, mainly due to manufacturing imperfections such as variations in capacitor size. These defects make some cells to cross the read threshold after repeated activation of nearby rows. Such vulnerability is rare but persistent over the chip's lifetime. To mitigate this, we introduce a row-based offset mapping scheme. We first profile vulnerable cells using the method in [26]. A naive strategy would be to precompute the significance of all lower-order bits and avoid placing critical bits on vulnerable cells. While feasible in principle [14,26], exhaustive ranking is impractical due to the millions of bits in typical DNNs and the lack of a clear boundary between "significant" and "insignificant" lower bits.

Instead, guided by Observation 2, we note that bit flips reducing a weight's value usually have little effect on predictions, while flips increasing a weight's value are more harmful. As shown in Fig. 5, during weight loading, if a '0' bit in a positive weight would be placed on a known vulnerable cell (with unsigned quantization, negative weights no longer apply), we apply a small random row offset to relocate it. This lowers the chance that vulnerable cells align with critical bit positions, reducing the risk of harmful bit flips (Fig. 6).

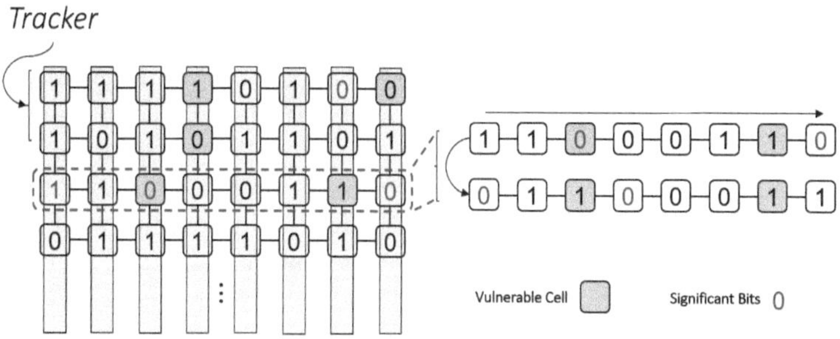

Fig. 6. Row-based Offset-Aware Randomized Weight Mapping

This randomization strategy reduces the likelihood that high-impact bits align with vulnerable DRAM cells. It also increases attack uncertainty, making it harder for adversaries to target critical parameters using Row Hammer. As a result, it reduces both the bit flip rate and the attack's effectiveness while preserving DNN accuracy with minimal protection cost.

Model Acceleration: Bit-Sliced Neuron Computation. As shown in Eq. 6, the dominant operation in Deep Neural Network (DNN) inference is the multiply-and-accumulate (MAC) between the intermediate activations x of the previous layer and the current layer's weights w. These operations require frequent and efficient weight loading. In traditional row-based mapping, all bits of a weight reside in a single row, allowing retrieval with one access. In contrast, column-based mapping scatters each weight's bits across multiple rows, requiring multiple accesses to reconstruct a weight. This fragmented pattern increases both energy consumption and access latency, which is critical since memory access dominates DNN energy cost [23, 24].

$$y = \hat{w} \cdot x = w_1 \cdot x_1 + x_2 \cdot x_2 + ... + w_n \cdot x_n \tag{6}$$

To mitigate this overhead, we propose a bit-slice-oriented neuron computation method that aligns inference flow with the memory layout. Equation 7 illustrates the quantization representation of an 8-bit weight, where b_i denotes the bit at position index i, and b_7 corresponds to the most significant bit (MSB).

$$w_q = b_7 \cdot 2^7 + b_6 \cdot 2^6 + \cdots + b_0 \cdot 2^0 \tag{7}$$

For a given layer l, all weights share the same quantization scale, so the same bit position across different weights has the same scaling factor $(b_i \cdot 2^i \cdot s)$. This property makes it feasible to process these bit positions independently but consistently across the network. Hence, the neuron computation $y = x \cdot w$ can be reformulated from Eqs. 6 to 8:

$$x \times w_q = x \times \left(\sum_{i=0}^{7} b_i \cdot 2^i \right) = \sum_{i=0}^{7} (x \times b_i) \cdot 2^i \tag{8}$$

Based on this, we design a bit-sliced computation flow that processes higher-order and lower-order bits separately (Fig. 7). Each MAC unit generates partial results, which are summed to produce the final activation. This format preserves computational correctness, supports pipelined or parallel execution, and achieves energy efficiency on bit-serial or bit-slice DNN accelerators [1,2]. By reusing partially loaded weights, reducing redundant fetches, and aligning with hardware-friendly computation, this method improves inference throughput while retaining the benefits of selective protection.

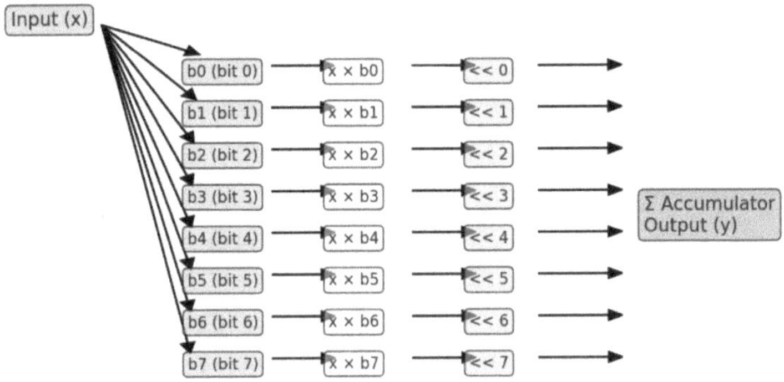

Fig. 7. Bit Slice based Computation Flow

Model Monitoring: Weight Checksum-Based Row Hammer Detection. NNHammerGuard leverages the inherent fault tolerance of DNN models to mitigate Row Hammer attacks, but bit flips can still occur in weakly protected rows. Once flipped, a weight bit remains corrupted until the weights are reloaded, and accumulated errors over time can degrade accuracy. To prevent this, NNHammerGuard introduces a periodic checksum mechanism. It reuses existing MAC units to compute a checksum by setting the input $x = 1$, effectively summing the stored weights. A mismatch from the expected checksum signals potential corruption, triggering a reload of correct weights from trusted storage.

4 Evaluation

4.1 Experimental Setup

Dataset and Benchmark: To thoroughly evaluate the performance of NNHammerGuard, we utilize two widely used image classification datasets that represent small and large-scale tasks: CIFAR-10 [11], and ImageNet [3]. We evaluate our approach on four diverse DNN architectures selected from publicly available pre-trained models: AlexNet [12], MobileNetV2 [19], VGG16 [21], and ResNet18 [7]. These models vary in depth, parameter count, and computational complexity, allowing us to assess NNHammerGuard's effectiveness and scalability across a broad range of network designs. **BFA Configuration:** As introduced in Sec 3.1, we adopt the same vulnerable bit search configuration as prior work [17,26]. Specifically, we relax the constraint that limits the number of bit flips to 100 in order to emulate the capabilities of a strong adversary, while acknowledging that inducing more than 100 flips is impractical in real-world systems. For each dataset and network configuration, we generate 10 independent sets of bit-flip patterns using different random seeds. Each set is crafted to degrade the model's inference accuracy to the level of random guessing, thereby ensuring both broad coverage and robustness in evaluating the effectiveness of the attack and the resilience of our defense mechanism.

4.2 Performance Evaluation:

Accuracy: Figure 8 shows the accuracy results on CIFAR-10, where #0, #5, ..., and #100 denote the number of bit flips. Without any protection, the model accuracy quickly drops to the random-guess level with only a few bit flips. When applying only the Column-based Mapping (denoted as CM), accuracy is largely preserved since the most significant bits are protected. However, accuracy degradation still occurs due to accumulated flips in less significant locations (with the incremental of flipped bits). With the full NNHammerGuard protection scheme (denoted as HG), the accuracy loss is reduced to less than 1%, which is negligible. This improvement comes from NNHammerGuard's ability to protect not only the most significant bits in the higher-order portions of the weights but also to reduce the probability of flips affecting relatively significant bits in lower-order portions. Similar results are observed on ImageNet, as shown in Fig. 9.

We also measure the impact of our checksum based Row Hammer detection, which the sum of the weights (layer by layer) will be calculated periodically. Without our Row Hammer detection, the bit flips can be accumulated through time and final degrade model's accuracy. The results are summarized in Fig. 10.

Hardware Overhead and Throughput Improvement: The major hardware components of NNHammerGuard can be categorized into four aspects: a) the counter and table design for rows that store significant bits, b) the bits used to record row offsets, and c) the overhead required to verify weight sums. We adopt the counter-table design from Graphene. However, unlike Graphene,

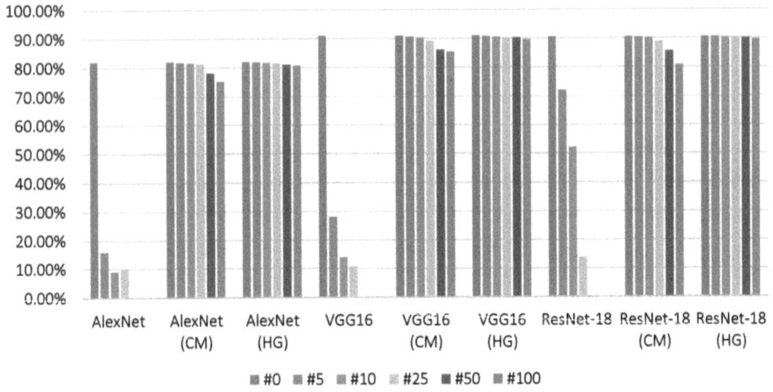

Fig. 8. Accuracy Results on Cifar-10

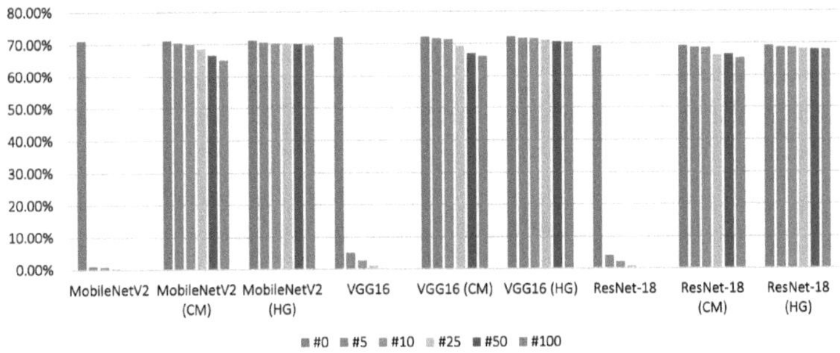

Fig. 9. Accuracy Results on ImageNet

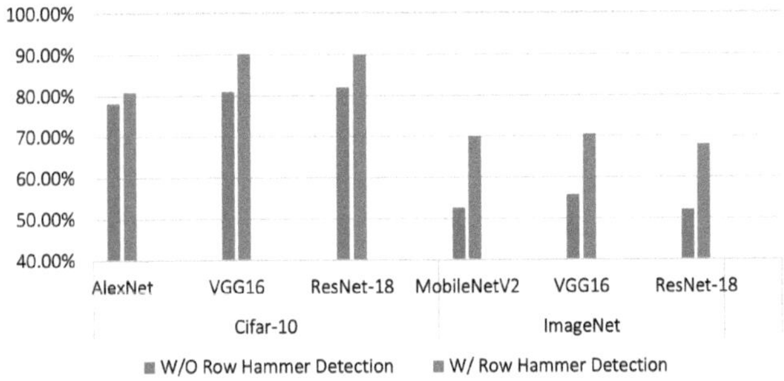

Fig. 10. Impact of Row Hammer Detection

NNHammerGuard only requires counters for rows storing the most significant bits (the two leftmost bits). As a result, NNHammerGuard needs only 1.1 KB of SRAM for the counter table, reducing the storage cost by approximately 70% compared to Graphene. The additional hardware requirements include the bits to record row offsets which incur about 0.4% DRAM overhead and 5% energy overhead. To evaluate performance, we adopt the FPGA accelerator design from

[1] and implement our bit-slicing–based dataflow on a Virtex-7 VC709 FPGA platform. Compared to the original row-based dataflow, the experimental results show that the bit-slicing design reduces memory accesses by a factor of 7.8 × and decreases computation delay by 7.2 ×.

5 Conclusion

In this work, we presented NNHammerGuard, a lightweight and efficient defense mechanism against Row Hammer attacks on DNNs. Leveraging the intrinsic fault tolerance of DNNs and the properties of vulnerable memory cells, NNHammerGuard introduces a significance-aware column-based DRAM mapping scheme, Row-based Offset-Aware Randomized Weight Mapping, and Weight Checksum for Row Hammer Detection to protect critical weight bits with minimal overhead. In addition, we proposed a bit-slice–based computation flow to optimize dataflow efficiency under the proposed memory mapping method. Our design integrates seamlessly with bit-sliced accelerators, delivering high throughput. Experimental evaluation on CIFAR-10 and ImageNet demonstrates that NNHammerGuard provides strong resilience against Row Hammer attacks while incurring negligible performance and memory overhead.

Acknowledgment. This research is partially supported by NSF grants CCF-2130688, CNS-2107057 and RISE support program.

References

1. Bilaniuk, O., Wagner, S., Savaria, Y., David, J.P.: Bit-slicing FPGA accelerator for quantized neural networks. In: 2019 IEEE International Symposium on Circuits and Systems (ISCAS), pp. 1–5 (2019). https://doi.org/10.1109/ISCAS.2019.8702332
2. Choi, I., Yoon, Y.S., Yang, J.S.: Bit-slice architecture for DNN acceleration with slice-level sparsity enhancement and exploitation. In: 2025 IEEE International Symposium on High Performance Computer Architecture (HPCA) (2025)
3. Deng, J., Dong, W., Socher, R., Li, L.J., Li, K., Fei-Fei, L.: Imagenet: a large-scale hierarchical image database. In: 2009 IEEE Conference on Computer Vision and Pattern Recognition, pp. 248–255 (2009). https://doi.org/10.1109/CVPR.2009.5206848
4. Dong, J., et al.: One-bit flip is all you need: when bit-flip attack meets model training. In: Proceedings of the IEEE/CVF International Conference on Computer Vision, pp. 4688–4698
5. Eccles, B.J., Rodgers, P., Kilpatrick, P., Spence, I., Varghese, B.: DNNshifter: an efficient DNN pruning system for edge computing. Futur. Gener. Comput. Syst. **152**, 43–54 (2024)
6. Frigo, P., et al.: Trrespass: exploiting the many sides of target row refresh. In: 2020 IEEE Symposium on Security and Privacy (SP), pp. 747–762 (2020). https://doi.org/10.1109/SP40000.2020.00090

7. He, K., Zhang, X., Ren, S., Sun, J.: Deep residual learning for image recognition. In: Proceedings of the IEEE Conference on Computer Vision and Pattern Recognition, pp. 770–778 (2016)
8. He, Z., Rakin, A.S., Li, J., Chakrabarti, C., Fan, D.: Defending and harnessing the bit-flip based adversarial weight attack. In: Proceedings of the IEEE/CVF Conference on Computer Vision and Pattern Recognition, pp. 14095–14103 (2020)
9. Kim, J.S., et al.: Revisiting rowhammer: an experimental analysis of modern dram devices and mitigation techniques. In: 2020 ACM/IEEE 47th Annual International Symposium on Computer Architecture (ISCA), pp. 638–651 (2020)
10. Kim, Y., et al.: Flipping bits in memory without accessing them: an experimental study of dram disturbance errors, pp. 361–372. ISCA '14, IEEE Press (2014)
11. Krizhevsky, A., Hinton, G.: Learning multiple layers of features from tiny images (2009)
12. Krizhevsky, A., Sutskever, I., Hinton, G.E.: Imagenet classification with deep convolutional neural networks. In: Pereira, F., Burges, C., Bottou, L., Weinberger, K. (eds.) Advances in Neural Information Processing Systems, vol. 25, Curran Associates, Inc (2012)
13. Li, J., et al.: Defending bit-flip attack through DNN weight reconstruction. In: 2020 57th ACM/IEEE Design Automation Conference (DAC), pp. 1–6. IEEE (2020)
14. Mutlu, O., Kim, J.S.: Rowhammer: a retrospective. IEEE Trans. Comput. Aided Des. Integr. Circuits Syst. **39**(8) (2019)
15. Orosa, L., et al.: A deeper look into rowhammer's sensitivities: experimental analysis of real dram chips and implications on future attacks and defenses. In: MICRO-54: 54th Annual IEEE/ACM International Symposium on Microarchitecture (2021)
16. Park, Y., Kwon, W., Lee, E., Ham, T.J., Ho Ahn, J., Lee, J.W.: Graphene: strong yet lightweight row hammer protection. In: 2020 53rd Annual IEEE/ACM International Symposium on Microarchitecture (MICRO), pp. 1–13 (2020)
17. Rakin, A.S., He, Z., Fan, D.: Bit-flip attack: crushing neural network with progressive bit search. In: Proceedings of the IEEE/CVF International Conference on Computer Vision, pp. 1211–1220 (2019)
18. Saileshwar, G., Wang, B., Qureshi, M., Nair, P.J.: Randomized row-swap: mitigating row hammer by breaking spatial correlation between aggressor and victim rows. In: Proceedings of the 27th ACM International Conference on Architectural Support for Programming Languages and Operating Systems. ASPLOS '22 (2022)
19. Sandler, M., Howard, A., Zhu, M., Zhmoginov, A., Chen, L.C.: Mobilenetv 2: inverted residuals and linear bottlenecks (2019)
20. Saxena, A., Saileshwar, G., Nair, P.J., Qureshi, M.: Aqua: scalable rowhammer mitigation by quarantining aggressor rows at runtime. In: 2022 55th IEEE/ACM International Symposium on Microarchitecture (MICRO), pp. 108–123 (2022)
21. Simonyan, K., Zisserman, A.: Very deep convolutional networks for large-scale image recognition (2015)
22. Tmamna, J., et al.: Pruning deep neural networks for green energy-efficient models: a survey. Cogn. Comput. **16**(6), 2931–2952 (2024)
23. Wan, Q., Wang, L., Wang, J., Song, S.L., Fu, X.: NAS-SE: designing a highly-efficient in-situ neural architecture search engine for large-scale deployment. In: 2023 56th IEEE/ACM International Symposium on Microarchitecture (MICRO) (2023)
24. Wan, Q., Xia, H., Zhang, X., Wang, L., Song, S.L., Fu, X.: Shift-BNN: highly-efficient probabilistic Bayesian neural network training via memory-friendly pat-

tern retrieving. In: MICRO-54: 54th Annual IEEE/ACM International Symposium on Microarchitecture. MICRO '21 (2021). https://doi.org/10.1145/3466752.3480120
25. Wang, L., Wan, Q., Wang, J., Chen, M., Peng, L., Fu, X.: Nani: energy-efficient neuron-aware hardware noise injection for adversarial defense using undervolting. J. Syst. Architect. **165**, 103424 (2025). https://doi.org/10.1016/j.sysarc.2025.103424
26. Yao, F., Rakin, A.S., Fan, D.: DeepHammer: depleting the intelligence of deep neural networks through targeted chain of bit flips. In: 29th USENIX Security Symposium (USENIX Security 20). USENIX Association (2020)
27. Yağlıkçı, A.G., et al.: Spatial variation-aware read disturbance defenses: experimental analysis of real dram chips and implications on future solutions (2024)
28. Zhang, C., et al.: Efficient one-shot neural architecture search with progressive choice freezing evolutionary search. Neurocomputing **598**, 127702 (2024). https://doi.org/10.1016/j.neucom.2024.127702

SpoofTrackBench: Interpretable AI for Spoof-Aware UAV Tracking and Benchmarking

Van Le[1(✉)] and Tan Le[2]

[1] Virginia Polytechnic Institute and State University, Blacksburg, VA 24061, USA
vanl@vt.edu

[2] Hampton University, Hampton, VA 23669, USA
tan.le@hamptonu.edu
https://sites.google.com/site/thanhtantp

Abstract. SpoofTrackBench is a reproducible, modular benchmark for evaluating adversarial robustness in real-time localization and tracking systems (RTLS) under radar spoofing. Leveraging the Hampton University Skyler Radar Sensor dataset, we simulate drift, ghost and mirror-type spoofing attacks and evaluate tracker performance using both Joint Probabilistic Data Association (JPDA) and Global Nearest Neighbor (GNN) architectures. Our framework separates clean and spoofed detection streams, visualizes spoof-induced trajectory divergence, and quantifies assignment errors via direct drift-from-truth metrics. Clustering overlays, injection-aware timelines, and scenario-adaptive visualizations enable interpretability across spoof types and configurations. Evaluation figures and logs are auto-exported for reproducible comparison. SpoofTrackBench sets a new standard for open, ethical benchmarking of spoof-aware tracking pipelines, enabling rigorous cross-architecture analysis and community validation.

Keywords: UAV spoofing · Trajectory analysis · Tracking algorithms · AI/ML · Radar tracking · Adversarial machine learning

1 Introduction

Radar-based tracking systems play a vital role in surveillance, autonomous navigation, and aerial asset monitoring. As these systems become increasingly integrated into civilian and defense infrastructure, they face rising threats from adversarial spoofing. Techniques such as drift, ghost and mirror injection can distort detection streams, mislead assignment logic, and destabilize tracker performance—posing safety, security, and operational risks [1–3]. In high-stakes environments like UAV coordination or border surveillance, even minor assignment errors can lead to mission failure or critical misinformation [4].

Despite growing awareness of spoofing risks, there is a lack of standardized, reproducible benchmarks that rigorously evaluate radar trackers under adversarial conditions. Most existing trackers are tested in clean environments with

H. R. Arabnia et al. (Eds.): CSCI 2025, CCIS 2969, pp. 412–427, 2027.
https://doi.org/10.1007/978-3-032-28460-0_29

synthetic clutter or noise, but without deliberate, labeled spoof injections. Comparisons across architectures—such as classical probabilistic trackers like Joint Probabilistic Data Association (JPDA) and graph-based trackers like modified Global Nearest Neighbor (GNN)—remain fragmented and unreproducible due to inconsistent scenario setups, injection logic, and evaluation metrics.

While progress has been made in spoof detection for domains such as GNSS [5,6] and Wi-Fi [7], radar-based tracking/RTLS benchmarks remain modality-specific and lack unified testing protocols. Existing work rarely incorporates visual interpretability (such as cluster overlays, trajectory continuity plots, or spoof annotation tools), which are crucial for understanding failure modes and reasoning about adversarial robustness. Moreover, most evaluations focus solely on tracker accuracy without considering reproducibility under randomized seeds, separation of clean and spoofed logs, or modularity for rapid scenario switching.

We present **SpoofTrackBench**, a modular, reproducible, and visually interpretable benchmark framework for radar-based tracking/RTLS in adversarial scenarios. SpoofTrackBench simulates radar detection streams with controlled injection of spoof types, offering labeled spoof logs, toggleable 2D/3D visualization modes, and tracker-agnostic evaluation harnesses. It supports both JPDA and GNN architectures, allowing bidirectional comparison under identical spoof conditions. The framework is built around robust logging, randomized scenario reinitialization, and interpretability overlays—making it suitable for both academic benchmarking and system-level diagnostic analysis.

Our proposed framework offers the following core contributions: 1) **Modular Spoof Injection Logic:** Parameterized drift, ghost and mirror spoofing, supporting rapid injection cycling and labeled detection streams. 2) **Quantitative Drift and Assignment Metrics:** Calculation of trajectory divergence, assignment misalignment, and spoof-induced cluster disruption for both JPDA and GNN trackers. 3) **Cross-Tracker Comparison Harness:** Unified metric logging and scenario reinitialization enabling fair, reproducible comparison.

1.1 Related Work and Literature Background

Radar Spoofing and Adversarial Tracking: Growing interest in adversarial resilience has spurred research into spoof detection across radar, GNSS, and wireless modalities. Das et al. [8] examined Kalman-based GNSS trackers under spoofed satellite signals, revealing assignment drift but lacking control over synthetic scenario variability. Gupta et al. [9] proposed interpretability techniques for Wi-Fi-based spoof detection, though their system did not track spatial cluster continuity or quantify assignment divergence in radar domains. Li and Mohapatra [10] advanced spoof classification metrics for single-modality pipelines but omitted cross-architecture evaluations or trajectory distortion metrics. These efforts highlight the fragmentation in current literature: most studies focus on spoof detection, not on spoof-aware tracking performance, and rarely support reproducible comparison across tracker types or spoof injection schemas.

Benchmarks in Related Domains: Benchmarks such as KITTI [11] and MOTChallenge [12] have shaped evaluation practices in computer vision and autonomous tracking. However, these datasets assume clean detection environments or focus on occlusion, leaving adversarial spoofing unaddressed. Few benchmarks simulate labeled injection scenarios, and fewer still include control over spoof parameters, separation of clean/spoofed logs, or interpretability overlays critical for understanding tracker failure modes. Radar-specific benchmarks remain scarce, especially those that support multiple spoof types, scenario modularity, and quantitative drift-from-truth metrics.

Need for Reproducible and Interpretable Adversarial Benchmarks: Across these domains, a recurring limitation is the absence of benchmarks that provide: 1) Modular spoof injection for diverse attack types (drift, ghost, mirror), 2) Tracker-agnostic evaluation with standard logging interfaces, 3) Visual overlays for interpretability and cluster inspection, 3) Reproducible scenario cycling and randomized seed management. SpoofTrackBench addresses these gaps through a unified framework designed for scientific rigor, extensibility, and practical clarity.

1.2 Contributions

We present SpoofTrackBench in Fig. 1, a reproducible benchmarking framework for radar-based tracking under adversarial spoofing conditions. It enables scientific comparison of tracker architectures, spoof injection types, and interpretability routines through modular simulation and standardized metrics. Specifically, our contributions include:

1) Modular Spoof Injection Logic: A configurable system supporting drift, ghost, and mirror spoof types with scenario-adaptive controls. Spoofs are injected using reproducible functions, enabling batch simulation and consistent labeling for interpretability and diagnostics.
2) Reproducible Drift Quantification and Evaluation: Separation of clean and spoofed detection streams, randomized seed cycling, and structured logging across time-steps. We

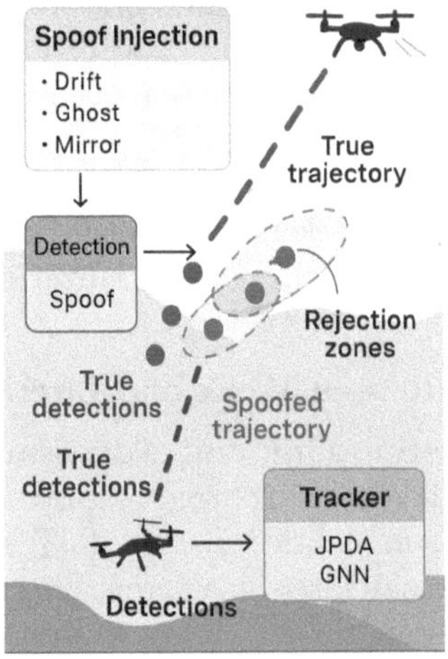

Fig. 1. Schematic comparison of spoof-aware trajectory tracking. JPDA and GNN respond differently to adversarial spoofing within gating regions: JPDA dilutes spoofed detections probabilistically, while GNN applies threshold-based rejection. The illustration highlights trajectory drift, spoof infiltration, and gating logic under adversarial conditions.

quantify drift-from-truth, assignment divergence, and cluster misalignment under controlled adversarial scenarios.
3) Tracker-Agnostic Benchmark Harness: A unified evaluation pipeline comparing JPDA and GNN trackers under identical spoof conditions.
4) Dataset Integration and Extensible Design: SpoofTrackBench supports modular scenario definitions and logging for reusability. The framework is designed for extensibility into future spoof types, multi-sensor fusion, and intelligent tracker architectures.

2 System Model

We consider a multi-sensor aerial surveillance architecture designed for spoof-aware tracking in contested environments [13,14]. The system comprises distributed sensing modalities feeding into a centralized AI-based Fusion Center, which performs adversarially resilient assignment, trajectory reconstruction, and interpretability logging.

2.1 Sensor Network and Data Ingestion

Sensor modalities include multiple sensors and/or detectors. 1) **Skyler Radar Sensors**: Provide high-resolution detection of non-cooperative aerial targets through micro-Doppler signatures. Particularly sensitive to platform dynamics and spoof-induced anomalies. 2) **ADSB (Automatic Dependent Surveillance–Broadcast)**: Capture positional broadcasts from cooperative aircraft, enabling spoof contrast through ID-grounded trajectories. 3) **LTE/4G Signal Detectors**: Record RF emissions and packet timings from airborne platforms, supporting detection continuity in signal-rich environments and complementary spoof resolution. Each sensor independently logs detections at fixed intervals. These detection logs are streamed to a central fusion node, where sensor-origin tags are preserved, malformed entries are filtered, and missing timesteps are padded using modular error-handling routines.

2.2 Complementarity and Robustness Through Fusion

Each sensor contributes distinct signatures. Firstly, radar enables platform dynamics and spoof traceability. Next, ADSB anchors cooperative identity and continuity. Finally, LTE/4G signals offer temporal density and spoof correlation through RF behavior. Fusion across these channels yields greater resilience to spoof injection. Disagreement between sensors (e.g. radar-based spoof detection without ADSB corroboration) triggers interpretability overlays and trajectory warnings. The centralized fusion logic exploits sensor complementarity to maximize assignment integrity and spoof-awareness.

3 Benchmark Design

SpoofTrackBench demonstrated in Fig. 1 is designed to support reproducible, spoof-aware benchmarking of radar tracking architectures under modular and adversarial conditions. The benchmark is structured into distinct components spanning data ingestion, spoof injection, tracker evaluation, and interpretability-driven logging [15].

3.1 Radar Dataset and Scenario Setup

We utilize the Hampton University Skyler Radar Sensor dataset, which provides volumetric micro-Doppler returns from airborne targets under varied maneuvering conditions. Detection frames are sampled across timesteps at configurable update rates (e.g. 0.5–1.0 s), allowing realistic simulation of both cooperative and non-cooperative aerial behavior.

Simulation scenarios are modularized as MATLAB structures, enabling parameter toggling for:

- Number and type of platforms (e.g. UAV, fixed-wing)
- Sensor coverage and detection uncertainty
- Scenario geometry, spoof entry points, and time durations

Each scenario instantiates synthetic ground truth with associated clean detection logs, which are subject to downstream spoof injection and tracking.

3.2 Spoof Injection Logic

SpoofTrackBench supports three primary spoof types, i.e. **Drift Spoof**, **Ghost Spoof** and **Mirror Spoof**. The first type of **Drift Spoof** is the result of temporal deviation of detection from true location, which simulates the gradual trajectory corruption. Next, **Ghost Spoof** is the case of false detections with no physical platform origin, which challengs the tracker resilience and assignment purity. Finally, **Mirror Spoof** is the reflection-based symmetry across geographic axes or sensor grid, which is designed to mislead motion models. We can present the derivations for these three spoof types as follows:

- **Drift Spoof**: Let the true detection at time t be $\mathbf{x}_t = [x_t, y_t]^\top$. Drift spoofing introduces a temporal deviation such that the spoofed detection becomes $\tilde{\mathbf{x}}_t = \mathbf{x}_t + \Delta_t$, where $\Delta_t = \alpha t \cdot \hat{\mathbf{v}}$, $\hat{\mathbf{v}}$ = unit drift direction. This simulates gradual trajectory corruption over time.
- **Ghost Spoof**: Ghost spoofing injects detections $\tilde{\mathbf{x}}_t$ at time t with no corresponding platform or track origin: $\tilde{\mathbf{x}}_t \notin \{\mathbf{x}_t^{(i)} \mid i = 1, \ldots, N\}$, and $\text{label}(\tilde{\mathbf{x}}_t) = \text{spoof}$. These detections challenge tracker assignment logic and clutter rejection.
- **Mirror Spoof**: Given a true detection $\mathbf{x}_t = [x_t, y_t]^\top$, mirror spoofing reflects it across a defined axis (e.g. $x = x_0$) or grid symmetry: $\tilde{\mathbf{x}}_t = [2x_0 - x_t, y_t]^\top$. This misleads motion models by simulating symmetric but false trajectories.

Scenario parameters dictate spoof type, injection rate, and platform targets. The function reinitializes detection logs per run, applying randomized seeds for controlled variability. Clean and spoofed detection streams are stored separately with spoof flags and timestamps for traceability.

3.3 Tracker Architectures

SpoofTrackBench supports tracker evaluation across JPDA and GNN (see Fig. 2). Here, JPDA supports gating thresholds, detection clustering and assignment scoring, whilst detections in GNN are nodes and edges encode spatial-temporal affinity.

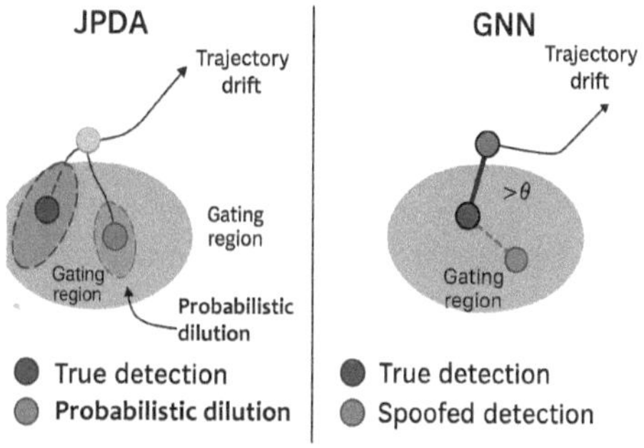

Fig. 2. Comparative schematic of JPDA and GNN tracking under spoofing conditions. JPDA handles spoofed detections via probabilistic dilution within the gating region, while GNN applies threshold-based rejection to isolate spoofed inputs. The illustration highlights differences in trajectory drift response and spoof filtering mechanisms.

Joint Probabilistic Data Association (JPDA) Tracker: The JPDA tracker performs soft data association by computing the probability that each detection originates from a given track. At each time step t, for track j, the update is a weighted sum over all gated detections:

$$\hat{\mathbf{x}}_t^{(j)} = \sum_{i=1}^{M} \beta_{ij} \cdot \mathbf{z}_t^{(i)} \tag{1}$$

where $\mathbf{z}_t^{(i)}$ is the i-th detection at time t, $\beta_{ij} = P(\mathbf{z}_t^{(i)} \mid \text{track } j)$ is the association probability, M is the number of detections within the gating region, and $\hat{\mathbf{x}}_t^{(j)}$ is the updated state estimate for track j. JPDA accounts for missed detections and clutter by normalizing over all feasible association events. It is sensitive to spoofing when false detections fall within the gate, as they dilute the association probabilities and degrade track confidence.

Global Nearest Neighbor (GNN) Tracker: The GNN tracker performs deterministic data association by assigning each detection to the nearest track based on a cost metric, typically Mahalanobis distance. At each time step t, the tracker solves:

$$\min_{\sigma \in \mathcal{S}} \sum_{i=1}^{N} d(\mathbf{z}_t^{(i)}, \mathbf{x}_t^{(\sigma(i))}) \tag{2}$$

where $\mathbf{z}_t^{(i)}$ is the i-th detection at time t, $\mathbf{x}_t^{(j)}$ is the predicted state of track j, $d(\cdot, \cdot)$ is the association cost (e.g. Mahalanobis distance), σ is a permutation mapping detections to tracks, and $\mathcal{S}$ is the set of feasible assignments. GNN uses gating to reject unlikely associations and solves the assignment problem using algorithms like the Hungarian method. It is robust to clutter and spoofing when spoofed detections fall outside gating thresholds.

Both architectures share input formats and evaluation pipelines, ensuring consistent comparison. We need to configure the parameters of gating distance or gate threshold, clutter density and birth probability P_{birth}, and tracker confidence thresholds via scenario definitions, which is then passed into tracker initialization modules. This enables robust tuning across spoof conditions.

3.4 Enhancement with AI-Based Target Tracking

We derive the simulation for object tracking by using the radar sensor in the radar surveillance system. Then, we develop the algorithms of multiple target tracking using the global nearest neighbor. In the algorithm, we utilize the clustering the region and address the process of data association. It means that we will fit the single most likely hypothesis at each scan. There are many challenges in the cluster-based multi-target tracking. The consecutive errors would come from the cause of the wrong measurements for the specific target or the false alarm. This usually happens when we consider the scenario of tracking multiple objects. Also, the clustering methods may create the wrong region of object movement and hence cause the incorrect acquired data. All of these lead to false association between the previous known targets and measurements. Hence, the system cannot keep track on the target (i.e. loosing track or breaking track) due to the wrong assignment of measurements to track. Specifically, the cluster must be designed correctly so that we must avoid the case of high density clusters. In this case, the computation for tracking cause a heavy burden to the computing system and may lead to serial numbers of false tracks. Furthermore, we need to design the fair clusters with the same/similar number of targets to be tracked. By doing so, we can balance the computing resource allocation to every cluster and hence improve the CPU time (i.e. making the designed tracking system to be suitable for real time implementations). In addition, we can also reduce the errors of assigned measurements to wrong tracks (i.e. dealing with data association problem successfully). Furthermore, both GNN and JPDA operate effectively under nominal conditions, these methods are susceptible to spoofing attacks that inject falsified detections or manipulate trajectory continuity.

To address these unique challenges, we firstly propose to use the Graph Convolutional Neural Network for assignment learning in the GNN, and we call this tracking mechanism as trackerGNN. Interested readers can find more detailed designation of Graph Neural Network in [3,16]. Deep learning approaches, including Graph Neural Network and Convolutional Neural Network-based trackers, offer adaptability and learning-based generalization, especially for GNN [3,17–20]. In the following, we briefly describe the algorithm to the multiple target tracking mechanism, which is used to the JPDA. In fact, we employ the data correlation processing to assign the best measurements to track associations. Firstly, we will use the Kalman filter for predicting the possible future position, given the predetermined associated covariance matrix of data. We will evaluate the valid observation/measurements belonging to the region/cluster or not in probabilistic manner. The second step is to associate the measurements with the tracks. This is because 1) the measurement is in region of one cluster with multiple tracks or in the border of multiple clusters with multiple tracks and 2) multiple measurements are in the same region of target track. To address this critical issue, we develop the Hungarian matching algorithm [21] to determine the optimal assigned pairs of observation and target track.

3.5 Logging and Reproducibility

SpoofTrackBench emphasizes reproducibility through: 1) **Detection Stream Separation**: Clean and spoofed logs are stored distinctly, preserving origin and labeling integrity. 2) **Seed Management**: Each scenario initializes random seeds with logging of seed IDs per run, enabling exact reconstruction. 3) **Snapshot Evaluation**: At each timestep, tracker outputs, assignment matrices, and drift metrics are saved as MATLAB structs. 4) **Auto-Export Mechanism**: Upon scenario completion, the benchmark auto-generates comparison plots (assignment accuracy, drift overlays), summary statistics, and scenario logs under organized directories. Visual exports include 1) assignment history overlays with spoof highlighting, 2) cluster purity and trajectory divergence plots, and 3) 2D/3D toggled views (dimMode) with tracker ID labels. The benchmark is designed for extensibility: future tracker modules, spoof types, and multi-sensor fusion configurations can be integrated with minimal code changes using scenario template inheritance and modular helper functions.

4 Evaluation Strategy

SpoofTrackBench employs a layered evaluation framework to quantify tracker integrity, interpret spoof resilience, and compare architectural behavior under controlled adversarial conditions. The evaluation strategy spans metric logging, visual overlays, and bidirectional tracker diagnostics [22–24].

4.1 Performance Metrics

To robustly assess tracking fidelity and spoof impact, we determine the following performance metrics.

- **Drift-from-Truth Quantification**: For each tracked platform, we log mean and maximum Euclidean error between reconstructed trajectory and ground truth, per timestep.
- **Assignment Divergence**: Measures mismatch between assigned detections and true platform associations using normalized confusion scores and switch frequencies.
- **Cluster Misalignment**: Evaluates clustering purity and probability-weighted error across temporal windows, particularly sensitive to ghost spoof insertions.
- **Spoof Detection Flags**: Each detection is labeled, enabling downstream statistics on spoof inclusion, recovery rates, and false positive attribution.

4.2 Cross-Tracker Comparison

SpoofTrackBench supports bidirectional comparison of tracking architectures through an automated evaluation module. This function computes **Numerical Benchmarking**, **Visual History Comparison**, and Spoof-Type Segmentation. **Numerical Benchmarking** aggregates drift errors, assignment accuracy, spoof recovery rate, and false association ratios across both tracker outputs. **Visual History Comparison** generates multi-timestep overlays (assignment, spoof recovery, cluster transitions) with side-by-side renderings. Results of **Spoof-Type Segmentation** are bucketed by spoof types (i.e. drift, ghost and mirror), enabling conditional analysis and architecture-specific vulnerability scoring. Exported examples include 1) drift error heatmaps across platforms and timesteps, 2) cluster purity timelines and divergence plots, and 3) assignment history and switch rate graphs.

5 Experiments, Results and Discussion

We present simulation results designed to evaluate the spoof resilience, assignment fidelity, and interpretability of tracking architectures using SpoofTrackBench. Experiments span varied spoof types, tracker configurations, and visualization diagnostics.

5.1 Simulation Setup

We simulate the scenario of drones flying and get the data for the radar measurements. This dataset would be used for our experiment on target tracking in the subsequent sections.

Scenario Setup: For the initialization of the scenario, simulation time is run until objects halt movement. Three moving and one stationary objects are initialized each with a class ID. In addition to the dimensions of moving object being the same, the radar cross section signature (RCS), whose parameters decide the intensity of the reflected radar signal from the target of each moving object are the same as well. Used parameters for the RCS signature include signature fluctuation model, the sampled RCS pattern, pattern frequency, Azimuth angles and elevation angles. The trajectories of each moving object vary in course, ground speed, climb rate, auto-pitch and auto-bank for a diverse readings. Experiments were orchestrated using a cyclic spoof evaluation loop, where each run initializes a unique spoof configuration: 1) **Spoof Type Cycling**: Drift, ghost, and mirror spoof categories are applied across multiple scenario seeds using the benchmark's orchestration module; 2) **Parameter Variation**: Detection noise levels, spoof injection timing, and agent counts were varied across runs to assess robustness under dynamic stress conditions; 3) **Platform Configuration**: Both cooperative and non-cooperative platforms were simulated. Cooperative agents use ADSB overlays, while radar-only detections were reserved for spoof-prone entities; 4) **Runtime Logging**: Each evaluation run generates a structured folder tree containing tracker outputs, spoof injection parameters, and interpretability exports. Fusion center inputs are normalized across sensors, maintaining dimensional consistency and detection field integrity across both clean and spoofed streams.

5.2 GNN vs. JPDA Without Spoofing Attacks

In the following, we present some essential results. Firstly, we consider the scenario of trajectories, where each moving object vary in course, ground speed, climb rate, auto-pitch and auto-bank for a diverse readings. The mentioned scenario can be seen in Fig. 3(a), which displays the detections (history) along with detection trajectories in the solid black lines and true target trajectories in dotted grey lines. In Figs. 4(a) and 4(b), we evaluate the GNN-based and JPDA-based multi-object tracking algorithms, where detections in colored dots along with their probability distribution surrounding them are displayed with predicted trajectories in solid colored lines. It can be seen that when objects are sparsely spaced, i.e. track one and two from -600 to 0 m along the x-axis, trajectory predictions are well defined. From 0 to 600 along the x-axis, however, tracked objects are cluttered and as a result trajectories and assignments

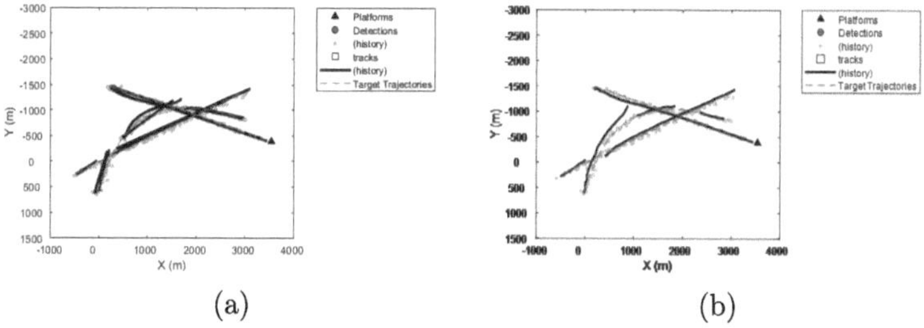

Fig. 3. Trajectory tracking methods: (a) The GNN trajectory tracking, and (b) The JPDA trajectory tracking.

becomes disorganized. While Fig. 3(a) maps trajectories similar to Fig. 3(b), it can be seen that it fails at overlapped tracks. In a multi-object tracking algorithm, detections must assigned to a tracked object without the interference of other objects detection, meaning the prediction of one object should not be affected by the observations of another. GNN provides an ease of computation, however, its performance is limited with cluttered objects. JPDA can overcome the GNN's cluster issues by assigning weights to the closest detection of each tracked object and using the weight combinations for assignment. Therefore, when a cluster of objects is established, each detection is not discretely assigned to each object, but is collaboratively performed with the adaptive weight. Hence, JPDA has a higher accuracy of assignment when comparing GNN's assignment to the graph of the original object trajectories.

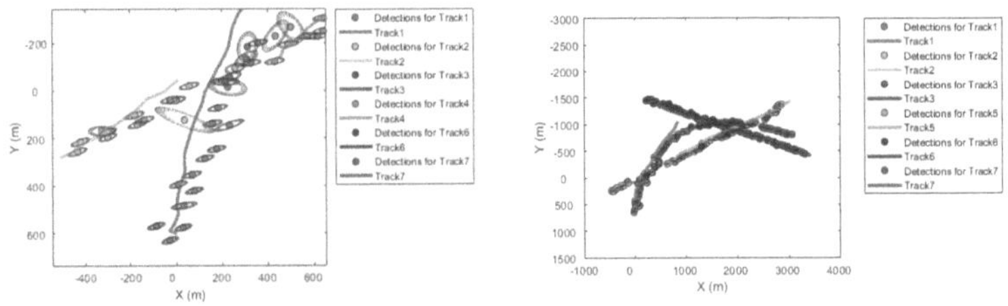

(a) (b)

Fig. 4. Evaluation of trackers: (a) Evaluation of GNN trajectory tracker, and (b) Evaluation of JPDA trajectory tracker.

5.3 GNN vs. JPDA Under Spoof Pressure

We firstly evaluate tracking performances of the GNN and JPDA under the consideration of spoofing attacks in Figs. 5(a) and 5(b), respectively. In Ghost

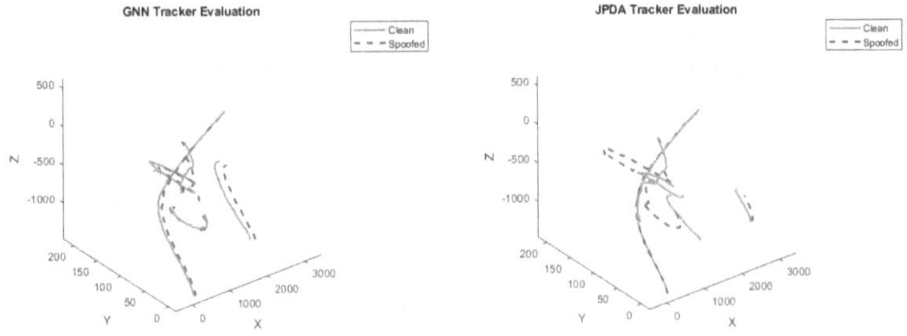

(a) (b)

Fig. 5. Evaluation of trackers under Ghost Spoofing: (a) Evaluation of GNN trajectory tracker, and (b) Evaluation of JPDA trajectory tracker.

Spoofing Attacks, the spoofing attacks cause significant distortions in the formulated trajectories of both GNN and JPDA trackers, which manifest as degraded track continuity, increased positional drift, and reduced assignment confidence. These effects are especially pronounced in JPDA due to its probabilistic association mechanism, which is more susceptible to spoof-induced ambiguity within the gating region. GNN exhibits stronger performance in spatial reasoning, especially under ghost and mirror spoof types. Its message-passing architecture supports cross-detection affinity scoring, enabling partial immunity to clustering confusion and ghost proliferation. GNN uses hard gating and assigns each detection to the nearest track if it falls within a predefined threshold. Spoofed detections outside the gate are ignored. Even if spoofed detections are inside the gate, only the closest one is selected, making GNN more resilient to clutter and spoof bursts. However, GNN trackers exhibit: 1) Increased sensitivity to irregular temporal cadence (e.g. intermittent spoof injection), 2) Reduced explainability due to latent node embeddings and opaque decision boundaries. In contrast, JPDA uses soft gating and computes association probabilities β_{ij} for each detection-track pair. When spoofed detections fall within the gate, they dilute the probability mass, reducing confidence in true associations. This can lead to track drift, fragmentation, or premature termination. In JPDA, the association probability β_{ij} is computed as $\beta_{ij} = \frac{P(\mathbf{z}_t^{(i)}|\text{track } j)}{\sum_{k=1}^{M} P(\mathbf{z}_t^{(k)}|\text{track } j)}$, where M is the number of detections in the gate. When spoofed detections are injected, the denominator increases, reducing β_{ij} for the true detection. So, the tracker may update the track with a weighted average that includes spoofed positions. Over time, this leads to confidence decay, trajectory corruption, and false track coalescence. Therefore, JPDA offers the following benefits depending on the considered environment. 1) Transparent assignment logic with interpretable gating and likelihood updates. 2) Greater resilience under gradual drift scenarios through Kalman filter continuity. 3) Weaker discrimination under dense spoof scenarios due to probabilistic dilution and clutter overlap. Figures 6(a), 6(b), and 6(c) provide detailed visualizations of JPDA tracker performance under three spoofing

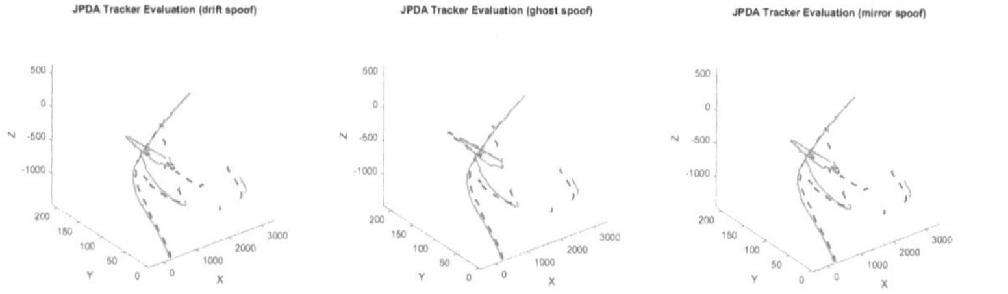

(a) Drift Spoof (b) Ghost Spoof (c) Mirror Spoof

Fig. 6. JPDA tracker performance under spoof injection scenarios (solid green lines represent estimated UAV trajectories, while dashed red lines illustrate spoofed paths affected by gradual drift injection). (a) Drift spoof, (b) Ghost spoof, (c) Mirror spoof. (Color figure online)

Table 1. Spoof Impact Benchmark: Drift (m) and Normalized Impact (%)

Tracker	Spoof Type	Drift (m)	Normalized Impact (%)
GNN	drift	77.10	15.42
GNN	ghost	66.25	13.25
GNN	mirror	72.10	14.42
JPDA	drift	66.10	13.22
JPDA	ghost	77.05	15.41
JPDA	mirror	75.38	15.08
Average (GNN)	–	72.48	14.36
Average (JPDA)	–	72.84	14.90
Average (drift spoof)	–	71.60	14.32
Average (ghost spoof)	–	71.65	14.33
Average (mirror spoof)	–	73.74	14.75
Overall Average	–	76.24	15.25

scenarios. In Fig. 6(a), solid green lines represent the estimated UAV trajectories, while dashed red lines illustrate spoofed paths affected by gradual drift injection, highlighting JPDA's ability to preserve track continuity despite adversarial corruption. Figure 6(b) shows tracked UAV paths in green and false detections introduced by ghost spoofing in red, demonstrating JPDA's resilience in filtering out non-physical targets. In Fig. 6(c), green lines depict estimated UAV trajectories, while red dashed lines reflect spoofed paths generated by symmetry-based mirror injection. Collectively, these visualizations underscore JPDA's robustness in maintaining trajectory integrity, rejecting deceptive inputs, and resisting spoof-induced distortions across diverse attack types. Finally, these findings suggest complementary deployment strategies, i.e. JPDA for low-spoof environments with clearer temporal continuity, and GNN for adversarial scenarios requiring dynamic assignment discrimination.

5.4 Spoof-Type Impact on Cluster Integrity

Table 1 demonstrates the performances of GNN and JPDA trackers across all spoof types, each exhibiting distinct clustering behaviors. 1) **Drift Spoofs**: Induce gradual trajectory bending, resulting in delayed cluster separation and subtle assignment drift. 2) **Ghost Spoofs**: Generate phantom detections that infiltrate clusters, triggering early confusion and tracker instability. 3) **Mirror Spoofs**: Create symmetrical echoes that challenge spatial models, often producing mirrored clusters with high affinity misgroupings. These observations highlight the unique failure modes introduced by different spoofing strategies and underscore the importance of spoof-aware evaluation in resilient tracking pipelines. GNN's affinity-based graph edges partially resolve ghost confusion but can misassign in drift and mirror cases. JPDA maintains cluster purity longer during drift but is prone to ghost contamination. This underscores the need for spoof-specific tracker evaluation rather than generic benchmarking.

5.5 Spoof Detection Flags: UAV Trajectory Overlay Under Drift Injection

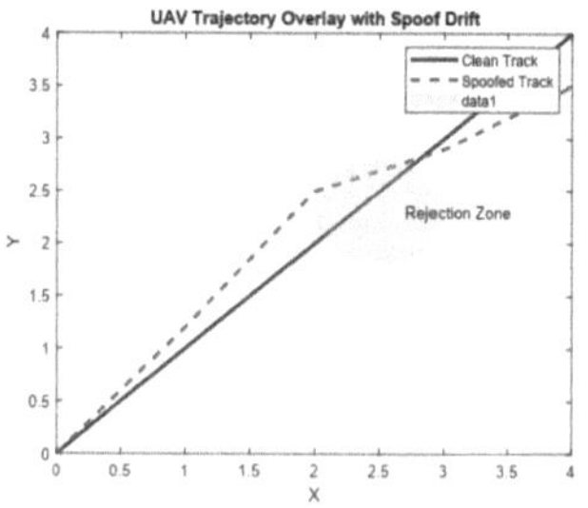

Fig. 7. UAV Trajectory Overlay with Spoof Drift. The clean track (solid blue) represents the true UAV trajectory, while the spoofed track (dashed red) illustrates the adversarial drift injection. The yellow rejection zone highlights regions where spoofed detections are filtered by the tracking system. (Color figure online)

Figure 7 illustrates the UAV trajectory overlay under spoof drift injection, where the clean track (solid blue) and spoofed track (dashed red) diverge across the X-Y plane. The yellow rejection zone delineates regions where spoofed detections are actively filtered by our proposed detection scheme. This mechanism enhances tracking resilience by isolating corrupted inputs and preserving trajectory integrity. By leveraging spatial divergence and rejection logic, our spoof-aware detection framework mitigates the risk of tracking failure under adversarial drift attacks. It enables robust trajectory recovery, improves downstream tracker performance, and supports interpretable diagnostics for spoof injection events. This capability is critical for mission-sensitive UAV operations in contested environments.

6 Conclusion and Future Work

SpoofTrackBench establishes a reproducible, architecture-agnostic benchmark for radar-based tracking under adversarial spoof conditions. By integrating modular spoof injection, tracker comparison, and interpretability overlays, it provides the community with an open framework for evaluating assignment resilience and trajectory fidelity. Key contributions include 1) a flexible spoof injection pipeline supporting drift, ghost, and mirror types with configurable parameters; 2) architecture-neutral benchmarking modules for classical JPDA and GNN trackers, enabling quantitative and qualitative comparison under identical spoof scenarios; and 3) automated logging, evaluation snapshots, and interpretability overlays to support scientific reproducibility and visual storytelling.

To further advance spoof resilience and tracking interpretability, we envision the following roadmap: 1) **Integration of Deep Neural Spoof Classification (SpoofNet)**: We will embed SpoofNet with recent AI/ML algorithms [3,16,17,19,20,25–27], a real-time spoof detection module trained on injection-labeled examples (drift, ghost and mirror), to classify detection integrity on-the-fly. This module will enhance interpretability overlays with spoof confidence scores and support dynamic tracker behavior based on spoof severity. 2) **Hybrid Deep Quantum Neural Network (DQNN) Tracker Architecture**: Inspired by JPDA's probabilistic foundations and GNN's spatial modeling, we propose a DQNN tracker integrating SpoofNet's outputs with quantum gate-inspired assignment logic [3]. This architecture will probabilistically reason about detection ambiguity while maintaining stable tracking in high-spoof environments.

Acknowledgment. This work was supported in part by the Commonwealth Cyber Initiative (CCI) Experiential Learning Program and the South Big Data Innovation Hub Partnership Nucleation program.

References

1. Kerns, A.J., Shepard, D.P., Bhatti, J.A., Humphreys, T.E.: Unmanned aircraft capture and control via GPS spoofing. J. Field Rob. **31**(4), 617–636 (2014)
2. Restivo, R.D., et al.: GPS spoofing on UAV: a survey. In: IEEE INFOCOM 2023-IEEE Conference on Computer Communications Workshops (INFOCOM WKSHPS), pp. 1–6. IEEE (2023)
3. Tan, L., Van, L., Sachin, S.: Privacy-aware framework of robust malware detection in indoor robots: hybrid quantum computing and deep neural networks. TechRxiv (2025)
4. Sathaye, H., Strohmeier, M., Lenders, V., Ranganathan, A.: An experimental study of GPS spoofing and takeover attacks on UAVs. In: USENIX Security Symposium (2022)
5. Psiaki, M.L., Humphreys, T.E.: GNSS spoofing and detection. IEEE Aeros. Elec-tron. Syst. Maga. (2016)
6. Radoš, K., Brkić, M., Begušić, D.: Recent advances on jamming and spoofing detection in GNSS. Sensors **24**(13), 4210 (2024)
7. Chen, Y., Trappe, W., Martin, R.P.: Detecting and localizing wireless spoofing attacks. In: IEEE International Symposium on Security and Privacy (2009)
8. Yue, S., Xu, R., Yan, Q., Liu, J.: Extended kalman filter-based GNSS signal tracking method to counter spoofing attacks. Lect. Notes Electr. Eng. **537**, 439–447 (2019)
9. Machaj, J., Safon, C., Matúška, S., Brída, P.: Detection of access point spoofing in the wi-fi fingerprinting based positioning. Sensors **24**(23), 7624 (2024)
10. Kamble, M.R., Sailor, H.B., Patil, H.A., Li, H.: Advances in anti-spoofing: from the perspective of asvspoof challenges. APSIPA Trans. Signal Inf. Process. **9**, e19 (2020)

11. Geiger, A., Lenz, P., Urtasun, R.: Are we ready for autonomous driving? The kitti vision benchmark suite. In: Conference on Computer Vision and Pattern Recognition (CVPR), pp. 3354–3361 (2012)
12. Dendorfer, P., et al.: Motchallenge: a benchmark for single-camera multiple target tracking. Int. J. Comput. Vision **129**(4), 845–881 (2021)
13. Lefèvre, S., Vasquez, D., Laugier, C.: A survey on trajectory prediction methods for autonomous driving. IEEE Trans. Intell. Transp. Syst. **17**(2), 512–529 (2016)
14. Bar-Shalom, Y., Willett, P., Kirubarajan, T.: Tracking of people and vehicles using multiple cameras. IEEE Aerosp. Electron. Syst. Mag. **20**(4), 38–45 (2001)
15. Xiang, Y., Alahi, A., Savarese, S.: Learning to track: online multi-object tracking by decision making. In: Proceedings of the IEEE International Conference on Computer Vision, pp. 4705–4713 (2015)
16. Le, T., Le, V.: DPFAGA-dynamic power flow analysis and fault characteristics: a graph attention neural network. In: The 2025 International Conference on the AI Revolution: Research, Ethics, and Society (AIR-RES 2025) (2025)
17. Wang, Q., Tan, L.T., Hu, R.Q., Qian, Y.: Hierarchical energy efficient mobile edge computing in IoT networks. IEEE Internet Things J. (2020)
18. Wang, Q., Tan, L.T., Hu, R.Q., Wu, G.: Hierarchical collaborative cloud and fog computing in IoT networks. In: 2018 10th International Conference on Wireless Communications and Signal Processing (WCSP), pp. 1–7 (2018)
19. Zahin, A., Tan, L.T., Hu, R.Q.: Sensor-based human activity recognition for smart healthcare: a semi-supervised machine learning. In: Han, S., Ye, L., Meng, W. (eds.) AICON 2019. LNICST, vol. 287, pp. 450–472. Springer, Cham (2019). https://doi.org/10.1007/978-3-030-22971-9_39
20. Zahin, A., Tan, L.T., Hu, R.Q.: A machine learning based framework for the smart healthcare monitoring. In: 2020 Intermountain Engineering, Technology and Computing (IETC) (2020)
21. Tan, L.T., Le, L.B.: Channel assignment with access contention resolution for cognitive radio networks. IEEE Trans. Veh. Technol. **61**(6), 2808–2823 (2012)
22. Shabtai, A., Moskovitch, R., Elovici, Y., Dolev, S.: Detection of malicious code using behavioral-based machine learning. J. Comput. Virol. Hack. Techn. **12**(4), 231–247 (2016)
23. Zheng, Z., Zhong, Q., Wang, Y., Zhan, W.: Deep learning-based object detection for autonomous driving: survey. IEEE Trans. Intell. Transp. Syst. **22**(10), 6148–6165 (2021)
24. Rudenko, A., Palmieri, L., Herman, M., Kitani, K.M., Gavrila, D.M., Arras, K.O.: Human motion trajectory prediction: a survey. Int. J. Rob. Res. **39**(8), 895–935 (2020)
25. Le, T., Shetty, S.: Artificial intelligence-aided privacy preserving trustworthy computation and communication in 5G-based IoT networks. Ad Hoc Netw. **126**, 102752 (2022)
26. Tan, L.T., Hu, R.Q., Hanzo, L.: Heterogeneous networks relying on full-duplex relays and mobility-aware probabilistic caching. IEEE Trans. Commun. **67**, 5037–5052 (2019)
27. Tan, L.T., Hu, R.Q.: Mobility-aware edge caching and computing in vehicle networks: a deep reinforcement learning. IEEE Trans. Veh. Technol. **67**(11), 10190–10203 (2018)

LP-RFL: Enabling Low-Cost Robust Federated Learning via Label-guided Weight Gradient Pruning to Against Backdoor Attacks

Lening Wang[1,2(✉)], Saide Zhu[3], Hanxiang Mu[4], and Xin Fu[2]

[1] Department of Computer Science, Prairie View A&M University, Prairie View, USA
lenwang@pvamu.edu
[2] Department of Electrical and Computer Engineering, University of Houston, Houston, USA
[3] Department of Information Science Technology, Penn State Berks, Reading, USA
[4] Liberal Arts and Science Academy, Austin, USA

Abstract. Federated Learning (FL) is a distributed machine learning paradigm that allows multiple clients to collaboratively train a shared model without sharing their raw data. However, deploying FL on mobile clients presents two major challenges. First, FL is vulnerable to backdoor attacks because malicious clients can insert hidden triggers or backdoors into the global model through their local updates. Second, mobile devices often have limited communication and computational resources, which restrict their ability to perform complete training iterations. In our research, we observed that weight gradients in FL are highly sparse and exhibit label specific sparse patterns. We further found that gradients from malicious clients attempting backdoor attacks show clear inconsistencies when compared with the sparse pattern of their targeted incorrect labels. Based on this observation, we propose a label based gradient pruning approach to mitigate backdoor insertions. On the server side, we profile a critical set of weight gradients (the sparse pattern) for each label using clean images and a pretrained initial model. During local training, clients prune gradients that fall outside the critical set for the corresponding label. This pruning based strategy blocks backdoor attempts and reduces redundant computations, which significantly lowers the computation and communication costs on mobile clients. Through our experiments, our proposed method effectively reduces the Attack Success Rate of backdoor attacks from 90% to 10% without noticeable accuracy loss on clean images. Additionally, our framework improves computation efficiency by an average of 3.2× (up to 3.9×).

Keywords: Federated Learning · Backdoor Attack · Pruning · Security

H. R. Arabnia et al. (Eds.): CSCI 2025, CCIS 2969, pp. 428–443, 2027.
https://doi.org/10.1007/978-3-032-28460-0_30

1 Introduction

Deep Neural Networks (DNNs) has achieved remarkable success in various applications. DNNs often requires extensive training data and iterations, leading to significant training costs. Typically, training tasks are performed on a powerful centralized server. This centralized approach raises concerns about data privacy. To address these concerns, Federated Learning (FL) has emerged as a distributed machine learning approach enables multiple clients to train a centralized model without sharing their for maximum data privacy.

However, deploying FL on mobile clients presents two significant limitations. Firstly, FL has been shown to be vulnerable to backdoor attacks [2]. In such attacks, an adversary disrupts the training process by introducing a targeted misclassification functionality into the model. This is typically achieved through data or model poisoning [22]. Since the data is decentralized in FL and the centralized server lacks access to individual clients' local data, malicious clients can inject poisoned training data into their updates. When these updates are aggregated with other clients' updates, they introduce backdoors into the model, leading to model poisoning. Secondly, on the client side, mobile devices typically have limited computational and communication resources compared to powerful centralized servers [23]. This poses a significant challenge for mobile clients to perform complex training iterations, making it difficult for them to efficiently execute the demanding training processes.

Previous works in addressing backdoor attacks in FL mainly leveraged the vote mechanism to optimize the aggregation function to influence the direction of weight gradients [14]. While these methods can be effective in certain scenarios, they may face limitations when dealing with a large number of malicious clients. Moreover, in a non-IID (independent and identically distributed) setting, where the data distribution across clients is uneven, the direction of weight gradients can vary significantly, making the voted results unreliable. On the other hand, it's important to note that the aforementioned aggregation solutions do not directly contribute to improving the training efficiency on local clients. Considering these limitations, we propose a different approach in our algorithm for robustness training against backdoor attacks in FL. We focus on the weight gradients, which are crucial parameters exchanged in FL. By exploiting the sparse nature of weight gradients, we seek to provide a practical and efficient solution for deploying FL on mobile devices, effectively addressing the challenges posed by backdoor attacks and resource limitations.

When training the clean images, it is widely recognized that weight gradients exhibit high sparsity when computed for each training example [8]. This implies the majority of near-zero weight gradients can be skipped without affecting the training convergence. By disregarding these negligible weight gradients, the training cost for FL can be significantly reduced. Furthermore, we make a key observation that the sparse pattern in weight gradients is closely correlated with the ground truth (correct) labels. In other words, training examples from the same class tend to have similar sets of effectively updated weights. When poisoned training examples are introduced to maliciously insert backdoors, the calculated significant weight gradients (i.e., weight gradients with most contributions) differ from the significant weight gradients associated with the target (incorrect) labels. These findings provide valuable insights into the characteristics of weight gradients in FL, and they form the basis for our proposed approach to combat backdoor attacks by leveraging the sparse patterns in weight gradients. To be specific, by extracting a weight gradient set for each label, we can effectively identify and remove the significant malicious weight updates caused by backdoored training images. As a result, the model's resilience against backdoor attacks is greatly enhanced and meanwhile, the overhead of calculating and transmitting large amounts of insignificant weight gradients is minimized.

In this study, we propose a low-cost **R**obust **FL** framework to against backdoor attacks by conducting the **L**abel-guided weight gradient **P**runing (LP-RFL). We begin by profiling a weight gradients set for each label using the pre-trained initial model and clean data. During the local training step in FL, only the weight gradients within the corresponding set are calculated, while the remaining weight gradients are pruned. For the clean images, the significant weight gradients are likely align in the corresponding weight gradient set to ensure model's accuracy. On the other hand, in adversarial images, the significant weight gradients take on a different nature. These weight gradients are considered most malicious and are likely aligned outside the corresponding weight gradient set. By removing them, we can prevent backdoor insertions. From our experiments, LP-RFL effectively reduces the Attack Success Rate (ASR) of backdoor attacks from 90% to 10% without noticeable accuracy loss on clean images. Moreover, unlike other pruning methods, which require expensive high-precision calculations to identify insignificant parameters, LP-RFL leverages provided labels to locate pruned weight gradients, which are known in advance during the training tasks. Based on our experiments, LP-RFL improves computation efficiency by an average of 3.2× (up to 3.9×). Overall, LP-RFL demonstrates its effectiveness in mitigating backdoor attacks, preserving model accuracy, and improving computation efficiency.

2 Background and Related Works

Federated Learning is a decentralized machine learning approach that allows multiple parties to collaboratively train a shared model without exchanging their raw data. In FL, a centralized server will distribute an initialized DNN model trained on public dataset to selected mobile clients, who will then train the model using their own data. After a predefined interval, which is defined as one communication round, each mobile client will send its calculated weight gradients back to the server. The sever then will aggregate the weight gradients from all the clients to produce one global model. This process repeats until the training process finished.

Backdoor Attack and Data/Model Poisoning in FL. During FL, a malicious client can deliberately manipulate training to introduce a hidden vulnerability, known as a Backdoor Attack. Such attacks aim to produce a model that behaves normally on clean inputs but misbehaves when given inputs containing a specific trigger. In FL, backdoors are usually implanted through data poisoning: the attacker injects trigger patterns into certain images (e.g., adding a trigger to cat images and labeling them as dogs), training the model to misclassify triggered inputs during inference. Backdoor attacks are especially concerning in FL because the central server has no direct access to client data and treats all clients equally, making it hard to distinguish malicious from honest participants [14]. To defend backdoor, most of methods are proposed to optimize the aggregation process. For example, [3,11,18] are pre-aggregation methods which measure distance between benign and malicious updates to identify adversarial clients. The in-aggregation method operates while the server is aggregating local models, using techniques such as differential privacy [12], robust learning rates [15], smoothness and perturbation [13], and local validation [1] to mitigate the effects of backdoors. [24] introduced the post-aggregation defense strategy of identifying and removing neurons with low activation when presented with benign samples.

3 Label Guided Robustness FL Training Through Weight Gradient Pruning

Prior to moving forward, we first introduce two categories of weight gradients.

Significant/Insignificant Weight Gradients: Significant weight gradients refer to the weight gradients that has the most contributions to minimize the loss. Usually, the magnitude of these gradients indicates their significances. On the other side, insignificant weight gradients refer to gradients that have very small magnitudes. In our study, unless explicitly specified, we consider the top 20% weight gradients in each layer as the significant weight gradients. However, this threshold can be adjusted based on specific requirements.

Malicious/Honest Weight Gradients: Malicious weight gradients refer to the weight gradients computed using the poisoned or backdoored examples. These weight gradients capture the influence of the trojan patterns. On the other hand, honest weight gradients are the weight gradients calculated using the clean data. These gradients represent the true and valid updates to the model. Based on this distinction, the significant weight gradients that are computed using the poisoned data can be referred to as significant malicious weight gradients. These weight gradients play a crucial role of backdoor insertion.

3.1 Observation

We first present two key observations that are relevant to FL and backdoor attacks in FL. These observations lay the foundation for our proposed algorithm for robustness training to counter backdoor attacks in FL. Before demonstrating our key observations, we briefly introduce the settings of our FL framework.

The centralized server is capable for model distribution and aggregation. We assume the initial distributed model is already well-trained on public dataset. After that, the mobile clients are using the local data to keep updating the model. We assume the attackers are not able to access the centralized server and the public dataset used at the server side are safe and clean. On the client side, the attackers are only able to access/modify their local data to perform backdoor insertions. Those local clients with poisoned data are defined as the malicious clients. On the other hand, the centralized server is always honest but without any knowledge of the malicious clients and their poisoned data. During the training, the ranking of significant weight gradient is calculated using the Fisher Information Matrix [19].

Observation I: During the FL training, each individual training class, represented by its ground truth label, exhibits a unique sparse pattern in weight gradients.

Observation Validation: To verify this observation, we first assign 10 honest clients and assume their local training data are clean. The initial model (VGG-11) is pre-trained on public dataset 1K MNIST/Cifar-10 and the client using rest images in both IID and non-IID setting. It is well-known high sparsity exist in calculated weight gradients. This means that for each training iteration, only a small portion of weight gradients are effectively updated, while the rest majority remain close to zero. We record the significant weight gradients for each individual training example and transfer the weight gradients matrix into a bit matrix ($BM_{sample/class}$) where the significant weight gradients are marked as 1s and the insignificant weight gradients are marked as 0s. We find that the sparse pattern (the location of 1s) in the weight gradients exhibits high similarity across the training examples within the same class but limited similarity between different classes. The similarity is measured the overlapping ratio of 1s between two bit matrices using Eq. 1, where i, j represents different examples or different classes. Figure 1 shows the similarity between the significant weight gradients' from the same class is around 70% on average and can reach up to 86% for certain classes (e.g., class 1 MNIST). We use a heat map in Fig. 2 to represent the sparse pattern (of significant weight gradients) similarity between two different classes, where same indexes from both the horizon/vertical blocks represent the same class. We calculate the sparse pattern similarity of arbitrary two different classes using the averaged significant weight gradients calculated using 1000 examples of the same class. The average similarity is only 24% on average.

$$Similarity_{i,j} = \frac{number\ of\ 1\,s\ in\ BM_i \wedge BM_j}{number\ of\ 1\,s\ in\ BM_i \vee BM_j} \tag{1}$$

Insights: In FL, the initial model distributed by the centralized server is typically pre-trained on a large, public dataset. This pre-training allows the initial model to learn general features and patterns that are applicable to a wide range of tasks and provide a good starting point for the FL training process. Once the initial model is distributed to the mobile clients, each client utilizes its local data to

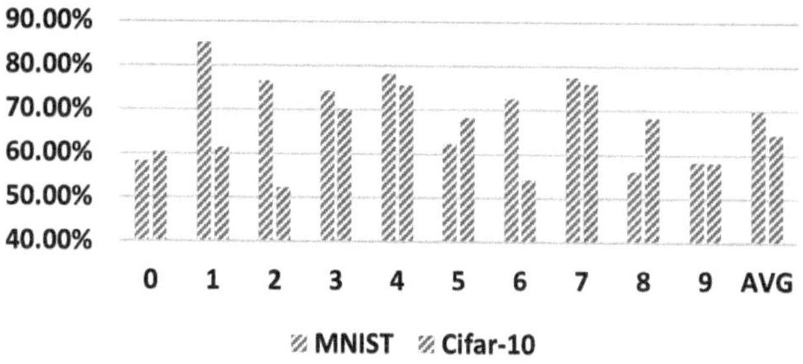

Fig. 1. Weight gradient sparse pattern similarity between different examples within the same class, X axis represents the class type.

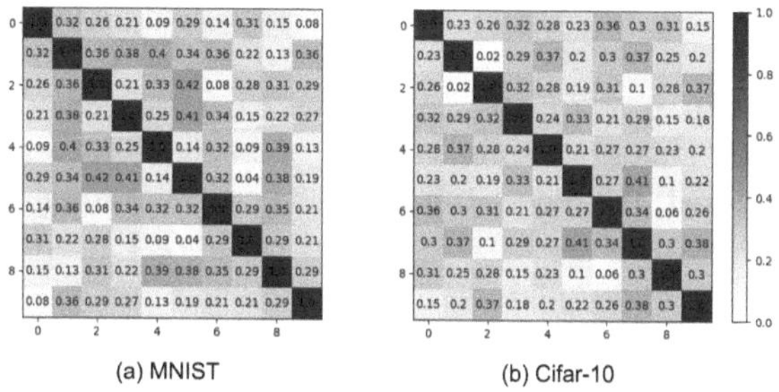

Fig. 2. weight gradient sparse pattern similarity between different classes.

further improve the model. During the local training iterations, for each class, the forward pass will activate a set of neurons (i.e., activations with large values), while those activated neurons are defined as the significant neurons and usually represent the learnt features of the corresponding input image [16]. According to Equation $\Delta W_l = \delta_{l+1} {}^* I_l$ from weight gradient calculation step, the weight gradient is the summation of multiplication results between forward activations and backward errors, which means the calculated weight gradients are mostly determined by the large activations and large errors. The large activations are the significant activations from forward pass. Because of the use of the RELU function, the derivation of the significant activations will be kept, and likely to produce the large errors as well. *Hence the significant neurons are the major contributors to the significant weight gradients.* For the training examples from the same class, they suppose to share a similar feature space which have the same features (i.e., significant neurons) to represent the corresponding class. Hence, the sparse pattern of weight gradients will be consistence for the same class too. On the other hand, the weight gradients' sparse patterns are different across different classes as different classes are likely to have different significant neurons.

Observation II: There is a mismatch between the significant weight gradients calculated using a backdoored training example and the significant weight gradients associated with the given incorrect label.

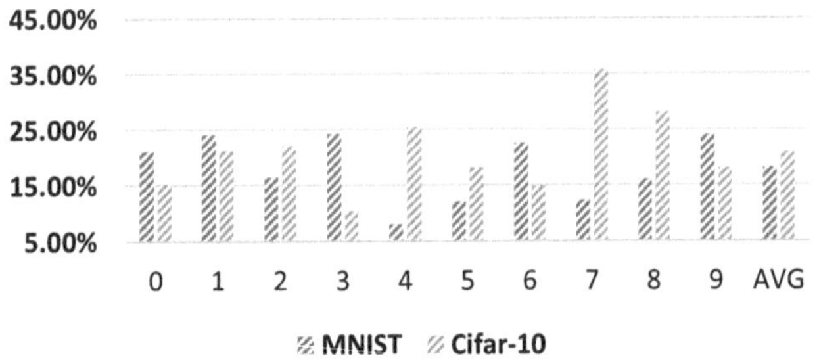

Fig. 3. Weight gradient sparse pattern similarity between significant weight gradients using backdoor example and significant weight gradients associated with the ground truth (correct) label, X axis represents the class type.

Observation Validation: We further blend some backdoored training examples into four clients' local data and profile significant weight gradients for each class using the initial model. We repeat the above experiments to measure the weight gradient sparse pattern similarity between the calculated significant weight gradients using backdoor inputs and the significant weight gradients associated with the given incorrect label (profiled using given incorrect label's clean images). As shown in Fig. 3, the similarity score is only about 21% on average.

Insights: In previous studies, researchers have utilized the discrepancy between the calculated significant neurons during the inference and the significant neurons associated with the output label to detect adversarial inputs [16]. When it comes to backdoored images with inserted triggers, the fundamental insights remain the same. The significant neurons of the poisoned image encompass the features corresponding to the ground truth correct label and the additional features introduced by the inserted triggers. However, they do not contains the features corresponding to the provided incorrect label since the incorrect label is unrelated to the ground truth correct label. As a result, the significant neurons calculated for the poisoned image will differ from the significant neurons associated with the given incorrect label. Recall that, *the significant neurons are the major contributors to the significant weight gradients*, the calculated significant weight gradients using given incorrect label will be different from the significant weight gradients associated with the ground truth correct label.

3.2 Methodology

It is well-known high sparsity exists in the weight gradients and only a few weights (indicated by the significance of corresponding weight gradients) are effectively updated for every iteration [26]. Conversely, the remaining weight gradients are considered insignificant and can be pruned without negatively impact the training accuracy. As a result, we aim to develop a pruning scheme for to achieve cost-effective robust FL against backdoor attacks. Revealed by Observation I, both the significant and insignificant weight gradients are highly correlated with the associated labels and can be pre-profiled using the initial pre-trained model from the server side. Observation II suggests that backdoored training examples tend to induce substantial significant updates (calculated using incorrect label) on the weights original should be insignificant if trained using clean images belong to the incorrect label.

Motivated by Observation I and II, we first profile a set of significant weight gradients for each label using the pre-trained model and clean images **on the server side** using pre-trained model and clean public data. We represent the weight gradient set using a bit mask, where each bit m_{loc} indicates whether the corresponding weight gradients are in the set. These sets are then synchronized with all clients. During each training iteration **on the client side**, only the weight gradients within this set are updated. For clean images, the calculated significant weight gradients align with the significant weight gradients of the provided ground truth correct label, ensuring the benign accuracy. Conversely, for backdoored images, the significant weight gradients (which are malicious) tend to fall outside the significant weight gradient set for the given incorrect label. Consequently, these weight gradients can be pruned to mitigate the impact of backdoored images. The weight gradients that are retained are typically insignificant and do not hinder convergence.

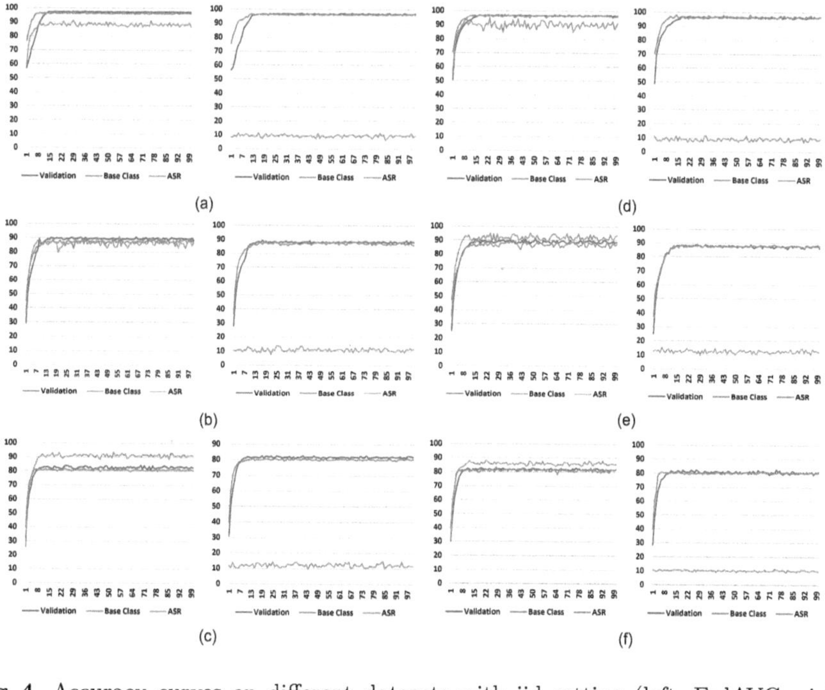

Fig. 4. Accuracy curves on different datasets with iid setting (left: FedAVG; right: LR-RFL): (a) LeNET-MNIST; (b) VGG16-Fashion MNIST; (c) MobileNetV2-Cifar10. Accuracy curves on different datasets with non-iid setting (left: FedAVG; right: LR-RFL): (d) LeNET-MNIST; (e) VGG16-Fashion MNIST; (f) MobileNetV2-Cifar10.

4 Evaluation

4.1 Experimental Setup

We evaluate on both MNIST [4], Fashion MNIST [25] and Cifar-10 [9] datesets using 3 most common neural networks (LeNET [10], VGG16 [20], MobileNetV2 [17]). We compare our work with four aggregation based methods: (i) FedAVG [21] , (ii) FoolsGold [5], (iii) Robust Learning Rate (RLR) [15] and (iv) Pruning Neurons [24]. FoolsGold is a pre-aggregation method to assign individual learning rates to each device. RLR is a in-aggregation method to assign the direction of learning rate per dimension based on the majority weight gradient direction and Pruning Neurons is a post-aggregation method to remove suspicious neurons after the aggregation stage.

We implemented LP-RFL in PyTorch. We simulate federated learning for 100 rounds involving 20 agents and 10 epoch local training each aggragation, of which F fraction are deemed malicious. Here Poisoned Model Rate (PMR) $= F$, and we set $F = 10\%$ as default. The objective of the backdoor task is to manipulate the model to misclassify instances from a base class as the target class, leveraging trojan patterns [6]. In other words, the model with the backdoor will classify instances which contain trojan patterns from the base class to the target class. To achieve backdoor injection, we assume the attacker can only perform the backdoor injection through local data corruption. An adversary can corrupt the local datasets of the malicious agents by introducing a trojan pattern to a fraction K of instances belonging to the base class and relabeling them as the target class, where K is set to 40% in default. However, apart from these actions, the adversary is unable to observe or modify the updates made by honest agents, nor can they influence the computations performed by honest agents or the aggregation server. In each round, the server selects participating clients and generating the weight gradient set (represented by bit mask) for each class using $p_{wg} = 20\%$ where p_{wg} means the percentage of kept weight gradients. These clients then conduct local training for $E = 10$ epochs, using a batch size of $B = 32$, before transmitting their updates to the server. 100 aggregation rounds are performed in total. Upon receiving and aggregating the updates, we evaluate the model's performance on the validation data using three key metrics: validation accuracy, base class accuracy, and backdoor attack success rate (ASR). We measured the computation reduction in terms of the averaged reduction of floating-points operations (FLOPs) counts. More importantly, since our method is orthogonal to the aggregation based method, we discuss the possibility of combing our method with other robust aggregation methods and also discuss the selection of parameters.

4.2 Results

Accuracy on both IID Settingand Non-IID Setting. We initially consider a setting where the data is distributed in an i.i.d. fashion. Each agent is assigned an equal number of samples from the training data through uniform sampling. We then transit to a more realistic setting for federated learning, where the

Table 1. Final backdoor and validation accuracy for different methods on non-iid with default setting.

	MNIST		Fashion MNIST		Cifar10	
	ASR	Validation	ASR	Validation	ASR	Validation
FedAVG	89.07%	97.51%	90.07%	89.11%	91.15%	81.27%
FoolsGold	8.18%	97.50%	8.08%	88.25%	8.65%	80.12%
RLR	7.01%	97.28%	8.01%	89.21%	9.16%	80.05%
Prune Neurons	8.22%	97.25%	12.22%	88.22%	9.99%	80.59%
LP-RFL	7.15%	97.27%	9.10%	88.45%	9.52%	80.99%

Table 2. Final backdoor and validation accuracy for different methods on non-iid with different Fs (Fashion MNIST).

	$F = 10\%$		$F = 20\%$		$F = 40\%$	
	ASR	Validation	ASR	Validation	ASR	Validation
FedAVG	90.07%	89.11%	92.01%	88.15%	94.10%	86.47%
FoolsGold	8.08%	88.25%	10.20%	87.25%	12.56%	86.05%
RLR	8.01%	89.21%	14.07%	86.01%	19.02%	82.01%
Prune Neurons	12.22%	88.22%	14.51%	84.25%	20.78%	80.17%
LP-RFL	9.10%	88.45%	9.28%	87.94%	9.10%	86.77%

data is distributed in a non-i.i.d. fashion. In this scenario, different classes are distributed among the clients, and each client may have a different distribution over the classes. We randomly select 40% to 100% training classes on each client and the number of training examples on each client are randomly sampled as well. To evaluate the performance of our proposed LP-RFL scheme, Fig. 4 plots the training curves for both FedAvg and LP-RFL in both the i.i.d. and non-i.i.d. settings. The horizontal axis represents the aggregation rounds, while the vertical axis represents the accuracy. We conduct these experiments using three different networks with different datasets.

In these figures, the left sub-figures depict the training curves obtained using the FedAvg, while the right sub-figures illustrate the training curves using our LP-RFL scheme. The blue curve represents the validation accuracy, the orange curve represents the base class accuracy and the grey curve represents the Attack Success Rate (ASR). We observe that there is no noticeable drop in accuracy when comparing LP-RFL to FedAvg. Furthermore, our LP-RFL scheme provides convergence guarantees on clean images under both the i.i.d. and non-i.i.d. settings. These experimental results demonstrate that LP-RFL is capable of achieving comparable accuracy to FedAvg. The convergence guarantee on clean images reinforces the effectiveness of our approach in maintaining the integrity and accuracy.

Table 3. Final backdoor and validation accuracy for different methods on non-iid with different Ks (Fashion MNIST).

	$K = 20\%$		$K = 40\%$		$K = 60\%$	
	ASR	Validation	ASR	Validation	ASR	Validation
FedAVG	85.07%	90.22%	90.07%	89.11%	91.15%	88.07%
FoolsGold	7.55%	88.81%	8.08%	88.25%	8.17%	87.52%
RLR	7.07%	89.18%	8.01%	89.21%	9.33%	87.58%
Prune Neurons	7.95%	89.25%	12.22%	88.22%	10.08%	87.09%
LP-RFL	7.28%	89.77%	9.10%	88.45%	9.54%	87.69%

ASR Degrading on Backdoored Examples. We further evaluate the defending quality of LP-RFL. As shown in Fig. 4, using LP-RFL can degrade the ASR from 90% to 10% on average compared to FedAVG, the most popular FL framework. We further compared LP-RFL with three state-of-the-art robust FL strategies under the non-iid setting. Table 1 reports the final accuracy results under different benchmarks. As it shows, LP-RFL achieves the state-of-the-art accuracy on both benign accuracy and the reducing of backdoor ASR. Note here, we only show the results in non-IID setting in following sections not only because of the page limits, but also the IID setting is over-idealized and not practical in real work situations. Our LP-RFL can achieve the state-of-the-art accuracy and defending qualities in both non-IID and IID settings as well.

Defending Against Different Amount of Malicious Clients/Images. We first analyze the defending quality of LP-RFL with different amount of malicious clients. By increasing Fs, we can add more malicious clients into the FL process. Table 2 shows the final accuracy results under different Fs on VGG16-Fashion MNIST. LP-RFL achieves better accuracy level than others with larger F because the other aggregation based methods rely on the vote scheme to determine the learning rate or updating direction. With the increment of malicious clients, the more wrong updating knowledge and the less correct updating knowledge will be collected by the server, which will fool the server to make the incorrect aggregation decisions.

We then fix F and increase the K to analyze the defending quality of LP-RFL with different amount of malicious images per client. Table 3 shows the final accuracy results under different Ks on VGG16-Fashion MNIST benchmark. The ASR of FoolsGold and RLR are not sensitive to the number of Ks because the whole malicious clients are likely to be removed to enhance the model's robustness. LP-RFL achieves better validation accuracy on smaller Ks compared to FoolsGold. FoolsGold assign a learning rate per client, as a result, the useful information from malicious clients will be ignored as well, which will hurt the convergence. LP-RFL prunes the malicious weight gradients in an example-per-example basis, it can achieve the balanced accuracy in both small Ks and large Ks.

4.3 Defending Against Adaptive Attack

Apart from the data pollution attack, we further assess LP-RFL's defensive capabilities in a scenario where an attacker could entirely manipulate the malicious clients, disregarding our updating rules, which is also known as the adaptive attack. Here, our primary objective involves implementing a Server Verification (SV) step to filter the malicious clients on the server side, leveraging received weight gradients from each client. Employing a sign-based method [15], we determine the majority updating direction (negative or positive) for each individual weight gradient. If a client contains more than $\lambda\%$ (weight gradient dissimilarity rate) of weight gradients opposing the majority updating direction, then the corresponding client is removed from participating FL. Our experiments reveal that the averaged Weight Gradient Dissimilarity Rate (WGDR) from malicious clients is 76.11% while the averaged WGDR from honest clients is 45.75%, hence we set $\lambda\% = 70\%$ in our experiments. The results are shown in Table 4. Notably, validation accuracy experiences some losses with the RV step, primarily due to misclassifying honest clients as malicious. Given that attackers find it more challenging to fully control the learning algorithm and that successful backdoor injection primarily occurs through data pollution [7], implementing the Server Verification step isn't mandatory in using LP-RFL to counter backdoor attacks.

Table 4. Final backdoor and validation accuracy when combine LP-RFL with RLR (Fashion MNIST).

	Mask Ignore Attack		Gradient Scaling Attack	
	ASR	Validation	ASR	Validation
LP-RFL	90.07%	89.11%	11.77%	75.02%
LP-RFL+SV	1.09%	88.02%	1.18%	87.67%

4.4 Combining with Robustness Aggregation Methods

Lastly, it is essential to highlight that our method is orthogonal to existing aggregation-based methods. This implies that our defense mechanism can be combined with other robust aggregation methods to enhance the overall defense against backdoor attacks while preserving the benign accuracy. We using an example of combing RLR and LP-RFL together, we can achieve low ASR and high validation accuracy in both low F and high F settings, which will make the FL more robustness to various of backdoor attack scenarios. The results are shown in Table 5. Since LR-RFL itself is not sensitive to the selection of Ks, combine RLR and LP-RFL will also be robust to the change of Ks. Combing LP-RFL with aggregation-based methods is also an efficient way to filter out the malicious clients which update malicious weight gradients to the server because the aggregation step is performed at the server side and the server is always honest.

Table 5. Final backdoor and validation accuracy when combine LP-RFL with RLR (Fashion MNIST).

	$F = 10\%$		$F = 40\%$	
	ASR	Validation	ASR	Validation
RLR	8.01%	89.21%	19.20%	82.01%
LP-RFL	9.10%	88.45%	9.10%	86.77%
RLR+LP-RFL	7.92%	89.20%	9.00%	86.67%

Computational Reductions and Communication Overhead To measure the impact of LP-RFL on the computational cost, we collect the number of floating point operations (FLOPs) per epoch required on the baseline model. Results of these experiments are listed in Table 6, as it shows, since we can know the pruned weight gradients in advance before calculating them, the proposed design has reduced the computational cost up to 3.9× and 3.2× on average. The FLOPs reductions are similar across different networks because they are determined by the desired pruning rate. If we use a similar pruning rate (i.e., p_{wg}) for all benchmarks, the reductions are similar as well. **Communication Overhead**: LP-RFL uses 1-bit to generate the mask to guide the weight gradient pruning, thus the overhead is negligible to the overall communication overhead (e.g. 3%).

Table 6. Computation Reduction on Client

	FedAVG	LP-RFL	FLOPs Reduction
LeNET	1.01T	259G	3.9×
VGG16	28.05T	9.35T	3.0×
MobileNetV2	17.27T	6.40T	2.7×

4.5 Defense Quality Analyzing via Parameters Tuning

In order to explain the effectiveness of our defense mechanism and provide empirical justification, we want to prove that the influence of malicious weight gradients must overcome the influence of honest weight gradients to successfully embed the backdoor into the model. To further explore and quantify the influence of agents on the model, we use a simple experiment from [15] to determine whether the influence of the adversary surpasses the combined influence of the honest clients. After each round, we compute the empirical Fisher Information Matrix (FIM) to identify the 500 most significant weight gradients for both adversarial and honest mappings. Then we denote the set of weight gradients updated to minimize the loss for adversarial mapping as S1 and for honest mapping as S2. Similarly, the set of weight gradients updated to maximize the loss for adversarial mapping is denoted as S3, and for honest mapping as S4. Then, we quantify the net adversarial influence as $I_{adv} = \|S_1 \backslash S_2\|_2 - \|S_3 \backslash S_4\|_2$ and the net honest influence as $I_{hon} = \|S_2 \backslash S_1\|_2 - \|S_4 \backslash S_3\|_2$. Finally, the net influence is then given by $I_{hon} - I_{adv}$.

Figure 5(a) shows our result. The horizon axis represent the aggregation rounds and the vertical axis represents the net influence score. We can see that, without LP-RFL, the malicious updatings will overcome the honest updatings and give negative influence score to insert the backdoor triggers. On the other side, with LP-RFL, the honest updatings will overcome the malicious updatings and give positive influence score to prevent the backdoor triggers.

Second, we measure the backdoor loss of different p_{wg}s. Recall that the fundamental ideas is to remove the significant malicious weight gradients, the higher p_{wg} is the more chance the significant malicious updatings will be retained. The results are shown in Fig. 5(b), where the horizon axis represent the aggregation rounds and the vertical axis represents the backdoor loss. With the FedAVG (p_{wg} = 100%), the backdoor loss drops quickly to get high ASR, on the other side, with the decreasing of p_{wg}, the backdoor loss is failed to minimize. The results prove the effectiveness of our defense mechanism. It is alreaday be proven that the weight gradients can be aggressively pruned without impacting the training convergence on clean inputs, and we choose $p_{wg} = 20\%$ based on our experiences.

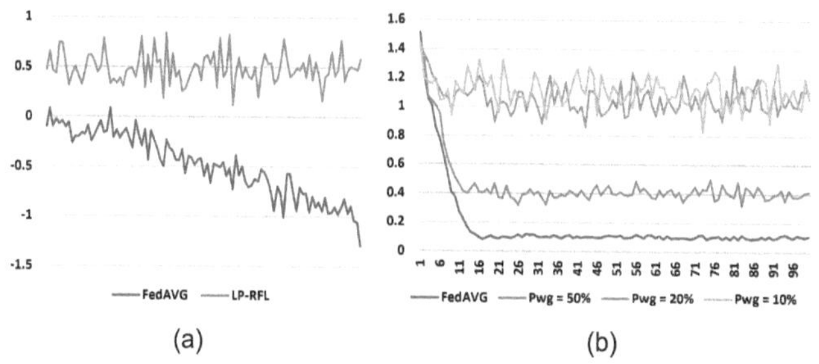

Fig. 5. (a) Net influence score of FedAVG and LP-RFL; (b) The impact of p_{wg} on backdoor loss.

5 Conclusion

In this work, we study FL from an adversarial perspective, and construct a simple defense mechanism to against backdoor attacks. We propose a label-based gradient pruning approach for FL to mitigate the impact caused by malicious clients. We profile a critical set (i.e., the sparse pattern) of weight gradients for each label on the server side using clean images and pre-trained initial model. Since training labels are known in advance, we can easily prune weight gradients that do not belong to the critical set. Moreover, our pruning-based training strategy efficiently eliminates redundant computations on local clients, resulting in significant savings in on-client computation and communication resources. Through experiments, our proposed method effectively reduces the Attack Success Rate of backdoor attacks from 90% to 10% and without noticeable accuracy loss on clean images. Additionally, our framework improves computation efficiency by an average of 3.2× (up to 3.9×), while still achieving the state-of-the-art accuracy.

Acknowledgment. This research is partially supported by NSF grants CCF-2130688, CNS-2107057 and RISE support program.

References

1. Andreina, S., Marson, G., Mollering, H., Karame, G.: Baffle: backdoor detection via feedback-based federated learning. In: 2021 IEEE 41st International Conference on Distributed Computing Systems (ICDCS). IEEE Computer Society (2021). https://doi.org/10.1109/ICDCS51616.2021.00086
2. Bhagoji, A.N., Chakraborty, S., Mittal, P., Calo, S.: Analyzing federated learning through an adversarial lens. In: Chaudhuri, K., Salakhutdinov, R. (eds.) Proceedings of the 36th International Conference on Machine Learning. Proceedings of Machine Learning Research, vol. 97, pp. 634–643. PMLR (2019). https://proceedings.mlr.press/v97/bhagoji19a.html
3. Blanchard, P., El Mhamdi, E.M., Guerraoui, R., Stainer, J.: Machine learning with adversaries: byzantine tolerant gradient descent. In: Guyon, I., et al. (eds.) Advances in Neural Information Processing Systems, vol. 30. Curran Associates, Inc. (2017)
4. Deng, L.: The mnist database of handwritten digit images for machine learning research. IEEE Signal Process. Mag. **29**(6), 141–142 (2012)
5. Fung, C., Yoon, C.J.M., Beschastnikh, I.: Mitigating sybils in federated learning poisoning (2020)
6. Goldblum, M., et al.: Dataset security for machine learning: data poisoning, backdoor attacks, and defenses (2021)
7. Gosselin, R., Vieu, L., Loukil, F., Benoit, A.: Privacy and security in federated learning: a survey. Appl. Sci. **12**(19) (2022). https://doi.org/10.3390/app12199901
8. Hong, Z., Yue, C.P.: Efficient-grad: efficient training deep convolutional neural networks on edge devices with gradient optimizations. ACM Trans. Embed. Comput. Syst. **21**(2) (2022). https://doi.org/10.1145/3504034
9. Krizhevsky, A., Hinton, G., et al.: Learning multiple layers of features from tiny images (2009)

10. Lecun, Y., Bottou, L., Bengio, Y., Haffner, P.: Gradient-based learning applied to document recognition. Proc. IEEE **86**(11), 2278–2324 (1998). https://doi.org/10.1109/5.726791
11. Muñoz-González, L., Co, K.T., Lupu, E.C.: Byzantine-robust federated machine learning through adaptive model averaging (2019)
12. Naseri, M., Hayes, J., Cristofaro, E.D.: Toward robustness and privacy in federated learning: experimenting with local and central differential privacy. CoRR arxiv:2009.03561 (2020)
13. Nguyen, T.D., et al.: Flame: taming backdoors in federated learning (extended version 1) (2023)
14. Nguyen, T.D., Nguyen, T., Nguyen, P.L., Pham, H.H., Doan, K., Wong, K.S.: Backdoor attacks and defenses in federated learning: survey, challenges and future research directions (2023)
15. Ozdayi, M.S., Kantarcioglu, M., Gel, Y.R.: Defending against backdoors in federated learning with robust learning rate (2021)
16. Qiu, Y., Leng, J., Guo, C., Chen, Q., Li, C., Guo, M., Zhu, Y.: Adversarial defense through network profiling based path extraction. In: Proceedings of the IEEE/CVF Conference on Computer Vision and Pattern Recognition (CVPR) (2019)
17. Sandler, M., Howard, A.G., Zhu, M., Zhmoginov, A., Chen, L.: Mobilenetv2: inverted residuals and linear bottlenecks. In: 2018 IEEE Conference on Computer Vision and Pattern Recognition, CVPR 2018, Salt Lake City, UT, USA, 18–22 June 2018, pp. 4510–4520 (2018). https://doi.org/10.1109/CVPR.2018.00474
18. Shen, S., Tople, S., Saxena, P.: Auror: defending against poisoning attacks in collaborative deep learning systems. In: Proceedings of the 32nd Annual Conference on Computer Security Applications. ACSAC '16 (2016). https://doi.org/10.1145/2991079.2991125
19. Shoham, N., et al.: Overcoming forgetting in federated learning on Non-IID data. arXiv preprint arXiv:1910.07796 (2019)
20. Simonyan, K., Zisserman, A.: Very deep convolutional networks for large-scale image recognition. In: International Conference on Learning Representations (2015)
21. Sun, T., Li, D., Wang, B.: Decentralized federated averaging (2021)
22. Tolpegin, V., Truex, S., Gursoy, M.E., Liu, L.: Data poisoning attacks against federated learning systems. In: Chen, L., Li, N., Liang, K., Schneider, S. (eds.) ESORICS 2020. LNCS, vol. 12308, pp. 480–501. Springer, Cham (2020). https://doi.org/10.1007/978-3-030-58951-6_24
23. Wang, L., Sistla, M., Chen, M., Fu, X.: Bs-pfl: enabling low-cost personalized federated learning by exploring weight gradient sparsity. In: 2022 International Joint Conference on Neural Networks (IJCNN), pp. 1–8 (2022). https://doi.org/10.1109/IJCNN55064.2022.9892137
24. Wu, C., Yang, X., Zhu, S., Mitra, P.: Mitigating backdoor attacks in federated learning (2021)
25. Xiao, H., Rasul, K., Vollgraf, R.: Fashion-mnist: a novel image dataset for benchmarking machine learning algorithms (2017)
26. Ye, X., et al.: Accelerating CNN training by pruning activation gradients. In: European Conference on Computer Vision (2019). https://api.semanticscholar.org/CorpusID:220647488

Detecting Botnet-Infected IoT Devices Using Variational Autoencoders

Jorge Proaño[1(✉)], Miguel Flores[2,3], and Roberto Andrade[4]

[1] Departamento de Matemáticas, Escuela Politécnica Nacional, Quito, Ecuador
jorge.proanio01@epn.edu.ec

[2] MODES, Departamento de Matemáticas, Escuela Politécnica Nacional, Quito, Ecuador
miguel.flores@epn.edu.ec

[3] Escuela Superior de Ingeniería y Tecnología, Universidad Internacional de La Rioja, Logroño, Spain

[4] Colegio de Ingenierias, Universidad San Francisco de Quito, Quito, Ecuador
randradep@usfq.edu.ec

Abstract. The exponential growth of the Internet of Things (IoT) has led to an increase in cyberattacks, particularly through the use of botnets that compromise vulnerable devices to perform malicious activities such as distributed denial-of-service (DDoS) attacks. Traditional intrusion detection systems often require labeled datasets, which are difficult to obtain in real-world IoT environments. This study proposes a novel unsupervised approach for identifying botnet-infected IoT devices using Variational Autoencoders (VAEs). The model is trained solely on legitimate network traffic to learn a baseline of normal behavior, enabling the detection of anomalies that may indicate compromised devices. To enhance the model's ability to detect anomalies, embeddings are used to effectively represent categorical variables, preserving the semantic relationships between categories. The proposed method is evaluated using public datasets and achieves high accuracy, precision, recall, and F1-score, demonstrating its effectiveness in distinguishing between benign and malicious flows. The findings suggest that VAEs offer a promising and scalable solution for anomaly-based intrusion detection in IoT networks, especially when labeled attack data is scarce or unavailable.

Keywords: VAEs · Cyberattacks · Anomaly Detection

1 Introduction

The rapid expansion of the Internet of Things (IoT) has introduced a wide range of innovative applications across various domains, including smart homes, industrial systems, healthcare, and transportation. However, this unprecedented proliferation of connected devices has also exposed significant security challenges. Many IoT devices are deployed with minimal security configurations, lack firmware updates, or use default credentials, making them easy targets for

H. R. Arabnia et al. (Eds.): CSCI 2025, CCIS 2969, pp. 444–453, 2027.
https://doi.org/10.1007/978-3-032-28460-0_31

malicious actors. As the number of these vulnerable devices grows, so does the potential surface for cyberattacks.

Among the most prevalent threats to IoT ecosystems are botnets, networks of compromised devices controlled by attackers. One of the most notorious examples is the Mirai botnet, which in 2016 infected thousands of IoT devices to launch large-scale distributed denial-of-service (DDoS) attacks. Mirai exploited basic vulnerabilities, such as open Telnet ports and hard-coded credentials [1]. Since then, various Mirai variants and other botnets have emerged, continuing to exploit the limited security posture of IoT infrastructures.

Detecting botnet activity in IoT networks remains a difficult task. Traditional Intrusion Detection Systems (IDS) often rely on signature-based techniques or supervised learning models that require large labeled datasets of both benign and malicious traffic [2]. However, obtaining accurate labels, especially for emerging threats, is resource intensive and not always feasible in real-world IoT environments, where traffic patterns are heterogeneous and evolve rapidly.

In response to these limitations, this study explores the use of variational autoencoders (VAEs), a class of unsupervised deep learning models, for anomaly detection in network traffic [3]. VAEs are capable of learning the distribution of benign traffic and identifying deviations that may correspond to malicious behavior. Unlike supervised models, VAEs do not require labeled attack data, making them particularly attractive for detecting novel or evolving threats in dynamic environments such as IoT networks [4].

The core idea is to train the VAE using only legitimate traffic, allowing it to learn a compact latent representation of normal behavior [13]. Once trained, the model can reconstruct benign flows with high fidelity, while traffic generated by infected devices will likely yield higher reconstruction errors, thus signaling a possible compromise [5]. This methodology focuses on detecting anomalies at the flow level, where statistical and time-based features of communication between devices are analyzed.

The main objective of this work is to develop and evaluate a VAE-based detection system capable of identifying IoT devices infected by botnets through analysis of network traffic. The proposed approach is validated using publicly available datasets that simulate realistic IoT network environments. Performance is assessed using common evaluation metrics such as accuracy, precision, recall, and F1 score. The results demonstrate the potential of VAEs to support scalable, low-overhead, and label-free detection strategies in IoT security.

2 Theoretical Framework

2.1 Internet of Things (IoT)

The Internet of Things (IoT) refers to a vast network of interconnected physical devices, such as sensors, cameras, home appliances, wearables, and industrial machines, that are capable of collecting, exchanging, and processing data over the Internet. Although IoT offers tremendous benefits in terms of automation and data-driven decision making, it also introduces serious security risks.

Their limited computational resources and real-time requirements often preclude the use of traditional cybersecurity tools, making IoT environments particularly susceptible to exploitation. Insecure configurations, hardcoded credentials, and unpatched firmware are among the most common vulnerabilities exploited by attackers.

2.2 Botnets

A botnet is a network of devices connected to the Internet that have been compromised by malware and are remotely controlled by an attacker, often without the knowledge of the device owner. Once infected, these devices, referred to as 'bots' or 'zombies', can be orchestrated to carry out a variety of malicious activities, including spamming, data exfiltration, credential stuffing, click fraud, and Distributed Denial of Service (DDoS) attacks [1].

Botnets continue to pose a critical threat due to their scalability, stealth, and the ease with which attackers can repurpose vulnerable IoT devices into weapons. Their decentralized nature and rapid mutation also challenge traditional detection approaches, which often rely on known signatures or IP blacklists.

2.3 Machine Learning and Unsupervised Anomaly Detection

Machine learning (ML) has gained prominence in cybersecurity due to its ability to model complex behaviors and detect patterns that can indicate malicious activity. In particular, unsupervised learning methods are useful in situations where labeled data is scarce or incomplete, common issues in the context of IoT traffic analysis. These models aim to discover the structure of the data without relying on predefined categories, making them well suited for anomaly detection [6]. In the context of network traffic, anomaly detection involves learning the typical patterns of communication and flagging flows that deviate from this learned behavior. Such anomalies may correspond to intrusions, misconfigurations, or malware infections [7]. One powerful technique for this task is the use of Autoencoders, which are neural networks trained to reconstruct their input data. When trained on benign data, autoencoders develop a compact latent representation that allows high-fidelity reconstruction [8]. Anomalous inputs, which differ significantly from the training distribution, tend to produce high reconstruction errors.

3 Variational Autoencoder Model

Variational Autoencoders (VAEs) extend this concept by introducing a probabilistic framework [8]. Rather than encoding inputs to fixed vectors, VAEs learn a distribution over the latent space, allowing them to generalize better and capture uncertainty. This makes VAEs particularly suitable for detecting subtle or emerging threats in dynamic environments such as IoT networks [9]. By modeling the distribution of normal traffic, VAEs can effectively identify flows that fall outside expected patterns, indicating potential botnet activity. Unlike traditional autoencoders that map an input $\mathbf{x}$ deterministically to a latent code $\mathbf{z}$, VAEs model the latent space as a distribution and learn parameters for both the encoder and decoder probabilistically.

3.1 Latent Space Encoding

Given an input sample $\mathbf{x}$, the encoder outputs the parameters of a probability distribution over the latent variable $\mathbf{z}$, typically a Gaussian:

$$q_\phi(\mathbf{z}|\mathbf{x}) = \mathcal{N}(\mathbf{z}; \boldsymbol{\mu}_\phi(\mathbf{x}), \boldsymbol{\Sigma}_\phi(\mathbf{x})) \tag{1}$$

Here, $\boldsymbol{\mu}_\phi(\mathbf{x})$ and $\boldsymbol{\sigma}_\phi(\mathbf{x})$ are the encoder's outputs representing the mean and standard deviation vectors of the approximate posterior distribution.

To allow gradient-based optimization, the *reparameterization trick* is used:

$$\mathbf{z} = \boldsymbol{\mu}_\phi(\mathbf{x}) + \boldsymbol{\sigma}_\phi(\mathbf{x}) \odot \boldsymbol{\epsilon}, \quad \boldsymbol{\epsilon} \sim \mathcal{N}(0, \mathbf{I}) \tag{2}$$

This separates the stochastic sampling from the model parameters, enabling backpropagation.

3.2 Decoder and Reconstruction

The decoder learns to reconstruct the original input $\mathbf{x}$ from the latent variable $\mathbf{z}$ by modeling the likelihood $p_\theta(\mathbf{x}|\mathbf{z})$, often assumed to be Gaussian for continuous inputs:

$$p_\theta(\mathbf{x}|\mathbf{z}) = \mathcal{N}(\mathbf{x}; \boldsymbol{\mu}_\theta(\mathbf{z}), \boldsymbol{\Sigma}_\theta(\mathbf{z})) \tag{3}$$

3.3 Loss Function: Evidence Lower Bound (ELBO)

Training a VAE involves maximizing the marginal log-likelihood of the data $\log p(\mathbf{x})$. Since the true posterior $p(\mathbf{z}|\mathbf{x})$ is intractable, VAEs optimize a tractable lower bound known as the Evidence Lower Bound (ELBO):

$$\log p(\mathbf{x}) \geq \mathbb{E}_{q_\phi(\mathbf{z}|\mathbf{x})}[\log p_\theta(\mathbf{x}|\mathbf{z})] - \mathrm{KL}(q_\phi(\mathbf{z}|\mathbf{x})||p(\mathbf{z})) \tag{4}$$

The ELBO has two terms:

- **Reconstruction Loss:** Measures how well the decoder can reconstruct $\mathbf{x}$ from $\mathbf{z}$
- **KL Divergence:** Regularizes the learned posterior $q_\phi(\mathbf{z}|\mathbf{x})$ to stay close to the prior $p(\mathbf{z})$, typically a standard normal $\mathcal{N}(0, \mathbf{I})$

To improve the classification, we use β-ELBO, that is a variant of the ELBO used in variational autoencoders to balance the trade-off between reconstruction accuracy and the regularization of the latent distribution.

However, the β-ELBO introduces a hyperparameter β that adjusts the weight of the regularization term:

$$\beta\text{-ELBO} = \mathbb{E}_{q_\phi(z|x)}[\log p_\theta(x|z)] - \beta * \text{KL}(q_\phi(z|x) \parallel p(z)) \tag{5}$$

- When $\beta > 1$, more weight is given to regularization, promoting a more structured latent space, although it may reduce reconstruction quality.
- If $\beta < 1$, reconstruction accuracy is prioritized over regularization, allowing more flexibility in the latent space.
- With $\beta = 1$, the standard ELBO formulation is recovered.

3.4 Reconstruction Error

The first term of the ELBO, $\mathbb{E}_{q(z|x)}[\log p(x|z)]$, represents the expected log-likelihood of reconstructing the input x from its latent representation z. When modeling $p(x|z)$ as a normal distribution, this term is directly related to the mean squared error (MSE), indicating that reconstruction quality can be evaluated using this metric.

Assuming that $p(x|z)$ follows a normal distribution with mean $\hat{x}$ (the reconstruction) and variance σ^2, we can write:

$$p(x|z) = \mathcal{N}(x; \hat{x}, \sigma^2) \tag{6}$$

The logarithm of this probability is given by:

$$\log p(x|z) = -\frac{1}{2\sigma^2}\|x - \hat{x}\|^2 - \frac{1}{2}\log(2\pi\sigma^2) \tag{7}$$

Taking the expectation with respect to $q(z|x)$ yields:

$$\mathbb{E}_{q(z|x)}[\log p(x|z)] = -\frac{1}{2\sigma^2}\mathbb{E}_{q(z|x)}[\|x - \hat{x}\|^2] - \frac{1}{2}\log(2\pi\sigma^2) \tag{8}$$

This shows that minimizing the reconstruction error is equivalent to maximizing the expected log-likelihood under the Gaussian assumption.

4 Methodology

4.1 Dataset and Feature Selection

To evaluate the proposed detection approach, this study uses the APOSEMAT-IOT23 dataset, which contains 23 network captures 20 of benign traffic and 3 of malicious traffic simulating realistic conditions in an IoT environment [10]. The dataset was generated using a testbed that integrates various IoT devices and common network protocols, capturing different types of attacks, such as reconnaissance, DoS, DDoS, and data theft. For this work, only traffic related to IoT devices is considered. From the dataset, a subset of flow-level features was selected, focusing on statistics that can reflect device behavior without relying on payload inspection. These features include packet and byte counts, flow duration, number of TCP flags, rate of packets per second, source and destination port numbers, and time-based indicators, obtaining 83 variables finally.

4.2 Data Preprocessing

The dataset preprocessing was divided into quantitative and categorical variable treatments. For the 63 quantitative features, no missing values were detected, so no imputation was required. However, due to varying scales and distribution biases, all numerical variables were normalized using the min-max scaling technique to the $[0, 1]$ range.

For the 20 categorical variables, two embedding strategies were employed. Contextual features such as IP addresses, ports, and protocols were encoded using Word2Vec [12], following the IP2Vec approach [13], where each variable was treated as part of a sentence in a flow-based corpus (See, Fig. 1). Non-contextual features were embedded using Keras' `Embedding` layer, with embedding dimensions set as $min(50, n/2)$, where n is the number of categories.

This preprocessing yielded a total of 440 features: 63 numeric variables and 377 embedding dimensions derived from the 83 original variables.

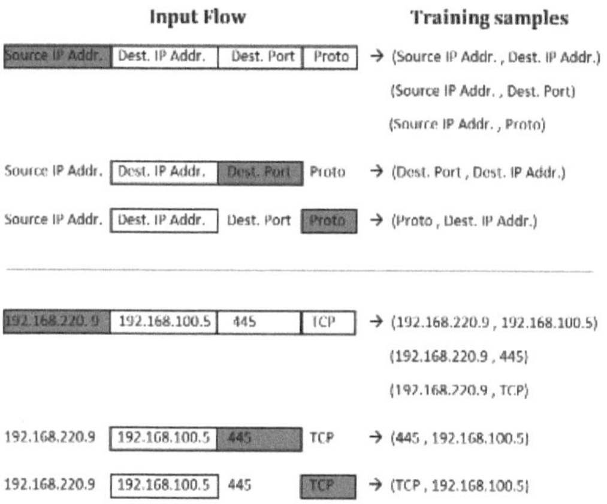

Fig. 1. Transforming a Network Flow into Training Samples for Word2Vec.

4.3 Variational Autoencoder Architecture

The proposed detection model leverages a Variational Autoencoder (VAE), composed of two primary components: the encoder and the decoder. The encoder maps the input features to a latent space defined by a mean vector and a standard deviation vector, allowing the model to capture uncertainty in the representation. A latent vector is then sampled from this distribution and passed to the decoder, which aims to reconstruct the original input.

The VAE is trained by minimizing a composite loss function comprising: (1) the reconstruction loss, typically the mean squared error between the input and its reconstruction, and (2) the Kullback–Leibler (KL) divergence, which encourages the latent distribution to approximate a standard normal distribution. This regularization promotes generalization and stability.

To perform anomaly detection, the model is trained exclusively on benign traffic, learning the normal behavior patterns. Anomalous inputs, such as malicious traffic, typically yield higher reconstruction errors due to their deviation from the learned distribution [14]. Therefore, a threshold on the reconstruction error, derived empirically from a validation set of benign samples, is used to classify inputs as normal or anomalous.

5 Results

The proposed Variational Autoencoder (VAE) model was implemented using Pyro and Spark. The selected dataset was preprocessed as described, and the VAE was trained exclusively on benign traffic. The model architecture included an encoder with two hidden layers (440 and 150 neurons) and a latent dimension of 30. The decoder mirrored this structure. Training was conducted over 20 epochs using a batch size of 10,000, a learning rate of 0.001 and a regularization parameter $\beta = 1.1$. After training, the model was evaluated using a separate test set containing both benign and malicious flows. The reconstruction error was computed for each instance in the test set, and a threshold was determined based on the 1st percentile of the reconstruction error on a validation subset of benign flows (See, Fig. 2).

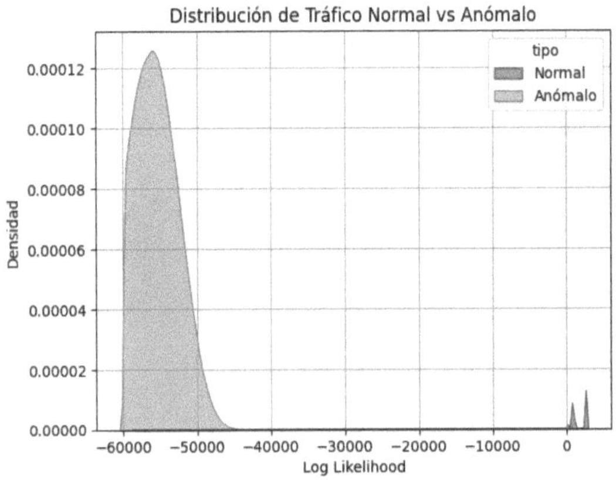

Fig. 2. Reconstruction error distribution from benign traffic.

Using the reconstruction error *threshold* = -2758.31, each test flow was classified as either benign or anomalous. The classification results were compared against the ground truth labels to compute the following performance metrics shown in Table 1.

Table 1. Evaluation Metrics for VAE-based Botnet Detection

Metric	Value
Accuracy	0.9916
Precision	0.9490
Recall	1.0000
F1-Score	0.9739

These results demonstrate that the VAE model is highly effective at distinguishing between benign and botnet-infected flows. The high precision suggests that most of the detected anomalies were truly malicious, while the recall indicates that a substantial portion of actual anomalies were successfully identified.

To further assess the model's discriminative capability, the Receiver Operating Characteristic (ROC) curve was generated by varying the reconstruction error threshold. The Area Under the Curve (AUC) was calculated to be 0.99, indicating excellent overall performance and confirming that the VAE is capable of separating benign and anomalous traffic with high reliability (See, Fig. 3).

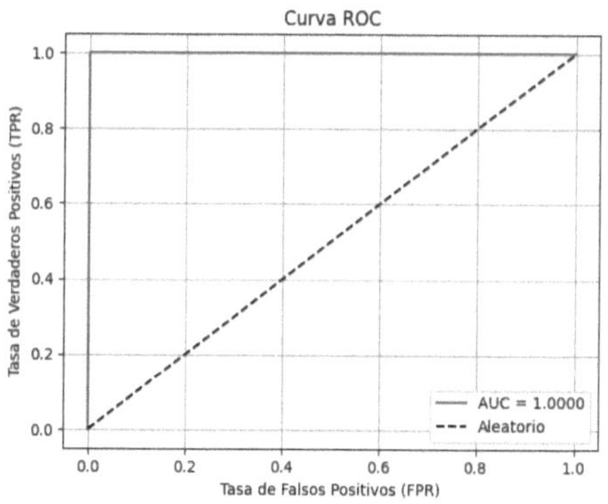

Fig. 3. Receiver Operating Characteristic (ROC) curve.

5.1 Comparative Analysis

To contextualize the effectiveness of the VAE model, its performance was compared to traditional machine learning classifiers such as Random Forest (RF), as well as to a VAE variant that did not include categorical variable embeddings. The RF model was trained in a supervised setting using labeled data and achieved competitive accuracy. However, it requires a substantial amount of labeled malicious traffic for training a limitation not shared by the VAE, which operates in an unsupervised manner [15]. These results underscore the advantage of unsupervised VAEs in environments with limited labeled data, such as IoT networks. The use of embeddings for categorical variables significantly improved performance, as the variant without them was less effective at detecting anomalies.

6 Discussion

The experimental results shows that Variational Autoencoders (VAEs) offer a powerful solution for detecting anomalous traffic in IoT networks, even in the absence of labeled attack data. By modeling only benign behavior during training, the VAE learns to reconstruct expected traffic patterns and effectively flags deviations that may correspond to botnet activity. This unsupervised learning paradigm is particularly suited to environments where new threats frequently emerge and labeled datasets are incomplete or unavailable. Unlike signature-based detection or supervised machine learning models, VAEs do not require continuously updated rule sets or annotated data. This allows for greater adaptability and reduces the operational burden on network administrators, who can deploy the model in real-time systems without extensive configuration. In addition, the use of flow-level features enables the system to remain lightweight and efficient, making it feasible for integration into resource-constrained IoT infrastructures. Since the model operates on statistical and temporal characteristics of the traffic—rather than deep packet inspection—it also avoids potential privacy concerns and processing overhead.

Despite its strengths, the approach is not without limitations. The model's effectiveness is strongly dependent on the quality of the benign training data. If this dataset contains undetected malicious samples or is not representative of the true diversity of normal traffic, the model may either miss threats or generate false positives. Furthermore, the selection of the reconstruction error threshold is critical to performance and requires empirical tuning to balance precision and recall in different deployment contexts. Another challenge is the interpretability of anomalies detected by the VAE. Although the model can signal deviations from the learned behavior, it does not inherently explain the nature of the anomaly or its potential impact. This limits its immediate usefulness in forensic analysis or incident response, where explainability is crucial. However, the promising results open avenues for future work. One possibility is to combine VAEs with semi-supervised learning or ensemble methods to enhance robustness and interpretability. Additionally, integrating feedback mechanisms from human analysts could refine the threshold dynamically and improve trust in the system.

7 Conclusions and Future Work

The results demonstrated high performance across key evaluation metrics, including accuracy, precision, recall, F1-score, confirming the model's potential in securing IoT networks against botnet threats. The approach's main strengths lie in its independence from labeled data, adaptability to evolving threats, and low computational cost, making it well-suited for dynamic and resource-constrained environments. The use of embeddings helps maintain relationships between network traffic characteristics, which improves the model's ability to learn from data. However, the success of the model depends heavily on the quality of the training data and the selection of the detection threshold. Addressing these challenges requires further exploration into adaptive thresholding techniques, feedback-driven calibration, and methods to improve the interpretability of detected anomalies.

References

1. Antonakakis, M.,et al.: Understanding the Mirai Botnet. In: Proceedings of the 26th USENIX Security Symposium (SEC'2017), USENIX Association, USA, pp. 1093–1110 (2017)
2. Xing, Y., Shu, H., Zhao, H., Li, D., Guo, L.: Survey on Botnet Detection Techniques: Classification, Methods, and Evaluation. Math. Problems Eng. **2021**(1), 6640499 (2021)
3. Doersch, C.: Tutorial on Variational Autoencoders (2021). arXiv:1606.05908
4. Ganguly, A., Earp, S.W.F.: An Introduction to Variational Inference. arXiv:2108.13083 (2021)
5. Aouini, Z., Pekar, A.: NFStream: a flexible network data analysis framework. Comput. Netw. **204**, 108719 (2022)
6. An, J., Cho, S.: Variational Autoencoder based Anomaly Detection using Reconstruction Probability (2015)
7. Bishop, C.M.: Pattern Recognition and Machine Learning. Springer, New York (2006)
8. Kingma, D.P., Welling, M.: Auto-Encoding Variational Bayes (2022). arXiv:1312.6114
9. O. K. C. U., Pranavi, D., Laxmi, B.R.A., D. R.: Variational Autoencoder for IoT Botnet Detection. In: Using Computational Intelligence for the Dark Web and Illicit Behavior Detection, pp. 74–88, IGI Global (2022)
10. Garcia, S., Parmisano, A., Erquiaga, M.J.: IoT-23: a labeled dataset with malicious and benign IoT network traffic (2020)
11. Fielding, R.T., Reschke, J.: Hypertext Transfer Protocol (HTTP/1.1): Semantics and Content. RFC 7231, IETF (2014)
12. Mikolov, T., Sutskever, I., Chen, K., Corrado, G.S., Dean, J.: Distributed Representations of Words and Phrases and their Compositionality. In: Advances in Neural Information Processing Systems, vol. 26 (2013)
13. Ring, M., Dallmann, A., Landes, D., Hotho, A.: IP2Vec: Learning Similarities Between IP Addresses. In: 2017 IEEE International Conference on Data Mining Workshops (ICDMW), pp. 657–666 (2017)
14. Zavrak, S., İskefiyeli, M.: Anomaly-Based Intrusion Detection From Network Flow Features Using Variational Autoencoder. IEEE Access, **8**, 108346–108358 (2020)
15. Blei, D.M., Kucukelbir, A., McAuliffe, J.D.: Variational Inference: A Review for Statisticians. J. American Stat. Assoc. **112**(518), 859–877 (2017). arXiv:1601.00670

Forensics and Logs Analysis: LLMs-Based Log Analysis and Correlation for Cyber Forensics

Sharina Eisa Aljallaf(✉), Mohammed Hussain, and Monther Aldwairi

College of Technological Innovation, Zayed University, Dubai, UAE
{M80008688,mohammed.hussain,monther.aldwairi}@zu.ac.ae

Abstract. Logging is a core security practice that records system activities for monitoring and forensic analysis. However, the unstructured nature of logs, the volume of alerts, and the complexity of modern attacks limit the effectiveness of traditional analysis methods. This paper presents an intelligent framework for log analysis that employs large language models, more specifically ChatGPT and Gemini AI. The proposed framework applies natural language processing and auto-correlation techniques to detect anomalies, correlate events, and identify potential security breaches in log data. By leveraging Big Data processing and visualization, the framework reduces false positives and enhances the accuracy of forensic investigations. Experimental evaluation demonstrates that the approach effectively uncovers suspicious activities and provides structured insights for cyber forensics analysts.

Keywords: Auto-Correlation · Big Data · ChatGPT · Gemini · AI · LLMs · Logs · Log Forensics

1 Introduction

Security breaches and cyber threats have become a critical concern for individuals and organizations [1]. System and application logs are an essential source of evidence for detecting attacks, tracing their origin, and preventing future incidents. Tools such as Intrusion Detection Systems (IDS), firewalls, and Active Directory monitoring continuously generate large volumes of log data. However, the scale, unstructured format, and heterogeneous nature of logs make manual inspection time-consuming and error-prone. This paper investigates the power of deep learning—specifically large language models (LLMs) such as ChatGPT and Gemini, to enhance and automate log analysis. These AI models can interpret and connect unstructured data from various systems, making it easier to detect patterns, identify security breaches, and respond to threats faster and more accurately. They can make cybersecurity tools smarter and more efficient, ultimately helping organizations stay ahead of cyberattacks.

2 Background

The utilization of Large Language Models (LLMs) presents a powerful approach to analyzing and correlating different logs. By harnessing the natural

H. R. Arabnia et al. (Eds.): CSCI 2025, CCIS 2969, pp. 454–465, 2027.
https://doi.org/10.1007/978-3-032-28460-0_32

language processing capabilities of LLM models specially ChatGPT and Gemini researchers can extract valuable insights for identifying security breaches in logs using advanced technique.

- ChatGPT: Developed by OpenAI, is a large language model trained to understand and produce text that mimics the language of human. It can be used to analyze text-based logs to identify patterns, anomalies, and potential correlations between events. By inputting log data, ChatGPT can process the data to identify possible connections [2].
- Gemini: Is a Google AI assistant for writing, planning, and more efficient summary of search results. We find it particularly helpful in log analysis and correlation, as Gemini possesses the ability to handle and examine large volumes of log data from several sources. To do this, algorithms would need to be utilized to identify patterns, anomalies, and correlations between various log entries and systems [3]. The combination of ChatGPT and Gemini can provide a more robust analysis by combining natural language understanding with specialized log analysis techniques.

3 Literature Review

Log analysis and correlation involved parsing, filtering, and combining log data from various sources to identify patterns, anomalies, and potential security incidents [4]. Unlike traditional approaches to this process, which were significantly limited by their inability to handle large and complex log data, machine learning strategies facilitated the efficient analysis of such data. Another noteworthy advantage of deep and machine learning techniques is their ability to recognize anomalies or patterns from extensive data and make predictions from massive log datasets [5]. This section reviewed ML, DL, and other approaches

3.1 Machine Learning-Based Approach

The increasing sophistication of cyberattacks led researchers to apply ML methods for intrusion detection and log correlation. For example, K.A Tait et al. [6] reviewed intrusion detection techniques and compared ML algorithms for binary and multiclass attacks. While binary classification achieved higher accuracy, multiclass classification with k-nearest neighbor showed better adaptability to attack categories. The study demonstrated the promise of ML for IDS log correlation but remained limited to a single dataset. Aldwairi and Alsaadi [7] developed FLUKES, a JavaScript tool that used regular expressions to extract threats from ASCII-based logs such as FTP, Snort, and IIS server logs. FLUKES enabled visualization and forensic analysis but struggled with scalability and required frequent updates to its signature set. Sarker et al. [8] provided an overview of data science and ML applications in cybersecurity. Their work emphasized how ML techniques could extract actionable insights from heterogeneous security data, though challenges included the lack of labeled datasets and limited adaptability across different log formats. Unlike more specialized tools, this work highlighted the broader application of ML to cybersecurity problems rather than focusing solely on log correlation. Similarly, Skopik, Landauer, and Wurzenberger [9,23] reviewed log data analysis with ML, covering clustering, classification, and anomaly detection. They concluded that ML could automate log analysis, but context remained crucial—some anomalies can only be identified relative to their operational environment. Pavlov and Voloshina [10] investigated IDS alert correlation techniques, highlighting clustering and classification methods to group alerts. They identified a persistent problem of false positives and a dependence on high-quality IDS data. Chuah et al. [4] surveyed log-correlation tools for failure diagnosis in clustered systems, providing a taxonomy of existing methods for spotting trends, outliers, and possible failures. Abdalla and Jumaa [11] expanded this with a comprehensive survey of supervised and unsupervised ML techniques for log analysis, noting two open problems: (i) the need for ML models trained on real-world datasets to reduce false positives/negatives, and (ii) the challenge of maintaining effective alert correlation in dynamic environments.

3.2 Deep Learning-Based Approach

Deep learning offered advanced capabilities for capturing patterns in unstructured logs. Sarker [12] reviewed DL techniques across supervised, unsupervised, and hybrid learning, highlighting their real-world applications and challenges in building robust models. Pan et al. [13] introduced a Transformer-based model that learned system behavior from normal logs to detect anomalies in an unsupervised manner. This reduced the reliance on labeled data but remained computationally intensive. Brazell et al. [14] applied convolutional neural networks for log correlation at scale, demonstrating improved detection accuracy, though results depended heavily on data quality and labeling. Khoei et al. [5] offered a systematic overview of DL techniques and future research needs, pointing out the difficulty of building adaptable models across diverse real-world scenarios. Karlsen et al. [15] benchmarked LLMs for log analysis, providing valuable insights into model performance across diverse datasets but highlighting issues of noisy data and generalization. Qi et al. [16] proposed Adanomaly, an adversarial DL framework combining log parsing, feature extraction, and adversarial training for anomaly detection. While effective, it was computationally demanding. Nam et al. [17] developed a hybrid BERT-CNN model for predicting VM failures, which showed strong predictive accuracy but required high-quality data and computational resources.

3.3 Other Log Analysis Approach

Big data technologies were integrated into log analysis. Shendi et al. [18] examined predictive analytics for real-time intrusion detection using big data, focusing on scalability but noting gaps in integration across diverse devices. Zhang and Zhao [19] proposed a polygon-based visualization method for IDS log correlation, enabling better interpretation of multidimensional log attributes but limiting its effectiveness due to attribute selection constraints. LogMine [20] introduced an efficient log parsing system for real-time analytics, automating field extraction and correlation. Similarly, Chen et al. [21] presented a scalable distributed framework for log collection, analysis, and visualization, demonstrating its utility for large-scale deployments despite complexity in configuration and potential false positives.

4 Proposed Log Analysis and Correlation Methodology

The methodology adopted in this study applies natural language processing (NLP) techniques based on deep learning algorithms to analyze log data, detect anomalies, and establish correlations between different log entries. The proposed solution leverages large language models for intelligent log analysis, with particular focus on the capabilities of ChatGPT and Gemini AI. Log files collected from a publicly available dataset are used as input for the analysis. The LLMs process the unstructured log data to:

1. Parse and interpret log entries into meaningful structured representations.
2. Identify patterns and anomalies that indicate potential security incidents.
3. Correlate alerts across multiple sources to provide a contextualized view of system activity.
4. Generate threat scenarios that summarize suspicious or malicious behavior for further investigation.

This approach enables a forensic investigator or security analyst to detect abnormal activity, understand system behavior more effectively, and enhance decision-making during incident response. By relying on LLMs, the proposed methodology reduces the dependency on manual analysis and rule-based systems, improves adaptability to diverse log formats, and enhances accuracy in identifying security threats.

4.1 Research Design

To conduct our experiment on LLMs based on log analysis and correlation we went through different stages. The main framework is as follows:

1. Data collection: We started by searching for a comprehensive dataset of Intrusion Detection Systemlogs. The data for this research is primarily sourced from network traffic logs recorded by IDS. These logs are rich repositories of both normal and potentially malicious activities, providing a real-world context for anomaly detection. These logs contain information about potential security threats, such as network traffic anomalies and suspicious activities. The log dataset we used was maintained by the Austrian Institute of Technology (AIT) those logs demonstrate a small enterprise network comprising of mail server, file share, WordPress server, VPN, firewall, etc. [22]. This data

source was collected over four to six days from 2022-01-14 and ends on 2022-01-18. Logs files from the various machines in the network were collected, including, DNS, webserver, authentication server, firewall, and system logs. The network traffic was also captured as pcap files [22].

2. Data analysis: For our analysis, we are using two different LLMs to generate the output response of a specific prompt that we came up with and send it to the LLMs to analyze the data and detect anomalies as shown in Figs. 1 and 2. To analyze the data, random logs were selected from multiple hosts. We tested the ability of LLMs to analyze different logs, by providing LLMs with excerpts from various log files. Then, the below prompt was used to request LLMs to analyze the log. This process was repeated many times, and we recorded the response of the LLMs. Correlate alerts across multiple sources to provide a contextualized view of system activity. Generate threat scenarios that summarize suspicious or malicious behavior for further investigation.

Prompt

I want you to act as forensic investigator in cyber security field, I want you to check and investigate for me if there is a security breach that could be identified in the logs. You can also check if there is any abnormal or suspicious activity in those logs and analyze for me what type of threat is being generated. and provide me with a scenario of what was happening on this system.

[Log]

Fig. 1. Prompt Used for Analysis

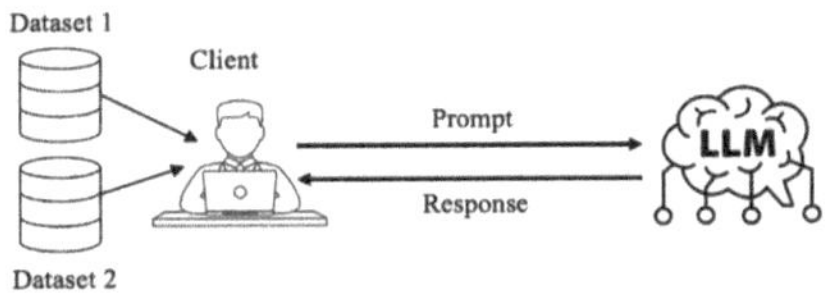

Fig. 2. Process of Analysis Using LLMs to Generate Output

3. Data correlation: Using LLMs techniques to correlate and identify patterns and relationships within the logs. Correlation helps determine if there is a correlation between the current log entry and previous log entries, which can provide insights into recurring security threats or attack patterns. To test whether there is a correlation or not we will send a specific prompt to the LLMs to check if there is a relationship between both datasets.

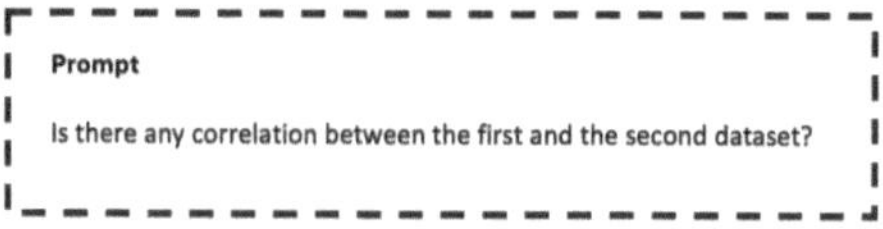

Fig. 3. Prompt Used for Correlation

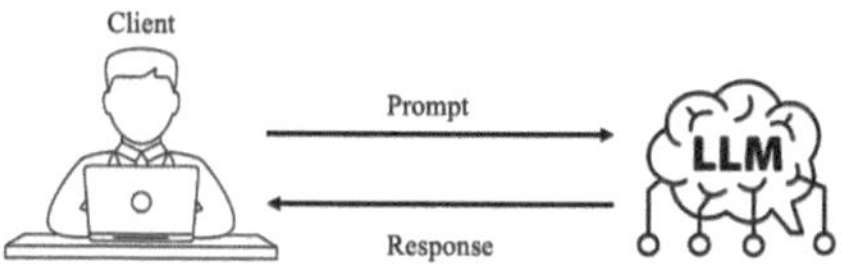

Fig. 4. Process of Correlation Using LLMs to Generate Output

4.2 Experiment and Implementation

For the experiment and implementation phase, we propose an intelligent solution to analyze and correlate log files using ChatGPT and Gemini AI. We analyze log files by providing LLMs with excerpts from various log files. Which will be acting as a forensics investigator to detect any abnormal or suspicious activity and analyze the log file to identify which type of threat is being generated to provide the users with a scenario of what was happening on their system and checks if there is any correlation between the datasets. This process was repeated many times, and we recorded the response of the LLMs (Figs. 3, 4).

1. Logs : The dataset utilized comprises network logs sourced from a diverse range of hosts within a single network environment. This dataset encompasses various hosts, including the iNet firewall, iNet DNS server, mail server, Monitoring server, and intranet server. Each host contributes distinct log entries reflecting its specific functionalities and interactions within the network to gain a better understanding, refer to Fig. 5 which demonstrates the structure of the dataset. By analyzing and correlating these heterogeneous logs, the experiment aims to uncover patterns, anomalies, and potential security threats within the network infrastructure, demonstrating the effectiveness of LLM-based techniques in ChatGPT and Gemini for cyber forensic investigations in complex network environments.
2. LLM-Based Log Analysis: ChatGPT and Gemini were used to analyze log files and identify abnormal activities based on contextual understanding and

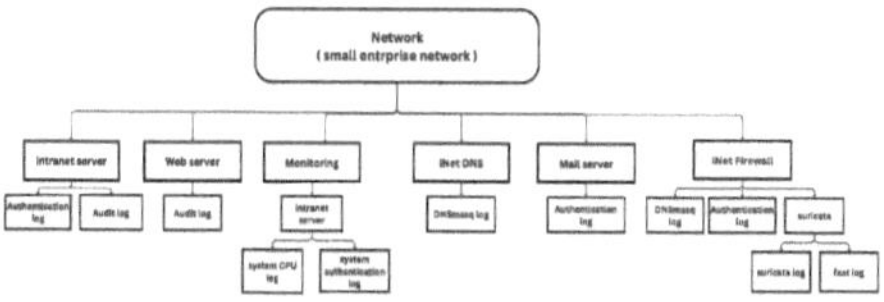

Fig. 5. Dataset Structure

anomaly detection capabilities. The log file content that was copied from the dataset was selected after a manual exploration as we wanted to see if the large language models can detect correlation. For the testing, we are using two different logs and asking both LLMs to generate a response based on the prompt that we developed earlier. The results indicate that both ChatGPT and Gemini successfully produced outputs demonstrating their ability to analyze log data.

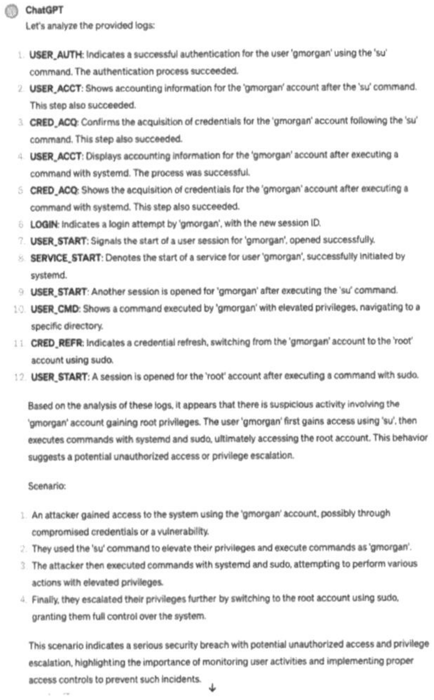

Fig. 6. ChatGPT analysis results

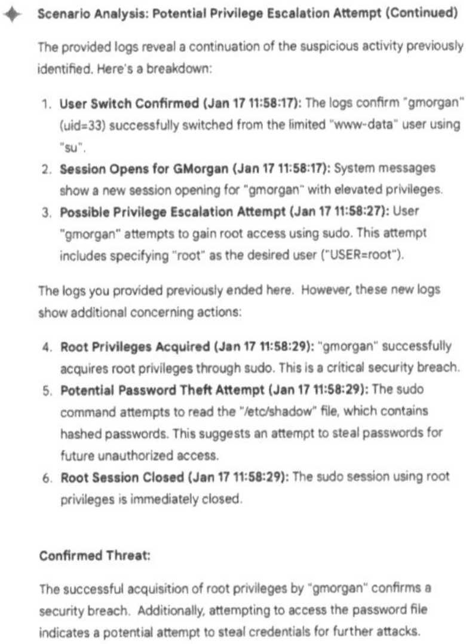

Fig. 7. Gemini analysis results

3. Data Correlation : To test whether there is a correlation or not we will send a specific prompt to the LLMs to check if there is a relationship between both datasets. ChatGPT and Gemini were used to indicate if there is a correlation between the two log files, which were encompassed from a single network. This is used to check if there is any relationship between the first and the second Log source. Both LLM's successfully demonstrate meaningful log correlation capabilities, to analyze and understand multiple log entries. Correlations between unrelated events or activities that are documented in logs can be found using ChatGPT and Gemini. Together, they can provide a more robust analysis by combining natural language understanding with log analysis techniques. Moreover, their capability to correlate logs from several systems makes it easier to identify connections and dependencies, giving enterprises the flexibility to prevent future problems, reduce risks, and strengthen

their cybersecurity. The ability of these tools to analyze logs from multiple systems allows for a more comprehensive view of the environment. This means that they can detect interactions and dependencies between different systems that might contribute to the overall security posture or operational state of an enterprise. One of the examples of ChatGPT and Gemini's response is shown below.

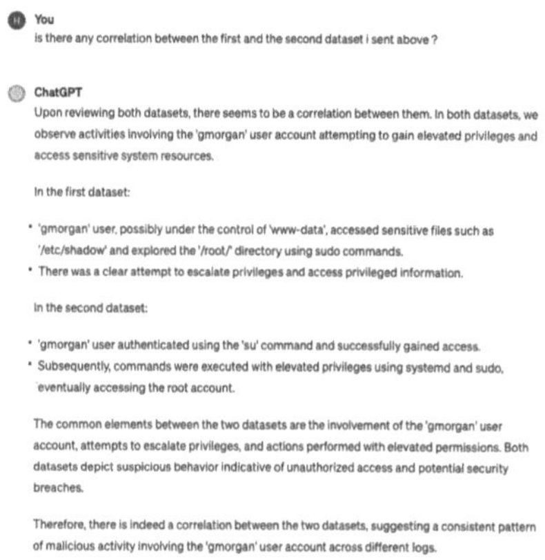

Fig. 8. ChatGPT correlation results

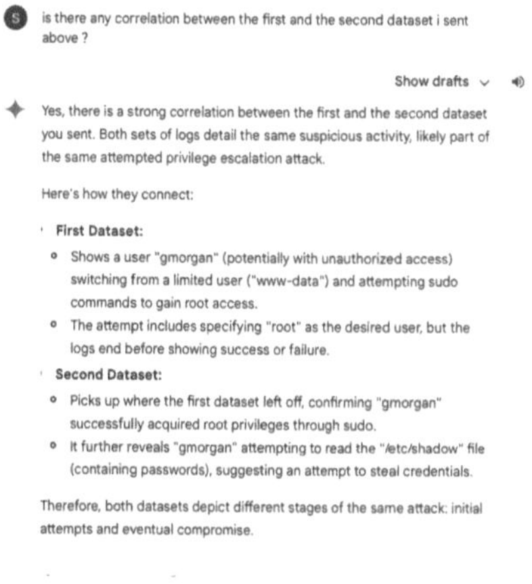

Fig. 9. Gemini correlation results

5 Results

The results from ChatGPT and Gemini analysis and correlation were compared to evaluate their performance in detecting abnormal activities across the different case studies. The confusion matrices provided detailed insight into how each model classified normal versus abnormal log entries. The evaluation was conducted on a dataset of 15 total samples cases but for demonstration purposes we only viewed 8 case studies. Although 15 samples offer more reliability than extremely small sets, the dataset is still considered limited, which means the results should be interpreted with caution due to higher sensitivity to misclassification and reduced generalizability (Figs. 6, 7, 8 and 9). Although both models were evaluated on a small dataset of 15 samples, the comparison offers initial insights into their performance. Gemini demonstrated higher accuracy and a stronger F1 score compared to ChatGPT, suggesting better consistency in anomaly detection for this dataset. However, larger and more diverse datasets are required to validate these findings, reduce performance variability, and ensure that the models generalize effectively across different real-world scenarios.

5.1 ChatGPT Confusion Matrix

ChatGPT achieved an accuracy of 0.6667 and an F1 score of 0.7059. The results show that ChatGPT identified most abnormal activities (TP) but generated a noticeable number of false positives, indicating a tendency to misclassify normal cases as suspicious. This behavior could lead to excessive alerting in operational environments. Additionally, the presence of false negatives suggests that ChatGPT occasionally failed to detect actual abnormal events, which is a critical concern in cybersecurity applications (Fig. 10) (Table 1).

Table 1. ChatGPT confusion matrix

	Predicted Negative	Predicted Positive
Actual Negative	4	3
Actual Positive	2	6

- ∘ **True Negatives (TN)** = 4
- ∘ **False Positives (FP)** = 2
- ∘ **False Negatives (FN)** = 3
- ∘ **True Positives (TP)** = 6

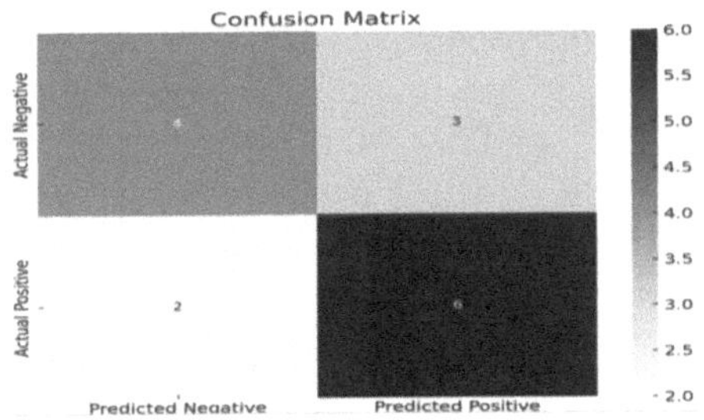

Fig. 10. ChatGPT confusion matrix correlation results

- ∘ **Accuracy** = 0.6667
- ∘ **F1 score** = 0.7059

5.2 Gemini Confusion Matrix

Gemini achieved an accuracy of 0.8000 and an F1 score of 0.8235, outperforming ChatGPT in both metrics. Gemini generated fewer false positives and fewer false negatives, indicating stronger reliability in distinguishing between normal and abnormal activities. This is especially important in security environments where detection accuracy directly influences incident response (Fig. 11) (Table 2).

Table 2. Gemini confusion matrix

	Predicted Negative	Predicted Positive
Actual Negative	5	2
Actual Positive	1	7

- **True Negatives (TN)** = 5
- **False Positives (FP)** = 1
- **False Negatives (FN)** = 2
- **True Positives (TP)** = 7

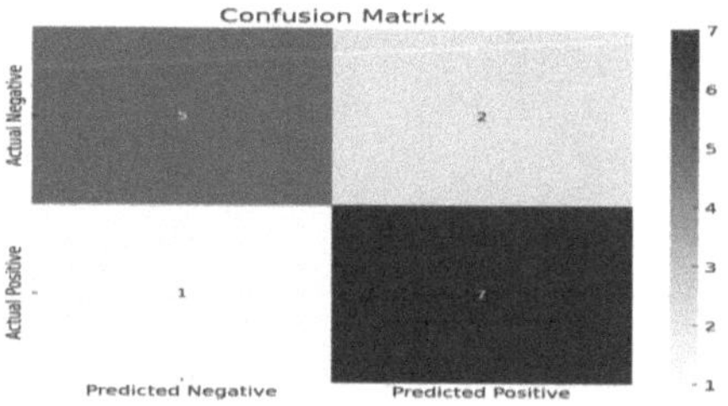

Fig. 11. Gemini confusion matrix correlation results

- **Accuracy** = 0.8000
- **F1 score** = 0.8235

6 Comparsion Between ChatGPT and Gemini

Both ChatGPT and Gemini function in a similar way, they have a similar user interface as the user type in the prompt to get a response. However, the majority of their differences lie in their language, training models, and data privacy [3]. In terms of the language models, they both offer more advanced and faster models by subscribing. The data sources that ChatGPT and Gemini employ to train their LLMs are the primary distinction between them. GPT-3.5 uses predetermined data that ends in January 2022. Gemini uses real-time data that is retrieved from the internet. It is set up to choose information from sources that correspond to particular subjects [3]. Both LLMs have privacy concerns as they store users' data in their systems. When it comes to Log analysis and correlation Gemini is a better choice as it is specifically designed for log data analysis and is optimized for efficiency offering high-performance capabilities for tasks such as anomaly detection, pattern recognition, and data analysis.

7 Conclusion

As organizations increasingly rely on digital systems and processes to manage and store sensitive information, the risk of security breaches becomes ever more significant. Log data generated by these systems can be a valuable source of information for detecting and responding to security breaches, but the sheer volume and complexity of this data can make it difficult to extract meaningful insights. The utilization of Large Language Models (LLMs) such as ChatGPT and Gemini holds a promising result for analyzing and correlating different datasets. Our exploration into the capabilities of ChatGPT and Gemini has demonstrated their effectiveness in dealing with different data sources which can provide valuable insights into network security incidents. Additionally, visualizing log data can enhance the understanding of patterns and anomalies, ultimately improving the detection and response capabilities of intrusion detection systems.

References

1. Romeo, M.D.S.S.: Intrusion detection system (IDS) in internet of things (IoT) devices for smart home. Int. J. Psychosocial Rehabilitation **23**(4), 1217–1227 (2019). https://doi.org/10.37200/IJPR/V23I4/PR190448
2. Brown, T.B., et al.: Language Models are Few-Shot Learners (2020)
3. Raffo, D.: Gemini vs. ChatGPT: what's the difference?, TechTarget (2024). https://www.techtarget.com/searchenterpriseai/tip/Geminivs-ChatGPT-Whats-the-difference
4. Chuah, E., Jhumka, A., Malek, M., Suri, N.: A survey of log-correlation tools for failure diagnosis and prediction in cluster systems, IEEE Access, vol. 10, pp. 133487–133503 (2022). https://doi.org/10.1109/ACCESS.2022.3231454
5. Chuah, E., Jhumka, A., Malek, M., Suri, N.: A survey of log-correlation tools for failure diagnosis and prediction in cluster systems. IEEE Access **10**, 133487–133503 (2022). https://doi.org/10.1109/ACCESS.2022.3231454
6. Khoei, T.T., Slimane, H.O., Kaabouch, N.: Deep learning: systematic review, models, challenges, and research directions. Neural Comput. Appl. **35**(31), 23103–23124 (2023). https://doi.org/10.1007/s00521-023-08957-4
7. Tait, K.-A., et al.: Intrusion Detection using Machine Learning Techniques: An Experimental Comparison (2021)
8. Aldwairi, M., Alsaadi, H.H.: FLUKES, In: Proceeding International Conference Future Networks and Distributed Systems, ACM, pp. 1–6 (2017) https://doi.org/10.1145/3102304.3102337.
9. Sarker, I.H., Kayes, A.S.M., Badsha, S., Alqahtani, H., Watters, P., Ng, A.: Cybersecurity data science: an overview from machine learning perspective. J. Big Data **7**(1), 1–29 (2020). https://doi.org/10.1186/s40537-020-00318-5
10. Skopik, F., Landauer, M., Wurzenberger, M.: Online log data analysis with efficient machine learning: a review. IEEE Security Privacy **20**(3), 80–90 (2022). https://doi.org/10.1109/MSEC.2021.3113275
11. Pavlov, A., Voloshina, N.: Analysis of IDS alert correlation techniques for attacker group recognition in distributed systems, pp. 32–42 (2020). https://doi.org/10.1007/978-3-030-65726-0_4.

12. Abdalla, R.R., Jumaa, A.K.: Log file analysis based on machine learning: a survey. UHD J. Sci. Technol. **6**(2), 77–84 (2022). https://doi.org/10.21928/uhdjst.v6n2y2022.pp77-84
13. Sarker, I.H.: Deep learning: a comprehensive overview on techniques, taxonomy, applications and research directions. SN Comput. Sci. **2**(6), 1–20 (2021). https://doi.org/10.1007/s42979-021-00815-1
14. Pan, J.: AI-based log analyser: a practical approach (2022)
15. Brazell, S., et al.: A machine-learning-based approach to assistive well-log correlation, pp. 469–479 (2019). https://doi.org/10.30632/PJV60N4-2019a1.
16. Karlsen, E., et al.: Benchmarking large language models for log analysis, security, and interpretation (2023)
17. Qi, J., et al.: Adanomaly: adaptive anomaly detection for system logs with adversarial learning, In: Proceeding IEEE/IFIP NOMS, pp. 1–5 (2022). https://doi.org/10.1109/NOMS54207.2022.9789917.
18. Nam, S., et al.: VM failure prediction with log analysis using BERT-CNN model, In: Proceeding IEEE CNSM, pp. 331–337 (2022). https://doi.org/10.23919/CNSM55787.2022.9965187.
19. Shendi, M., Elkadi, H., Khafagy, M.: A study on big data log analysis: goals, challenges, issues, and tools (2019)
20. Zhang, J., Zhao, L.J.: Visualization of IDS logs based on regular polygon of even sides. Appl. Mech. Mater. **602–605**, 2157–2162 (2014). https://doi.org/10.4028/www.scientific.net/AMM.602-605.2157
21. Hamooni, H., et al.: LogMine, In: Proceeding 25th ACM International Conference Information and Knowledge Management, pp. 1573–1582 (2016). https://doi.org/10.1145/2983323.2983358.
22. Chen, C.L., et al.: A scalable architecture for log analysis and intrusion detection, unpublished
23. Landauer, M., et al.: Maintainable log datasets for evaluation of intrusion detection systems. IEEE Trans. Dependable Secure Comput. **20**(4), 3466–3482 (2023). https://doi.org/10.1109/TDSC.2022.3201582

Enhancing Cyber Defense with a Modular Machine Learning Framework for Real-Time Phishing Detection

Dejene Dedefa(✉), Agustina Lorda, and Adeoluwa Afolabi

Department of Computer Science, Lincoln University, Pennsylvania, PA 19352, USA
ddedefa@lincoln.edu, {agustina.lorda, adeoluwa.afolabi}@lions.lincoln.edu

Abstract. This paper presents a modular anomaly detection framework for real-time phishing email classification using live Gmail data and machine learning. Grounded in Research Software Engineering (RSE) principles, the system supports reproducible workflows, user-driven experimentation, and dynamic model retraining. Built on a five-layer architecture, it integrates semi-manual feature engineering and TF-IDF vectorization. Experiments on Kaggle datasets 40K to 50K emails of which 30–40% are phishing. All classification algorithms show precision and recall above 95%, with Random Forest reaching 99.17% precision and Logistic Regression giving 97.21% recall. The GUI enables algorithm switching, performance visualization, and CSV export. This adaptable framework bridges academic research and practical cyber defense deployment.

Keywords: Cyber Defense · Phishing Detection · Classification · ML · RSE

1 Introduction

Phishing attacks have surged in recent years, exploiting both technical vulnerabilities and human behavior. According to the Anti-Phishing Working Group (APWG), phishing incidents rose by nearly 40% in 2023 alone. The attack rose 22% in 2024, and continued to rise by 13% in 2025 with email-based attacks accounting for a significant portion of data breaches and financial losses [1]. Traditional detection systems often rely on static rules or blacklists, which struggle to adapt to novel attack patterns.

This paper introduces a modular layered anomaly detection framework system that empowers users to experiment with machine learning models on live Gmail data. Our modular architecture supports real-time prediction, model switching, and performance visualization, making it suitable for both research and instructional use. As Roshanaei et al. [5] outline, integrating AI and ML into cybersecurity frameworks is essential for adaptive threat detection and strategic resilience. This work is informed by RSE principles, which prioritize maintainable codebases, reproducible workflows, and collaborative development - a key to bridging academic research and deployable cybersecurity tools, emphasizing experiential learning and student empowerment. The work aims to equip users, learners and researchers with hands-on tools for navigating real-world cybersecurity challenges.

H. R. Arabnia et al. (Eds.): CSCI 2025, CCIS 2969, pp. 466–477, 2027.
https://doi.org/10.1007/978-3-032-28460-0_33

2 Related Work

Phishing detection is a rapidly evolving field within AI, fueled by increasingly sophisticated attacks and the proliferation of machine learning tools. Researchers have explored diverse approaches - from supervised learning and deep neural networks to ensemble and reinforcement methods - each contributing to smarter threat detection systems.

Kavya and Sumathi [2] review emerging ML techniques such as CNNs, RNNs, GANs, and graph-based models, emphasizing ensemble learning and adversarial robustness. Tamal et al. [12] introduce an Optimal Feature Vectorization Algorithm (OFVA) with supervised classifiers, achieving over 97% accuracy and underscoring the value of feature selection and hyperparameter tuning, both supported in our modular architecture. Similarly, Gandotra and Gupta [10], Shahrivari et al. [3], and Purbay and Kumar [9] highlight the effectiveness and variability of supervised models, reinforcing the need for flexible experimentation.

Jabbar and Al-Janabi [1] propose a reinforcement learning framework with high precision and low false positives on email datasets. Dutta [13] explores RNNs for phishing website detection, showcasing the power of sequence modeling, despite the article's later retraction. Beaman and Isah [6] focus on header-based anomaly detection, while our system expands feature extraction to full email content.

Quang et al. [16] identify scalability and reproducibility gaps in phishing detection, advocating for modular systems - a theme echoed by Bannert [4], Bannert et al. [14], Zhang et al. [7] and Zhang et al. [15], who promote RSE principles like modularity and collaborative development. These principles are embedded in our system, which supports live email access, dynamic model selection, and reproducible experimentation.

From a software engineering lens, Harman [8] discusses AI's impact on development practices, while, our prior work, Mamo and Dedefa [11] argue for layered, context-aware architectures to enhance scalability.

Unlike prior systems that focus on URL or header-based detection, our approach integrates full email content, supports dynamic model selection, and embeds RSE principles for reproducibility and modularity. This positions our system not only as a technical solution but also as a research and educational platform for AI-driven cybersecurity defense development.

3 Data Preparation

This chapter outlines the foundational steps taken to prepare the dataset and construct meaningful features for model training and evaluation. The process involved initial data inspection, semi-manual feature engineering, and text vectorization to ensure compatibility with machine learning algorithms.

3.1 Initial Analysis

The datasets were sourced from Kaggle and each contained between 40,000 and 50,000 email records.

- Cleaned and used multiple datasets of emails from Kaggle, labeled as 1 for phishing and -1 for legitimate. Due to conventions used by ML algorithms, 0 is also used to represent legitimate emails in some of the data sets, in processing and in the output.
- Verified structural integrity with columns including sender, subject, body, and the label.
- Conducted data quality checks such as null value detection and descriptive statistics to ensure consistency.

3.2 Feature Engineering

To enhance model performance and interpretability, a set of manually engineered features was extracted from the email content.

- Extracted quantitative features:

  Link count based on occurrences of "http" and "www".
  Suspicious keyword frequency using a curated list of over 20 phishing-related terms.
  Email body length to capture structural anomalies.

- Implemented sender verification:

  Detected public domains such as gmail.com and yahoo.com.
  Identified domain mismatches between sender address and claimed domain.
  Flagged unknown domains, noting that 99.8% of phishing emails originated from unfamiliar sources.

3.3 Text Vectorization

- Textual features were transformed into numerical representations to support machine learning algorithms.
- Combined subject and body content into a unified text field.
- Applied TF-IDF (Term Frequency / Inverse Document Frequency) vectorization, limited to the top 300 features to manage dimensionality.
- Merged the TF-IDF output with manually engineered features to produce the final modeling dataset.

4 System Architecture and Methodology

Designing a modular system for real-time phishing detection required navigating several technical and methodological challenges. Each stage of development demanded careful reasoning, iterative refinement, and strategic decision-making to ensure robustness, reproducibility, and instructional value. The following challenges help to shape the architecture and informed key design choices:

- Unbalanced Dataset: The training data contained significantly more legitimate emails than phishing samples. This imbalance made accuracy an unreliable metric. Instead, precision, recall, and F1 score were prioritized to better capture the system's sensitivity to phishing threats.

- False Positives in Unsupervised Models: Unsupervised algorithms such as Isolation Forest and One-Class SVM produced a high rate of false positives. These misclassifications reduced trust in the predictions and highlighted the limitations of anomaly detection without labeled data.
- Complex Feature Engineering: Extracting meaningful features from raw email content was nontrivial. Attributes like hyperlink count, suspicious keyword presence, and sender domain patterns required multiple iterations and domain-specific insight to define and validate.
- TF-IDF and Manual Feature: Integration Merging TF-IDF vectors with manually engineered features introduced challenges in dimensionality alignment and data formatting. Careful preprocessing and validation were necessary to ensure compatibility during model training and prediction.

The proposed system adopts a dynamic, five-layer modular architecture for email anomaly detection, integrating modular components that support real-time prediction, secure data access, and flexible model experimentation. At the top, the Interface Layer provides a Python based GUI to support real-time model selection and uploading of new training dataset during the run time. It enables users to select machine learning models, load new datasets, and initiate predictions with minimal friction. The Preprocessing Layer leverages pandas and TF-IDF vectorization to transform raw email content into structured, ML-compatible formats. We also adopt a semi-manual feature engineering approach, combining domain-informed feature selection with automated extraction techniques to enhance model performance. These features include hyperlink count, sender domain analysis, and security prone keywords which complement automated vectorization.

The Machine Learning Layer supports both supervised algorithms like Random Forest, Logistic Regression, etc., and unsupervised techniques like Isolation Forest, One-Class Support Vector Machine (SVM) with a built-in performance visualization to aid model evaluation. The External Service Layer integrates Gmail API access via OAuth2 credentials, ensuring secure retrieval of live email data. Finally, the Data Pipeline Layer orchestrates the end-to-end flow; parsing emails, preprocessing content, and generating predictions in real time. The five layers (Fig. 1) of the architecture are:

- Interface Layer: The interface layer consists of a Python-based graphical user interface that enables users to select classification models, load datasets, and initiate predictions. It integrates real-time confusion matrix visualization and CSV export functionality to support phishing detection, user feedback, and model evaluation. After the model is trained on a large dataset, users can trigger phishing predictions on newly arriving emails by clicking a designated event button, which fetches messages from a specified inbox and applies the trained model for dynamic threat assessment.
- Preprocessing Layer: The preprocessing layer transforms raw email content into structured formats suitable for machine learning. It uses pandas for data manipulation and TF-IDF vectorization to encode textual features from the email body, subject, and sender fields. This layer also incorporates semi-manually engineered features such as hyperlink count and sender domain analysis. The output is stored in a CSV file which serves as the input for model training and testing.

- Machine Learning Layer: The machine learning layer supports both supervised and unsupervised learning algorithms. Supervised models include Logistic Regression, Random Forest, and K Nearest Neighbor. Unsupervised models such as Isolation Forest and One-Class SVM can also be used for anomaly detection without labeled data. Trained models and vectorizers are saved as PKL files to ensure reproducibility. The layer also includes tools for visualizing performance metrics such as confusion matrices and ROC curves to guide model selection and evaluation.
- External Service Layer: The external service layer integrates the Gmail API using OAuth2 authentication to securely access live email data. It uses modules such as API_GoogleOAuth.py and a credential file (credentials.json) to establish authorized connections. This layer isolates authentication and data retrieval from the core processing pipeline, maintaining system security while enabling access to real-world email streams.
- Data Pipeline Layer: The data pipeline layer coordinates the flow of email classification. It manages the parsing of incoming emails, applies preprocessing routines, and executes predictions using the trained model. When triggered by the user, this layer fetches newly arrived emails from a specified inbox, transforms them into vectorized formats, and stores the classification output results in a CSV file. It also supports feedback and result visualization to help users assess model performance and refine detection strategies.

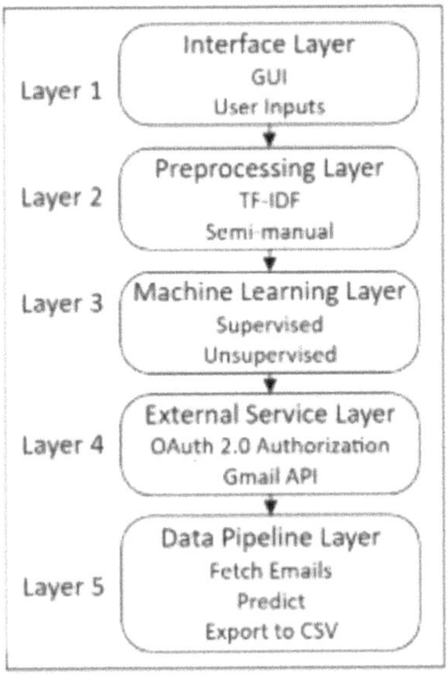

Fig. 1. Five Layers of the Architecture

Figure 2 illustrates the full architecture, highlighting the functional boundaries and data flow across the layers.

Each layer is modular and independently testable, enabling seamless expansion and customization. The five-layer modular architecture reflects core RSE principles such

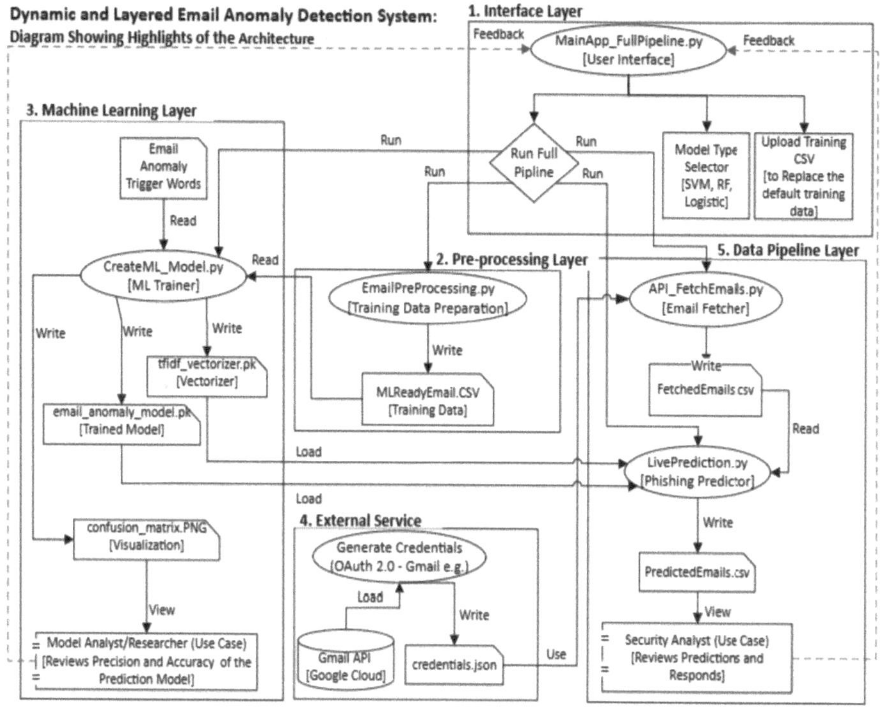

Fig. 2. The System Architecture

as separation of concerns (clear division of tasks), testability, and extensibility. Each module is verifiable in isolation and well positioned for collaborative development and future reuse. This design builds on the principles articulated in our prior work, Mamo and Dedefa [11], which advocates for layered, context-aware systems to promote modularity and scalability. The pipeline's modular structure further supports iterative debugging and flexible deployment across diverse operational contexts.

5 Implementation Details

Our system has a GUI that provides intuitive access to model selection and prediction. Backend modules ensure seamless communication between layers. The system is implemented in Python, using libraries such as scikit-learn, pandas, and matplotlib. Gmail integration is handled via the Google API client and OAuth2 credentials. The implementation follows RSE best practices, including modular code organization, version-controlled development, and inline documentation to support reproducibility and collaborative refinement. The GUI also includes dropdown menus, buttons for data fetch and prediction, and embedded plots for performance visualization.

Initial development was conducted in Google Collab, with later migration to VS-Code to be able to authenticate and use Gmail API on local machines to fetch life emails for prediction. A python library called joblib is used for serializing trained models and vectorizers, ensuring reproducibility.

Environment setup: Python 3.9 +, scikit-learn, pandas, matplotlib, google-auth, google-api-python-client. Figure 3 illustrates the GUI interface, highlighting the interface and a sample interactive screen with output.

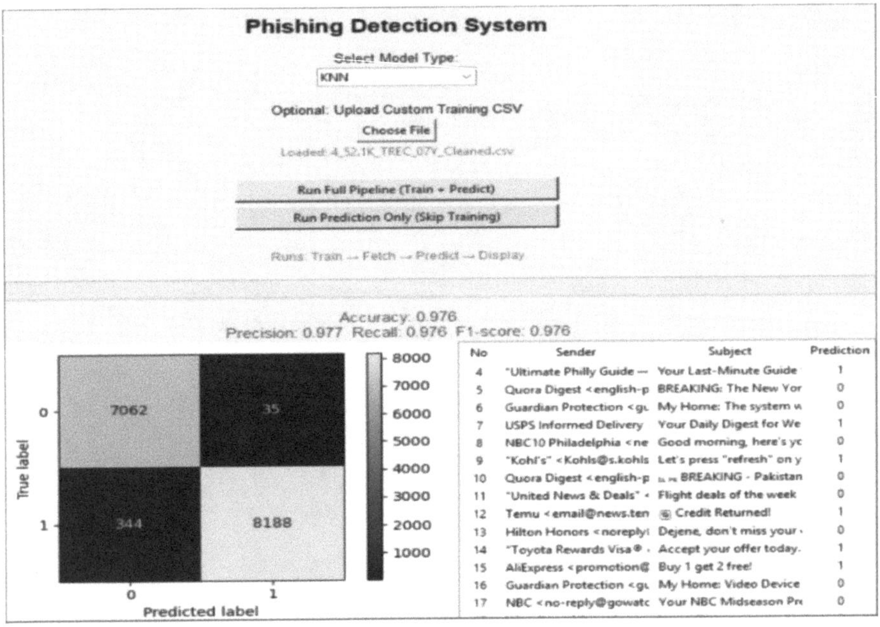

Fig. 3. The GUI Interface

6 Experimental Results

Using a labeled phishing email dataset, the system was evaluated across multiple models. Supervised models consistently outperformed unsupervised ones in accuracy, precision, recall, F1 and Area Under the ROC curve (AUC) score. AUC quantifies the ROC curve into a single number. AUC = 1.0 indicating perfect classification and AUC = 0.5 indication no better than random. Our initial model development & training involves:

- Implementation of 8 different algorithms (4 supervised, 4 unsupervised)
- Supervised models: KNN, Logistic Regression, Random Forest, SVM
- Unsupervised models: Isolation Forest, Local Outlier Factor, KMeans, One-Class SVM
- Split data into 70% training and 30% test sets

The experiments on the initial models are conducted using multiple publicly available phishing datasets from Kaggle (TREC, Nazario, Nigerian) with 3 features or predictors (sender, subject and body) containing 40 to 50 thousand emails each and the emails are labeled as phishing or benign.

Confusion matrices (Fig. 4) were used to guide model selection and refinement, helping identify misclassification patterns. A confusion matrix being a summary of the classification model's performance by comparing predicted labels with actual outcomes, highlighting true positives, true negatives, false positives, and false negatives. It provides a clear view of both accuracy and error types, making it essential for evaluating models in security contexts. In our phishing detection system, Logistic Regression achieves the highest recall, correctly identifying 6546 phishing emails while missing only 189, making it the most reliable for minimizing undetected threats. Random Forest, though more precise with just 52 false positives, misses 314 phishing cases, and K-Nearest Neighbor (KNN) shows a higher rate of both false positives and false negatives. Given our focus on security, where missing a phishing attempt can have serious consequences, recall is the most critical metric, and Logistic Regression stands out as the preferred model for comprehensive threat detection.

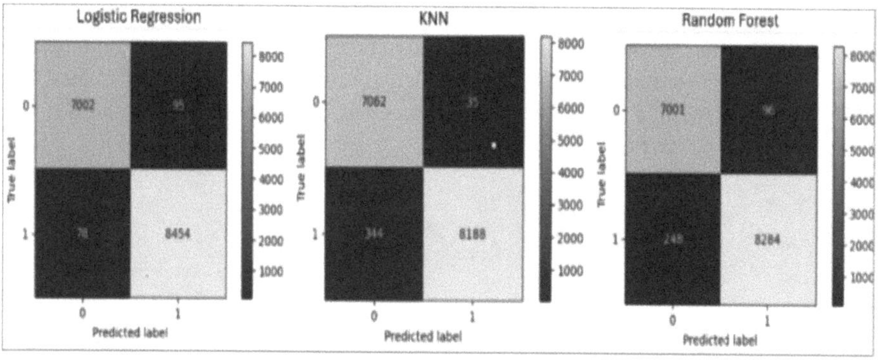

Fig. 4. Sample Confusion Matrix

Our results align with Purbay and Kumar [9], who observed divergent performance across supervised algorithms in phishing URL detection. Users can switch models and datasets dynamically, enabling comparative analysis and iterative refinement. Performance metrics are visualized within the GUI for transparency and interpretability.

Semi-manual feature engineering such as suspicious keyword detection, hyperlink count in the email body, and sender domain analysis have also been used to improve model accuracy. Combining TF-IDF with the features offers a hybrid approach that captures both linguistic and structural anomalies. The integration of semi-manual and TF-IDF features improves robustness and adaptability to varied phishing strategies. GUI usability enables non-technical users to interact with the system, broadening its accessibility for students, researchers and other users.

Table 1. Performance Comparison of Classification Algorithms

Classification Algorithm	Accuracy	Precision	Recall	F1 Score	AUC
Logistic Regression	97.73%	97.78%	97.21%	97.49%	1.00
Random Forest	99.17%	99.17%	95.54%	97.32%	1.00
KNN	97.27%	97.25%	96.73%	96.99%	0.99
SVM	97.47%	97.40%	97.01%	97.20%	1.00

Table 1 presents a comparative evaluation of four classification algorithms based on Accuracy, Precision, Recall, F1 Score, and AUC. Random Forest achieved the highest overall accuracy of 99.17% and precision of 99.17%, indicating strong predictive performance. However, its recall of 95.54% was slightly lower than the others, suggesting a modest trade-off in sensitivity. Logistic Regression and SVM offered balanced performance across all metrics, with F1 Scores of 97.49% and 97.20%, respectively. KNN showed competitive results but slightly lagged in precision and F1 Score. On the other hand, all unsupervised classifiers achieved accuracy, precision, recall values less than 80% and an average of about 50%. Comparison chart of the four measures for the four algorithms is shown in Fig. 5.

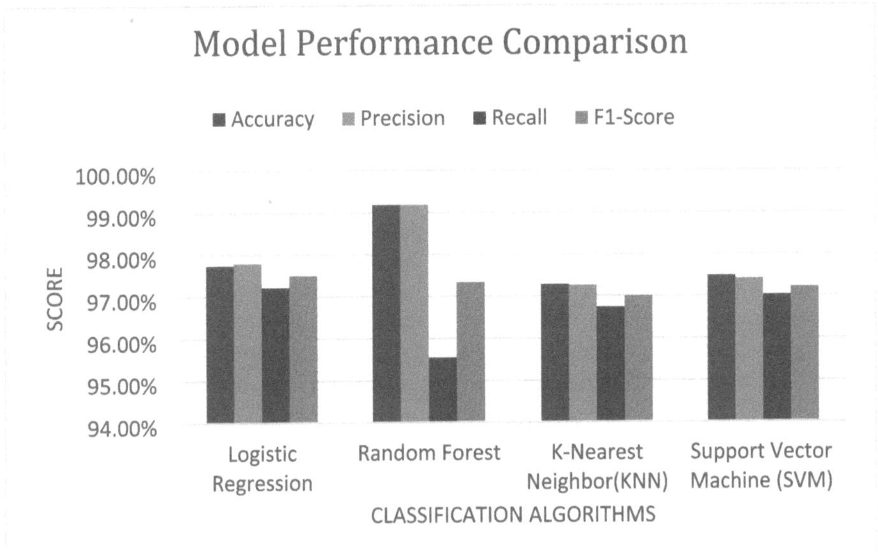

Fig. 5. Performance Comparison of Classifiers

Where F1 Score is a key evaluation metric in our classification tasks as it balances both precision (how many predicted positives are correct) and recall (how many actual positives are identified). It is especially valuable when dealing with imbalanced datasets, where relying on accuracy alone can be misleading. The F1 Score is defined as the harmonic-mean of precision and recall, ensuring that both metrics are given equal weight calculated as:

$$F1\,Score = 2x(Precision\,x\,Recall)/(Precision + Recall).$$

A high F1 Score indicates that the model is both accurate in its positive predictions and sensitive to detecting actual positives, making it ideal for applications like phishing classification where false negatives and false positives carry significant consequences.

ROC analysis (Fig. 6) further confirms these trends: Logistic Regression, Random Forest, and SVM each achieved a perfect AUC of 1.00, indicating flawless class separation across thresholds, while KNN followed closely with an AUC of 0.99. These results suggest that although Random Forest leads in average performance, Logistic Regression and SVM may offer more consistent generalization, particularly in applications where recall and class discrimination are critical.

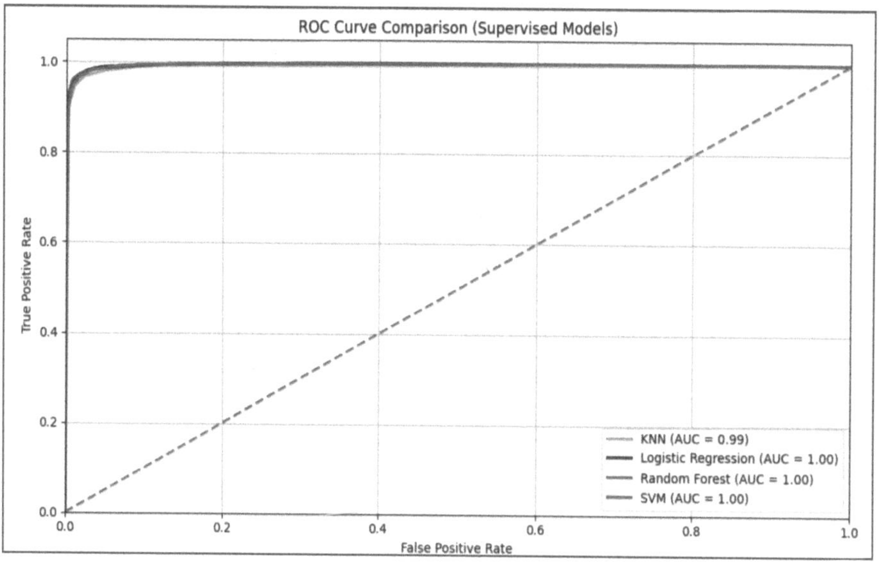

Fig. 6. ROC Curve

As shown in the confusion matrix, performance table and graph, and ROC curve, Logistic Regression consistently delivers strong classification results, particularly in identifying phishing emails with minimal false negatives. In security management, where missing a threat can lead to serious consequences, recall is the most critical metric. A high recall ensures that phishing attempts are reliably detected, reducing the risk of undetected attacks. Given its superior recall and balanced overall performance, Logistic Regression is the best performing classification algorithm for email phishing detection in this study. Accordingly, our system provides a dropdown menu that allows users to select from three classification algorithms: Logistic Regression, Random Forest, and K Nearest Neighbor. Logistic Regression is set as the default option, reflecting its status as the preferred model based on our study.

7 Discussion and Conclusion

The experimental results highlight the effectiveness of supervised models in phishing detection. Logistic Regression, Random Forest, and Support Vector Machine consistently achieved strong precision, recall, and F1 scores. Among these, Random Forest provided the highest overall accuracy and interpretability. These models benefit from labeled training data, which supports better generalization and reduces false positive rates. In contrast, unsupervised approaches such as Isolation Forest and One-Class SVM struggled to detect the subtle patterns found in phishing emails. These patterns often depend on linguistic cues and structural anomalies that are difficult to capture without labeled examples.

The integration of semi-manual feature engineering with TF-IDF vectorization improved model sensitivity. Features such as link count, suspicious keyword frequency, and domain verification complemented the textual encoding and contributed to more reliable classification outcomes.

During development, several challenges emerged. Managing feature dimensionality required careful preprocessing. Debugging Gmail API interactions demanded attention to authentication flow and data formatting. Ensuring smooth integration across system layers involves iterative testing and modular design. The performance table illustrates trade-offs between classifiers and emphasizes the importance of balancing precision, recall, and interpretability based on the intended deployment context. Limitations of the current system include the size of the dataset, reliance on Gmail-specific integration, and the absence of deeper contextual analysis.

This study presents a layered anomaly detection framework for real-time phishing detection. The architecture includes five modular components: interface, preprocessing, machine learning, external services, and data pipeline. Each layer supports dynamic model switching, live email access, and performance visualization. The graphical user interface makes the system accessible to nontechnical users. Its modularity allows for iterative refinement and future expansion.

From a pedagogical perspective, the system serves as a scaffold for students and researchers. It enables exploration, extension, and repurposing of components with minimal overhead. This supports the broader goal of building sustainable and research-ready software ecosystems. As Harman [8] observes, artificial intelligence is reshaping software engineering practices. This shift is reflected in the system's modular design and integration of machine learning techniques.

Future work may include cross-platform email access, integration of deep learning models, and the use of explainable AI techniques to enhance adaptability and transparency.

Funding Acknowledgment. This work was supported by the National Science Foundation under NSF 21-523 and NSF 23-521.

References

1. User Guide: scikit-learnevelopers, "Supervised Learning," scikit-learn (2025). https://scikit-learn.org/stable/supervised_learning.html

2. Google Workspace: "Gmail API," Google Developers (2025). https://developers.google.com/workspace/gmail/api/guides
3. Alam, N.A., et al.: Phishing Email Dataset, Phish No More: The Enron, Ling, CEAS, Nazario, Nigerian & SpamAssassin Datasets. Kaggle Datasets (2025). https://www.kaggle.com/datasets
4. Bannert, M.: Best Practices in Research Software Engineering. Chapman & Hall/CRC Data Science Series (2024). ISBN: 978-1032261270
5. Roshanaei, M., Khan, M.R., Sylvester, N.N.: Enhancing Cybersecurity through AI and ML: Strategies, Challenges, and Future Directions. J. Inf. Secur. **15**, 320–339 (2024). https://doi.org/10.4236/jis.2024.153019. https://www.scirp.org/journal/jis
6. Beaman, C., Isah, H.: Anomaly Detection in Emails using Machine Learning and Header Information. arXiv preprint arXiv:2203.10408 (2022). https://doi.org/10.48550/arXiv.2203.10408. https://arxiv.org/abs/2203.10408
7. Zhang, Y., et al.: Building an AI-ready RSE Workforce. arXiv preprint arXiv:2111.04916, (2021). https://doi.org/10.48550/arXiv.2111.04916. https://arxiv.org/abs/2111.04916
8. Harman, M.: The Role of Artificial Intelligence in Software Engineering. In: Proc. 1st Int. Workshop Realizing AI (2012). https://doi.org/10.1109/RAISE.2012.6296531. https://www.computer.org/csdl/proceedings-article/raise/2012/06227961/12OmNxzuMNt
9. Purbay, M., Kumar, D.: Split Behavior of Supervised Machine Learning Algorithms for Phishing URL Detection. Lect. Notes Electr. Eng. **683**, 451–460 (2021). https://doi.org/10.1007/978-981-15-6840-4_40
10. Gandotra, E., Gupta, D.: "An Efficient Approach for Phishing Detection using Machine Learning", in Algorithms for Intelligent Systems, Springer. Singapore (2021). https://doi.org/10.1007/978-981-15-8711-5_12
11. Mamoa, B., Ejigu Dedefa, D.: A Generic Layered Architecture for Context Aware Applications. In: Proc. Int. Symp. Emerging Internetworks, Communication and Mobility (EICM), vol. 34, pp. 619–624. Halifax, Canada (2014). https://doi.org/10.1016/j.procs.2014.07.085
12. Tamal, M., et al.: Optimal Feature Vectorization Algorithm for Phishing URL Detection Using Supervised Machine Learning. Frontiers in Comp. Sci. **6**, 1428013 (2024). https://doi.org/10.3389/fcomp.2024.1428013/full
13. Dutta, S.: Phishing Website Detection Using Recurrent Neural Networks. Int. J. Comput. Appl. **183**(45), 1–6 (2021). (Note: Retracted, but methodology remains influential.)
14. Bannert, C., et al.: Reproducible Machine Learning in Cybersecurity: A Research Software Engineering Perspective. J. Open Res. Softw. **11**(1), 1–12 (2023)
15. Zhang, Y., et al.: Modular AI Systems for Cybersecurity: Bridging Software Engineering and Machine Learning. IEEE Trans. Software Eng. **50**(2), 345–359 (2024)
16. Quang, H., et al.: Phishing Detection in the Big Data Era: A Systematic Literature Review. Comput. Secur. **124**, 102984 (2023)

Neural Networks for Ransomware Detection: A Comparative Evaluation of RNN, CNN, and ANN Models

Luice Khamboo, Ali Al-Sinayyid(✉), Samuel Adewumi, Vaishnavi Kottala, Sheetal Phuyal, and Zimren Drew Dixon

Department of Mathematics and Computer Science, West Virginia State University, Institute, USA
{lkhamboo,ali.alsinayyid,sadewumi,vkottala,sphuyal1, zdixon1}@wvstateu.edu

Abstract. Ransomware continues to be a significant cybersecurity threat, disrupting individuals and organizations by encrypting critical data and demanding payment for decryption. As ransomware variants continue to evolve, effective detection methods are essential to mitigate their impact. This research evaluates the performance of three deep learning models Convolutional Neural Networks (CNN), Recurrent Neural Networks (RNN), and Artificial Neural Networks (ANN) in detecting ransomware based on execution behavior and network activity. Among these models, CNN achieved the highest accuracy (0.9864), followed by RNN (0.9262) and ANN (0.8872). The findings suggest that CNNs are particularly effective in identifying ransomware patterns, while RNNs capture sequential dependencies, and ANNs contribute to feature learning. This study underscores the potential of deep learning in strengthening ransomware detection and cybersecurity defenses.

Keywords: Neural Network · CNN · ANN · RNN · Ransomware · Detection · AI Detection

1 Introduction

Ransomware is a rapidly evolving cyber threat that encrypts a victim's files and demands payment for the encryption. These attacks target individuals, businesses, and public institutions, often causing severe financial and operational disruptions [7]. By leveraging advanced encryption techniques, ransomware operators render critical data inaccessible, demanding ransom, typically in cryptocurrency, in exchange for the decryption key [12]. As cybercriminals refine their tactics, ransomware has become increasingly sophisticated, bypassing traditional security measures, and exploiting vulnerabilities in

H. R. Arabnia et al. (Eds.): CSCI 2025, CCIS 2969, pp. 478–491, 2027.
https://doi.org/10.1007/978-3-032-28460-0_34

modern digital infrastructures. Even in 2024, ransomware remains one of the most pressing cybersecurity challenges. Reports indicate a dramatic surge in ransomware incidents, with over 4,399 attacks recorded in 2023 [18] nearly double the previous year's count. Although the percentage of victims willing to pay ransoms has declined, the financial toll continues to rise, surpassing $1 billion in ransom payments in 2023 alone [18]. Ransomware variants are frequently modified to bypass traditional detection systems, making it clear that more advanced techniques are required to mitigate this growing danger. Machine learning (ML) and deep learning (DL), particularly through the use of neural networks, could be better implemented on mitigating the attack. These AI-powered approaches can analyze large volumes of data, detect intricate patterns, and identify anomalies that may indicate the presence of ransomware [8, 10, 11, 19].

This research explores the effectiveness of three deep learning models Recurrent Neural Networks (RNN), Convolutional Neural Networks (CNN), and Artificial Neural Networks (ANN) in detecting ransomware. These models were evaluated based on key performance metrics, including accuracy, precision, and recall, to assess their ability to correctly identify ransomware samples while minimizing false positives and negatives. By comparing these models, the study aims to identify the most efficient neural network architecture for ransomware detection, contributing valuable insights to the field of cybersecurity and offering potential solutions for improving defenses against ransomware attacks.

2 Literature Review

Rhode et al. [20] present an early-stage malware detection approach that shows how recurrent neural networks (RNNs) can accurately classify malicious files, including ransomware, within just a few seconds of execution. Unlike traditional dynamic analyses that rely on long execution traces or categorical API calls, they use compact, continuous machine-activity metrics (CPU use, memory consumption, network traffic, process creation) that generalize better and are harder to evade. Their experiments reveal that GRU-based RNNs outperform classical classifiers and achieve around 94% accuracy within one second on previously unseen ransomware, highlighting the practical value of early detection before encryption begins. In their conclusion, the authors emphasize that neural networks are particularly effective because they capture temporal dependencies in evolving runtime behavior, making them more robust than static methods. This insight is significant for ransomware research, as it shows why dynamic features paired with neural architectures are essential for proactive detection. This conclusion is critical because it highlights that the strength of neural networks lies not merely in classification power but in their ability to model evolving runtime behavior, which aligns directly with the growing need for adaptive ransomware detection systems that rely on dynamic features like network activity and execution logs [20].

Artificial Intelligence (AI) has become an important tool for ransomware detection in response to the growing complexity and sophistication of attacks. Traditional approaches such as signature-based systems often fail against modern ransomware, which frequently employs polymorphic techniques to bypass detection [3–6]. AI-based methods, particularly machine learning (ML) and deep learning (DL), provide enhanced detection capabilities by identifying patterns in network behavior and detecting anomalies before ransomware can complete data encryption. Supervised learning algorithms, including Support Vector Machines (SVM) and Random Forest (RF), have achieved high accuracy in classifying ransomware behavior, as demonstrated by Ferdous et al. [2] (2024). In addition, unsupervised approaches such as clustering and anomaly detection allow AI models to recognize previously unseen ransomware variants. This adaptability highlights the potential of AI-driven solutions to complement traditional methods, as they can continuously learn from new data and strengthen defenses against evolving threats [2, 13, 14].

Sneha et al. [9] emphasize the escalating severity of ransomware threats, focusing on variants such as Locky and Crypto that have caused widespread operational and financial disruption, as seen in major outbreaks like WannaCry and GrandCrab. They note that traditional detection methods, including static and dynamic analysis, often fall short against modern ransomware, which leverages polymorphism and encryption to bypass defenses. The financial toll of these attacks is illustrated by the 2019 Norsk Hydro incident, where recovery costs reached $71 million despite the company's refusal to pay the ransom. In addition, global damages from ransomware were projected to rise to $20 billion by 2021 (Mindsight, 2019), reflecting the continued exploitation of vulnerabilities and highlighting the urgent need for stronger and more adaptive security strategies [9].

The rapid evolution of ransomware has prompted a shift from traditional signature-based detection methods, which fail against novel and obfuscated attacks, to machine learning and neural network approaches that can capture complex behavioral patterns. Previous studies have shown that Artificial Neural Networks (ANNs) effectively distinguish benign from malicious executables with accuracies above 97%, while Convolutional Neural Networks (CNNs) achieve similarly high performance by treating malware as image data, though at the risk of feature extraction errors. Recurrent Neural Networks (RNNs) have further advanced detection by modeling sequential API calls, with attention-based LSTM variants surpassing 90% accuracy in capturing recurrent ransomware behaviors [21]. These works collectively demonstrate that different neural network architectures offer distinct advantages ANNs excel at general classification, CNNs at feature extraction, and RNNs at sequence modeling yet each also faces limitations in scalability, interpretability, or dataset diversity [1].

F. Manavi and A. Hamzeh(2020) [16] introduced a static ransomware detection method that leverages Portable Executable (PE) headers and Convolutional Neural Networks (CNNs) to address the shortcomings of traditional techniques. Their approach converts the first 1024 bytes of the PE header into 32×32 grayscale images, allowing CNNs to capture the structural differences between benign and malicious files through automated feature extraction. This method achieved high detection accuracy, remained resistant to packing, and proved more efficient than dynamic or opcode-based methods by eliminating the need for program execution. Moreover, the CNN model successfully categorized ransomware into families such as Cerber, Locky, Wannacry, Sage, and TeslaCrypt, demonstrating its robustness and flexibility for ransomware detection [16].

3 Methodology

3.1 Problem Definition and Data Processing

To effectively detect ransomware, we employ a deep learning-based classification approach that processes network activity and file execution logs. Ransomware detection involves identifying malicious file execution behavior and network anomalies, where each event is represented as a feature vector capturing encryption patterns, file modifications, process injections, and packet anomalies. Data is collected from endpoint monitoring systems and honeypots, with features extracted and normalized for model compatibility. The dataset is split into training (70%), validation (15%), and testing (15%) subsets using stratified sampling. We evaluate three deep learning architectures: RNN, which captures temporal dependencies in ransomware execution; CNN, which extracts spatial patterns from network traffic; and ANN, which classifies ransomware activity based on extracted features. The detection pipeline follows a structured process, beginning with feature extraction and preprocessing, followed by model selection and training using labeled data. Model performance is evaluated using accuracy, precision, and recall, and upon deployment, files are classified based on a threshold score, where those identified as ransomware are quarantined while benign files are allowed to execute.

3.2 Pseudocode

Ransomware Detection Pipeline

```
 WHILE system_running DO
  IF new_file_detected() THEN
    // Feature extraction
    features = extract_features(file)

    // Data preprocessing
    processed_data = preprocess(features)

    //  Automated Model Selection
    model = select_model() // Returns pre-configured optimal model

    // Prediction
    malware_prob = model.predict(processed_data)

    // Decision making
    IF malware_prob >= 0.5 THEN
      quarantine_file(file)
      send_alert("Malware detected!")
    ELSE
      allow_execution(file)
      log_entry(file, "Clean")
    END IF

    // Update system records
    update_database(file, malware_prob, model.name)
  END IF
END WHILE
```

3.3 Mathematical Model

Each model is trained independently to evaluate its effectiveness in detecting ransomware Each model is trained independently to evaluate its effectiveness in detecting ransomware patterns.

Recurrent Neural Network (RNN) Model

Training the RNN Model.

Let the input feature vector at time step τ be χ_τ, the hidden state be γ_τ, and the output be φ_τ. The RNN learns from sequences by updating its hidden state:

$$\gamma_\tau = \phi(\chi)\big(\omega_\gamma \gamma_{\tau-1} + \omega_\chi \chi_\tau + \beta_\gamma\big) \tag{1}$$

$$\varphi_\tau = \sigma\left(\omega_\varphi \gamma_\tau + \beta_\varphi\right) \tag{2}$$

where:
ω_{Δ} is the weight matrix for hidden state transitions.
ω_{χ} is the weight matrix mapping inputs to hidden states.
ω_{φ} is the weight matrix mapping hidden states to output probabilities.
β_{γ}, β_{φ} are bias terms.
σ is the sigmoid activation function.
$\phi(\chi)$ is the hyperbolic tangent activation function.

Loss Function and Backpropagation:
Using binary cross-entropy, the loss is minimized to improve classification accuracy:

$$\Gamma = -\frac{1}{\forall}\sum_{i=1}^{\forall}\left[\phi_i log\left(\hat{\phi}\_i\right) + (1-\phi_i)loglog\left(1-\hat{\phi}_i\right)\right] \tag{3}$$

Backpropagation Through Time (BPTT) updates weights based on temporal dependencies.

Final Probability Equation:
The probability of a file being ransomware is given by:

$$\delta_{RNN} = \varphi_{\tau} = \sigma\left(\omega_{\varphi}\gamma_{\tau} + \beta_{\varphi}\right) + \sum_{\tau=1}^{T}\frac{e}{\sum_{\zeta} e^{\zeta}} + \frac{||||^2 + ||\chi||^2}{2} \tag{4}$$

If $TM_{RNN} \geq 0.5$, classify as malware; otherwise, classify as benign.

Convolutional Neural Network (CNN) Model
Training the CNN Model:
Convolutions capture hierarchical relationships in input features:
For an input feature map χ and a filter ω:

$$\xi_{\varepsilon,\iota} = \sum_{\mu}\sum_{\eta}\omega_{\varphi_{\mu,\eta}}\chi_{\varepsilon-\mu,\iota-\eta} + \beta_{\varphi} \tag{5}$$

Rectified Linear Unit (ReLU) activation introduces non-linearity:

$$\Lambda_{\varepsilon,\iota} = max(0, \xi_{\varepsilon,\iota}) \tag{6}$$

Pooling reduces dimensionality while preserving essential information:

$$\wp_{\varepsilon,\iota} = max(\Lambda_{\mu,\eta}) \tag{7}$$

ω_{ϕ} is the convolutional filter applied to extract spatial features.
β_{φ} is the bias term in the convolutional layer.
$\Lambda\varepsilon,\iota$ is the activation map after applying ReLU.
$\wp\varepsilon,\iota$ is the pooled feature map

Fully Connected Layer for Classification:
The extracted features are flattened and passed through a dense layer:

$$\Psi = \sigma\left(\omega_{\mu}\wp + \beta_{\mu}\right) \tag{8}$$

where:
ω_μ is the weight matrix of the fully connected layer mapping convolutional features to final classification.
β_μ is the bias term for the dense layer.

Final Probability Equation:
The probability of ransomware for CNN is given by:

$$\delta_{CNN} = \sigma\left(\omega_\mu \wp + \beta_\mu\right) + \frac{1}{\forall}\sum_{\varepsilon=1}^{\forall} loglog\left(1 + e^{-}\right) + \frac{||\omega_{\mathtt{C}}||^2}{2} \tag{9}$$

where $\omega_{\mathtt{C}}$ transforms extracted features into probability outputs.
If $TM_{CNN} \geq 0.5$, classify as malware; otherwise, classify as benign.

Artificial Neural Network (ANN) Model
Training the ANN Model:
Feedforward layers transform input features:
Let χ be the input vector, and ω, β the weight matrix and bias:

$$\xi = \omega_\chi + \beta \tag{10}$$

Applying activation functions (ReLU or Sigmoid) introduces non-linearity:

$$\Lambda = f() \tag{11}$$

Final *classification* layer:

$$\Psi = \sigma(\omega_o \Lambda + \beta_o) \tag{12}$$

ω_o is the weight matrix connecting the hidden layer to the output layer.
β_o is the bias term for the output layer.

Final Probability Equation:
The probability of ransomware for ANN is given by:
Where:

$$\delta_{ANN} = \sigma(\omega_o \Lambda + \beta_o) + \sum_{\varepsilon=1}^{\forall} loglog\left(\frac{1}{1 + e^{-}}\right) + \frac{||\omega_o||^2}{2} \tag{13}$$

where ω_o maps learned feature representations to output probabilities.
If $TM_{ANN} \geq 0.5$, classify as malware; otherwise, classify as benign.

3.4 Advantages of the Proposed Detection Framework

This pipeline ensures high accuracy by leveraging deep learning techniques such as RNNs, CNNs, and ANNs, each contributing to detecting ransomware from different perspectives. The feature selection process improves interpretability and efficiency, while data preprocessing minimizes biases. Scalability is ensured through integration with real-time streaming platforms. The use of automated alerts and quarantine mechanisms enhances security by providing proactive defense against evolving ransomware threats. By analyzing behavioral patterns and network activity, this approach effectively detects ransomware attacks before they can cause significant damage, making it a crucial tool in modern cybersecurity.

3.5 System Architecture and Data Flow

This flow represents the structured decision-making process in a ransomware detection system, breaking down the data flow and model selection:

1. Data Intake: The process starts with detecting new data, such as network traffic, file executions, or system logs.
2. Feature Extraction: Key characteristics of the data are extracted, such as file modifications, encryption patterns, or network behavior.
3. Preprocessing: The extracted features undergo normalization and encoding to make them suitable for deep learning models.
4. Model Selection: The system evaluates multiple models (CNN, RNN, and ANN) in parallel to classify the data.
5. Classification: The models determine the probability that the input represents ransomware.
6. Decision Making: If the malware probability exceeds 0.5, the file is quarantined; otherwise, execution is allowed.
7. Logging and Alerts: Actions are logged, alerts are sent if needed, and databases are updated for future analysis.

This structured approach ensures a robust and real-time response to ransomware threats, leveraging deep learning for accurate detection (Fig. 1).

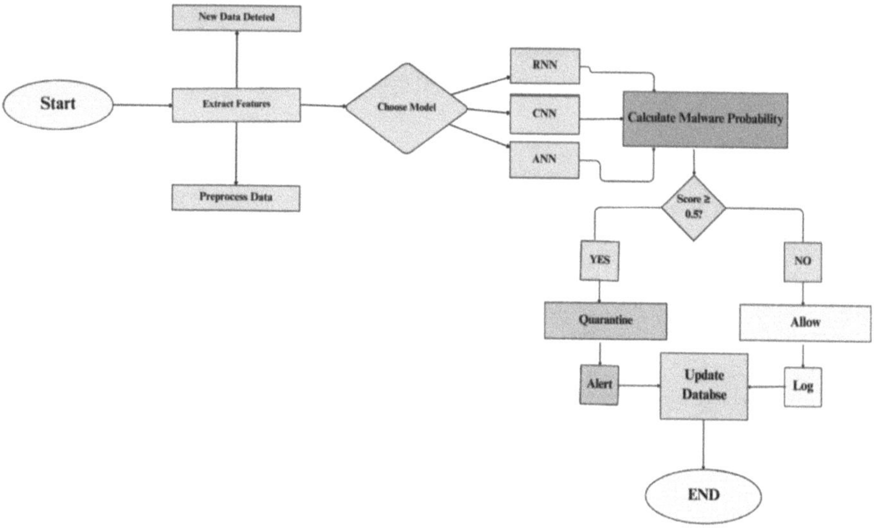

Fig. 1. Ransomware Detection Model Workflow

3.6 Comparative Analysis of Model Performances

Dataset Overview

The dataset used for this study consists of 21,752 samples, evenly divided between 10,876 malicious and 10,876 benign files. It includes 26 distinct malware families, with a strong emphasis on 11 major ransomware families such as Cerber, DarkSide, Dharma, GandCrab, LockBit, Maze, Phobos, REvil, Ragnar Locker, Ryuk, Shade, and WannaCry. The dataset is structured into four main categories, making it a robust resource for training machine-learning models in ransomware classification and detection [17].

Convolutional Neural Network (CNN) Model

The Convolutional Neural Network (CNN) model was trained on the dataset using 18 selected features critical for distinguishing ransomware behavior, including file system activities (files_text, files_malicious, files_suspicious, files_unknown), process execution patterns (total_processes, processes_monitored, processes_malicious, processes_suspicious), network communications (network_dns, network_connections, network_http, network_threats), registry modifications (registry_total, registry_read, registry_write, registry_delete), and system call interactions (apis, dlls_calls). Over 5 epochs, with each epoch consisting of 109 iterations, the model continuously adjusted its weights to minimize the loss function and enhance classification accuracy. The validation accuracy steadily improved from 0.7535 in Epoch 1 to 0.9920 in Epoch 5, demonstrating the model's ability to learn and generalize from these discriminative feature patterns.

The final evaluation results showed that the CNN model achieved an accuracy of 0.9864, with precision of 0.9774 and a recall of 0.9959. This performance profile built upon the model's analysis of critical features spanning file operations, process behaviors, network activities, registry changes, and system calls indicates that the model is not only highly accurate overall but also exceptionally reliable, correctly identifying almost all true malware while maintaining very few false positives. These results, demonstrating near-perfect detection capabilities, establish this CNN model as the most effective solution for ransomware detection in this study.

Artificial Neural Network (ANN) Model

The ANN model was trained using 100 iterations, where each iteration refined the model's weights through a fully connected network of neurons. Despite this longer training period, the ANN model attained an accuracy of 0.8872, with a precision of 0.8926 and recall of 0.8872. These results indicate that while the ANN model was capable of detecting ransomware, its overall performance was lower than that of the CNN model, particularly in its ability to generalize to unseen data.

Training durations were optimized per architecture to ensure peak performance. The ANN used early stopping, halting at 100 iterations to prevent overfitting despite a 1000-iteration safety limit. The more complex CNN and RNN required fuller training for 5 epochs (109 iterations per epoch, 545 total) to avoid underfitting.

Recurrent Neural Network (RNN) Mode

The RNN model, trained over 5 epochs, leveraged sequential data to recognize patterns in ransomware behavior. Each epoch comprised 109 iterations, during which the model progressively refined its understanding of the dataset. Accuracy improved from 0.7889 in Epoch 1 to 0.9262 in Epoch 5, with recall increasing from 0.7168 to 0.9628, indicating enhanced ransomware detection capability over time. The final evaluation yielded an accuracy of 0.9262, precision of 0.8972, and recall of 0.9628, showcasing its ability to identify ransomware effectively.

The evaluation of CNN, RNN, and ANN models for ransomware detection highlights the superiority of CNN, which achieved the highest accuracy (0.9864), precision (0.9774), and recall (0.9959). Its ability to effectively learn spatial patterns made it the most reliable model for distinguishing between malicious and benign files. The RNN model, leveraging sequential dependencies, performed well with an accuracy of 0.9262 and a recall of 0.9628, indicating strong detection capabilities but slightly lower precision. In contrast, the ANN model, despite undergoing 100 training iterations, had the lowest accuracy (0.8872), suggesting that a fully connected architecture was less effective in capturing ransomware patterns. These results justify the choice of CNN as the optimal model for ransomware detection due to its superior performance in identifying and classifying threats. However, incorporating RNNs and ANNs still adds value, as they provide alternative perspectives on feature representation.

To further validate the effectiveness of these models, we computed their F1 scores, a crucial metric that combines both precision and recall into a single value. The F1 score provides a balanced assessment of model performance by considering both false positives and false negatives, making it particularly useful in evaluating classification models where an imbalance between precision and recall could impact real-world applicability. This metric is especially relevant for ransomware detection, where both accurate identification and minimal misclassification are critical [15].

The F1 score is calculated using the following equation:

$$F1 = 2 \times \frac{(Precision \times Recall)}{(Precision + Recall)} \tag{14}$$

To assess the effectiveness of the proposed ransomware detection models, Table 1 presents a comparative analysis of the Artificial Neural Network (ANN), Convolutional Neural Network (CNN), and Recurrent Neural Network (RNN) based on key evaluation metrics, including accuracy, precision, recall, and F1 score. This tabular representation provides a clear numerical comparison, highlighting the superior performance of the CNN model across all metrics.

Table 1. Comparison of Performance Metrics Across Models

Model	Accuracy	Precision	Recall	F1 Score
CNN	0.9864	0.9774	0.9959	0.9866
RNN	0.9262	0.8972	0.9628	0.9288
ANN	0.8872	0.8926	0.8872	0.8899

Furthermore, Fig. 2 provides a visual representation of these performance metrics, allowing for a clearer comparison of the models' respective strengths and weaknesses. The grouped bar graph effectively highlights the variations in accuracy, precision, recall, and F1 score across the three models, reinforcing the numerical findings presented in Table 1. Notably, the CNN model demonstrates superior precision and recall values, underscoring its strong capability to accurately detect ransomware while minimizing false positives and false negatives.

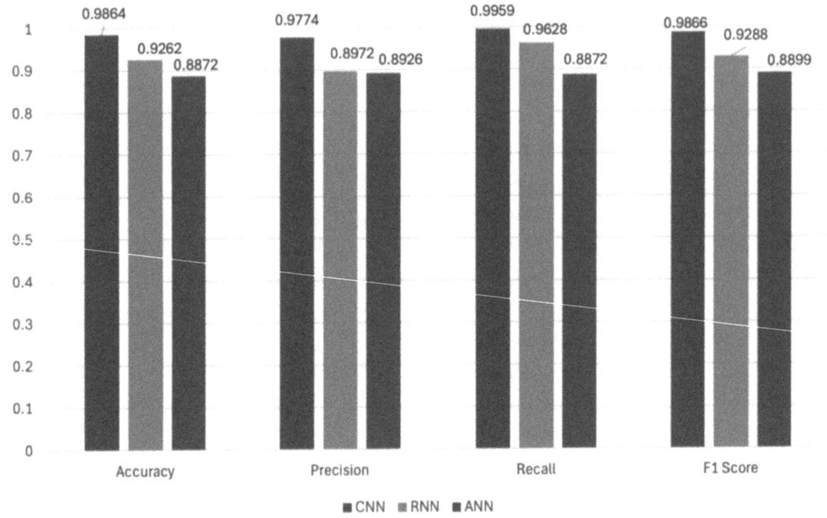

Fig. 2. Model Performance Comparison

Furthermore, Fig. 2 provides a visual representation of these performance metrics, allowing for a clearer comparison of the models' respective strengths and weaknesses. The grouped bar graph effectively highlights the variations in accuracy, precision, recall, and F1 score across the three models, reinforcing the numerical findings presented in Table 1. Notably, the CNN model demonstrates superior precision and recall values, underscoring its strong capability to accurately detect ransomware while minimizing false positives and false negatives.

3.7 CNN Performance in Ransomware Detection

The CNN model demonstrated exceptional efficiency and effectiveness, achieving superior accuracy in just 5 epochs. Its convolutional layers allowed it to extract complex patterns from the dataset, enabling the model to learn and adapt to ransomware behaviors quickly. The ability of CNN to process data in parallel and apply convolutional filters made it highly efficient in feature extraction, significantly reducing the number of parameters and avoiding overfitting. Unlike the ANN model, which required 100 iterations yet still struggled with lower accuracy, CNN was able to refine its weights and improve classification performance rapidly. The validation accuracy increased from 0.7535 in Epoch 1 to an impressive 0.9920 by Epoch 5, showcasing its rapid learning and ability to generalize well to unseen data. On the other hand, while the RNN model also improved over the epochs, its sequential nature made training slower, and the model struggled with long-term dependencies due to vanishing gradients, which impacted its performance. CNN's structured learning approach, efficient weight updates, and ability to capture intricate features of ransomware data allowed it to outperform both ANN and RNN models, making it the most powerful and reliable choice for ransomware detection in this study.

Table 2. CNN Training and Validation Metrics for Each Epoch

Epoch	Train Accuracy	Train Precision	Train Recall	Val Accuracy	Val Precision	Val Recall	Time/Epoch
1	0.9098	0.9304	0.8841	0.7535	0.9836	0.5325	1.0s
2	0.977	0.9711	0.9829	0.8506	0.9878	0.7202	0.5s
3	0.9815	0.9752	0.9877	0.9	0.9873	0.8173	0.5s
4	0.9839	0.9767	0.9912	0.9888	0.9884	0.99	0.5s
5	0.9848	0.9773	0.9923	0.992	0.9858	0.9989	0.5s

Table 2 presents the training and validation metrics recorded across five epochs. This table provides insight into how the model's accuracy, precision, and recall evolved during training. The steady improvement in both training and validation performance highlights the effectiveness of CNN in learning complex ransomware patterns. The time taken per epoch remained consistent, demonstrating an efficient training process (Fig. 3).

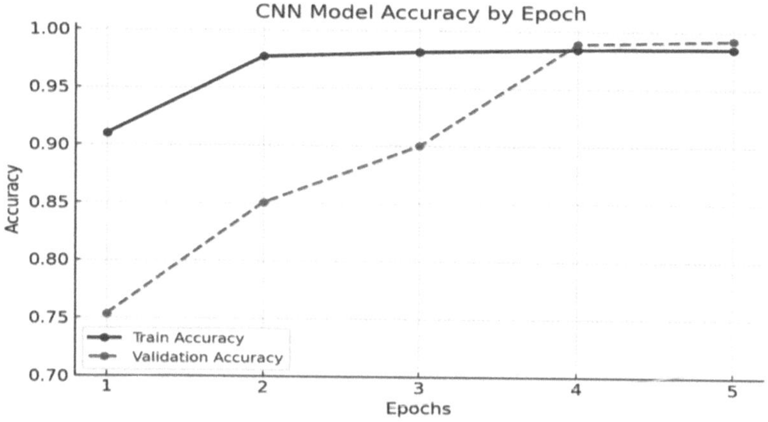

Fig. 3. CNN Model Accuracy by Epochs

3.8 Future Work

In the future, ransomware detection can be significantly improved by adopting hybrid deep learning approaches that combine the strengths of CNNs and RNNs. While CNNs excel at extracting spatial features, RNNs are effective in capturing sequential patterns, making a combined model more robust in detecting complex ransomware behaviors. Additionally, implementing adversarial training techniques can enhance the resilience of detection models against sophisticated evasion tactics employed by advanced ransomware variants. By training models with adversarially generated samples, security systems can become more resistant to attacks designed to bypass traditional detection mechanisms.

4 Conclusion

This research demonstrates the potential of deep learning techniques for detecting ransomware by analyzing network activity and execution behavior. Among the evaluated models, CNNs achieved the highest accuracy (0.9864), precision (0.9774), and recall (0.9959) within the scope of this study, while RNNs also performed well by capturing sequential dependencies (accuracy of 0.9262), and ANNs provided insights into feature relationships despite slightly lower performance. These results suggest that deep learning can offer advantages over traditional rule-based and heuristic methods by enabling automated extraction of ransomware indicators; however, the effectiveness of any given model remains dependent on the type of features and datasets used. The study also underscores the importance of careful feature selection, data preprocessing, and considerations for real-time deployment in building effective ransomware defense mechanisms.

References

1. Chaithanya, B.N., Brahmananda, S.H.: AI-enhanced defense against ransomware within the organization's architecture. J. Cyber Secur. Mobil. (2022). https://doi.org/10.13052/jcsm2245-1439.1146
2. Ferdous, J., Islam, R., Mahboubi, A., Islam, M.Z.: AI-based ransomware detection: a comprehensive review. IEEE Access **12**, 136666–136695 (2024). https://doi.org/10.1109/ACCESS.2024.3461965
3. Jegede, A., Fadele, A., Onoja, M., Aimufua, G., Mazadu, I.J.: Trends and future directions in automated ransomware detection. J. Comput. Soc. Inf. **1**(2), 17–41 (2022). https://doi.org/10.33736/jcsi.4932.2022
4. Lee, K., Lee, S.-Y., Yim, K.: Machine learning based file entropy analysis for ransomware detection in backup systems. IEEE Access **7**, 110205–110215 (2019). https://doi.org/10.1109/access.2019.2931136
5. Oz, H., et al.: A survey on ransomware: evolution, taxonomy, and defense solutions. ACM Comput. Surv. (2022). https://doi.org/10.1145/3514229
6. Poudyal, S., Dasgupta, D.: AI-powered ransomware detection framework. In: 2020 IEEE Symposium Series on Computational Intelligence (SSCI) (2020). https://doi.org/10.1109/ssci47803.2020.9308387
7. Razaulla, S., et al.: The age of ransomware: a survey on the evolution, taxonomy, and Research Directions. IEEE Access **11**, 40698–40723 (2023). https://doi.org/10.1109/access.2023.3268535
8. Smith, D., Khorsandroo, S., Roy, K.: Machine learning algorithms and frameworks in ransomware detection. IEEE Access **10**, 117597–117610 (2022). https://doi.org/10.1109/access.2022.3218779
9. Sneha, M., Arya, A., Agarwal, P.: Ransomware detection techniques in the dawn of Artificial intelligence: a survey. In: 2020 The 9th International Conference on Networks, Communication and Computing, vol. 7, pp. 26–33 (2020). https://doi.org/10.1145/3447654.3447659
10. Talabani, H.S., Abdulhadi, H.M.T.: Bitcoin ransomware detection employing rule-based algorithms. Sci. J. Univ. Zakho **10**(1), 5–10 (2022). https://doi.org/10.25271/sjuoz.2022.10.1.865
11. Alraizza, A., Algarni, A.: Ransomware detection using machine learning: a survey. Big Data Cogn. Comput. **7**(3), 143 (2023). https://doi.org/10.3390/bdcc7030143

12. Anghel, M., Racautanu, A.: A note on different types of ransomware attacks (2019). https://eprint.iacr.org/2019/605.pdf
13. Bae, S.I., Lee, G.B., Im, E.G.: Ransomware detection using machine learning algorithms. Concurr. Comput. Pract. Exp. **32**(18) (2019). https://doi.org/10.1002/cpe.5422
14. Bello, I., et al.: Detecting ransomware attacks using intelligent algorithms: recent development and next direction from deep learning and big data perspectives. J. Ambient. Intell. Humaniz. Comput. **12**(9), 8699–8717 (2020). https://doi.org/10.1007/s12652-020-02630-7
15. Bold, R., Al-Khateeb, H., Ersotelos, N.: Reducing false negatives in ransomware detection: a critical evaluation of machine learning algorithms. Appl. Sci. **12**(24), 12941 (2022). https://doi.org/10.3390/app122412941
16. Manavi, F., Hamzeh, A.: A new method for ransomware detection based on PE header using convolutional neural networks. In: 2020 17th International ISC Conference on Information Security and Cryptology (ISCISC), Tehran, Iran, pp. 82–87 (2020). https://doi.org/10.1109/ISCISC51277.2020.9261903
17. Amjad Hussain, A.H.: Ransomware Dataset 2024. 1.0, Zenodo (2024). https://doi.org/10.5281/zenodo.13890887
18. Ransomware survey: Ransomware stats. Ransomware.org. (2024). https://ransomware.org/ransomware-survey/
19. Djenna, A., Bouridane, A., Rubab, S., Marou, I.M.: Artificial intelligence-based malware detection, analysis, and mitigation. Symmetry **15**(3), 677 (2023). https://doi.org/10.3390/sym15030677
20. Rhode, M., Burnap, P., Jones, K.: Early-stage malware prediction using recurrent neural networks. Comput. Secur. **77**, 578–594 (2018). ISSN 0167–4048, https://doi.org/10.1016/j.cose.2018.05.010
21. Madani, H., Ouerdi, N., Boumesaoud, A., et al.: Classification of ransomware using different types of neural networks. Sci. Rep. **12**, 4770 (2022). https://doi.org/10.1038/s41598-022-08504-6
22. Alsinayyid, A., Kadiyala, S.: SVM enhanced detection of volume-based attacks in IoT networks (2025). https://doi.org/10.13140/RG.2.2.14330.81600
23. Al-Sinayyid, A., Mannuru, V., Ali Jewel, M.J., Sasidhar, K.: Analytical study for cybersecurity attacks and defenses characteristics. In: 2023 International Conference on Computational Science and Computational Intelligence (CSCI), Las Vegas, NV, USA, pp. 853–860 (2023). https://doi.org/10.1109/CSCI62032.2023.00143

An Analytical Study of Machine Learning Methods for Cyberattack Detection in Industrial Control Systems

Vishnu Vardhan Reddy Sheri(✉), Ali Al-Sinayyid, Rohith Reddy Battula, Jaya Chandra Gundeboina, Samuel Adewumi, and Shikhar Pandey

Department of Mathematics and Computer Science West, Virginia State University, Institute, USA

{vsheri,ali.alsinayyid,rbattula,jgundeboina,sadewumi, spandey}@wvstateu.edu

Abstract. Industrial Control Systems (ICS), which is the backbone of critical infrastructure such as energy, water, and manufacturing, are under constant threat from cyberattacks. Hence, this comparative study highlights how state-of-the-art machine learning and deep learning methods, particularly Random Forest, Long Short-Term Memory (LSTM), Convolutional Neural Networks (CNN), along with hybrid models, can be useful for the detection of ICS-specific attacks. This allows analysis of tradeoffs in detection accuracy, computational complexity and dataset performance, such as ensemble models with high accuracy, sequence-based models for temporal anomalies, and lightweight methods for resource-constrained environments. The study sets a benchmark for determining appropriate techniques enabling researchers and practitioners to maximize ICS resilience with respect to evolving cyber threats, allowing a critical assessment of the strengths and limitations of these algorithms across multiple datasets and attack categories.

Keywords: Industrial Control Systems (ICS) · Machine Learning (ML) · Deep Learning (DL) · Random Forest (RF) · Long Short-Term Memory (LSTM) · Convolutional Neural Networks (CNN)

1 Introduction

In the contemporary world, it is Industrial Control Systems (ICS) that play a crucial role in the day-to-day operations of transportation, energy, water purification, and factories. It is these infrastructure needs that underpin the operation of industry, power plants, smart grids, and water treatment facilities. In the past, ICS were constructed for reliability, availability, and resilience, usually working in isolated environments without any direct connections to external networks. However, with the evolution of Information Technology (IT) solutions, increasing interconnectivity has resulted in extensive exposure to cybersecurity threats [5]. Although moving to a set of interconnected systems has been beneficial in many respects, new threats emerge in the growing use of Internet of Things (IoT) devices within industrial networks.

H. R. Arabnia et al. (Eds.): CSCI 2025, CCIS 2969, pp. 492–505, 2027.
https://doi.org/10.1007/978-3-032-28460-0_35

These enhanced connections have markedly increased exposure to cyberattacks such as ransomware, DDoS. Conventional cybersecurity measures, with their associated intrusion detection systems (IDS) or firewalls, were usually implemented with IT networks in view and are incompatible with the requirements of ICS environments [15]. Thus, cybercriminals have developed more sophisticated attacks against Industrial Control Systems, which have included ransomware attacks on systems to disrupt operations, encrypt critical files, and extort businesses [13]. High-profile incidents like the Stuxnet worm attack on Iran's nuclear facilities and the Triton malware attack on a Saudi Arabian petrochemical plant illustrate how seriously these attacks affect both infrastructure and national security [4].

Cyberattacks, including Ransomware, are among the most perilous of the various types of cyber assaults for ICS, as it has the ability to incapacitate critical services that rely on these systems. Industrial Control System administrators face the challenge of identifying and responding to ransomware attacks in real time to mitigate the risk of system failures and minimize delay, which can have catastrophic consequences for their systems. This challenge is further complicated by the design features of Industrial Control Systems, which necessitate high availability. These systems are unable to sustain false positives or service disruptions, rendering traditional security methodologies unsuitable for these environments [15].

Recently, machine learning (ML) has gained traction as a strong candidate to augment ICS security for use in ransomware and advanced cyber threats detection and response. Machine learning algorithms are able to process much more data in a way that traditional intrusion detection systems [11] cannot, such as examining a wide batch of network traffic and system data, finding patterns of activity that suggest malicious activity, while uncovering anomalies in a similar manner [5]. Through continuous data learning, machine learning intrusion detection systems (IDS) embrace new attack types like zero-day threats and malware evolution without continuous manual interaction [8]. In this paper, the aim is to assess the performance of machine learning based algorithms for detecting cyber-attacks and other malicious activities in ICS. Namely, the paper will evaluate and compare the performance of multiple popular machine learning models such as Random Forest, Support Vector Machines (SVM), K-Nearest Neighbor (KNN), Decision Trees and Neural Networks [31]. This research will help define which algorithms are the best at preventing cyber-attacks in critical infrastructure by analysing their accuracy, low false positive rates, and fast response times.

The purpose of this study is to make more useful and adaptable intrusion detection solutions for Industrial Control Systems. In determining the most robust machine learning methods, the research will contribute towards strengthening the resilience of Industrial Control Systems against the increasing risk of cyberattacks, ultimately enhancing the security and stability of the critical infrastructures that support modern society.

2 Literature Review

In a study of AI-Augmented Threat Response Systems with Real-Time Adaptive Defense presented by Rajashekhar Reddy Kethireddy et al. [2], a deep learning-based reinforcement learning (RL) model proposes an AI framework for a threat security paradigm, proposing improvements on threat detection and real-time adaptive defense based on these models. For detection, a high reliability (98.5% accuracy) CNN-based network is employed with an RL agent dynamically adapting the defense strategies, cutting down response time by 30% after threat detection. The study uses the CIC-IDS2017 dataset for training and testing; the model has proved autonomously capable of threat detection in real-time and response time is faster than what static defense systems, such as the traditional way, are able to do.

This paper investigates the application of machine learning (ML) algorithms for the detection of attacks in Industrial Control Systems (ICS) [5], especially those used in Industrial Internet of Things (IIoT) settings. With IoT, robotics, artificial intelligence, and blockchain, IIoT technologies are widely used and entering into Industrial systems. While these technologies foster systems that operate self-optimizing and self-configurating, allowing systems to perform operations that do complex tasks, they raise new security challenges, especially considering ICS.

James Graham et al. (2023) published the paper Improving Cybersecurity for Industrial Control Systems. Specifically, in [1], cybersecurity vulnerabilities relating to Industrial Control Systems (ICS), more specifically supervisory control and data acquisition (SCADA) systems with a central control unit (MTU) and remote-control units (RTUs), are discussed. The majority of the critical challenges Graham listed are the need to secure legacy systems that have lengthy lifecycles, the absence of cybersecurity professionals trained specifically in that area, software patch applications being delayed, and a low level of industry awareness towards cybersecurity issues. The paper refers to a few ways that ICS security can be improved, including more conventional IT solutions (network segmentation and firewalls), but also maintains that "specialized solutions will be required" to secure field devices. For example, the author describes the SCADA-Guard™ Security Preprocessor (SPP). With smaller kernel size, compartmentalized software, and an access control scheme, the device provides greater security for both legacy and modern front-end ICS setups.

With this research, Ijaz Ahmad, Atif Alamri and Mushfiqur Rouf [10] proposed a novel approach of CNN-LSTM hybrid intrusion detection system for enhancing cybersecurity in the Internet of Things (IoT) networks. Utilization in real-time is performed with the proposed model, which works both for the collection of temporal and spatial data and the detection of multiple cyberattacks with high accuracy. Studies on lightweight anomaly detection frameworks and adaptive intrusion prevention mechanisms prove the relevance and effectiveness of the proposed approach of the authors. They are experts designing intelligent security systems and deep learning applications in the Internet of Things (IoT).

In this paper, the novelty value of artificial intelligence (ML) techniques for detecting ICS anomalies has been considered [15]. Different techniques are also assessed for intrusion detection on SCADA datasets, such as Random Forest techniques, K-Nearest Neighbors, Naive Bayes, and Logistic Regression. The methods of finding anomalies are mainly classification models, feature extraction, and data cleaning. Measures such as accuracy, recall, and F1-score are employed to measure performance. The Random Forest classifier had the highest accuracy (96%), followed by Naive Bayes (88%), K-Nearest Neighbors (94%), and Logistic Regression (92%), based on the results. In Spyder

Python, network traffic is analyzed here, focusing on TCP and UDP protocol profiling and anomaly detection. While false positives still remain issues, the study establishes a superiority of machine learning intrusion detection systems (IDS) in order to realize a performance boost in ICS. Further research will focus on improving detection based on online behavior analysis and using hybrid models.

It is found through the literature review that to secure Cyber-Physical Systems, a sophisticated form of intrusion detection must be made, which is often hindered through traditional methods. A pioneering approach has been taken by Siddique et al. [3], who put forward a new GAN-LSTM hybrid model that got 87% accuracy on SWaT and 91% accuracy on WADI datasets, outperforming MAD-GAN by a mile. Previous research found this model improves precision and recall for structured attacks. However, more recent studies have proved that persistent challenges remain in detecting subtle, complex intrusions and optimizing static thresholds.

Shalyga et al. [29] proposed a neural network (NN) based forecasting method for anomaly detection (AD) for industrial control systems based on the Secure Water Treatment (SWaT) testbed dataset. Genetic algorithms were used by the researchers to automatically optimize the NN architecture, and they ultimately found that the best model for the same type of monitoring is a multilayer perceptron (MLP) that contains encoder, analyzer, and decoder blocks. This optimized MLP model had an F1-score of 0.812 and a NAB score of 69.612 for detecting 25 out of 34 physical attacks in the SWaT dataset. The model also pays more attention to the NAB metric than the F1-score, as it values early detection more highly and is less sensitive to the long duration of certain anomalies.

In this regard, Jiang Tian et al. [14] proposed a new anomaly detection framework combining One-Class Support Vector Machines (OC-SVM) with Particle Swarm Optimization (PSO) to increase detection accuracy for anomaly detection in cybersecurity applications. The proposed model uses OC-SVM. With the help of PSO to dynamically adjust hyperparameters, OC-SVM classification performance is enhanced for anomaly detection on large volumes of complex data with more advanced classification methods. It can be observed that this hybrid methodology is more accurate and computationally less costly than the currently implemented OC-SVM; this approach demonstrates that it can serve as the optimal option for identifying cyber threats to vital systems.

In this paper, Xu et al. [17] introduce a lean secure federated learning scheme for anomaly detection on wireless Industrial Control Systems (ICS) that are susceptible to eavesdropping, tampering, and injection attacks. Based on a residual multibeam self-attention CNN for spatial-temporal feature extraction coupled with dynamic parameter pruning (reducing communication costs by 89.6%) and adaptive AES encryption with elliptic curve key negotiation, their end-edge collaborative technique resulted in >99% accuracy on the Smart Meter, NSL-KDD, and UNSW-NB15 datasets. The proposed framework proved robust against parameter tampering (50% attack rate) and malicious data injection, providing a scalable, low-latency method to secure resource-constrained ICS environments.

Sun et al. [30] advanced MTS-DVGAN, an unsupervised dual variational generative adversarial network for multivariate time series anomaly detection in Cyber-Physical Systems. This model implements feature center loss and contrastive constraints to achieve greater stability and better distinguish between normal and anomalous data in the embedding space. The architecture was thoroughly evaluated over three public benchmarks, namely against the industrial control system datasets SWAT and WADI. On the SWAT dataset, MTS-DVGAN provided best results against competitors with an accuracy of 94.84% and F1-score of 79.87%. These results showed high accuracy, with the optimum F1-score being 97.84% on the WADI dataset, The results indicate that performance is faster and more stable compared to previous GAN-based systems.

According to this study by Sardar Muhammad Ali et al. [16], cybersecurity vulnerabilities of Industrial Control Systems (ICS) and cyber threats of cyber-attacks in the field of Critical Infrastructure are examined. They analyze numerous frameworks, such as ISO 27001, NIST, and CIS in reviewing these security models for the purpose of protecting ICS against cyber-attacks. In mitigating risks, they emphasize the role of Information Security Management Systems (ISMS) and decision-making models like the OODA loop and PDCR cycle. The research highlights the importance of AI and recommender systems in providing more security through risk detection and compliance improvement. In addition, the study identifies research gaps, which require more practical validation of cybersecurity frameworks and implementing AI technologies to protect ICS from cyber-attacks. This research highlights the critical importance of targeted security mechanisms and predictive risk management practices regarding ICS and security against new cyber threats such as ransomware.

Considering such problems with false positive rates and high computational burden of the existing methods, this work describes Improved Autoencoder (ImpAE) to handle the anomaly identification of Industrial Control Systems (ICS) [24]. The proposed model performed better than previous ICS security detection in accuracy, precision, recall, and F1-Score as evidenced by the precision of 0.993, 96% accuracy, recall of 0.673, and F1-Score of 0.771 with respect to the SWaT dataset generated over a real water treatment plant.

In this paper, Mokhtari et al. [18] introduced a measurement intrusion detection system (MIDS) with its supervised machine learning model to detect the stealthy ICS attacks that circumvent the conventional NIDS. Using a hardware-in-the-loop (HIL) testbed, they created a labeled dataset (HAI) targeting sensor and control data. Random Forest reached a 99.76% accuracy, indicating the capability of the model to identify threats that are difficult to detect via network traffic alone and are scalable for ICS security. It also tackles insider threats and encrypted data manipulation. These are important drivers for incorporating measurement-based detection and network-level IDS for critical infrastructure.

Kim et al. [27] performed a comprehensive comparative study on 5 state-of-the-art time series anomaly detection models for Industrial Control Systems in order to clarify the mode of selection for real-world scenarios. Using public datasets made available via SWaT and HAI, their evaluation paid attention to detection accuracy and training/testing times. The best model was data-dependent, wherein InterFusion provided the highest F1-score for SWaT (90.7%), whereas RANSynCoder performed much better for HAI (82.9% F1-score). Despite this good accuracy over InterFusion, RANSynCoder proved a more practical option, as the efficiency behind that choice is superior, and it was critically observed that employing only about 40% of the training dataset can provide comparable performance, yielding a huge reduction in overhead, according to the study.

Xue et al. [28] investigated deep neural networks for industrial anomaly detection, establishing the task as a self-supervised autoregressive model to learn normal system dynamics. They employed the multivariate time-series measurements from the Secure Water Treatment (SWaT) testbed dataset with both LSTM and CNN based backbones and DeepSVDD. Importantly, researchers proposed a joint loss approach that could be adopted by LSTM models in order to support a successful combination of continuous and discrete features existing in the dataset. The simple Neural Network (NN) model attained the highest F1-score out of the architectures compared with 0.82831, and the LSTM model with the joint loss achieved the best overall AUC performance of 0.881.

3 Classification

In this study, an overview on classification-based assessment of various attack detection approaches is carried out in Industrial Control Systems (ICS). The emphasis is also the one that focuses on calculating basic metrics like detection accuracy, computational complexity, and algorithmic fit. Examples include the success rates of Random Forest models, which can consistently meet high accuracy requirements at the expense of substantial computational overhead, making real-time applications difficult. Classifiers such as Naïve Bayes and K-Nearest Neighbors (KNN) reduce latency and achieve competitive accuracy and may be used in time-sensitive environments. Therefore, to decide which model is appropriate, performance must be balanced with the constraints of the system.

Hybrid model-based methods in the form of decision trees, neural networks, and optimization models like Particle Swarm Optimization (PSO) are a great compromise between accuracy and the resources. Deep learning architectures such as Convolutional Neural Networks (CNN) and Long Short-Term Memory (LSTM) networks that provide good adaptability to dynamic threats are expensive in computing power. And so their actual deployment becomes similarly influenced by the limitations of the system and the nature of the dataset.

Classification analyzes in detail a variety of Industrial Control Systems (ICS) attack scenarios and their accuracy, computational cost, latency, and suitability for a specified dataset. Here, the performance of multiple assault methods is critically assessed, focusing on the complexity of these attacks, their detection accuracy, and how they respond to the detection mechanism. The analysis provides insight as to the pros and cons of some detection techniques in the context of ICS by focusing on crucial aspects such as attack mitigation and operational practicality (Table 1).

3.1 Analysis Study

Methodology. This study is a systematic review of machine learning methods of detecting cyberattacks in Industrial Control Systems (ICS). Peer-reviewed articles and reputable preprints were included if (i) algorithms were applied to recognized ICS security datasets (e.g., SCADA, SWaT, WADI, HAI, ICS-Network), (ii) empirical results with at least one standard performance metric such as Accuracy, Precision, Recall, or F1-score were reported, and (iii) the attack types and algorithm characteristics were evaluated.

Table 1. Comparative Analysis of ICS Attack Detection Methods.

Ref	Algorithm	Complexity	Accuracy	Dataset	Attacks faced
[5]	Random forest	O (m × n log n)	99.84%	SCADA	DoS, Reconnaissance
[6]	Random forest	O (m × n log n)	99.50%	ICS anomaly	DoS, Man-in-the-Middle attacks
[7]	Long Short-Term Memory	O (n × d × h)	96.67%	SWaT	Distributed denial of service attacks
[9]	Naïve Bayes Classifier	O (n × m)	96.00%	SCADA	DoS, Man-in-the-Middle attacks
[10]	CNN, CB_SMOTE	O (k × n × m)	97.00%	ICS Custom	Protocol Exploits, Command Injection
[12]	DT + NB with MLP	O (k × n × m)	92.70%	SCADA	DoS, Man-in-the-Middle attacks
[14]	OCSVM + PSO	O (n^2 × m)	95.00%	Generic ICS	Generic Anomalies
[15]	Hybrid J48 + Bayes Network	O (n × m)	96.18%	SCADA	DoS, MITM, Modbus/TCP
[16]	Random forest, KNN	RF:O (m × n log n) KNN: O (n^2)	81.00%	General-purpose ICS	Ransomware, Phishing, Social engineering
[17]	RMS-CNN-LSFL	O (n × m)	99.00%	Smart Meter / NSL-KDD	Tampering Attacks, Injecting Attacks
[18]	Random forest	O (k × n × d log n)	99.76%	HAI	PCL, Stealthy Attacks, Primitive attacks
[19]	CNN	O (k × n × m)	97.80%	IoT/ICS Hybrid	Ransomware, DDoS
[20]	OCSVM	O (n^2) to O (n^3)	98.80%	SCADA	MitM, SYN Flooding, Honeypot interaction
[21]	DNN with Random Forest	O (n log n)	90.60%	Adversarial Dataset	White-box and Black-box attacks
[22]	RF with Federated Learning	O (n log n)	93.51%	Federated ICS	DoS, Man-in-the-Middle attacks

(*continued*)

Table 1. (*continued*)

Ref	Algorithm	Complexity	Accuracy	Dataset	Attacks faced
[23]	Decision Tree	O (log n)	93.00%	SCADA	DoS, Distributed denial of service attacks
[2]	CNN	O (k × n × m)	98.50%	CIC-IDS2017	APT, Ransomware
[24]	ImpAE	O (n^2)	96.00%	SWaT	APT, DDoS, DoS
[25]	Random forest	O (n log n)	93.44%	ICS Network Data	Intrusion Detection (Network and Anomaly)
[26]	KNN	O (n^2)	99.20%	Phishing Dataset	Phishing Attacks

The algorithm used, dataset, attack scenario, and performance metrics were extracted from each paper. These results were normalized for consistency in tabular and graphical comparison. Figures and tables reflect these extracted results to highlight relative strengths and limitations of different approaches across datasets.

3.2 Accuracy

Determination of Industrial Control Systems (ICS) attack detection is necessary during which various aspects such as accuracy, computational complexity, and application against a variety of attack cases needs to be assessed. These conditions directly correlate with the capability of detection in an architecture to both identify and alleviate security threats without blocking operation in areas such as energy grids, water treatment plants and processing stations. Real-time detections, with low false alarms are key since a potential downtime or false positive in a model can have disastrous safety and economic effects.

As shown in Fig. 1, It shows that the performance patterns of the machine learning algorithms differ due to the design of the model and the applicability of the training to different threats in identifying Industrial Control Systems (ICS) cyberattacks. Compared to all other approaches, Random Forest outperforms with an accuracy of 99.84% with respect to Denial-of-Service (DoS) attacks on SCADA datasets through the use of the ensemble to analyze structured traffic patterns.

LSTM models perform strongly on Man-in-the-Middle (MITM) attacks due to their ability to model temporal dependencies (96.67% accuracy on SWaT datasets). Such performance results can be partially attributed to their ability to analyze sequential communication data. On the other hand, algorithms like K-Nearest Neighbors (KNN) and untested hybrids, such as J48 + Bayes Network, perform badly when dealing with ICS applications; KNN's accuracy dropped to 81% on general-purpose datasets since it was unsuitable for high-dimensional ICS data.

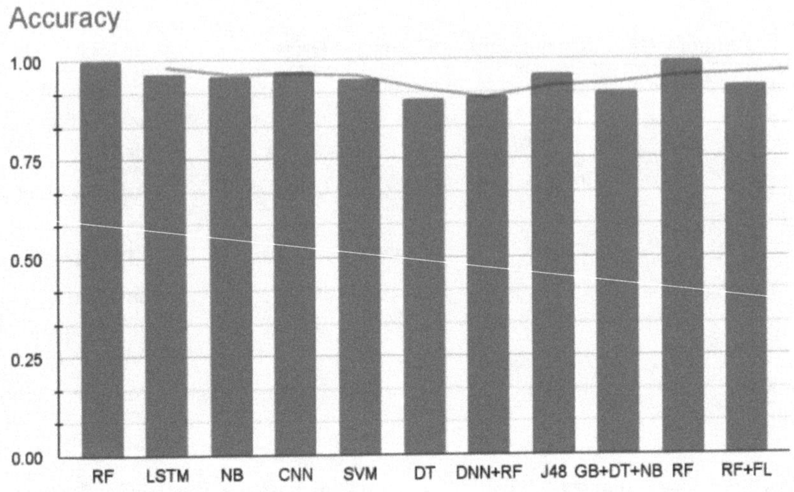

Fig. 1. Represents Accuracy for each Algorithm.

3.3 Accuracy vs Attacks

The accuracy of different cyberattacks against critical infrastructure that affect it (probably especially ICS type) can be seen on the Accuracy vs. Attacks graph. Denial-of-Service (DoS), DDoS, and Modbus TCP attacks have almost perfect accuracy, and for each type of attack, there is a highly reliable detection rate, indicating a well-optimized detection system. Other threats, such as man-in-the-middle attacks, corrupted packets, and protocol exploitation, have also shown excellent detection accuracy, albeit at a lower level (Fig. 2).

Variability in accuracy suggests that while most assault types are eventually detected, there are a few that could still be challenging to identify, particularly if they include complex avoidance measures. This underlines the ongoing need for improved anomaly detection, intrusion detection systems (IDS), and adaptive cybersecurity techniques to ensure robust ICS protection. DoS/DDoS, MITM/Replay, Command/Response/SQL Injection, Ransomware/Phishing, and Protocol Exploits/False Data Injection (FDI) are some of the key cyber-attacks against ICS.

3.4 Complexity

As shown in Fig. 3, the complexities related to ICS detection methods are broadened to various classes. The most frequent are O(k × n × m) (19%), followed by O(n log n), O(m × n log n), and O(n × m), each contributing approximately 14.3%. The few categories that contribute less are O(log n), O(n2 × m), O(n2–O(n3)), and O(n × d × h), which account for 4.8% of the distribution. This diversity suggests that although moderate-complexity algorithms prevail, a sizeable subset of higher-order algorithms continue to

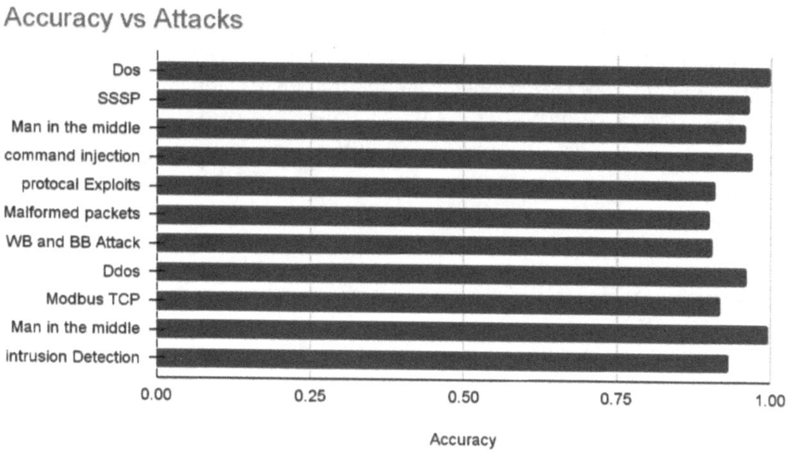

Fig. 2. Represents Accuracy of algorithms for each Attack faced.

be meaningful, particularly for detecting sophisticated and evolving cyberattacks despite their costlier computation.

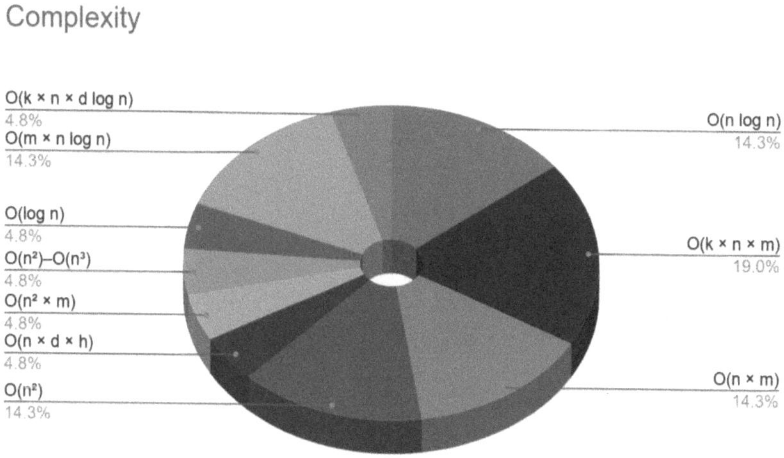

Fig. 3. Represents percentage division of complexities for various algorithms.

Complexities are stated according to the original papers. Given differences in dataset features, dimensionality of features, preprocessing steps taken, and parameterization, a single algorithm may be presented with different complexities (e.g., the same algorithm can have different complexities). For instance, Random Forest has been reported as O(m × n log n) [5, 6, 16], O(k × n × d log n) [18], and O(n log n) [21, 22, 25], since it varies based on the number of features (m), trees (k), or depth of trees (d) on various datasets. Also, CNN-based models look something similar; O(k × n × m) [2, 10, 12, 19], although runtime actually depends on convolutional layers and the scale of the set. LSTM is reported as O(n × d × h) [7], and depth (d) and hidden state size (h) depend on configuration. One-Class SVM will have O(n2 × m) [14] or O(n2)–O(n3) [20], depending on kernel expansion. On the other hand, simpler classifiers like Decision Trees (O(log n) [23]) or Naïve Bayes (O(n × m) [9, 15, 17]) have stable low-to-moderate complexity. These differences highlight that complexity values should be treated as scaling metrics of the dataset, as well as experimental conditions, and not as absolute constants.

3.5 Algorithm vs Dataset

Figure 4 describes the performance of algorithm–dataset in terms of accuracy in a heatmap. Each individual cell shows the detection performance of a given algorithm on a dataset, darker colors represent higher accuracy, while lighter colors indicate lower performance. Readers are then able to compare which algorithms generalize well across datasets and which ones are more context dependent.

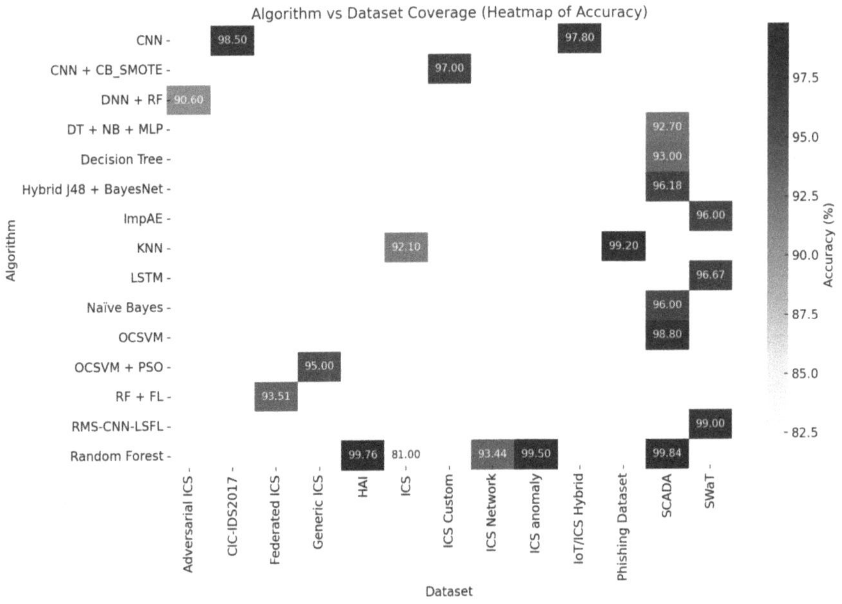

Fig. 4. Represents a heatmap of accuracy between algorithms and datasets.

As illustrated in Fig. 4, Random Forest achieves the maximum accuracy of 99.76% on HAI, but also still achieves stable performance on SCADA and SWaT. LSTM achieves 99.20% on ICS anomaly datasets, highlighting its predictive ability of sequential dependencies. CNN-based methods consistently exceed 97% across adversarial and custom ICS datasets and are adaptable for complex input structures. In contrast, DNN + RF (90.60%) and ICS-Custom models (81.00%) achieve lower results, demonstrating dataset imbalance and less generalization. Overall, these heatmap highlights how tree-based models are superior in organized settings versus sequence-oriented and convolutional architectures over temporal and adversarial environments.

3.6 Identifying the Best Algorithm

In ICS algorithms, selection should be sensitive to the type of attack and to the operational limitations of settings. Ensemble architectures, for instance, Random Forest, with 99.84% accuracy on SCADA datasets, can be highly effective against DoS and protocol manipulation through modeling structured traffic behavior and filtering it through multiple decision trees. Time series-based models such as Long Short-Term Memory (LSTM) networks, which achieved 96.67% accuracy on the SWaT dataset, are well suited for detecting temporal anomalies such as replay, and Man-in-the-Middle (MITM) attacks that may involve abnormal command sequences and spurious sensor readings. Likewise, XGBoost is better prepared in handling correlated industrial features, reaching 99.76% accuracy on HAI for command injection detection, far outperforming simpler classifiers like Naïve Bayes.

For resource constrained ICS edge and security environments, lightweight algorithms like LightGBM can complement these methods, as it achieves a 92% F1-score on phishing and ransomware detection using rapid inference and the lower resource load. With Hybrid methods, detection coverage is also increased, while combinations from LightGBM to Random Forest could obtain 99% accuracy on S7 traffic for Modbus/TCP spoofing in DPI. There is no one optimal algorithm, Random Forest can use a lot of memory for example, LSTM is costly for training, and boosting models like XGBoost raise issues as to interpretability. Consequently, the "best" algorithm is inherently situation specific so the tradeoff must be found between accuracy and computational complexity, and scalability for real-world ICS applications.

4 Conclusion

Industrial Control Systems (ICS) face increasingly high-impact cyberattacks that compromise critical infrastructure reliability through the manipulation of system availability and protocol vulnerabilities. This study compared multiple machine learning models on large-scale datasets (SCADA, SWaT, HAI) for detection accuracy and complexity. The study showed that Random Forest had an excellent accuracy on structured traffic (99.84% accuracy on SCADA), but presented a substantial computational cost when compared with the LSTM (96.67% accuracy on SWaT) architectures that were more effective to handle temporal anomalies, including replay and Man-in-the-Middle attacks.

CNN-methods demonstrated strong adaptability for protocol exploits, whereas KNN or Naïve Bayes are also capable of achieving competitive performance in latency-sensitive conditions. Taken together, the analysis indicates that no single algorithm is universally effective; the choice depends on the type of attack, system limits, and characteristics of dataset. By integrating accuracy and complexity, the analysis provides a benchmark for the optimal choice of detection methods on real-world ICS. Future studies should extend comparative testing to more extensive datastores, refine hybrid models and integrate interpretability, to ensure that machine learning–based anomaly detection is strong and feasible in protecting industrial environments from dynamic cyber threats.

References

1. Graham, J., Hieb, J., Naber, J.: Improving cybersecurity for industrial control systems. In: 2016 IEEE 25th International Symposium on Industrial Electronics (ISIE) (pp. 618–623). IEEE (2016)
2. Kethireddy, R.R.: AI-Augmented threat response systems with real-time adaptive defense. Intelligence **1**(1), 62–71 (2023)
3. Siddique, M.S., Khan, M.A.R., Ahammad, I., Nath, N., Das, J.R., Rahman, F.: An Intelligent Intrusion Detection System for Cyber-Physical Systems Using GAN-LSTM Networks. Franklin Open 100281 (2025)
4. Formby, D., Durbha, S., Beyah, R.: Out of control: ransomware for industrial control systems. In: RSA conference (Vol. 4, p. 8) (2017)
5. Arora, P., Kaur, B., Teixeira, M.A.: Evaluation of machine learning algorithms used on attacks detection in industrial control systems. J. Inst. Eng. (India): Ser. B **102**(3), 605–616 (2021)
6. Dehlaghi-Ghadim, A., Moghadam, M.H., Balador, A., Hansson, H.: Anomaly detection dataset for industrial control systems. IEEE Access **11**, 107982–107996 (2023)
7. Perales Gómez, Á.L., Fernández Maimó, L., Huertas Celdrán, A., García Clemente, F.J.: Madics: a methodology for anomaly detection in industrial control systems. Symmetry **12**(10), 1583 (2020)
8. Psychogyios, K., Papadakis, A., Bourou, S., Nikolaou, N., Maniatis, A., Zahariadis, T.: Deep learning for intrusion detection systems (IDSs) in time series data. Future Internet **16**(3), 73 (2024)
9. Eigner, O., Kreimel, P., Tavolato, P.: Attacks on industrial control systems-modeling and anomaly detection. In: Proceedings of the 4th International Conference on Information Systems Security and Privacy (pp. 581–588). SCITEPRESS-Science and Technology Publications (2018)
10. Gu, H., Lai, Y., Wang, Y., Liu, J., Sun, M., Mao, B.: DEIDS: a novel intrusion detection system for industrial control systems. Neural Comput. Appl. **34**(12), 9793–9811 (2022)
11. Bernieri, G., Conti, M., Turrin, F.: Evaluation of machine learning algorithms for anomaly detection in industrial networks. In: 2019 IEEE International Symposium on Measurements & Networking (M&N) (pp. 1–6). IEEE (2019)
12. Varghese, S.A., Ghadim, A.D., Balador, A., Alimadadi, Z., Papadimitratos, P.: Digital twin-based intrusion detection for industrial control systems. In: 2022 IEEE International Conference on Pervasive Computing and Communications Workshops and Other Affiliated Events (PerCom Workshops) (pp. 611–617). IEEE (2022)
13. Butt, U.J., Abbod, M., Lors, A., Jahankhani, H., Jamal, A., Kumar, A.: Ransomware Threat and its Impact on SCADA. In: 2019 IEEE 12th International Conference on Global Security, Safety and Sustainability (ICGS3) (pp. 205–212). IEEE (2019)
14. Tian, J., Gu, H.: Anomaly detection combining one-class SVMs and particle swarm optimization algorithms. Nonlinear Dyn. **61**(1), 303–310 (2010)
15. Mubarak, S., Habaebi, M.H., Islam, M.R., Rahman, F.D.A., Tahir, M.: Anomaly detection in ICS datasets with machine learning algorithms. Comput. Syst. Sci. Eng. **37**(1) (2021)
16. Ali, S.M., Razzaque, A., Yousaf, M., Shan, R.U.: An automated compliance framework for critical infrastructure security through Artificial Intelligence. IEEE Access (2024)
17. Xu, C., Du, X., Li, L., Li, X., Yu, H.: End-edge collaborative lightweight secure federated learning for anomaly detection of wireless industrial control systems. IEEE Open J. Ind. Electron. Soc. **5**, 132–142 (2024)
18. Mokhtari, S., Abbaspour, A., Yen, K.K., Sargolzaei, A.: A machine learning approach for anomaly detection in industrial control systems based on measurement data. Electronics **10**(4), 407 (2021)

19. Kavitha, D., Thejas, S.: AI enabled threat detection: leveraging artificial intelligence for advanced security and cyber threat mitigation. IEEE Access (2024)
20. Maglaras, L.A., Jiang, J.: Intrusion detection in SCADA systems using machine learning techniques. In: 2014 Science and Information Conference (pp. 626–631). IEEE (2014)
21. Ding, Y., Wang, L., Zhang, H., Yi, J., Fan, D., Gong, B.: Defending against adversarial attacks using random forest. In: Proceedings of the IEEE/CVF Conference on Computer Vision and Pattern Recognition Workshops (p. 0) (2019)
22. Markovic, T., Leon, M., Buffoni, D., Punnekkat, S.: Random forest based on federated learning for intrusion detection. In: IFIP International Conference on Artificial Intelligence Applications and Innovations (pp. 132–144). Cham, Springer International Publishing (2022)
23. Azam, Z., Islam, M.M., Huda, M.N.: Comparative analysis of intrusion detection systems and machine learning-based model analysis through decision tree. IEEE Access **11**, 80348–80391 (2023)
24. Aslam, M.M., Tufail, A., De Silva, L.C., Haji Mohd Apong, R.A.A., Namoun, A.: An improved autoencoder-based approach for anomaly detection in industrial control systems. Syst. Sci. Control Eng. **12**(1), 2334303 (2024)
25. Deshpande, C., Gens, D., Franz, M.: Stackbert: machine learning assisted static stack frame size recovery on stripped and optimized binaries. In: Proceedings of the 14th ACM Workshop on Artificial Intelligence and Security (pp. 85–95) (2021)
26. Lin, Y., et al.: Phishpedia: A hybrid deep learning based approach to visually identify phishing webpages. In: 30th USENIX Security Symposium (USENIX Security 21) (pp. 3793–3810) (2021)
27. Kim, B., Alawami, M.A., Kim, E., Oh, S., Park, J., Kim, H.: A comparative study of time series anomaly detection models for industrial control systems. Sensors **23**(3), 1310 (2023)
28. Xue, F., Yan, W., Wang, T., Huang, H., Feng, B.: Deep anomaly detection for industrial systems: a case study. In: Annual Conference of the PHM Society (Vol. 12, No. 1, p. 8) (2020)
29. Shalyga, D., Filonov, P., Lavrentyev, A.: Anomaly detection for water treatment system based on neural network with automatic architecture optimization. arXiv preprint arXiv:1807.07282
30. Sun, H., Huang, Y., Han, L., Fu, C., Liu, H., Long, X.: MTS-DVGAN: anomaly detection in cyber-physical systems using a dual variational generative adversarial network. Comput. Secur. **139**, 103570 (2024)
31. Al-Sinayyid, A., Battula, R.R., Kadiyala, S., Mannuru, V., Sanford, T., Sanchez, A.: Analytical study on advanced persistent threat detecting, defending and mitigating. In: World Congress in Computer Science, Computer Engineering & Applied Computing (pp. 355–370). Cham: Springer Nature Switzerland (2024)

Evaluating Random Forest, Decision Tree and Naive Bayes for Real Time DDoS Detection in 5G Environments

Vaishnavi Kottala(✉), Ali Al-Sinayyid, Venkatesh Mannuru, Luice Khamboo, Samuel Adewumi, and Joshua Gilliland

Department of Mathematics and Computer Science West, Virginia State University, Institute, USA

{vkottala,ali.alsinayyid,vmannuru,lkhamboo,sadewumi, jgilliland}@wvstateu.edu

Abstract. DDoS attacks are a major security concern in 5G networks, capable of overwhelming bandwidth and disrupting essential services. In this paper we assess three classifiers based on machine learning algorithms which are Random Forest (RF), Naive Bayes (NB), and Decision Tree (DT) using statistical and protocol-level features to detect malicious traffic. We find experimental results indicating RF had the highest accuracy at 0.982, NB at 0.932, and DT at 0.864. These results emphasize RF's suitability for modeling complex traffic patterns, NB's efficiency in lightweight detection, and DT's role in interpretability through rule-based classification. The entire investigation shows that the use of machine learning in DDoS detection can improve the security posture of next generation 5G networks.

Keywords: 5G Networks · DDoS Detection · Cybersecurity · Machine Learning · Random Forest · Intrusion Detection · Real-Time Classification · Artificial Intelligence

1 Introduction

Fifth generation (5G) networks allow ultra-low latency, huge bandwidth and massive connectivity, providing services like smart healthcare [11], industrial automation and autonomous systems. Despite giving rise to new chances, these features also increase the attack surface, leaving networks open to sophisticated security threats [10]. Of these, Distributed Denial-of-Service (DDoS) attacks are of special concern, when traffic is deliberately overflowing in a systematic manner, resulting in the depletion of bandwidth and computing resources, and in extreme cases serious service degradation or outages. With 5G's reliance on virtualization, edge computing, and network slicing, these attacks grow more disruptive and more difficult to mitigate.

Extended service interruptions in sensitive fields like healthcare and transportation, where uninterrupted communication is critical, can have a serious impact. Conventional defenses, such as signature- or rule-based mechanisms, are insufficient against modern

H. R. Arabnia et al. (Eds.): CSCI 2025, CCIS 2969, pp. 506–517, 2027.
https://doi.org/10.1007/978-3-032-28460-0_36

DDoS attacks [1] in terms of both velocity and magnitude. This has prompted two studies on data-driven methodologies for detecting hidden patterns and adapting to changing attack strategies. Machine learning (ML) offers a novel approach for analyzing large amounts of traffic data, detecting anomalies, and classifying malicious flows [2].This paper examines the effectiveness of Random Forest (RF), Naive Bayes (NB), and Decision Tree (DT) classifiers in identifying DDoS adversarial traffic in 5G. The comparative effectiveness of these models in terms of detectability, computational efficiency, and interpretability is investigated, and findings on how these can be used to build effective scalability and reliability in security for next-generation networks are discussed [3].

2 Literature Review

Artificial intelligence improves the efficiency of 5G networks by improving MIMO detection, precoding, and channel coding, using models such as DetNet, OAMPNet, QWF, LAMS, and NOMS. Zhang et al. (2020) [4] highlights the dual role of AI in improving performance and strengthening cybersecurity resilience. AI and machine learning address interference in dynamic 5G environments, such as un- manned aerial vehicles, the Internet of Things, and vehicular networks, by enabling real-time detection and secure resource allocation. In contrast to conventional methods, AI adjusts to diverse conditions, effectively handling both jamming and unintentional interference.

Arjoune and Faruque (2020) [5] emphasize AI's importance and highlight ongoing challenges. Tian et al. (2019) [6] also propose an AI-driven security framework that includes a knowledge graph, automated attack/defense tools, and a validation testbed. The framework improves threat detection, accuracy, and real-time response in 5G systems by integrating dynamic vulnerability mining, game theory, and reinforcement learning. Network slicing enables 5G to provide isolated, QoS-optimized networks, but there are limitations for complex services like V2X and IIoT (Wijethilaka Liyanage, 2021) [7].

B5G improves emergency response in smart cities with reliable, low latency communication and secure slicing (Rafique et al., 2024) [8]. It enables real-time services such as rescue drones and safeguards critical infrastructures. Sullivan et al. (2021) [9] identify security vulnerabilities in 5G, especially at the physical layer, and suggest employing PLS techniques to alleviate risks like eavesdropping. However, the development of technology and inconsistent standards still pose significant problems.

NIST, FAA, and CISA have identified threats like aviation disruptions and malware in supply chains. Mohan et al. (2022) [12] point to the need for strong cybersecurity frameworks. Olimid and Nencioni (2020) [13] also state that the detailed security evaluation of 5G slicing is limited by standards development and implementation complications. Maintaining confidentiality and isolation in multi-purpose environments poses significant challenges. 3 Secure MANO necessitates the deployment of AI-powered policy enforcement. Efficient trust management and device protection are critical to preventing breaches. The identified issues emphasize the need for further investigation into the security of 5G slicing.

Ahmad et al. [14] conducted a review of critical security measures for 5G and beyond, emphasizing the importance of layered approaches across the physical, network, and application layers, as well as strong authentication, privacy, and advanced cryptography. Dutta and Hammad [15] conducted a systematic review of security challenges and potential solutions in 5G networks, focusing on systemic vulnerabilities. Ahmed et al. [16] conducted a survey on the privacy and security requirements, as well as the challenges that come with secure 5G-enabled IoT. Benzaid and Taleb [17] emphasized AI's dual role as both a defensive tool and a potential cybersecurity threat in contexts beyond 5G.

Soldani [18] investigated the future of 5G security in the context of ICT, focusing on emerging threats and mitigation strategies. Sun et al. [19] created an automated framework for 5G security that integrates dynamic offensive and defensive strategies. Mathew [20] discussed network slicing in 5G and the security implications. Jover and Marojevic [21] looked into potential vulnerabilities in 5G protocols. Fang and Qian [22] proposed adaptable mechanisms to enhance security and privacy in 5G wireless networks. Suomalainen et al. [23] identified machine learning as a potential threat to 5G security and highlighted how adversaries exploit it.

Kim [24] identified security threats and classified attacks aimed at the 5G core network from a protocol perspective. Al-Sinayyid et al. [25] investigated phishing defense and attribution analysis techniques. Al-sinayyid [26] shared insights on detecting and mitigating advanced persistent threats. Al-sinayyid and Kadiyala [27] demonstrated the successful detection of IoT volume-based attacks using SVM techniques.

3 Design of an Algorithm

To design a suitable algorithm to recognize DDoS attacks in 5G networks, we first need to clearly define the problem and then take the dataset into a structured manner. This guarantees that the next steps of model training and evaluation are based in formulated goals and quality input data.

3.1 Problem Statement and Data Preprocessing

In 5G networks, Distributed Denial-of-Service (DDoS) traffic detection relies on statistical and protocol-level data to differentiate malicious from legitimate flows. The data used to represent each traffic flow itself is an example of a feature vector providing metrics like flow duration, packet size distribution, protocol type, and TCP flag counts. Hence, the key challenge is to handle large volumes of traffic efficiently, as well as building models to be able to generalize across different attacks. The dataset is preprocessed to be compatible with machine learning classifiers. Normalization of continuous features through standard scaling and conversion of categorical values into numerical columns are performed.

The Synthetic Minority Oversampling Technique (SMOTE) is also used to generate synthetic samples of minority classes to eliminate a class imbalance, that is between benign and attack flows. After preprocessing, the dataset is divided into training (80%) and testing (20%) samples using stratified sampling which is to maintain the ratio of attack to normal traffic. This allows the Random Forest, Decision Tree, and Naive Bayes classifiers to be trained on even-seeded data, and to be able to evaluate them in a consistent manner, measured based on accuracy, precision, and recalls.

3.2 Pseudocode

The pseudocode above represents the training of the Random Forest model, identifying relevant features, and real-time detection of DDoS traffic in 5G networks.

```
Require:
• D = (X, y): Network traffic dataset
• τ ∈ (0,1): test split ratio (default 0.2)
• k ∈ ℕ: top features to keep (default 10)
Ensure:
• ŷ: attack predictions on Xtest
• Acc, Prec, Rec: performance metrics

function DETECTATTACKS(D, τ, k)
  (Xtrain, Xtest, ytrain, ytest) ← Split(D, τ, stratified=TRUE, shuffle=TRUE)
  scaler ← Fit(StandardScaler, Xtrain)
  Xtr ← Transform(scaler, Xtrain);  Xte ← Transform(scaler, Xtest)
  RF ← TrainRandomForest(Xtr, ytrain)
  imp ← FeatureImportance(RF)
  Fsel ← TopK(imp, k)
  ŷ ← Predict(RF, Xte[:, Fsel])
  (Acc, Prec, Rec, F1) ← Evaluate(ŷ, ytest)
  return (ŷ, Acc, Prec, Rec)

function REALTIMEDETECTION(pkt, RF, scaler, Fsel)
  f ← ExtractFeatures(pkt)
  f ← Transform(scaler, f)
  predicted ← Predict(RF, f[Fsel])
  if predicted = "Attack" then TriggerResponse()
end if
end function
```

This strong architecture of the model is perfect for high-stakes cybersecurity issues like real-time intrusion detection which demands fast and dependable decision-making. It is highly scalable and therefore suitable for deployment across platforms due to its ability to handle complex data. The solution's scalability allows it to be integrated into intelligent decision-making systems, making it suitable for dynamic environments with large datasets that require real-time processing and analysis.

3.2.1 Input Data

Input of algorithm is described as follows:

$$X = \{x1,\ x2,\ \ldots,\ xk\}, Y = \{y_1, y_2, \ldots, y_k\} \quad (1)$$

where X represents the set of input features and Y represents the corresponding target labels.

3.2.2 Random Forest Stage

The Random Forest stage merges the individual features X with features ψ denoted by:

$$F = [X, \psi] \tag{2}$$

where:

1. X is the set of original features.
2. ψ is the additional features generated by the model.

Next, we train the Random Forest on X and Y, and then select important features or reduce noise. This can be formalized as the following optimization problem.

$$\text{Min} \sum_{i=1}^{n} f(\min \varsigma), \text{ subject to } \forall\theta \geq \in \tag{3}$$

$$\hat{y} = \frac{1}{T} \sum_{i=1}^{T} h_t(x) \tag{4}$$

where $h_t(x)$ denotes the prediction from the t-th decision tree in the forest, and T is the total number of trees.

Consequently, we have the following optimization function:

$$\Phi = f(x) = \begin{cases} \min(\frac{1}{2T}||\psi||^2 + C\sum_{i=1}^{T} h_t(x) \\ (\sqrt{\frac{1}{T}\sum_{i=1}^{T} h_t(x)})^2 \end{cases} \tag{5}$$

where:

1. ψ is the feature set,
2. $h_t(x)$ is the prediction function of the t-th tree,
3. T is the number of trees in the forest,
4. $\hat{y}$ is the aggregated prediction outcome

The program launches an automated classification process in Python, starting with labeled training data from a CSV file. The Random Forest trains multiple decision trees with different samples of data. After training, metrics like accuracy, precision, and recall evaluate the model. Once deployed, the model makes real-time predictions, with each decision tree casting a vote, initiating responses, and providing assistance.

3.3 System Architecture and Data Flow

The Random Forest (RF) based DDoS detection system follows the workflow illustrated in Fig. 1. It consists of data preprocessing, training, and classification stages, which work together to identify and mitigate malicious traffic in real time.

The process of the proposed system starts with loading raw network traffic data and proceeds with preprocessing such as normalization, feature extraction, and encoding. A Random Forest classifier (RF) is learned from the annotated traffic data, and if the performance requirements are not met, the training is performed again with an adjusted set of parameters or additional feature input, and finally an optimized model is established.

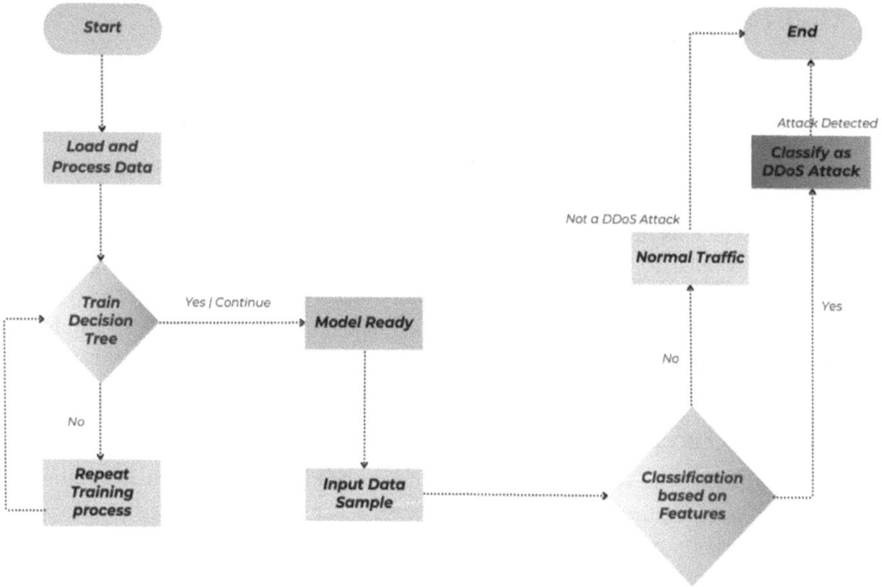

Fig. 1. Represents the process of DDoS attack detection using RF model

After it is settled, the model is used to classify the incoming real-time or batch traffic samples based on some selected statistical features, namely flow duration, packet size, and flag counts.

The RF model computes predictions from multiple decision trees for normal and malicious traffic. Normal traffic may proceed without any interference, while traffic found as being of Distributed Denial-of-Service (DDoS) nature can generate alerts and mitigation attempts such as rate limiting, blocking, or quarantining. Finally, the output segment labels each traffic flow as benign or attack traffic and carries on the automatic detection and response in the 5G security.

3.4 Application in Python(Random Forest)

In Python-based machine learning, after training the Random Forest model, it is capable of predicting class labels or probabilities for new data. A prediction that indicates malicious activity and exceeds a predetermined confidence threshold classifies the input as a potential threat. Packet size, flow duration, and flag counts can all help identify abnormal patterns. Decisions are aggregated across multiple boundaries in decision trees, allowing for more accurate classification. Random Forest demonstrates high reliability for real-time threat detection in a dynamic 5G network.

4 Model Evaluation and Testing

4.1 Dataset Overview

The dataset used in this study includes 6,171,315 samples as well as 12 columns of network traffic flows containing examples of benign and DDoS attacks captured using a 5G network slicing testbed. Each flow is described using statistical and protocol-level characteristics such as flow duration, packet length distribution, port numbers, protocol type, and flag counts. To ensure diversity in traffic conditions, the dataset is divided into multiple slices and collection days before being unified. This dataset is extremely stable after preprocessing with normalization, encoding, and label binarization for training and evaluating machine learning models applicable to DDoS detection in 5G environments.

4.1.1 Random Forest Algorithm Evaluation

The Random Forest (RF) model was trained on the 5G DDoS dataset using statistical and protocol-level features to differentiate between benign and malicious traffic. The properties included flow duration, packet size distribution, protocol type, port numbers, and TCP flag counts. The model formed an ensemble of decision trees in which each tree was trained on a bootstrapped subset of data with randomly chosen features at each split to capture complex patterns in traffic behavior.

During this time, the RF model focused on fine-tuning decision boundaries and used the majority voting method among the trees for improving the reliability of their classification. The final assessment results indicated that the Random Forest model achieved an accuracy of 0.9820 (98%), with perfect precision, recall on the test set. As demonstrated in this performance profile, the model classified all malicious flows, obtaining zero false positives, so that this model can be used for real-time detection of DDoS. In this paper, Random Forests proved to be the best classifier in 5G network security, displaying a near-perfect detection ability which achieved perfect accuracy.

4.1.2 Decision Tree Evaluation

The Decision Tree (DT) classifier was trained on the dataset utilizing essential traffic features, thereby creating a hierarchical decision rule for distinguishing between normal and attack flows. Each of the nodes corresponded to a condition, expressed in values of the features, rendering a transparent and interpretable way to classify.

Our performance evaluation had shown that Decision Tree with an accuracy of 0.864 (86%) had precision and recall than the ensemble-based Random Forest. While decision trees provide clarity and ease of understanding, their reduced accuracy suggests a tendency for misclassifying certain flows, highlights their vulnerability to overfitting when utilized in isolation.

4.1.3 Naive Bayes Evaluation

The Naive Bayes (NB) model was trained on the 5G DDoS dataset with statistical and protocol-level features like flow duration, packet size distribution, protocol type, port numbers, and TCP flag counts. NB applied a probabilistic framework which assumes that features have conditional independence, which allowed the training for the NB to quickly calculate class probabilities providing lightweight classification models for high-throughput network environments. The straightforward nature of the system made for rapid inference and hence is considered a viable model in real-time detection cases.

In the final evaluation, the Naive Bayes model achieved an accuracy of 0.932 (93%) and was able to achieve balanced precision, recall, as well as F1-score values. This is an indication that the model could reliably identify most malicious flows. NB did not achieve the exact performance of Random Forest, but it was a robust and efficient technique to detect DDoS in 5G networks (Fig. 2).

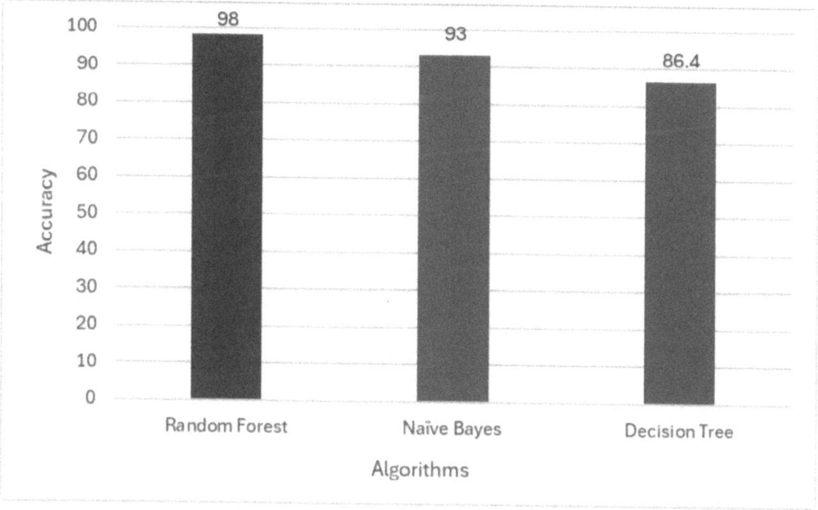

Fig. 2. Represents Accuracy comparison of Random Forest, Decision Trees, and Naive Bayes algorithms.

The comparison shows that the Random Forest classifier reached 98% accuracy, proving that it worked for traffic with many types. The Naive Bayes is known to be an effective naive statistical model with an accuracy of 93%. The Decision Tree achieved 86% overall accuracy, indicating it is easy to explain, but it is prone to overfitting compared to ensemble methods. This means that ensemble techniques such as Random Forest perform better than other methods in detecting DDoS attacks that occur in the 5G domain.

In order to validate the performance of the proposed models, we calculated their F1-scores, an important value that integrates precision and recall. The F1-score is a useful result for achieving a balanced performance evaluation of the model by balancing the number of false positives and false negatives, and therefore this F1-score is crucial when a classification model is tested for network security tasks. That is especially important for 5G based DDoS detection because accurate identification of malicious traffic and minimal misclassification of benign flows are critical to maintaining service reliability.

The F1-score is computed using the below equation.

$$F1 = 2 \times \frac{(Precision \times Recall)}{(Precision + Recall)} \tag{6}$$

To determine the effectiveness of proposed DDoS detection models. Table 1 presents an analysis of RF, DT, NB classifiers by evaluation metrics such as accuracy, precision, recall, and F1-score There is a clear numerical comparison in this tabular representation, showing that the Random Forest model outperforms their accuracies across all metrics.

Table 1. Comparison of Metrics across models

Model	Accuracy	Precision	Recall	F1-Score
RF	0.982	0.981	0.982	0.98
DT	0.86	0.85	0.86	0.86
NB	0.93	0.931	0.932	0.93

Moreover, Fig. 3 presents metrics for performance, allowing an easier comparison of the models' relative advantages and disadvantages. The grouped bar chart shows the accuracy, precision, recall, and F1-score of the three classifiers, supporting the numerical information shown in Table I. However, the Random Forest (RF) model demonstrates superior performance across all metrics, with accurate detection of DDoS attacks.

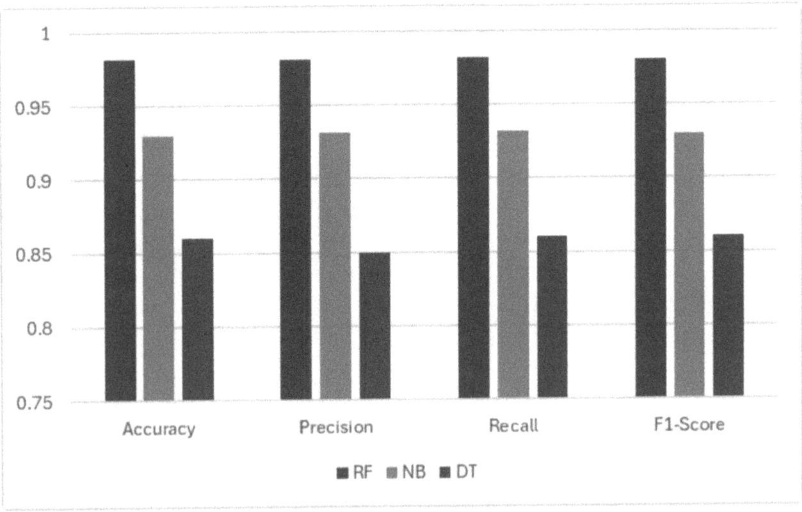

Fig. 3. Performance comparison of RF, NB, and DT models for DDoS detection in 5G networks.

5 Final Algorithm Evaluation

Comparing the three classifiers allowed for more specific distinctions regarding their respective fit for detecting DDoS in 5G networks. Random Forest performed significantly well, achieving 98% accuracy and stable classification metrics, demonstrating its suitability for deployment in high-assurance environments. More importantly, Decision Tree attained an impressive 86% accuracy, making it very informative, because of its hierarchy of rules, which gives full transparency to the users of the network, but also showed less accuracy compared to ensemble-based methods whose predictive power is less consistent.

This yields a performance level of 93% accuracy for Naive Bayes, which achieves the trade-off between computational efficiency and detection, thus being particularly useful in resource-constrained environments like edge deployments. Combined these results make the position of Random Forest as the preferable model for robust detection, while Naive Bayes provides a lightweight alternative and Decision Tree adds critical interpretability. Its complementary strengths point to the possibility of a hybrid or layered strategy to further increase the resilience to more variety and changing patterns of DDoS attacks in the 5G environment.

6 Future Work

Future advancements in DDoS detection within 5G networks may be significantly enhanced through the application of hybrid learning methods that integrate machine learning and deep learning techniques. Ensemble models such as Random Forest effectively manage various traffic features; however, deep learning models like CNNs and RNNs excel at capturing intricate spatial and temporal dependencies within network flows. A combined framework enhances adaptability to evolving attack patterns. Using federated and online learning methods can also improve models' real-time adaptability by allowing them to keep up with new traffic patterns without having to be retrained in a central location. Using these new technologies, future detection systems can become more scalable, resilient, and efficient in preventing advanced DDoS attacks in next-generation networks.

7 Conclusion

This research provides a clear structure of the performance levels of these models in the evaluation for DDoS detection in 5G networks. Random Forest showed excellent results with highest accuracy with 98%, which shows the robustness in model performance for capturing complex traffic patterns and reliability for real-time deployment Naive Bayes achieved 93% accuracy while maintaining effectiveness and computational power. In a low-capability detection environment, Decision Tree achieved 86% accuracy and provided interpretability.

The findings show that ensemble-based methods perform best, whereas simpler models excel in scenarios requiring speed and transparency. The primary contribution of this study is the ability to identify potential nodes for various machine learning models in 5G security architectures. Random Forest is a reliable choice for high-assurance detection, Naive Bayes provides a lightweight representation when resources are limited, and Decision Tree enhances explainability for operational insight. Taken together, these findings emphasize the importance of striking a balance between algorithmic characteristics and deployment scenarios when developing scalable, adaptive, and resilient defensive measures against next-generation DDoS threats. This research advances our understanding of model trade-offs for 5G security and lays the groundwork for adaptive 5G solutions that can protect critical services in the future.

Acknowledgment. The authors would like to acknowledge the support of the US Department of Education under sub-award grant number R2301364. The DoE support made it possible to conduct this research and to disseminate the findings. The authors are grateful for the DoE's commitment to supporting cutting-edge research.

References

1. Kavitha, D., Thejas, S.: AI enabled threat detection: leveraging artificial intelligence for advanced security and cyber threat mitigation. IEEE Access (2024)
2. Benzaid, C., Taleb, T.: AI for beyond 5G networks: a cyber-security defense or offense enabler? IEEE Network **34**(6), 140–147 (2020)
3. Kaur, J., Khan, M.A., Iftikhar, M., Imran, M., Haq, Q.E.U.: Machine learning techniques for 5G and beyond. IEEE Access **9**, 23472–23488 (2021)
4. Zhang, C., Ueng, Y.L., Studer, C., Burg, A.: Artificial intelligence for 5G and beyond 5G: implementations, algorithms, and optimizations. IEEE J. Emerging Sel. Top. Circu. Syst. **10**(2), 149–163 (2020)
5. Lohan, P., Kantarci, B., Ferrag, M.A., Tihanyi, N., Shi, Y.: From 5G to 6G networks, a survey on AI-Based jamming and interference detection and mitigation. IEEE Open J. Commun. Soc. (2024)
6. Tian, Z., Sun, Y., Su, S., Li, M., Du, X., Guizani, M.: Automated attack and defense framework for 5G security on physical and logical layers (2019). arXiv preprint arXiv:1902.04009
7. Wijethilaka, S., Liyanage, M.: Survey on network slicing for Internet of Things realization in 5G networks. IEEE Commun. Surv. Tutorials **23**(2), 957–994 (2021)
8. Rafique, W., Barai, J., Fapojuwo, A.O., Krishnamurthy, D.: A survey on beyond 5g network slicing for smart cities applications. IEEE Commun. Surv. Tutorials (2024)
9. Sullivan, S., Brighente, A., Kumar, S.A., Conti, M.: 5G security challenges and solutions: a review by OSI layers. IEEE Access **9**, 116294–116314 (2021)
10. Fang, L., Zhao, B., Li, Y., Liu, Z., Ge, C., Meng, W.: Countermeasure based on smart contracts and AI against DoS/DDoS attack in 5G circumstances. IEEE Network **34**(6), 54–61 (2020)
11. Moglia, A., et al.: 5G in healthcare: from COVID-19 to future challenges. IEEE J. Biomed. Health Inform. **26**(8), 4187–4196 (2022)
12. Mohan, J.P., Sugunaraj, N., Ranganathan, P.: Cyber security threats for 5G networks. In: 2022 IEEE International Conference on Electro Information Technology (eIT), (pp. 446–454). IEEE (2022)
13. Olimid, R.F., Nencioni, G.: 5G network slicing: a security overview. IEEE Access **8**, 99999–100009 (2020)
14. Ahmad, I., Shahabuddin, S., Kumar, T., Okwuibe, J., Gurtov, A., Ylianttila, M.: Security for 5G and beyond. IEEE Commun. Surv. Tutorials **21**(4), 3682–3722 (2019)

15. Dutta, A., Hammad, E.: 5G security challenges and opportunities: a system approach. In: 2020 IEEE 3rd5G World Forum (5GWF) (pp. 109–114). IEEE (2020)
16. Ahmed, S.F., et al.: Toward a secure 5G-enabled internet of things: a survey on requirements, privacy, security, challenges, and opportunities. IEEE Access **12**, 13125–13145 (2024). https://doi.org/10.1109/ACCESS.2024.3352508
17. Soldani, D.: 5G and the future of security in ICT. In: 2019 29th International Telecommunication Networks and Applications Conference (ITNAC) (pp. 1–8). IEEE (2019)
18. Sun, Y., Tian, Z., Li, M., Zhu, C., Guizani, N.: Automated attack and defense framework toward 5G security. IEEE Netw. **34**(5), 247–253 (2020)
19. Mathew, A.: March. network slicing in 5G and the security concerns. In 2020 Fourth International Conference on Computing Methodologies and Communication (ICCMC) (pp. 75–78), IEEE (2020)
20. Jover, R.P., Marojevic, V.: Security and protocol exploit analysis of the 5G specifications. IEEE Access **7**, 24956–24963 (2019)
21. Fang, D., Qian, Y.: 5G wireless security and privacy: architecture and flexible mechanisms. IEEE Veh. Technol. Mag. **15**(2), 58–64 (2020)
22. Suomalainen, J., Juhola, A., Shahabuddin, S., M¨ammel¨a, A., Ahmad, I.: Machine learning threatens 5G security. IEEE Access **8**, 190822–190842 (2020)
23. Kim, H.: 5G core network security issues and attack classification from a network protocol perspective. J. Internet Serv. Inf. Secure. **10**(2), 1–15 (2020)
24. Al-Sinayyid, A., Ali Jewel, M.J., Mannuru, V., Sasidhar, K.: Defending characteristics and attribution analysis for phishing attacks. In: 2023 International Conference on Computational Science and Computational Intelligence (CSCI), pp. 868–874 (2023). https://doi.org/10.1109/CSCI62032.2023.00145
25. Alsinayyid, A.: Analytical study on advanced persistent threat detecting, defending and mitigating. https://doi.org/10.13140/RG.2.2.15169.67687
26. Al-Sinayyid, M.J.A.J., Mannuru, V., Sasidhar, K.: Analytical study for cyber security attacks and defenses characteristics. In: 2023 International Conference on Computational Science and Computational Intelligence (CSCI), pp. 853–860 (2023). https://doi.org/10.1109/CSC I62032.2023.00143 . IEEE
27. Alsinayyid, A., Kadiyala, S.: SVM Enhanced Detection of Volume-Based Attacks in IoT Networks. https://doi.org/10.13140/RG.2.2.14330.81600

Zero Trust Architectures and Artificial Intelligence

Syed Noaman Ali[1(✉)], Joe Francom[1], and Alin Zamfiroiu[1,2,3]

[1] Utah Tech University, Saint George, UT 84770, USA
{syed.ali,joe.francom,alin.zamfiroiu}@utahtech.edu
[2] Bucharest University of Economic Studies, 010374 Bucharest, Romania
[3] National Institute for Reseach and Development in Informatics–ICI Bucharest, 011455 Bucharest, Romania

Abstract. This paper examines Zero Trust Architectures (ZTA) and the vital role of Artificial Intelligence (AI) in modernizing cybersecurity. The ZTA model, proposed by Kindervag in 2010, represents a fundamental shift from traditional perimeter-based security. The central premise is that no entity (user, device, or application) is inherently trustworthy, regardless of location, and trust must be continuously assessed.

AI and Machine Learning (ML) are essential for processing the vast volume of data generated by continuous ZTA monitoring. This integration (i-ZTA) transforms ZTA into a more dynamic and context-aware framework. AI-based behavioral analytics monitor activity and detect deviations, enabling continuous assessment of trustworthiness.

Major challenges include vulnerability to adversarial attacks, model bias, and lack of transparency (opacity) of AI decisions. ZTA + AI has proven effective in financial services (fraud detection) and healthcare (microsegmentation). The paper contributes a synthesized roadmap of AI-enabled ZTA deployment across multiple industries which tackles common implementation barriers and suggests mitigation strategies.

Keywords: ZTA · cybersecurity · AI · machine learning · challenges

1 Introduction

Traditional security models have been more permissive of assets within the organization and have focused on the prevention of outward threats as the primary prevention necessity. The primary tenets of zero trust move the focus from perimeter-based awareness and enforcement to ensuring that all devices, regardless of location, are authorized to perform intended actions (Kindervag, 2010). Kindervag further proposed that all sources must be verified and secured, access controls must be appropriately regulated, and network traffic should be logged and monitored (Kang, 2023). All these actions should be performed continuously.

H. R. Arabnia et al. (Eds.): CSCI 2025, CCIS 2969, pp. 518–531, 2027.
https://doi.org/10.1007/978-3-032-28460-0_37

We have identified the key components of a zero-trust model to include the following:

- Who. This could refer to a user, process, application, service, or other network flow that is trying to access a resource or data. Being able to identify who is making a data request is critical. Identity verification and authentication are critical pieces of any security model. Some traditional methods to accomplish this are through certificates, single sign on (SSO), or MFA (Kandula et. al).
- What: Data or actions that could manipulate data fit into this component. Operations like read, write, execute, update, delete need to be validated. Typically, this is accomplished to access controls (U.S. Department of Defense, 2022). A least-privilege mode of access is key. Data will need to be secured in transit and at rest (Cunningham).
- When: While not necessarily referring to a time stamp, this could also relate to whether an action fits within a set of expected actions. It can include behavioral context or policy evaluation time frames. This can include frequency of requests, whether access is taking place during an allowed access window, or unusual access patterns.
- Where: A key piece of the zero-trust model is to not assume trust based on location. This refers to the origin of the request, typically in the form of network or physical location. Traditional methods of control involve firewalls, VPN, network segmentation, or intrusion-detection systems (Darshane et. al). In a zero-trust model, the types of controls are enhanced through micro segmentation (Mavroudis), geolocation, and other context-aware policies.
- How: This piece addresses whether the action aligns with allowed workflows. An analysis of whether a particular action conforms to expected organizational policy is imperative.

While identity, action, and location considerations do exist in traditional models they exist as a once at time of initial access. The zero-trust model also mandates that these steps are not done in isolation, trustworthiness is evaluated constantly (Liu et. al). The visibility and continuous analysis of the above components are also critical (Cunningham, 2018).

While the above principles identify the core precepts of ZTA, organizations rely on structured frameworks to aid in implementation. Two of the most widely recognized frameworks are NIST SP 800-207 and Forrester's Zero Trust EXtended (ZTX).

2 Foundations of Zero Trust Architecture

The motivation behind the Zero Trust Architecture (ZTA) model is that no computing resource should be inherently trusted regardless of location inside or outside an organization. Trust should be continually evaluated (NIST, 2020).

NIST outlines several principles and definitions for the ZTA framework. Resources are defined as any digital device that can access organizational data. The framework requires secure communications irrespective of location and ensures that access to a resource is only granted on a per-session basis and according to the least privilege model. A dynamic policy will control access to resources with no assumption of any device being inherently trusted. All assets are continuously checked, monitored, and updated before being allowed access to organizational digital resources. Authentication is required for access to any and all resources. Digital asset data is continuously aggregated and evaluated to improve organizational security posture.

Forrester's Zero Trust eXtended (ZTX) framework expands on the ZTA model by suggesting 4 tenets that should be included within an organization to implement a good zero trust posture: strategy, capabilities, technology, and features (Cunningham). Organizations must first decide on what their security goals are and how it fits within their ecosystem. An entity must understand what capabilities exist and whether they align with policy to make appropriate security decisions. Tools can then be selected and different features enabled or evaluated.

ZTX prioritizes data as the primary asset to protect. All data access requests must be controlled by Policy Enforcement Points (PEPs) and Policy Decision Points (PDPs) using an identity and a metadata-centric policy model. (Garbis & Chapman, 2021). Each user, device and network flow is continuously authenticated and authorized, not just at the perimeter. Access is dynamically calculated based on a variety of factors such as identity, device health, location, and behavior. Access to organizational resources utilizes the least-privilege rule.

While both the NIST and ZTX models share the same assumption that no device or network should be inherently trusted, they differ on semantics. NIST emphasizes monitoring device integrity to ensure it has not been compromised. ZTX puts stronger emphasis on the data itself as the central asset. In practice, organizations might elect to follow principles from both frameworks. NIST's posture on securing devices provides strong protection against a compromised endpoint, while Forrester's model addresses the need to tightly control data access in cloud and hybrid environments.

In the table below we compare the traditional security model (Sarkar et al., 2022) with a ZTA model (Kindervag, 2010) (Table 1).

Table 1. Traditional Security model vs ZTA model (Kindervag, 2010)

Dimension	Traditional Security Model	Zero Trust Architecture (ZTA)
Core Assumption	Internal assets are trusted; protect from external threats	No device, user, or application is inherently trusted, regardless of location
Security Focus	Perimeter defense (firewalls, IDS/IPS)	Continuous verification of identity, device, and behavior
Access Control	Broad access once inside the network; limited re-authentication	Least-privilege access granted per session, dynamically adjusted
Verification Frequency	One-time, typically at login or entry point	Ongoing, continuous authentication and authorization

(continued)

Table 1. *(continued)*

Dimension	Traditional Security Model	Zero Trust Architecture (ZTA)
Network Design	Flat or perimeter-segmented networks	Microsegmentation, context-aware policies, and dynamic trust zones
Resource Protection	Protects the "castle" (network perimeter)	Protects individual assets (data, apps, devices)
Adaptability	Static policies; slower to adapt to cloud or remote work	Dynamic, designed for hybrid cloud, mobile, and IoT environments

The traditional security model is certainly easier and cheaper to implement and manage in the short run, but having a more secure organizational posture will be more beneficial in the long run. ZTA also allows for more flexibility, more dynamic adaptation, and heightened security. The advent of artificial intelligence (AI) could reduce the complexities and cost of the traditional ZTA architecture to make it more feasible for implementation.

3 Tools Used in AI and Zero Trust Architecture

The underlying premise of ZTA is that trust will be evaluated continuously. This generates a vast amount of security data from users, devices, applications, and network flows. Current AI and ML trends can be integrated as analytical tools to aid in the processing of this data and facilitate the implementation of a true ZTA (Hussain et al, 2024).

Ramezanpour et al. present an intelligent zero trust architecture model (i-ZTA) that integrates AI to make intelligent dynamic trust assessments. Attacks can be detected and mitigated by utilizing machine learning to parse the substantial amounts of data collected. By leveraging AI-driven analytics, ZTA can evolve from static policy enforcement to a more adaptive, context-aware security framework (Ramezanpour et al, 2022).

Behavioral analytics involves the analysis of user behavior trends and fluctuations. Typically, they center around predictable patterns such as login time or data access and usage trends to make a trust decision. The advent of AI and ML has made the analysis of this data much more feasible and easier to incorporate into the ZTA model (Sina, 2025).

AI-powered behavioral analytics make the ZTA requirement of continuous trustworthiness evaluation possible. They can monitor user and device activity, record baselines, and detect deviations. Machine learning models can identify abnormal login patterns, or unusual access requests.

AI models in ZTA are vulnerable to adversarial attacks, where small, carefully crafted input can cause misclassification or bypass security controls. These attacks can target both the model's predictions and its explanations, allowing attackers to manipulate or conceal malicious activity, or extract confidential data (Olutimehin et al.).

Bias within a system can be induced through model training data, or poor model design. These can lead to unfair or inaccurate decisions in ZTA which compromises model efficacy.

If AI models are too opaque, it will be difficult to assess the efficacy. When a trustworthiness decision is made, it is important to be able to quantify why access was granted or denied. In ZTA, it is important to make the correct decision, but it is also important to validate why a particular decision was made. Without this transparency, it is difficult to assess if the model is working as intended.

Several mitigation strategies have been proposed for mitigating AI threats. Training the AI model on adversarial examples can help the model become more robust when learning how to deal with legitimate threats (Olutmehin et al.) (Kumar, 2022). This method of training is expensive due to the need for a large dataset and compute power. Preprocessing using deep neural networks to transform the data before feeding it to the AI model is also suggested as a mechanism for mitigation. Craighero et al. suggest a detection mechanism framework that combines multiple detection methods to aid against adversarial attacks using various scoring functions. (Craighero et al., 2023).

Combining multiple defense mechanisms creates a layered and more robust security posture. These hybrid methods have been shown to be more successful than single-method defenses against current and evolving attack strategies. (Olutimehin et al.) (Petihakis et al.)

Dynamic trust fluctuates based on interactions and experiences. The integration of AI into security systems influences the amount of trust placed into the system. Dynamic trust is based on factors such as transparency, explainability, system performance, and contextual relevance. Trust in AI decisions increases over time as users gain experience with the system (Cabiddu et al., 2022).

The effective deployment of Zero Trust Architecture (ZTA) enhanced with AI relied not only on phased strategies but on the availability of robust tools, technologies, and reference architectures. These enablers translate conceptual principles into practical implementations by providing identity control, continuous monitoring, adaptive threat detection, and a scalable environment. The following table highlights the most common tools and reference architectures identified in literature and practice that enable ZTA + AI deployments (Table 2).

These technologies form the building block of Zero Trust deployments and illustrate the ecosystem available for AI Integration.

Table 2. Most common used tools

Category	Common Tools / Technologies	Role in ZTA + AI	References
Identity & Access Management (IAM)	Okta, Azure AD, Ping Identity; MFA; Privileged Access Management (PAM)	Centralizes identity, enforces phishing-resistant MFA, enables AI-driven adaptive authentication	(; Ebuka Mmaduekwe Paul et al., 2024)
AI-Driven Threat Detection	Darktrace, CrowdStrike Falcon, Vectra AI; Google Chronicle; Microsoft Sentinel	AI/ML-powered anomaly detection, insider threat detection, automated response	(Alevizos et al., 2022; Alnaim and Alwakeel, 2025)
Endpoint & Network Security	Endpoint Detection & Response (EDR/XDR); Next-Gen Firewalls; Secure Access Service Edge (SASE)	Monitors device posture, isolates compromised endpoints, enforces Zero Trust perimeters	(Bradatsch et al., 2023; Ebuka Mmaduekwe Paul et al., 2024)
Behavioral Analytics & UEBA	UEBA platforms integrated into SIEM; Continuous Authentication (keystroke, biometrics)	Builds behavioral profiles, detects anomalies, continuously verifies user/device identity	(Alnaim and Alwakeel, 2025; Ebuka Mmaduekwe Paul et al., 2024)
Microsegmentation & ZTNA	VMware NSX, Cisco ZTNA, BeyondCorp	Isolates network segments, enforces least-privilege access, blocks lateral movement	(Ajish, 2024)
SIEM / SOAR Platforms	Splunk, Microsoft Sentinel, Elastic Security	Centralizes telemetry, applies AI-driven correlation, automates incident response workflows	(; Bradatsch et al., 2023)
AI Governance & Explainability	SHAP, LIME (XAI tools)	Increases transparency and trust in AI-driven ZTA decision-making	(Adnan Alvi and Jalil, 2025; Galli et al., 2024)

(*continued*)

Table 2. (*continued*)

Category	Common Tools / Technologies	Role in ZTA + AI	References
Privacy & Trust Enhancers	Federated Learning, Blockchain Identity, Zero-Knowledge Proofs (ZKP), Trusted Execution Environments (TEE)	Ensures data privacy, secure model training, decentralized trust validation	(Jose Diaz Rivera et al., 2024; Kato et al., 2023)
Reference Architectures	NIST SP 800-207A; CISA ZTMM; Google BeyondCorp; Microsoft ZTA; Cisco ZTA	Provide structured frameworks for phased ZTA adoption with AI integration	(; Ajish, 2024)

4 Real-World Applications and Case Studies

4.1 Financial Services

The financial sector encounters the most sophisticated cyber threats, which include high-velocity fraud schemes and complex social engineering attacks. Security models based on VPNs and flat networks expose institutions to lateral movement risks after penetrating perimeter defenses (Rose et al., 2020). The adoption of Zero Trust and AI-driven controls faces significant challenges. Compliance pressure and complexity with regulatory requirements such as PCI DSS 4.0, GDPR, and SWIFT CSCF v2024 impose stricter authentication, encryption standards, and real-time anomaly detection requirements ("PCI DSS," 2024; "Palo Alto Networks," n.d.). The protection of digital identities for customers, employees, and third parties is becoming increasingly complex due to emerging threats, including deepfake-enabled fraud, session hijacking, and adversarial AI threats. The implementation of strict security controls faces two significant barriers: employee and organizational resistance, and insufficiently skilled AI professionals (Kapoor, 2025).

The combination of Zero Trust with AI, implemented in phases, provides organizations with the most effective defense mechanism. Securing critical assets begins by adhering to NIST 800-207 guidelines. The foundation of Identity and Access Management (IAM) with multi-factor authentication (MFA) and behavioral biometrics verifies all requests in real-time (Daah et al., 2024). Financial service organizations that implement ZTA and AI systems are reported to achieve quantifiable enhancements in their ability to detect fraud and maintain regulatory compliance. AI-powered anti-money laundering (AML) and fraud detection systems identify 2–4 times more suspicious activities while achieving a 60% reduction in false positives.

The experience across the sector demonstrates that Zero Trust operates as a program rather than a product, as full-scale deployment requires coordination between security, IT, compliance, and business units. The "start small" approach, securing the most critical assets first before scaling, is more sustainable than attempting a full rollout at once. The high cost of implementation, combined with existing infrastructure limitations, presents a significant obstacle to ZTA adoption. The combination of ZTA and AI creates an adaptive security system; however, organizations must maintain sustained investment, governance, and organizational alignment to achieve maximum benefits.

4.2 Healthcare

Health sector systems are increasingly digitalized and heavily depend on Electronic Health Records (EHRs), cloud services, and thousands of Internet of Medical Things (IoMT) devices. The healthcare sector faces significant challenges because hospitals still rely on outdated IT infrastructures and vendor-managed devices. The implementation of compliance standards, such as HIPAA, GDPR, and new U.S. HHS 405(d)/HICP best practices, requires healthcare organizations to implement stricter access controls, monitoring, and segmentation (Vukotich, 2023). A layered strategy that combines Zero Trust Architecture (ZTA) with AI-driven analytics offers enhanced resilience. Strong identity controls are essential for adaptive multi-factor authentication (MFA) for clinicians and administrators, as well as certificate-based workload authentication and device posture contextual checks. AI detects real-time anomalies through its analysis of traffic patterns and user/device activities. Hospitals can train AI models through federated. AI-enabled continuous dynamic trust scoring systems use user role, location, and urgency levels to grant permissions. Adopting ZTA with AI produces measurable improvements across healthcare environments. The implementation of microsegmentation and policy enforcement in hospitals leads to enhanced endpoint security, resulting in reduced lateral movement and faster isolation of compromised endpoints. Operational resilience is strengthened through speedier incident response, minimized downtime, and ensured continuity of clinical workflows. The research, along with pilot programs, yielded several critical lessons. First, the system must strike an optimal balance between security and usability, as clinicians require easy access to their workflows through single sign-on (SSO), biometrics, or adaptive authentication, rather than complex protocols. Second, AI systems require human oversight. Finally, cost and scalability issues persist in the healthcare sector, as the industry requires more skilled cybersecurity experts and faces high costs when replacing legacy systems with the latest technologies (Zakhmi et al., 2025).

4.3 Government and Defense

Government and defense networks experience particular security threats from nation-state adversaries, cybercriminals, and insider threats. Attackers target their classified networks, operational technology (OT), industrial control systems (ICS), and defense supply chains, often by exploiting identity compromise and lateral movement techniques. Traditional perimeter defense systems do not provide adequate protection for modern hybrid, multi-cloud, and coalition networks. The primary obstacle to the adoption of new technologies is that defense platforms and OT protocols were never designed with modern authentication or patching cycles in mind (Zakhmi et al., 2025). The broad spectrum of devices that encompass IoT and weapon support systems creates integration difficulties. Organizations resist implementing a "never trust, always verify" model.

Finally, AI-enabled threats (deepfakes, automated intrusion, adversarial evasion) create a new competitive environment between attackers and defenders.

Identity-centric controls implement phishing-resistant MFA, continuous decision attestation, and role-based microsegmentation to protect users, workloads, and data. The OT/ICS domains achieve secure remote maintenance access through identity-aware gateways and brokered encrypted sessions, which enable high-availability processes. AI enhances ZTA capabilities through its ability to run User and Entity Behavior Analytics (UEBA), real-time anomaly detection, and automated triage through SOAR platforms (Katiyar et al., 2024). The Thunderdome program of DoD protects defense operations through ZTA integration with AI and software-defined networking. The system implements automated policy enforcement to maintain uniformity across multiple cloud platforms and classified domains.

According to the research, the correct implementation of ZTA + AI results in measurable improvements in the resilience of defense systems.. AI-based detection systems and automated response protocols reduce the mean time to detect/respond, enabling organizations to combat credential theft, supply-chain attacks, and insider threats. It also ensures that vendors must demonstrate that their products follow "secure-by-design" controls to be eligible for procurement (Fluchs et al., 2023). ZTA + AI enables mission continuity through its ability to support safer remote operations, protect critical infrastructure uptime, and coalition data exchanges.

The government and defense sector provide essential learning experiences. First, ZTA success depends on treating it as a transformation process rather than a tool. Second, due to legacy and OT constraints, ZTA controls are typically applied at gateways or enclaves rather than being retrofitted onto fragile systems (Phiayura and Teerakanok, 2023). Third, balancing security with mission usability is critical; overly rapid policies can disrupt operations or frustrate users. Interoperability across siloed systems and coalition environments remains challenging, necessitating flexible architectures and standards-based integration. Finally, the DoD possesses the financial resources to support enterprise-wide orchestration; however, smaller agencies struggle to obtain both the budget and expertise. The main takeaway is that ZTA with AI provides strong defense resilience; however, organizations must sustain funding, cultural transformation, and flexible governance systems to achieve success.

4.4 Education and SMEs

Educational institutions and SMEs encounter similar difficulties because they operate under low budgets, outdated systems, and insufficient IT personnel. The combination of multiple identities across staff members, students, contractors, and personal devices through BYOD creates challenges in maintaining consistent security policies. SMEs face obstacles to implementing advanced controls because they lack sufficient funds and skilled personnel. The combination of cultural resistance and insufficient cyber literacy skills makes users disable MFA and disregard security protocols, which prevents ZTA implementation. The implementation of AI systems introduces new security risks, including biased operations, and difficulties in understanding system decisions.

The most efficient approach to security control implementation involves a cost-efficient deployment process that follows NIST SP 800-207 guidelines. The first set of accomplishments includes deploying phishing-resistant MFA, centralizing identity providers, and applying least privilege access controls.. The deployment becomes more feasible by incorporating Federated learning and cloud-based managed services without moving sensitive student or customer data (Kumar, 2025). SME can rely on open-source IAM solutions (like Open Policy Agent) or low-cost SaaS security tools that include AI-driven EDR and automated remediation. Strategic roadmaps that prioritize identity first, followed by device, network, application, and data, provide a structured way to phase in adoption while staying within the budget limits.

When implemented incrementally, ZTA + AI offers significant risk reduction for every dollar invested. AI-based anomaly detection systems enhance incident response operations by reducing false positives and streamlining the incident assessment process. SMEs achieve GDPR, PCI DSS, and HIPAA compliance through ongoing logging operations and systems that enable auditable identity management. Educational institutions gain from stronger data privacy protections that align with FERPA, safeguard student records, and support secure cloud adoption.

The first step in deploying a Zero Trust program should focus on critical applications, as this approach delivers quick results before expanding to other areas. AI models require human oversight because they are prone to drift, bias, and adversarial evasion. Security controls achieve success through balanced measures of cultural adoption and technological implementation, since staff training, leadership support, and precise security controls result in long-term, sustainable adoption (Wang et al., 2025). The research requires additional empirical studies that focus on SMEs and small schools because current ZTA frameworks were created for enterprises and need modifications for smaller organizations.

5 Challenges and Limitations

The shift to Zero Trust Architecture (ZTA) represents a fundamental paradigm shift that is used in cybersecurity. By this adoption we are moving away from traditional perimeter-based defenses to a model where no entity is implicitly trusted (Lee and Park, 2025).

The transition from traditional-based security to Zero Trust Architecture (ZTA) presents formidable challenges, particularly when organizations attempt to integrate it with their existing legacy systems that are based on the traditional mechanism (Mensah, 2024). These older systems, which support many critical operations, were not designed with modern cybersecurity paradigms in mind and often lack the flexibility and features necessary for seamless ZTA adoption. The complexity of this integration is a recurring theme across various industries, from general enterprise environments to critical infrastructure sectors.

Legacy systems are prevalent in many organizations, including critical infrastructure sectors such as energy, transportation, and manufacturing (Ojo, 2025). These systems often rely on outdated hardware and software, some developed decades ago, without modern cybersecurity features built in. The traditional "castle and moat" security model, which assumes implicit trust in the network perimeter, has made these systems highly vulnerable to modern cyber threats, especially insider threats and lateral movements.

The integration of ZTA with legacy systems also carries significant operational and resource challenges such as:

- **High Implementation Costs**: Deploying ZTA is technologically complex and resource-intensive. Initial setup costs can be substantial, particularly for organizations heavily reliant on extensive legacy systems. These expenses include purchasing and deploying new hardware, software, and security tools, as well as integrating them into existing IT and Operational Technology (OT) infrastructures (Abbas and Gul, 2025).
- **Resource Strain and Performance Bottlenecks**: The continuous monitoring, verification, and enforcement required by ZTA can strain existing system resources, especially in large-scale or resource-constrained environments. This can lead to performance bottlenecks and latency, impacting operational efficiency (Bradatsch, et al., 2022).
- **Skills, Expertise, and Training Gaps**: Implementing ZTA, especially in sectors like critical infrastructure where there is a traditional focus on OT over IT, requires specialized knowledge of both domains, which is often lacking internally. This leads to a steep learning curve and potential reliance on external consultants, further increasing costs and prolonging implementation timelines (Ojo, 2025).
- **Maintenance Costs and Security Debt**: Maintaining and securing aging infrastructure, particularly when retrofitting with new security measures, can be costly. Organizations may prioritize maintaining functionality, accumulating a "security debt" that makes future ZTA adoption even harder (Ojo, 2025).
- **Backward Compatibility Challenges**: ZTA deployments face challenges in maintaining backward compatibility, specifically in enforcing advanced security features without excluding or overwhelming older devices that lack support for modern security mechanisms (Li, et al., 2025).

The complexity of integrating ZTA with legacy systems is a multi-dimensional challenge, encompassing technical incompatibilities, significant resource demands, and the need for careful strategic planning. Successfully navigating these complexities requires a thorough understanding of legacy system limitations, a commitment to phased and adaptive implementation, and a proactive approach to managing the associated costs and skill gaps.

To address these challenges, organizations must adopt integrated strategies. Implementing robust data governance frameworks, such as the Zero Trust Data Quality Framework (ZTDQF), can ensure continuous data validation and verification (Mohammed, et al., 2021). Privacy-oriented AI techniques–such as Federated Learning, differential privacy, homomorphic encryption, and secure multi-party computation–can protect sensitive data. At the same time, robust aggregation mechanisms in Federated Learning, the use of Explainable AI to increase transparency, and the adoption of adversarial training can strengthen the resilience of models. Other solutions such as blockchain, Trusted Execution Environments (TEE), or Continuous Adaptive Trust systems add an additional layer of security and integrity.

Finally, the transition to ZTA should be carried out gradually, accompanied by employee training and awareness programs. Such an approach facilitates adaptation and cultivates a security-oriented organizational culture (Ajish, 2024).

While Zero Trust Architecture offers robust defenses against modern cyber threats, its adoption is deeply intertwined with organizational readiness and the capacity for cultural change. Successfully navigating the challenges of organizational resistance, fostering a proactive security mindset, and addressing critical skill gaps through strategic change management and comprehensive training are paramount for realizing the full potential of ZTA in safeguarding digital assets in today's complex environments.

6 Conclusions

Moving to a Zero Trust architecture, complemented by artificial intelligence, is no longer just an option, but a necessity in an increasingly vulnerable digital environment. We have seen that the advantages are clear: continuous verification, faster threat detection and better protection of critical data. At the same time, the challenges cannot be ignored–costs, complexity of integration with legacy systems and risks specific to AI models can slow down progress. However, experience from different sectors shows that success lies not only in technology, but also in people: in user education, in organizational openness to change and in responsible governance of new tools. Zero Trust with AI is not a "ready-to-install" solution, but a gradual transformation path, in which each step brings more security and trust. Organizations that choose to take this path step by step build not only a more secure infrastructure, but also a more resilient digital future.

References

Abbas, G., Gul, S.: Zero Trust Architecture: Revolutionizing Cybersecurity in the Era of Advanced Threats (2025)

Adnan Alvi, M., Jalil, Z.: XRGuard: a model-agnostic approach to ransomware detection using dynamic analysis and explainable AI. IEEE Access **13**, 53159–53170 (2025). https://doi.org/10.1109/ACCESS.2025.3553562

Ajish, D.: The significance of artificial intelligence in zero trust technologies: a comprehensive review. J. Electr. Syst. Inf. Technol. **11**(1), 30 (2024)

Alevizos, L., Eiza, M.H., Ta, V.T., Shi, Q., Read, J.: Blockchain-Enabled intrusion detection and prevention system of APTs within zero trust architecture. IEEE Access **10**, 89270–89288 (2022). https://doi.org/10.1109/ACCESS.2022.3200165

Alnaim, A.K., Alwakeel, A.M.: Zero trust strategies for cyber-physical systems in 6G networks. Mathematics **13**, 1108 (2025). https://doi.org/10.3390/math13071108

Architecture: Int. J. Eng. Technol. Res. (IJETR) **9**(2), 379–386 (2024)

Bradatsch, L., Haeberle, M., Steinert, B., Kargl, F., Menth, M.: Secure service function chaining in the context of zero trust security. In: 2022 IEEE 47th Conference on Local Computer Networks (LCN) (pp. 123–131). IEEE (2022)

Bradatsch, L., Miroshkin, O., Kargl, F.: ZTSFC: a service function chaining-enabled zero trust architecture. IEEE Access **11**, 125307–125327 (2023). https://doi.org/10.1109/ACCESS.2023.3330706

Cabiddu, F., Moi, L., Patriotta, G., Allen, D.: Why do users trust algorithms? A review and conceptualization of initial trust and trust over time. Eur. Manage. J. (2022). https://doi.org/10.1016/j.emj.2022.06.001

CISA Releases Part One of Zero Trust Microsegmentation Guidance. CISA [WWW Document] (2025). URL: https://www.cisa.gov/news-events/alerts/2025/07/29/cisa-releases-part-one-zero-trust-microsegmentation-guidance

Cunningham, C., Blankenship, J., Balaouras, S., Murphy, R., Cyr, M., Dostie, P.: "The zero trust eXtended (ZTX) ecosystem", Forrester Research: Cambridge. MA, USA (2018)
Daah, C., Qureshi, A., Awan, I., Konur, S.: Enhancing zero trust models in the financial industry through blockchain integration: a proposed framework. Electronics **13**, 865 (2024). https://doi.org/10.3390/electronics13050865
Fluchs, S., Taştan, E., Trumpf, T., Horch, A., Drath, R., Fay, A.: Traceable security-by-design decisions for Cyber-Physical Systems (CPSs) by means of function-based diagrams and security libraries. Sensors **23**, 5547 (2023). https://doi.org/10.3390/s23125547
Craighero, F., et al.: Unity is strength: improving the detection of adversarial examples with ensembleapproaches. Neurocomputing **554**, 126576 (2023). ISSN 0925-2312 https://doi.org/10.1016/j.neucom.2023.126576
Galli, A., La Gatta, V., Moscato, V., Postiglione, M., Sperlì, G.: Explainability in AI-based behavioral malware detection systems. Comput. Secur. **141**, 103842 (2024). https://doi.org/10.1016/j.cose.2024.103842
Ganescu, B.-M., Passerat-Palmbach, J.: Trust the Process: Zero-Knowledge Machine Learning to Enhance Trust in Generative AI Interactions (2024). https://doi.org/10.48550/arXiv.2402.06414
Garbis, J., Chapman, J.: Data protection. Zero Trust Secur. (2021). https://doi.org/10.1007/978-1-4842-6702-8_13
Health, C. for D. and R.: Cybersecurity. FDA (2025)
Jose Diaz Rivera, J., Muhammad, A., Song, W.-C.: Securing digital identity in the zero trust architecture: a blockchain approach to privacy-focused multi-factor authentication. IEEE Open J. Commun. Soc. **5**, 2792–2814 (2024). https://doi.org/10.1109/OJCOMS.2024.3391728
Kandula, S.R., Kassetty, N., Alang, K.S., Pandey, P.: Context-aware multi-factor authentication in zero trust architecture: enhancing security through adaptive authentication. Int. J. Glob. Innovations Solutions (IJGIS) (2024). https://doi.org/10.21428/e90189c8.f525ef41
Kang, H., Liu, G., Wang, Q., Meng, L., Liu, J.: Theory and application of zero trust security: a brief survey. Entropy **25**(12), 1595 (2023). https://doi.org/10.3390/e25121595
Kapoor, S.: The imperative of zero trust architecture in modern product security. J. Comput. Sci. Technol. Stud. **7**, 1096–1102 (2025). https://doi.org/10.32996/jcsts.2025.7.130
Katiyar, D., Tripathi, M., Kumar, M., Verma, M., Sahu, D., Saxena, D.: AI and cyber-security: enhancing threat detection and response with machine learning. Educ. Adm. Theory Pract. **30** (2024). https://doi.org/10.53555/kuey.v30i4.2377
Kato, F., Cao, Y., Yoshikawa, M.: Olive: oblivious federated learning on trusted execution environment against the risk of sparsification. Proc. VLDB Endow. **16**, 2404–2417 (2023). https://doi.org/10.14778/3603581.3603583
Kindervag, J.: No more chewy centers: introducing the zero trust model of information security; Forrester Research: Cambridge, MA, USA, Volume 3 (2010)
Kumar, P.: Zero trust architecture for SME cybersecurity: enhancing resilience in the digital transformation Era 05 (2025)
Lee, J.H., Park, J.: A case study: leveraging SASE technology for zero trust implementation in cloud environments. In: 2025 International Conference on Information Networking (ICOIN), (pp. 590–594). IEEE (2025)
Liu, C., Tan, R., Wu, Y., et al.: Dissecting zero trust: research landscape and its implementation in IoT. Cybersecurity **7**, 20 (2024). https://doi.org/10.1186/s42400-024-00212-0
Hussain, M., Pal, S., Jadidi, Z., Foo, E., Kanhere, S.: Federated zero trust architecture using artificial intelligence. IEEE Wirel. Commun. **31**(2), 30–35 (2024). https://doi.org/10.1109/MWC.001.2300405
Manzoor, J., Waleed, A., Jamali, A.F., Masood, A.: Cybersecurity on a budget: evaluating security and performance of open-source SIEM solutions for SMEs. PLoS ONE **19**, e0301183 (2024). https://doi.org/10.1371/journal.pone.0301183

Vasilios, M.: Zero-Trust Network Access (ZTNA) (2024). https://doi.org/10.48550/arXiv.2410.20611
Mensah, F.: Zero trust architecture: a comprehensive review of principles, implementation strategies, and future directions in enterprise cybersecurity. Int. J. Acad. Ind. Res. Innovations (IJAIRI) **10**, 339–346 (2024)
Mohammed, M., Talburt, J.R., Dagtas, S., Hollingsworth, M.: A zero trust model based framework for data quality assessment. In: 2021 International Conference on Computational Science and Computational Intelligence (CSCI), (pp. 305–307). IEEE (2021)
Ojo, A.O.: Adoption of Zero Trust Architecture (ZTA) in the protection of critical infrastructure. Path Sci. **11**(1), 5001–5009 (2025)
Olutimehin, A., Ajayi, A., Metibemu, O., Balogun, A., Oladoyinbo, T., Olaniyi, O.: Adversarial threats to AI-Driven systems: exploring the attack surface of machine learning models and countermeasures. J. Eng. Res. Rep. (2025). https://doi.org/10.9734/jerr/2025/v27i21413
Payment Card Industry Data Security Standard (2024)
Petihakis, G., Farao, A., Bountakas, P., Sabazioti, A., Polley, J., Xenakis, C.: AIAS: AI-ASsisted cybersecurity platform to defend against adversarial AI attacks. Proceedings of the 19th International Conference on Availability, Reliability and Security (2024). https://doi.org/10.1145/3664476.3669920
PfMO, Z.: DoD Zero Trust Strategy (2022)
Phiayura, P., Teerakanok, S.: A comprehensive framework for migrating to zero trust architecture. IEEE Access **11**, 19487–19511 (2023). https://doi.org/10.1109/ACCESS.2023.3248622
Pokhrel, S.R., Yang, L., Rajasegarar, S., Li, G.: Robust zero trust architecture: joint blockchain based federated learning and anomaly detection-based framework. In: Proceedings of the SIGCOMM Workshop on Zero Trust Architecture for Next Generation Communications (pp. 7–12) (2024)
Rose, S., Borchert, O., Mitchell, S., Connelly, S.: Zero trust architecture. Natl. Inst. Stand. Technol. (2020). https://doi.org/10.6028/NIST.SP.800-207
Sarkar, S., Choudhary, G., Shandilya, S.K., Hussain, A., Kim, H.: Security of zero trust networks in cloud computing: a comparative review. Sustainability **14**(18), 11213 (2022). https://doi.org/10.3390/su141811213
Ahmadi, S.: Autonomous identity-based threat segmentation for zero trust architecture. Cyber Secur. Appl. **3**, 100106 (2025). ISSN 2772-9184. https://doi.org/10.1016/j.csa.2025.100106
Tabassi, E., 2023. Artificial Intelligence Risk Management Framework (AI RMF 1.0) (No. NIST AI 100–1). National Institute of Standards and Technology (U.S.), Gaithersburg, MD. https://doi.org/10.6028/NIST.AI.100-1
Tariq, S., Baruwal Chhetri, M., Nepal, S., Paris, C.: Alert fatigue in security operations centres: research challenges and opportunities. ACM Comput. Surv. **57**(224), 1–224:38 (2025). https://doi.org/10.1145/3723158
U.S. Department of Defense. *Zero Trust reference architecture* (2022). https://www.defense.gov/Resources/DoD-Zero-Trust-Reference-Architecture/
Vukotich, G.: Healthcare and cybersecurity: taking a zero trust approach. Health Serv. Insights **16**, 11786329231187826 (2023). https://doi.org/10.1177/11786329231187826
Wang, X., Bhuse, V., Cheng, Y.: A Zero Trust Module for Cybersecurity Education 12
What is Zero Trust Architecture? [WWW Document], n.d. Palo Alto Networks. URL. https://www.paloaltonetworks.com/cyberpedia/what-is-a-zero-trust-architecture
Zakhmi, K., Ushmani, A., Ranjan Mohanty, M., Agrawal, S., Banduni, A., Kakatum Rao, S.S.: Evolving zero trust architectures for AI-Driven cyber threats in healthcare and other high-risk data environments: a systematic review. Cureus (2025). https://doi.org/10.7759/cureus.85446
Zero Trust Maturity Model. CISA [WWW Document] (2023). URL https://www.cisa.gov/zero-trust-maturity-model

A Hybrid Architecture Integrating Graph Neural Networks and Long Short-Term Memory Networks for Advanced Malware Detection and Analysis

Uma Kannan(✉) and Rajendran Swamidurai

Alabama State University, Montgomery, AL 36104, USA
{ukannan,rswamidurai}@alasu.edu

Abstract. The frequency and severity of cyberattacks are increasing. During a Senate hearing, senior intelligence officials expressed concern that cyberthreats would eventually surpass terrorist attacks as the biggest danger facing the nation. Cyberattacks continue to evade detection. Cybercriminals utilize polymorphic malware because it can evade antivirus protection with just a tiny change to the script or binary code. Through Trojanized programs, tools, and services, cybercriminals introduce extremely evasive malwares. To tackle dangers posed by evasive malware, a malware detection module is required that is efficient, robust, and scalable. Pre-execution rule-based malware technologies are useless and impractical. We require state-of-the-art security systems capable of handling massive amounts of data and protecting us from both existing and future malware threats. In order to ward off existing and future evasive malware attacks, we propose a two-pronged approach that integrates multiple artificial intelligence techniques, including graphical neural networks (GNN) and recurrent neural networks (LSTM and GRU). The empirical findings of our study demonstrate that the GNN-LSTM and GNN-GRU hybrid framework attains superior predictive accuracy compared to its single-modality counterparts, hence validating the synergy realized through feature fusion.

Keywords: Malware Detection and Analysis · Graph Neural Networks · Long Short-Term Memory Networks · Hybrid GNN-LSTM Model · Hybrid GNN-GRU Model

1 Introduction

Today, cyber networks (cyberspace plus the Internet) are as integral to the American homeland as cities, farms, mountains, and coastlines. Because they are where nearly all of our daily activities take place, including shopping, banking, working, playing, learning, connecting with family, etc. [1]. Cyberspace is susceptible to an increasing spectrum of threats posed by criminals and nation-state actors. The objectives of cyberattacks include identity theft, data theft, espionage, and the interruption of essential activities [1].

H. R. Arabnia et al. (Eds.): CSCI 2025, CCIS 2969, pp. 532–538, 2027.
https://doi.org/10.1007/978-3-032-28460-0_38

In the early cyber period, malware threats were few, and basic pre-execution rules often detected them. Rapid expansion of the Internet and malware made manually constructed detection criteria impractical, necessitating new, powerful protection solutions. [2] Cybersecurity threats of today continue to find ways to fly under the radar. Cybercriminals employ polymorphic malware because a minor modification to the binary code or script enables these threats to evade detection by conventional antivirus software. Today, cybercriminals deploy highly evasive malwares using Trojanized versions of legitimate applications, tools, and services. [3] To combat the exponential growth of evasive malware threats, an efficient, robust, and scalable malware detection module is required. The old malware tools that rely on pre-execution rules are ineffective and impractical. We require tools based on advanced protection technologies that are capable of processing vast amounts of data and delivering long-lasting defense solutions against current and future attacks. [2, 3].

There are 23 million small businesses in the United States [4] with one or two IT specialists, no cybersecurity experts, and inadequate funding to build a robust cyber defense program. 75% of attacked enterprises, according to the Verizon Data Breach Report [4], were low-profile small businesses with 100 or less employees. In addition, the industry struggles to recruit qualified cybersecurity employees. Approximately 40% of entry-level security roles and more than 50% of senior and manager positions are vacant. In many cases, there are no personnel who know how to do this and how to operate it [5]. To solve this essential issue, we propose to design a new AI-based model capable of automatically detecting logic flaws in web applications and protecting small enterprises from cyberattacks.

Malware are becoming increasingly difficult to combat due to their constant increase in complexity and volume. Using machine learning techniques to automatically learn models and patterns behind such complexity and to develop solutions to keep pace with malware evolution is one of the most prevalent approaches in the literature. [6].

Software application vulnerability is a design flaw or implementation bug when exploited, compromises the application. Vulnerability is an implementation flaw that allows the attackers to orchestrate an attack using the applications legitimate sequences of actions to induce unintended consequences [7]. Researchers used three types of techniques to address applications vulnerabilities and identify malwares: 1) Secure frameworks, 2) white-box techniques, and 3) black-box techniques, but most of these research and commercial vulnerability scanners focus on existing types of malwares and only very few research focus on the detection of evasive malwares. The following are the drawbacks or gaps in the existing vulnerability detection systems:

- *Secure frameworks:* They are only suitable for systems under development; therefore, they are not suitable for applications that are already deployed.
- *White-box Techniques:* 1) Source code is needed to develop the system – which is not available in most cases, because today most of the applications are developed by third-party vendors, 2) They are technology or programming language dependent. That is, a system developed for one programming language cannot be used for applications developed on other languages, and 3) Sharing a source code with a third-party to develop a vulnerability detection system will leads to the compromising the privacy of the application stakeholders.

- *Black-box Techniques:* 1) Most of the existing systems do not work well with applications with multiple web-page interactions (that is, data-flow systems), and 2) They focus only on the existing malware types unable to detect a malware of similar family with slight modifications.

Therefore, there is a demand for a system that can detect all types of evasive malwares in applications independent of the functionality of the application.

Effective malware analysis necessitates capturing two opposing aspects of malicious program behavior: the static, inherent structural dependencies of code components and the dynamic, time-dependent flow of execution during runtime. Single-modal deep learning models frequently fail to grasp this critical duality. Models based entirely on structural analysis may be misled by temporal delays or execution masking, whereas models based solely on sequential analysis may overlook critical structural integrity aspects that define a malevolent component. This duality in feature space needs a complete solution capable of studying both dimensions concurrently. Recurrent neural network (RNN) models like LSTM (Long Short-Term Memory) and GRU (Gated Recurrent Unit) network models can track long-term dependencies, which makes them a great tool for malware strain analysis. This paper investigates the theoretical foundation, architectural implementations, and empirical validation of hybrid GNN-LSTM and hybrid GNN-GRU models in advanced malware detection.

2 Background and Related Work

2.1 Graph-Based Representations for Structural Analysis

The application of GNNs in malware analysis relies on abstracting program execution into structured graphs. Control Flow Graphs (CFGs) are the most common representation, modeling program execution flow where nodes typically represent basic blocks or API calls, and edges denote control flow or data dependencies.[8, 9] GNNs, due to their inherent ability to process data represented as nodes and edges that depict relationships between elements, are uniquely suited to capture the structural relationships within malware samples.[9].

GNN architectures, such as Graph Convolutional Networks (GCNs), are optimized for graph-level classification, effectively capturing and aggregating information from nodes and edges to identify malicious structural patterns.[10] For example, some approaches implement GCNs where nodes represent API calls and edges represent their dependencies, achieving robust identification of malicious behavior.[9] More advanced models, such as the DEGCN, have attempted to internally integrate temporal context by dynamically adjusting node weights based on API call significance and updating edge weights according to their temporal sequence. This approach achieved a 98.3% detection rate on a dataset of 1,400 malware samples.[9] However, embedding temporal awareness internally often increases the complexity of the single architecture, which can complicate training and interpretability, supporting the rationale for a modular, two-stream hybrid design.

2.2 Sequential Deep Learning for Temporal Analysis

Temporal analysis focuses on the ordered execution of operations extracted from dynamic analysis logs, typically consisting of API call sequences or opcode streams.[9, 11] Recurrent Neural Networks (RNNs) are the traditional models for processing this sequential data.

Long Short-Term Memory (LSTM) networks are the preferred RNN variant due to their specialized architecture—memory cells and gating mechanisms—which effectively addresses the vanishing gradient problem encountered in standard RNNs.[12] This capability allows LSTMs to effectively model the complex, long-range dependencies inherent in multi-step malware execution sequences. Studies evaluating sequential detection models confirm the superiority of LSTMs: an LSTM model achieved the highest accuracy at 96%, surpassing the Gated Recurrent Unit (GRU) model at 95% and the standard RNN at 93%.[12] This empirical advantage highlights the LSTM's critical role in capturing time-dependent context necessary for detection. Furthermore, techniques borrowed from Natural Language Processing (NLP), such as word embedding and bidirectional LSTMs (Bi-LSTMs), are employed for sequence analysis. Bi-LSTMs capture context from both preceding and succeeding instructions, enriching the semantic understanding of opcode sequences within basic blocks.[11, 13].

2.3 Synthesis of Existing Hybrid Approaches

The transition to explicit GNN-LSTM hybridization is predicated on the observed limitations of single-modality systems and the success of precursor hybrid models (e.g., CNN-LSTM). Previous hybrid research, such as models combining CNNs and LSTMs, successfully utilized feature fusion to combine spatial characteristics (from image representation) and temporal data, achieving high accuracy (98.7% for CPL-Net) [14, 15]. This work established the precedent that capturing spatio-temporal dynamics is necessary for robust malware detection.

The move to GNN-LSTM signifies an architectural maturation, shifting the structural analysis from generalized image features to semantically relevant program behavior graphs (CFGs). While attempts to inject sequential cues into GNNs (like DEGCN) demonstrate the structural need for temporal information [9], the modular, two-stream parallel GNN-LSTM architecture is emerging as the dominant strategy. This modularity allows for a cleaner separation of feature responsibilities, thereby simplifying training, improving robustness against targeted obfuscation, and potentially making post-hoc Explainable AI easier to implement.

3 Methodology

Generating high-quality graph inputs, such as CFGs or Program Dependency Graphs (PDGs), is the first critical step. Advanced methodologies utilize dynamic CFG extraction to ensure the graph reflects actual execution paths, rather than potential static paths [8]. Nodes in these graphs, representing basic blocks or API calls, must be embedded into meaningful feature vectors. Hybrid encoding schemes are favored, combining rule-based vectorization (for deterministic instructions) with autoencoder-based embedding to capture higher-level semantic properties. This dual approach addresses the challenge of handling out-of-vocabulary instructions and generating informative, fixed-length node representations [8, 13].

Execution traces, often derived from API call sequences, are processed into a sequential format for the LSTM. This involves tokenizing the calls and then applying embedding layers, analogous to word embedding in NLP, to represent the sequence numerically [11]. Given the variability in malware execution length, sequence padding or truncation is required to maintain consistency for training recurrent networks (Fig. 1).

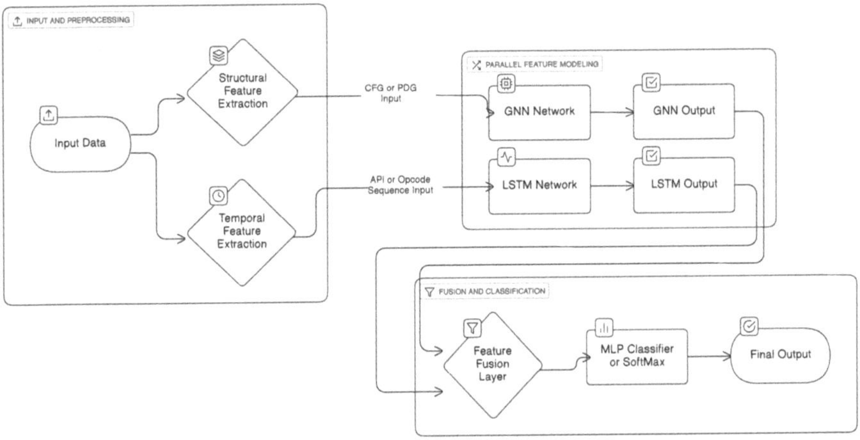

Fig. 1. Architectural Diagram.

Parallel Architecture Operation. In this setup, the GNN processes the graph input to generate a structural embedding vector, capturing global graph properties suitable for graph-level classification [4]. Simultaneously, the LSTM processes the sequential trace to produce a temporal embedding vector, which captures sequential dependency and chronological context [16].

Fusion Layer Design. The method by which these features are combined determines the final decision boundary:

- *Feature Fusion (Concatenation):* Concatenates the high-dimensional structural embedding from the GNN and the temporal embedding from the LSTM. The resulting combined feature vector is then input into a final classification layer (e.g., SoftMax or a multi-layer perceptron) [14].
- *Score Fusion:* Combines the derived risk scores (or probabilities) from each model. A weighted sum can be calculated for the final classification: $Final\ Score = \alpha \cdot LSTM\_Score + (1-\alpha) \cdot GNN\_Score$ [14], Score fusion often simplifies interpretability by explicitly quantifying the influence of each modality on the final decision.

4 Results

We tested the hybrid GNN-LSTM model using the balanced ClaMP malware dataset [17]. The hybrid GNN-LSTM model achieved an accuracy of 96.55% and an F1-score of 96.70%. This performance was significantly higher than the GNN-only baseline (90.82% accuracy, 90.94% F1-score) and the LSTM-only baseline (93.48% accuracy, 93.02% F1-score) (Table 1).

Table 1. Comparative Performance of Hybrid GNN-LSTM Models.

Model Architecture	Primary Feature Modality	Accuracy (%)	F1-Score (%)	Key Limitation Addressed
LSTM-only Baseline	Temporal (Sequential)	93.02	93.48	Lacks structural context
GNN-only Baseline	Structural (Graph)	90.94	90.82	Inability to model execution order
Hybrid GNN + LSTM (Fusion)	Structural & Temporal	96.55	96.70	Captures orthogonal information, reducing FPR/FNR

5 Conclusion

The hybrid GNN-LSTM framework represents a critical advancement in computational malware analysis, offering a robust solution for synthesizing the structural and temporal characteristics of cyber threats. By achieving significant performance gains over monomodal baselines, the architecture effectively combats heterogeneous obfuscation and supports complex analytical tasks. This structural integration demonstrates that concurrent capture of orthogonal feature spaces is a necessary requirement for achieving high robustness and accuracy in detecting modern, adaptive malware.

Acknowledgments. This research was funded by the National Science Foundation (NSF), grant number 2300405.

Disclosure of Interests. The authors have no competing interests to declare that are relevant to the content of this article.

References

1. Johnson, J.C.: Let's pass cybersecurity legislation. The Hill (2014). http://thehill.com/opinion/op-ed/217151-lets-passcybersecurity-legislation
2. Kaspersky: Machine learning methods for malware detection. Whitepaper (2020). www.kaspersky.com

3. Lazo, E.G.: Combing through the fuzz: Using fuzzy hashing and deep learning to counter malware detection evasion techniques. Microsoft 365 Defender Research Team. Microsoft Security (2021)
4. Byron Acohido: Why small businesses must tackle cybersecurity (2013). http://www.usatoday.com/story/cybertruth/2013/12/24/why-small-businesses-must-tackle-cybersecurity/4190325/. USA TODAY
5. Hackett, R.: For better cybersecurity, skip the shiny toy—invest in people and processes (2014). http://fortune.com/2014/09/09/for-better-cybersecurity-skip-the-shiny-toy-invest-in-people-and-processes/. Fortune
6. Ucci, D., Aniello, L., Baldoni, R.: Survey of machine learning techniques for malware analysis. Comput. Secur. (2018). https://doi.org/10.1016/j.cose.2018.11.001
7. Alkofahi, H.: Business Rules Discovery in Web Application through Process Mining. PhD Thesis. Auburn University (2020)
8. Dynamic CFG Construction and Explainability in GNN-based Malware Detection. Source: arXiv, https://arxiv.org/html/2504.16316v1
9. A Hybrid Malware Detection Method using Graph Neural Networks and Temporal Sequences (DEGCN Model). Source: SJSU ETD Projects. https://scholarworks.sjsu.edu/cgi/viewcontent.cgi?article=2268&context=etd_projects
10. Hierarchical Adversarial Attacks Against Graph Neural Network Based Detectors. Source: arXiv, https://arxiv.org/html/2401.05680v1
11. CNN-LSTM and Transfer Learning Models for Malware Classification based on Opcodes and API Calls. Source: SJSU Faculty Research. https://scholarworks.sjsu.edu/cgi/viewcontent.cgi?article=3685&context=faculty_rsca
12. Malware-SeqGuard: An Approach Utilizing LSTM and GRU for Effective Detection of Evolving Malware in Android Environments. Source: ResearchGate. https://www.researchgate.net/publication/393393523_Malware-SeqGuard_An_Approach_Utilizing_LSTM_and_GRU_for_Effective_Detection_of_Evolving_Malware_in_Android_Environments
13. Deep Learning for Zero-day Malware Detection and Classification: A Survey. Source: ACM Comput. Surv (2023)
14. A Hybrid Approach Using Graph Neural Networks and LSTM for Fraud Detection. Sourc.: Int. J. Res. Publ. Rev. https://ijrpr.com/uploads/V6ISSUE6/IJRPR48336.pdf
15. CPL-Net: A Malware Detection Network Based on Parallel CNN and LSTM Feature Fusion. Source: ResearchGate/MDPI, https://www.mdpi.com/2079-9292/12/19/4025
16. GCN with LSTM and Attention for Enhanced API Call Sequence Modeling in Malware Detection. Source: MDPI. https://www.mdpi.com/2079-9292/12/23/4817
17. https://github.com/urwithajit9/ClaMP/tree/master/dataset

Determinants of Trust in Information Security of Artificial Intelligence Assistant

Hyeon Jo[(✉)]

Department of Strategic Planning, RealSecu, Busan, Republic of Korea
sineoriz@gmail.com

Abstract. This study proposes a theoretical model to identify the factors influencing trust in the information security of artificial intelligence assistants (AIAs). Data were collected from 232 users with prior experience using AIAs. The research model was tested using partial least squares structural equation modeling (PLS-SEM). The findings indicate that system quality is significantly associated with trust. Privacy concerns regarding one's surroundings are also identified as a significant determinant of trust. In addition, utilitarian and hedonic motivations are found to be key antecedents of trust in information security. The study concludes with theoretical and practical implications.

Keywords: artificial intelligence assistant · information security · trust · system quality · privacy concerns · motivation

1 Introduction

The term Siri has become widely recognized among the general public. It is an artificial intelligence-based application that processes user voice input and supports decision-making (Gu et al., 2018). Siri represents a typical artificial intelligence assistant (AIA). Following its introduction, similar services such as Samsung Bixby, Naver Clova, Microsoft Cortana, Google Assistant, and Amazon Alexa have rapidly emerged in the market. AIAs are embedded in various electronic devices, including smartphones, smart speakers, and smart TVs, to enhance user convenience.

Despite their increasing adoption, the use of AIAs involves uncertainties related to trust, perceived risk, and privacy risk (Nguyen et al., 2021; Pal et al., 2021; Tanribilir, 2021). Users typically interact with AIAs through voice commands, during which personal information may be disclosed. This raises concerns that such information could be misused by the system or accessed by unauthorized third parties (Nguyen et al., 2019; Pal et al., 2021). Compared to text-based interfaces, voice-based AIAs are more susceptible to unintended exposure, such as being overheard by others, thereby intensifying privacy concerns. This potential for privacy invasion represents a critical drawback of AIAs. Accordingly, this study focuses on identifying the antecedents of trust in AIA information security.

System quality is widely recognized as a key determinant of information systems (IS) success (DeLone & McLean, 2003). It encompasses attributes such as stability,

H. R. Arabnia et al. (Eds.): CSCI 2025, CCIS 2969, pp. 539–546, 2027.
https://doi.org/10.1007/978-3-032-28460-0_39

reliability, and response speed, which collectively reflect overall system performance (Urbach & Müller, 2012). High system quality in AIAs is expected to enhance user trust in information security. Therefore, this study posits that system quality significantly influences trust.

Users may also experience privacy concerns during AIA usage (Nguyen et al., 2019). Prior research has primarily examined concerns related to the misuse of personal information or direct privacy violations (Lau et al., 2018; Pal et al., 2021; Tanribilir, 2021). More recent studies have extended this perspective to include concerns about others in the surrounding environment (Lutz & Newlands, 2021; Moorthy & Vu, 2015). Individuals with heightened sensitivity to privacy risks in public settings may exhibit lower trust in AIA security. Accordingly, this study proposes that privacy concerns regarding surroundings play a significant role in shaping trust in information security.

Furthermore, users engage with AIAs for both utilitarian and hedonic purposes (Moorthy & Vu, 2015; Tanribilir, 2021). Strong usage motivations may reflect greater trust in the system. Thus, this study posits that both utilitarian and hedonic motivations serve as important predictors of trust.

2 Related Work and Hypotheses

2.1 System Quality

System quality is defined as the reliability, response time, and availability of an information system (Gao & Waechter, 2017). When users perceive the overall system quality of an AIA to be high, they are more likely to evaluate its security positively. Therefore, this study proposes that system quality enhances trust in information security.

H1. System quality positively affects trust in information security.

2.2 Privacy Concerns on Surroundings

Privacy concerns regarding surroundings refer to the potential misuse, overhearing, or unintended exposure of information to nearby individuals. Users with higher levels of such concerns are likely to exhibit lower trust in AIA information security. Accordingly, this study posits that privacy concerns regarding surroundings negatively influence trust.

H2. Privacy concerns regarding surroundings negatively affect trust.

2.3 Utilitarian Motivation

Utilitarian motivation reflects the use of information systems for functional, goal-oriented tasks. Users who rely on AIAs for work or study purposes may be more willing to provide personal information, which may indicate higher trust in the system's security. Therefore, this study suggests that utilitarian motivation positively influences trust in information security.

H3. Utilitarian motivation positively affects trust in information security.

2.4 Hedonic Motivation

Hedonic motivation refers to the use of information systems for enjoyment and pleasure (Hyllegard et al., 2011). Users with strong hedonic motivation may engage in more personal or casual interactions with AIAs. Continued use despite the disclosure of private information may reflect trust in the system's security. Thus, this study proposes that hedonic motivation enhances trust in information security.

H4. Hedonic motivation positively affects trust in information security.

3 Methodology

3.1 Instrument

This study adapted measurement items from prior research and refined them to fit the AIA context. A pilot study was conducted to assess the reliability and validity of the constructs and to ensure an appropriate measurement structure. Prior to the main survey, experts in the information systems field reviewed the questionnaire to evaluate its clarity, logical flow, and potential ambiguity.

3.2 Subjects and Data Collection

This study employed a cross-sectional survey design. Data were collected through an online survey administered by a professional market research firm in South Korea in January 2021. To enhance data quality, attention-check items unrelated to the study were included. After excluding invalid or careless responses, a total of 232 responses were retained for analysis. Table 1 presents the demographic characteristics of the respondents.

Table1. Profile of the respondents

Demographics	Item	Subjects ($N = 232$)	
		Frequency	Percentage
Gender	Male	115	49.6%
	Female	117	50.4%
Age	Teens	8	3.4%
	20s	67	28.9%
	30s	87	37.5%
	40s	54	23.3%
	50s	16	6.9%
Usage Frequency	5–9 times	60	25.9%
	10–99 times	129	55.6%
	≥ 100 times	43	18.5%

4 Analysis and Results

The partial least squares (PLS) approach is widely employed in information systems research due to its suitability for complex, prediction-oriented models (Hair et al., 2017). This study analyzed the proposed theoretical framework using PLS-SEM with SmartPLS.

4.1 Reliability and Validity of the Data

This study assessed the measurement model by examining reliability, convergent validity, and discriminant validity. Internal consistency reliability was evaluated using composite reliability (CR) and Cronbach's alpha (Table 2). All CR values and Cronbach's alpha coefficients exceeded the recommended threshold of 0.70, indicating adequate reliability. Notably, the CR values were above 0.90, demonstrating strong internal consistency.

Convergent validity was assessed based on factor loadings and average variance extracted (AVE). All factor loadings exceeded the recommended threshold of 0.70 (Hair et al., 2006), ranging from 0.860 to 0.953, and were statistically significant at the $p <$ 0.001 level, thereby confirming satisfactory convergent validity (Bagozzi et al., 1991).

Table 2. Results of Reliability and Validity Assessment

Construct	Items	Mean	St. Dev	Factor Loading	Cronbach's Alpha	CR	AVE
System Quality	SYQ1	4.263	1.251	0.903	0.924	0.952	0.868
	SYQ2	4.332	1.220	0.946			
	SYQ3	4.310	1.238	0.944			
Privacy Concerns on Surroundings	PCS1	4.362	1.319	0.953	0.809	0.908	0.832
	PCS2	4.164	1.383	0.870			
Utilitarian Motivation	UTM1	4.547	1.299	0.860	0.894	0.934	0.825
	UTM2	3.983	1.432	0.940			
	UTM3	4.026	1.456	0.924			
Hedonic Motivation	HEM1	3.909	1.440	0.947	0.944	0.964	0.899
	HEM2	3.806	1.506	0.950			
	HEM3	3.737	1.513	0.948			

Discriminant validity was evaluated using the Fornell–Larcker criterion (Table 3). Specifically, the square root of the AVE for each construct exceeded the corresponding inter-construct correlations, indicating that each construct is empirically distinct from the others. These results confirm that the measurement model demonstrates satisfactory discriminant validity.

Table 3. Construct Correlations and Discriminant Validity

Constructs	1	2	3	4	5
1. System Quality	0.931				
2. Privacy Concerns on Surroundings	0.084	0.912			
3. Utilitarian Motivation	0.613	0.143	0.908		
4. Hedonic Motivation	0.469	0.126	0.334	0.948	
5. Trust	0.592	-0.237	0.464	0.434	0.946

Note: Diagonal values are the square root of AVE

4.2 Hypothesis Test

Structural equation modeling (SEM) using the PLS approach was conducted to examine the hypothesized relationships among the constructs. To assess the statistical significance of the path coefficients, a bootstrapping procedure with 500 resamples was employed. As illustrated in Fig. 1, all hypothesized relationships are supported.

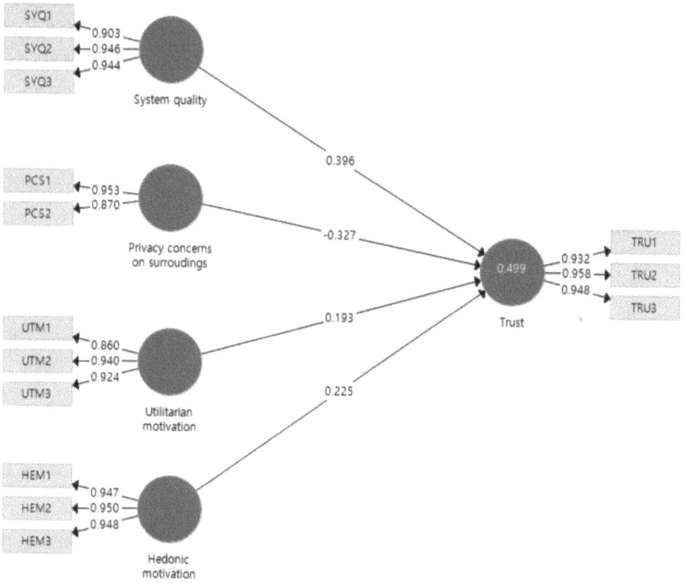

Fig. 1. SEM Results

System quality is positively and significantly associated with trust ($\beta = 0.396$, $t = 4.675$), supporting H1. Privacy concerns regarding surroundings exhibit a significant negative effect on trust ($\beta = -0.327$, $t = 5.257$), supporting H2. Utilitarian motivation has a significant positive effect on trust ($\beta = 0.193$, $t = 2.470$), supporting H3. Similarly, hedonic motivation exerts a significant positive influence on trust ($\beta = 0.225$, $t =$

3.896), supporting H4. Overall, the structural model explains approximately 49.9% of the variance in trust, indicating moderate explanatory power.

5 Conclusion

5.1 Summary

This study examined the determinants of trust in the information security of artificial intelligence assistants (AIAs). Specifically, system quality, privacy concerns regarding surroundings, utilitarian motivation, and hedonic motivation were proposed as key antecedents. Using cross-sectional survey data collected from 232 AIA users, the proposed model was analyzed via PLS-SEM following a two-step approach. The empirical results indicate that all hypothesized relationships are statistically significant.

5.2 Implications

This study offers several theoretical and practical implications. First, it extends the literature on AIA security by adopting a user-centered perspective. While prior research has primarily focused on technical and engineering aspects, this study identifies key behavioral and perceptual factors influencing trust among actual users. This perspective enables a more comprehensive understanding of security-related user behavior. Second, this study contributes to the literature by categorizing the antecedents of trust into system-related factors, privacy concerns, and motivational drivers. This conceptual differentiation provides a structured foundation for future research to further refine and validate these constructs.

From a practical standpoint, the findings suggest that developers should prioritize system quality and integrate privacy-sensitive design principles from the early stages of development. Additionally, marketers should consider users' privacy concerns and motivational factors when promoting AIA services.

5.3 Limitations and Future Research Directions

Despite its contributions, this study has several limitations. First, the analysis is limited to AIAs embedded in smartphones. Future research should examine AIAs across a broader range of devices to enhance the generalizability of the findings. Second, this study focuses exclusively on general users. Future studies should incorporate expert perspectives to provide a more comprehensive evaluation of AIA information security.

Appendix

System Quality (adapted from Nguyen et al. (2019)).

SYQ1 In terms of system quality, I would rate AI personal assistants highly.

SYQ2 Overall, AI personal assistants are of high quality.

SYQ3 Overall, I would give AI personal assistants a high-quality rating.

Privacy Concerns on Surroundings (adapted from Nguyen et al. (2019)).

PCS1 My personal information could be misused by people around me.

PCS2 I am concerned that people around me may use the information I provide to AI personal assistants for purposes of which I am not aware.

Utilitarian Motivation (adapted from Salehan et al. (2017)).

UTM1 I use AI personal assistants to obtain useful information.

UTM2 I use AI personal assistants to support my work.

UTM3 I use AI personal assistants to assist with my business or studies.

Hedonic Motivation (adapted from Salehan et al. (2017)).

HEM1 I use AI personal assistants for enjoyment or entertainment.

HEM2 I use AI personal assistants to escape from daily routines.

HEM3 I use AI personal assistants to feel excitement.

Trust (adapted from Nguyen et al. (2019)).

TRU1 I trust that my personal information will not be used for purposes other than those intended.

TRU2 I believe that my personal information is adequately protected.

TRU3 I am confident that my personal information is secure.

References

Bagozzi, R.P., Yi, Y., Phillips, L.W.: Assessing construct validity in organizational research. Adm. Sci. Q. **36**(3), 421–458 (1991). https://doi.org/10.2307/2393203

DeLone, W.H., McLean, E.R.: The DeLone and McLean model of information systems success: a ten-year update. J. Manag. Inf. Syst. **19**(4), 9–30 (2003)

Gao, L., Waechter, K.A.: Examining the role of initial trust in user adoption of mobile payment services: an empirical investigation. Inf. Syst. Front. **19**(3), 525–548 (2017). https://doi.org/10.1007/s10796-015-9611-0

Gu, D., Yang, X., Li, X., Jain, H.K., Liang, C.: Understanding the role of mobile internet-based health services on patient satisfaction and word-of-mouth. Int. J. Environ. Res. Public Health **15**(9), 1972 (2018)

Hair, J., Anderson, R., Tatham, B.R.: Multivariate data analysis, 6th edn. Prentice Hall (2006)

Hair, J., Hollingsworth, C.L., Randolph, A.B., Chong, A.Y.L.: An updated and expanded assessment of PLS-SEM in information systems research. Ind. Manag. Data Syst. **117**(3), 442–458 (2017). https://doi.org/10.1108/IMDS-04-2016-0130

Hyllegard, K.H., Ogle, J.P., Yan, R.-N., Reitz, A.R.: An exploratory study of college students' fanning behavior on Facebook. Coll. Stud. J. **45**(3), 601–617 (2011)

Lau, J., Zimmerman, B., Schaub, F.: Alexa, are you listening? Privacy perceptions, concerns and privacy-seeking behaviors with smart speakers. Proc. ACM on Hum.-Comput. Interact. **2**(CSCW), 1–31 (2018)

Lutz, C., Newlands, G.: Privacy and smart speakers: a multi-dimensional approach. Inf. Soc. **37**(3), 147–162 (2021)

Moorthy, A.E., Vu, K.-P.L.: Privacy concerns for use of voice activated personal assistant in the public space. Int. J. Hum.-Comput. Interact. **31**(4), 307–335 (2015)

Nguyen, D.M., Chiu, Y.-T.H., Le, H.D.: Determinants of continuance intention towards banks' Chatbot services in vietnam: a necessity for sustainable development. Sustainability **13**(14), 7625 (2021). https://www.mdpi.com/2071-1050/13/14/7625

Nguyen, Q.N., Ta, A., Prybutok, V.: An integrated model of voice-user interface continuance intention: the gender effect. Int. J. Hum.-Comput. Interact. **35**(15), 1362–1377 (2019). https://doi.org/10.1080/10447318.2018.1525023

Pal, D., Babakerkhell, M.D., Zhang, X.: Exploring the determinants of users' continuance usage intention of smart voice assistants. IEEE Access (2021). https://doi.org/10.1109/ACCESS.2021.3132399

Salehan, M., Kim, D., Kim, C.: Use of online social networking services from a theoretical perspective of the motivation-participation-performance framework. J. Assoc. Inf. Syst. **18**, 141–172 (2017). https://doi.org/10.17705/1jais.00449

Tanribilir, R.N.: Analysing antecedence of an intelligent voice assistant use intention and behaviour. F1000Research **10**(496), 496 (2021)

Urbach, N., Müller, B.: The updated DeLone and McLean model of information systems success. In: Dwivedi, Y.K., Wade, M.R., Schneberger, S.L. (eds.), Information Systems Theory: Explaining and Predicting Our Digital Society, Vol. 1 (pp. 1–18). Springer New York (2012). https://doi.org/10.1007/978-1-4419-6108-2_1

Poster Papers

Breaking the Guardrails: A Comprehensive Jailbreak Vulnerability Assessment of Local Large Language Models

Bikesh Regmi(✉), Binisa Giri, Chukwuemeka Duru, and Kofi Nyarko

Department of Computer Science,, Morgan State University, Baltimore, USA
{bireg1,bigir1,chdur8,kofi.nyarko}@morgan.edu

Abstract. This paper presents what is, to our knowledge, the first systematic study of jailbreak vulnerabilities in local large language models (LLMs). We evaluated 445 prompt–model combinations across five widely used models: Gemma2, Gemma3n, DeepSeek-R1, Command-R, and LLaMA 3.1. Our analysis focused on three attack categories—direct prompting, multi-step indirect attacks, and encoding-based methods. Results show clear differences in resilience: DeepSeek-R1 resisted all attempts (0% attack success rate), while Command-R was the most susceptible (14.6%). Multi-step indirect prompts were the most effective (12.6%), underscoring the challenge of defending against conversational exploitation. These findings expose critical security gaps in local deployments and provide quantitative benchmarks for AI safety improvement.

Keywords: Large Language Models · Jailbreak Attacks · AI Security · Vulnerability Assessment · Local Models

1 Introduction

Large language models (LLMs) are increasingly integrated into sensitive applications, from cybersecurity to education. Despite rapid progress in alignment, models remain vulnerable to *jailbreaking*—attacks that intentionally bypass safety guardrails to elicit harmful, deceptive, or restricted content [1]. Prior research has largely examined proprietary systems such as GPT-4 and Claude [2], leaving local deployments understudied.

Local LLMs offer privacy and flexibility but may lack robust filtering. This study bridges that gap by systematically evaluating five widely used local models to measure jailbreak resilience and analyze vulnerability transfer across architectures.

2 Methodology

2.1 Models Evaluated

We assessed five locally deployed LLMs: Gemma2, Gemma3n, DeepSeek-R1, Command-R, and LLaMA 3.1, chosen for architectural diversity and popularity.

H. R. Arabnia et al. (Eds.): CSCI 2025, CCIS 2969, pp. 549–552, 2027.
https://doi.org/10.1007/978-3-032-28460-0_40

All were tested using Ollama with fixed parameters (temperature 0.1, max tokens 512) to ensure fair comparison.

2.2 Attack Dataset and Taxonomy

Eighty-nine crafted jailbreak prompts were tested across:

- **Direct prompting**: explicit harmful requests.
- **Multi-step indirect**: multi-turn deceptive conversations.
- **Encoded/mixed**: role-play, encoding, or obfuscation.

This taxonomy captures common attack strategies used in practice [3].

2.3 Evaluation Metrics

We measured attack success rate (ASR), defense evasion rate (DER), severity (1–5 scale), and mean response time. Harmful outputs were verified through manual review following established safety taxonomies [4] (Table 1).

3 Results and Analysis

3.1 Overall Findings

The average ASR across all models was 7.4%, indicating that while most models blocked basic attempts, certain architectures remain vulnerable.

Table 1. Overall Security Metrics Across All Models

Metric	Mean	Std	Median	Max
ASR	0.074	0.262	0.000	1.000
DER	0.064	0.231	0.000	0.999
Severity	1.27	0.97	1.00	5.00
Resp. Time (s)	19.3	17.0	14.8	90.2

3.2 Model-Specific Vulnerabilities

Table 2 shows clear variation: DeepSeek-R1 achieved perfect defense, while Command-R exhibited 14.6% ASR.

3.3 Attack Type Effectiveness

As shown in Table 3, multi-step indirect attacks were the most successful (12.6%), validating prior work emphasizing conversational vulnerabilities [5].

Table 2. Model Comparison

Model	ASR	DER	Sev.
DeepSeek-R1	0.000	0.000	1.00
LLaMA 3.1	0.022	0.018	1.08
Gemma3n	0.090	0.074	1.32
Gemma2	0.112	0.099	1.43
Command-R	0.146	0.131	1.55

Table 3. Attack Effectiveness

Category	ASR	DER	Evals
Multi-step Ind.	0.126	0.111	135
Direct Prompt.	0.060	0.048	100
Encoded/Mix.	0.048	0.042	210

3.4 Cross-Model Transferability

Correlations between related models (e.g., Command-R and Gemma2, $r = 0.659$) indicate shared weaknesses that propagate across similar architectures. This implies systemic risk in families of models with overlapping training or fine-tuning datasets.

4 Implications and Defenses

Our findings reveal wide variation in model robustness, persistent weaknesses in multi-turn contexts, and vulnerability transfer across related architectures. We recommend layered defenses including context-aware moderation, role-play detection, ensemble validation, and adaptive safety tuning [6].

5 Related Work

Recent studies have increasingly explored the security and alignment of large language models, emphasizing both proprietary and open-weight systems. Wei et al. [7] analyzed safety fine-tuning strategies and demonstrated that even robust alignment training often fails to protect against specific formatting-based or "jailbreaking" attacks, identifying competing objectives and mismatched generalization as key failure modes. Similarly, Bhardwaj and Poria [8] introduced a red-teaming benchmark based on Chain of Utterances (CoU) prompting, showing that multi-turn conversational strategies can reliably bypass safety mechanisms in both closed- and open-weight models.

Other research, such as the work by Robey et al. [9], has investigated defensive architectures like SmoothLLM, which uses randomized input perturbations to detect and mitigate jailbreaking attempts. Compared to these general robustness and alignment studies, our work focuses specifically on local LLMs, offering a quantitative assessment of jailbreak transferability and model-level resilience in air-gapped or on-premise deployments.

6 Conclusion

This paper presents the first comprehensive evaluation of jailbreak vulnerabilities in local LLMs. While DeepSeek-R1 demonstrated complete resistance, other models remain susceptible, particularly to multi-step attacks. As open-weight models proliferate, dynamic and multi-layer safety mechanisms are urgently needed.

Acknowledgments. This research was supported by the *Center for Equitable AI & ML Systems (CEAMLS)*, Morgan State University. We thank the open-source AI safety community for their tools and evaluation frameworks.

References

1. Zou, A., Wang, Z., Carlini, N., Nasr, M., Kolter, J.Z., Fredrikson, M.: Universal and transferable adversarial attacks on aligned language models (2023). arXiv:2307.15043
2. Ji, J.: AI alignment: a comprehensive survey (2023). arXiv:2310.19852
3. Chao, P.: JailbreakBench: an open robustness benchmark for jailbreaking large language models (2024). arXiv:2404.01318
4. Inan, H.: Llama Guard: LLM-based input-output safeguard for human-AI conversations (2023). arXiv:2312.06674
5. Perez, E.: Red teaming language models to reduce harms (2022). arXiv:2209.07858
6. Greshake, K., et al.: Compromising real-world LLM-integrated applications with indirect prompt injection. Proc. AISec (2023)
7. Wei, A., Haghtalab, N., Steinhardt, J.: Jailbroken: How does LLM safety training fail? arXiv:2307.02483 (2023)
8. Bhardwaj, R., Poria, S.: Red-teaming LLMs using chain of utterances (2023). arXiv:2308.09662
9. Robey, A., Wong, E., Hassani, H., Pappas, G.J.: SmoothLLM: defending large language models against jailbreaking attacks (2023). arXiv:2310.03684

Author Index

H. R. Arabnia et al. (Eds.): CSCI 2025, CCIS 2969, pp. 553–554, 2027.
https://doi.org/10.1007/978-3-032-28460-0

FSC
www.fsc.org
MIX
Papier | Fördert
gute Waldnutzung
FSC® C083411

Kolibri 360 GmbH
Ferdinand-Jühlke-Straße 7
99095 Erfurt, Deutschland
produktsicherheit@kolibri360.de